Essentials of Applied Mathematics

Essentials of Applied Mathematics

Edited by Ada Hastings

Larsen & Keller
www.larsen-keller.com

Essentials of Applied Mathematics
Edited by Ada Hastings
ISBN: 979-8-88836-044-6 (Hardback)

Larsen & Keller
Published by Larsen and Keller Education,
5 Penn Plaza,
19th Floor,
New York, NY 10001, USA

Cataloging-in-Publication Data

Essentials of applied mathematics / edited by Ada Hastings.
p. cm.
Includes bibliographical references and index.
ISBN 979-8-88836-044-6
1. Mathematics. 2. Mathematical analysis. I. Hastings, Ada.
QA36 .E87 2023
510--dc23

For more information regarding Larsen and Keller Education and its products, please visit the publisher's website www.larsen-keller.com

Table of Contents

Preface VII

Chapter 1 **Quantum Implications of Non-Extensive Statistics** 1
Nana Cabo Bizet, César Damián, Octavio Obregón and Roberto Santos-Silva

Chapter 2 **On the Global Positivity Solutions of Non-Homogeneous Stochastic Differential Equations** 15
Farai Julius Mhlanga and Lazarus Rundora

Chapter 3 **Gibbs-Slice Sampling Algorithm for Estimating the Four-Parameter Logistic Model** 24
Jiwei Zhang, Jing Lu, Hang Du and Zhaoyuan Zhang

Chapter 4 **A Knowledge-Based Smart Trainer System for Transferring Knowledge from Coaches to Long Jump Students** 42
Teerawat Kamnardsiri, Pattaraporn Khuwuthyakorn, Sirinun Boripuntakul and Worawit Janchai

Chapter 5 **A Numerical Analysis of the Leakage Characteristics of an Embankment Dam Slope with Internal Erosion** 52
Yu Jia, Yu Ding, Xuan Wang, Jiasheng Zhang and Xiaobin Chen

Chapter 6 **On the (*k,s*)-Hilfer-Prabhakar Fractional Derivative with Applications to Mathematical Physics** 66
Muhammad Samraiz, Zahida Perveen, Gauhar Rahman, Kottakkaran Sooppy Nisar and Devendra Kumar

Chapter 7 **Waves that Appear from Nowhere: Complex Rogue Wave Structures and their Elementary Particles** 74
Nail Akhmediev

Chapter 8 **Numerical Analysis of Turbulent Flow and Heat Transfer in Internally Finned Tubes** 82
Zhanwei Liu, Yanwei Yue, Luchao She and Guangming Fan

Chapter 9 **A Primer on R for Numerical Analysis in Educational Research** 93
Tricia R. Prokop and Michael Wininger

Chapter 10 **Numerical Analysis of Zipper Fracturing using a Non-Planar 3D Fracture Model** 109
Zhen Wang, Lifeng Yang, Rui Gao, Guanshui Xu, Zhe Liu, Shaoyuan Mo, Meng Fan and Xin Wang

Chapter 11 **Probabilistic Assessment of Investment Options in Honey Value Chains** 120
Joshua Wafula, Yusuf Karimjee, Yvonne Tamba, Geoffrey Malava, Caroline Muchiri, Grace Koech, Jan De Leeuw, Josephat Nyongesa, Keith Shepherd and Eike Luedeling

Chapter 12 **Convergence Analysis and Approximate Optimal Temporal Step Sizes for Some Finite Difference Methods Discretising Fisher's Equation** 131
Koffi Messan Agbavon, Appanah Rao Appadu, Bilge Inan and Herve Michel Tenkam

Chapter 13 **Statistical Analysis of fNIRS Data: Consideration of Spatial Varying Coefficient Model of Prefrontal Cortex Activity Changes During Speech Motor Learning in Apraxia of Speech** **148**
Rachel Johnson, Jennifer Matthews, Norou Diawara and Rachel Carroll

Chapter 14 **Statistical Analysis of Multi-Relational Network Recovery** **163**
Zhi Wang, Xueying Tang and Jingchen Liu

Chapter 15 **A Mathematical Analysis of the Improving Sequence Effect for Monetary Rewards** **171**
Salvador Cruz Rambaud, María José Muñoz Torrecillas and Adriana Garcia

Chapter 16 **CPD-Structured Multivariate Polynomial Optimization** **179**
Muzaffer Ayvaz and Lieven De Lathauwer

Chapter 17 **Geostatistical Learning: Challenges and Opportunities** **203**
Júlio Hoffimann, Maciel Zortea, Breno de Carvalho and Bianca Zadrozny

Chapter 18 **Generation of a Dynamical Logic Gate from Unstable Dissipative Systems of Type 1** **218**
Roberto R. Rivera-Durón, Ricardo Sevilla-Escoboza and Qui-Ling Wang

Permissions

List of Contributors

Index

Preface

The purpose of the book is to provide a glimpse into the dynamics and to present opinions and studies of some of the scientists engaged in the development of new ideas in the field from very different standpoints. This book will prove useful to students and researchers owing to its high content quality.

Mathematics encompasses topics such as numbers, quantities and their changes, formulas and related structures, and spaces and shapes in which they exist. The application of mathematical principles, methods and concepts in different fields is known as applied mathematics. Some of the various fields in which applied mathematics is prominently used are computer science, biology, physics, finance, engineering, and medicine. Wider applications have been enabled with the advent of computers, such as studying new computer technology, and studying problems occurring in areas of science along with mathematical computation. A few of the major branches of mathematics, which are used within applied mathematics are differential equations and stochastics. Analytical and numerical methods are also used for providing real world solutions to different problems. This book includes some of the vital pieces of works being conducted across the world on applied mathematics. Those with an interest in this field would find it helpful.

At the end, I would like to appreciate all the efforts made by the authors in completing their chapters professionally. I express my deepest gratitude to all of them for contributing to this book by sharing their valuable works. A special thanks to my family and friends for their constant support in this journey.

Editor

1

Quantum Implications of Non-Extensive Statistics

Nana Cabo Bizet[1], César Damián[2], Octavio Obregón[1] and Roberto Santos-Silva[3]*

[1]Departamento de Física, División de Ciencias e Ingenierías, Universidad de Guanajuato, León, Mexico, [2]Departamento de Ingeniería Mecánica, Universidad de Guanajuato, Salamanca, Mexico, [3]Departamento de Ciencias Naturales y Exactas, CUValles, Universidad de Guadalajara, Jalisco, Mexico

***Correspondence:**
Nana Cabo Bizet
nana@fisica.ugto.mx

Exploring the analogy between quantum mechanics and statistical mechanics, we formulate an integrated version of the Quantropy functional. With this prescription, we compute the propagator associated to Boltzmann–Gibbs statistics in the semiclassical approximation as $K = F(T)\exp(iS_{cl}/h)$. We determine also propagators associated to different nonadditive statistics; those are the entropies depending only on the probability $S_\pm$ and Tsallis entropy S_q. For $S_\pm$, we obtain a power series solution for the probability vs. the energy, which can be analytically continued to the complex plane and employed to obtain the propagators. Our work is motivated by the work of Nobre et al. where a modified q-Schrödinger equation is obtained that provides the wave function for the free particle as a q-exponential. The modified q-propagator obtained with our method leads to the same q-wave function for that case. The procedure presented in this work allows to calculate q-wave functions in problems with interactions determining nonlinear quantum implications of nonadditive statistics. In a similar manner, the corresponding generalized wave functions associated to $S_\pm$ can also be constructed. The corrections to the original propagator are explicitly determined in the case of a free particle and the harmonic oscillator for which the semiclassical approximation is exact, and also the case of a particle with an infinite potential barrier is discussed.

Keywords: quantropy, nonlinear quantum systems, propagator, nonextensive entropies, path integrals

1 INTRODUCTION

Nonextensive entropies depending only on the probabilities have been obtained in [1]. They belong to a family of nonextensive statistical mechanics, relevant for nonequilibrium systems. Renowned examples are Tsallis (S_q) [2, 3] and Sharma-Mital [4]; all of them can be obtained within the framework of Superstatistics [5].

For the entropies depending only on the probability, there are two entropy functionals [1]:

$$S_+ = \sum_l \left(1 - p_l^{p_l}\right), \quad S_- = \sum_l \left(p_l^{-p_l} - 1\right),$$

where the index l runs over the states of the system and p_l denotes the probability of the state l. These expressions can be considered as building blocks for nonextensive entropies without parameters. For example, one can consider $S_1 = (S_+ + S_-)/2$. These entropies are noticeably distinct to Boltzmann–Gibbs (BG) entropy for systems with few degrees of freedom; however, when the number of degrees of freedom goes to the thermodynamic limit, they match perfectly with BG statistics Cabo [6]. This is when the probability p_l is small, the leading term in S_+ and S_- expansions is the BG entropy. Therefore, in this limit, BG statistics is recovered. They belong to the class of Superstatistics of [5] with an intensive parameter χ^2 distribution found in [1]. These entropies have also been studied in [6–10].

There is a universality of the Superstatistics family [5]. As it has been shown, for several distributions of the temperature, the Boltzmann factor essentially coincides up to the first expansion terms. This has as a consequence that also the entropies associated to these Boltzmann factors have all of them basically the same first corrections to the usual entropy. Furthermore, the three entropies listed here that depend only on the probability are expanded only on the parameter $y = p\ln p$; this is always smaller than 1 giving correction terms to the entropy which at any order are smaller than the previous ones. So, that any function of y proposed as another generalized entropy, depending only on this parameter, when expanded in y will basically coincide with one of the three ones studied here; clearly demanding that the first term in the expansion is $-y$ giving BG entropy. Thus, the entropies S_+, S_-, and their linear combinations can be considered as building blocks to compute any possible modified entropies depending only on the probability.

We are motivated by the concept of Quantropy developed by Baez and Pollard [11] and by nonlinear quantum systems with modified wave functions based on Tsallis statistics in [3, 12, 13]. For example, the work [13] developed a nonlinear quantum mechanics with q-mathematics motivated by Tsallis entropy. In recent years, there have also been other interesting developments in the connections of nonextensive entropies and quantum mechanics [14–25]. Also, the work [26] showed extensions of nonlinear quantum equations arising from an effective one particle treatment of many-body physics, such that the nonlinearity represents the interactions, obtaining wave function solutions that are q-distributions and including the harmonic oscillator potential. There exists as well a connection between nonlinear quantum equations and nonlinear diffusion and Fokker–Planck equations [18, 20] that also is noticed in [26]. Moreover, interesting applications of nonextensive entropies to compute statistical and thermodynamical properties of graphene and 2-dimensional quantum structures [27–30] have been developed. We develop a version of Quantropy in terms of the propagator of a quantum mechanical theory. Our generalized propagators could be connected to the appropriate quantum equations. Baez and Pollard's Quantropy is a functional of the amplitude on the path integral a, with the same functional form as the entropy in terms of the probability $Q = -\int_X a(x)\ln a(x)dx$. Giving the functional

$$\Phi_{\mathrm{BP}} = -\int_X a(x)\ln a(x)dx - \alpha\int_X a(x)dx \\ -\lambda\int_X a(x)S(x)dx, \tag{1}$$

where α and λ are Lagrange multipliers and x is a path in the space of all possible paths X. From the search of extrema of this functional, restricted to values of a normalized and an average of the action, Baez and Pollard obtained the relation $a = \exp(iS/\hbar - 1 - \alpha)$ with $\lambda = \frac{1}{i\hbar}$. Then, $a \sim \exp(iS/\hbar)$ with the normalization fixed by the Lagrange multiplier α. This scheme deepens on the relation between Quantum Mechanics and Statistical Physics, which has also been studied in different approaches [31, 32]. For example, in [33], the Fisher Information measure is employed to explore this relation.

In Baez and Pollard's approach, the energy is mapped to the action S and the temperature to $i\hbar$. We consider the same identification but instead we identify E with the classical action S_{cl}. Thus, we consider as the analog of the entropy a functional in terms of the propagator, instead of the amplitude a. This is an extrapolation of the Quantropy [11] to an integration over all classical paths. It is worth to mention that the analog to the microstate in statistical mechanics is a particle path in quantum mechanics. Such that as the partition function in statistical mechanics is the sum over all the microstates, the quantum mechanical analog is the sum over all the paths of the particle (Feynman path integral). The standard expression for the propagator is given semiclassically by $K \sim \exp(iS_{cl}/\hbar)$. We use this fact to define a kind of integrated Quantropy functional now in terms of the propagator for BG statistics, which we extend to the modified statistics S_+, S_-, and S_q.

This article is organized as follows. In **Section 2**, we obtain a series expansion for the probabilities versus βE for the generalized entropies depending only on the probabilities S_+ and S_-. In **Section 3**, we continue these expansions to the complex plane. In **Section 4**, we present a version of Quantropy for BG statistics, S_+ and S_- and S_q. In **Section 5**, we study in particular the case of the free particle propagator, obtained from the extrema of the Quantropy in the cases of S_+ and S_- and S_q for $q < 1$ and $q > 1$. We show that the K_q propagator results exactly in the q-exponential that defines the q-wave function for the free particle [3]. In a similar manner, we argue that the corresponding generalized propagators K_+ and K_- provide us with a procedure to construct Ψ_+ and Ψ_- for the free particle. Our method however gives the possibility to construct K_q, K_+, and K_- also for problems with interactions and by this mean to identify the corresponding wave functions. We also provide a way to perform the normalization inspired in Feynman and Hibbs work [34]. **Section 6** is devoted to the analysis of the harmonic oscillator, and we exemplify with the case corresponding to K_+. Finally, in **Section 7**, we study the K_+ propagator for the particle in an infinite potential barrier. In **Section 8**, we summarize our results and present the conclusion. In a **Supplementary Appendix** we present a numerical study of the propagators.

2 PROBABILITY DISTRIBUTIONS FOR SYSTEMS WITH MAXIMAL S_+ AND S_-

We start by developing a recurrent solution for the probability distribution of the generalized entropy S_+, introduced in [1]. On the contrary to BG statistics, for a system subjected to S_+ extremization, probability normalization, and energy conservation, there is not a simple inverse function of the probabilities p vs. the values of the energy state E. We overcome this difficulty by finding a series solution to the extremum equation. There are other possible series solutions, but we discuss here one that has a good convergence. At the end of the section, we give also the probability expansion for the entropy S_-, which is obtained by an equivalent Ansatz.

The functional to maximize the S_+ entropy subjected to probability normalization and averaged energy is given by [7, 8]:

$$\Phi_+ = \sum_l \left(1 - p_l^{p_l}\right) - \gamma \sum_l p_l - \beta \sum_l E_l p_l^{p_l+1}. \tag{2}$$

β and γ are Lagrange multipliers and E_l is the energy of the state l, with probability p_l. The average values of energy and the normalization value have been omitted for simplicity. The extrema of **(2)** given by $\frac{\delta \Phi_+}{\delta p_l} = 0$ gives a relation between the energy E and the probabilities p (we have omitted the index l):

$$\beta E = \frac{(-\gamma p^{-p} - 1 - \ln p)}{(1 + p + p \ln p)}. \tag{3}$$

Notice that we omit the subindex l from the quantities. Setting the Lagrange multiplier γ to −1, we first expand the previous equation around $p = 0$ that accounts to consider the expansion around $y = p\ln p = 0$. That is, for the exponential of minus **equation (3)**, one gets

$$\begin{aligned} e^{-\beta E} &= p - p^2 \ln p^2 + 1/2 p^3 \left(\ln p^2 + 2 \ln p^3 + \ln p^4\right) \\ &+ 1/6 p^4 \left(-3 \ln p^2 - 8 \ln p^3 - 9 \ln p^4 - 6 \ln p^5 - \ln p^6\right) \\ &+ 1/24 p^5 \left(12 \ln p^2 + 44 \ln p^3 + 70 \ln p^4 + 68 \ln p^5 + 42 \ln p^6 + 12 \ln p^7 + \ln p^8\right) + \dots \end{aligned} \tag{4}$$

We make the following Ansatz to solve **equation (4)**.

$$p = e^{-\beta E} \left(1 + \sum_{n=1} c_n e^{-n\beta E}\right), \tag{5}$$

where c_n can be functions of βE. Plugging (5) in **(4)**, to have the LHS equal to the RHS, the coefficients multiplying the powers of $e^{-n\beta E}$ with $n > 1$ have to vanish. This gives a recurrent expression for the coefficients c_n which for the first four coefficients is solved as

$$\begin{aligned} c_1 &= x^2, \\ c_2 &= 1/2 x^2 \left(-1 - 2x + 3x^2\right), \\ c_3 &= \frac{1}{6} x^2 \left(3 + 4x - 6x^2 - 24x^3 + 16x^4\right), \\ c_4 &= \frac{1}{24} x^2 \left(-12 - 16x + 60x^2 + 116x^3 + 30x^4 - 300x^5 + 125x^6\right). \end{aligned} \tag{6}$$

We have denoted βE as x. The coefficients **(6)** give the following approximate solution for the probabilities versus βE:

$$\begin{aligned} p_+ &= e^{-x} + e^{-2x} x^2 + \frac{1}{2} e^{-3x} x^2 \left(-1 - 2x + 3x^2\right) \\ &+ \frac{1}{6} e^{-4x} x^2 \left(3 + 4x - 6x^2 - 24x^3 + 16x^4\right) \\ &+ \frac{1}{24} e^{-5x} x^2 \left(-12 - 16x + 60x^2 + 116x^3 + 30x^4 - 300x^5 \right. \\ &\left. + 125x^6\right) + \dots . \end{aligned} \tag{7}$$

In **Figure 1**, we compare the exact value of p vs. βE with the power series solution **(7)** till 3rd order and with the Boltzmann distribution $e^{-\beta E}$.

Probability Expansion for the Entropy S_- -

Consider the other generalized entropy dependent only on the probabilities S_-. For this entropy, the functional to extremize Φ_- reads

$$\Phi_- = \sum_l \left(p_l^{-p_l} - 1\right) - \gamma \sum_l p_l - \beta \sum_l E_l p_l^{1-p_l}. \tag{8}$$

β and γ are Lagrange multipliers, and E_l is the energy of the state l and p_l its probability. Finding the extrema of **(8)** as $\frac{\delta \Phi_-}{\delta p_l} = 0$ and proposing the same Ansatz (5), we obtain a set of equations that can be solved to give the recursive probability solution:

$$\begin{aligned} p_- &= e^{-x} \left(1 - e^{-x} x^2 + \frac{1}{2} e^{-2x} x^2 \left(-1 - 2x + 3x^2\right) \right. \\ &\left. + \frac{1}{6} e^{-3x} \left(-3x^2 - 4x^3 + 6x^4 + 24x^5 - 16x^6\right)\right) \dots \end{aligned} \tag{9}$$

3 MODIFIED AMPLITUDE EXPANSIONS

In this section, we use the series solutions for the probabilities in terms of the energy obtained in the previous section, to perform an analytic continuation to the complex plane. Considering a as the amplitude of a path, this is a new complex variable substituting the probability p_l, and A as the action replacing βE_l. This identification will allow to study modified Quantropy functionals, for the definition of Baez and Pollard (equation **(1)**), as well as our definition **(18)**. The usual Quantropy solution will give an exponential $a \sim e^{\frac{iA}{\hbar}}$. In our approach, this would be the propagator $K \sim e^{\frac{iS_{cl}}{\hbar}}$. We want to analyze the new statistics S_+ and S_-. We will find a functional dependence of a vs. A (K vs. S_{cl}) that deviates from the exponential dependence.

The main idea is to complexify first the power expansion solution **(7)** since the amplitude is a complex number, such that we have a solution to the extrema of the modified Quantropy. The functional to extremize reads

$$\Phi_{BP,+} = \int \left(1 - a(x)^{a(x)}\right) dx - \alpha \int a(x) dx - \lambda \int A(x) a(x)^{a(x)+1} dx. \tag{10}$$

Finding the extrema of **(10)** w.r.t. a, i.e., solving $\frac{\delta \Phi_{BP,+}}{\delta a(x)} = 0$, one gets

$$\frac{A}{i\hbar} = \frac{(-\gamma a^{-a} - 1 - \ln a)}{(1 + a + a \ln a)} = F\left(a\left(\frac{A}{i\hbar}\right)\right). \tag{11}$$

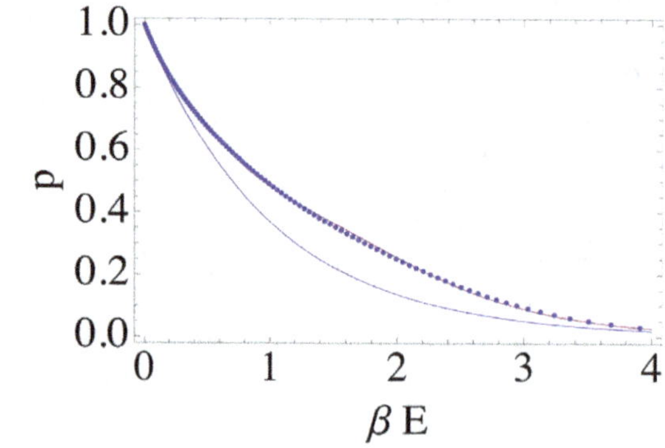

FIGURE 1 | Probability versus βE. The blue line represents the BG statistics distribution. The dots represent the exact dependence in **(3)**, S_+ statistics distribution, while the continuous red curve overlapping with the dotted line represents the power series solution **(7)** till order 3, i.e., up to the e^{-4x} correction.

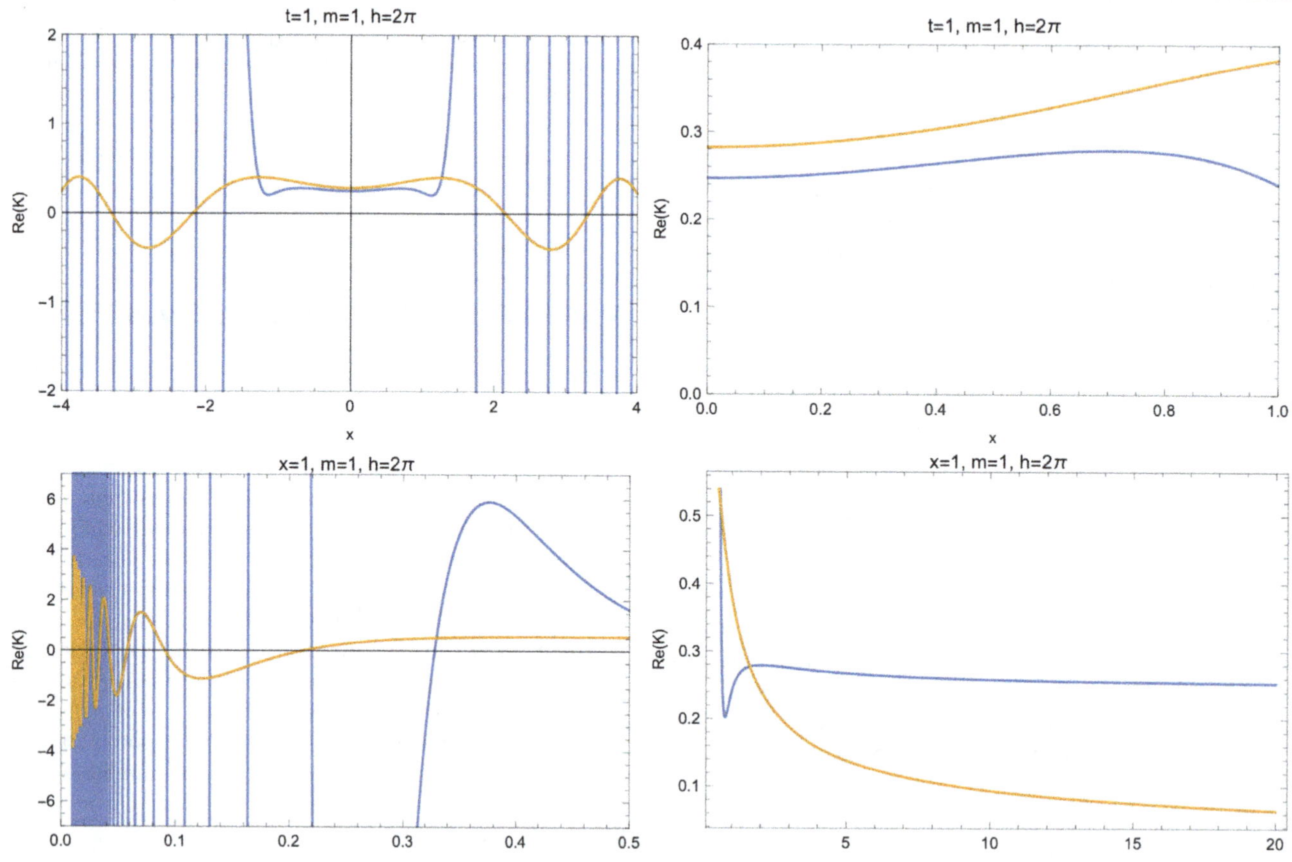

FIGURE 2 | Real parts of the modified propagator (blue line) vs. standard propagator (yellow line), for the free particle for the modified statistics S_+. We set the mass and the Planck constant to unity. The quantum regime is given by $S_{cl} \approx h$. Imposing $S_{cl} \lesssim h$ which translates for fixed $x = 1$ in $t \gtrsim 1/2$, for fixed $t = 1$ translates in $x^2 \lesssim 2$. Notice that in the quantum regime $S_{cl} \approx h$, there are differences between the standard and the modified propagator. In the classical regime, there are many oscillations caused by the series expansion that should sum up when computing more terms. When $S_{cl} \ll h$, then both results coincide. The difference between the results of the modified propagator and the standard free particle is that the modified propagator result could be interpreted as the particle with an effective potential; this would give a spatially bounded wave function.

The range of validity of the propagators expressions depends on the convergence of the imaginary series solution to this equation. The series is obtained by doing the replacement βE by $\frac{A}{i\hbar}$ and p by a in **(3)**. Also, the Lagrange multipliers have to be mapped: β to λ and α to γ. The minus sign gives the right sign after a rotation on the argument of the exponential (similar to a Wick rotation). As the series solution continuation of **(7)**, we obtain the following expression:

$$\begin{aligned} a_+\left(\frac{A}{i\hbar}\right) = e^{iA/\hbar}\Big(1 - \frac{A^2}{\hbar} e^{i(A/\hbar)} \\ -\frac{1}{2}\left(A^2/\hbar^2\right)\left(-1 + 2i(A/\hbar) - 3\left(A^2/\hbar^2\right)\right)e^{i2(A/\hbar)} \\ -\frac{1}{6}\left(A^2/\hbar^2\right)\left(3 - 4i(A/\hbar) + 6\left(A^2/\hbar^2\right) - 24i\left(A^3/\hbar^3\right) \right. \\ \left. +16\left(A^4/\hbar^4\right)\right)e^{3i(A/\hbar)} + \ldots\Big). \end{aligned} \tag{12}$$

Since A has units of action, the argument of the exponentials and the terms on the expansion are adimensional. Substituting this expression on the constraint **equation (11)**, we obtain the real and imaginary parts of $F\left(a\left(\frac{A}{i\hbar}\right)\right)$. In **Figure 2**, the relevant difference of the propagators K_+ obtained employing **(12)** with respect to the standard one can be observed in the region of $S_{cl} \approx \hbar$

Using the parameter $\lambda = \frac{1}{i\hbar}$ in **(12)**, the expression for the amplitudes becomes

$$\begin{aligned} a_+(A) = e^{-\lambda A}\Big(1 + (\lambda A)^2 e^{-\lambda A} \\ +\frac{1}{2}(\lambda A)^2\left(-1 - 2\lambda A + 3(\lambda A)^2\right)e^{-2\lambda A} \\ +\frac{1}{6}(\lambda A)^2\left(3 + 4\lambda A - 6(\lambda A)^2 - 24(\lambda A)^3 \right. \\ \left. +16(\lambda A)^4\right)e^{-3\lambda A} + \ldots\Big). \end{aligned} \tag{13}$$

It is not difficult to observe that all terms of this expansion can be written as derivatives with respect to the parameter λ. If we derive with respect to λ, the usual amplitude we obtain is $\frac{\partial}{\partial\lambda}e^{-\lambda A} = -Ae^{-\lambda A}$. Higher derivates can be written as

$$\left(\frac{\lambda}{n}\right)^m \frac{\partial^m}{\partial\lambda^m}e^{-n\lambda A} = (-1)^m(\lambda A)^m e^{-n\lambda A}, \tag{14}$$

where m and n are positive integers. Thus, we rewrite (3) as

$$\begin{aligned} a_+ = e^{-\lambda A} + \frac{\lambda^2}{4}\frac{\partial^2}{\partial\lambda^2}e^{-2\lambda A} \\ +\frac{1}{2}\left(-\frac{\lambda^2}{3^2}\frac{\partial^2}{\partial\lambda^2} + 2\frac{\lambda^3}{3^3}\frac{\partial^3}{\partial\lambda^3} + 3\frac{\lambda^4}{3^4}\frac{\partial^4}{\partial\lambda^4}\right)e^{-3\lambda A} \\ +\frac{1}{6}\left(3\frac{\lambda^2}{4^2}\frac{\partial^2}{\partial\lambda^2} - 4\frac{\lambda^3}{4^3}\frac{\partial^3}{\partial\lambda^3} - 6\frac{\lambda^4}{4^4}\frac{\partial^4}{\partial\lambda^4} + 24\frac{\lambda^5}{4^5}\frac{\partial^5}{\partial\lambda^5} + 16\frac{\lambda^6}{4^6}\frac{\partial^6}{\partial\lambda^6}\right)e^{-4\lambda A} + \ldots . \end{aligned} \tag{15}$$

One can compute the corrections to any order. Those corrections to the usual amplitude $a_0 = e^{iS/\hbar}$ can be interpreted as higher order interactions of the action at different frequencies[1].

Now, one can apply the same method to determine the distribution arising from the Quantropy with statistics S_-. We also have to perform the extension to the complex plane. The amplitude distribution for the modified Quantropy coming from S_- is given by the following:

$$\begin{aligned} a_- = e^{i(A/\hbar)}\Big(1 + (A/\hbar)^2 e^{i(A/\hbar)} \\ -\frac{1}{2}(A/\hbar)^2\left(-1 + 2i(A/\hbar) - 3(A/\hbar)^2\right)e^{2i(A/\hbar)} \\ +\frac{1}{6}\left(3(A/\hbar)^2 - 4i(A/\hbar)^3 + 6(A/\hbar)^4 \right. \\ \left. -24i(A/\hbar)^5 + 16(A/\hbar)^6\right)e^{3i(A/\hbar)} + \ldots\Big). \end{aligned} \tag{16}$$

4 QUANTROPY IN TERMS OF THE PROPAGATOR

In this section, we present as an alternative proposal a kind of integrated version of the Quantropy of [1]. First, we do it for the BG entropy and then for S_+, S_-, and S_q. The change in distribution probabilities which arise from modified entropies in statistics is now reflected in the quantum arena as modifications to the propagators. The propagators between points in space-time $(\mathbf{x}_a, t_a)$ and $(\mathbf{x}_b, t_b)$ in quantum mechanics determine the probability amplitude of particles to travel from certain position to another position in a given time. As modified entropies in statistical physics lead to modified probability distributions, distinct probabilities of propagation over all paths from $(\mathbf{x}_a, t_a)$ and $(\mathbf{x}_b, t_b)$ will arise from a modified Quantropy.

In the work [11], the Quantropy functional associated with BG statistics was formulated, and its maximization leads to the weight on the path integral $a \sim \exp(-\lambda S)$ with $\lambda = 1/i\hbar$. We propose another functional, which in a sense constitutes an integrated version of Quantropy. Its maximization leads to the propagator $K(x) \sim \exp(-\lambda S_{cl}(x))$. The Wentzel–Kramers–Brillouin (WKB) method [35] allows to compute the wave function in a semiclassical approximation. In a sense, this is linked to our approach, in which we maximize a functional which determines the propagator with a semiclassical approximation in terms of the classical action S_{cl}. This is exact for the free particle, the harmonic oscillator as well as for other cases [36, 37]. The procedure is applied to generalized entropy functionals, giving a modified propagator. For the Tsallis statistics, we obtain $K_q(x) \sim \exp_q(-\lambda S_{cl}(x))$. This structure is the same as the one of the wave function for the free particle of the Tsallis statistic $\Psi_q(x) = \exp_q(i(kx - wt))$, which has been proposed as solution of the nonlinear quantum equations of [13]. According to Feynman arguments, one can start with the free particle propagator and determine the corresponding wave function, as discussed in [34]. Thus, our procedure allows to find a propagator which can be identified with the wave function of interest. We should note that the propagator resulting from our procedure will describe not only the

[1]For example, for a massive particle those will be contributions from multiples of the particle mass. For the harmonic oscillator also, there will be contributions with a tower of masses and frequencies.

free particle but to a good approximation any other problem with its corresponding classical action. Our method should give the wave function solution for the problem of interest.

With the same method, we write functionals for S_+ and S_- and obtain probability distributions; we can write the corresponding Quantropies and obtain the propagators K_+ and K_- and correspondingly extrapolate them to the wave functions Ψ_+ and Ψ_-. This would give us the quantum behavior for a given action. The obtained propagators can be related to nonlinear quantum systems studied in the literature [26].

To define our functionals, we use the semiclassical limit to compute the propagator; this is $K(x) = F(a,b)e^{\frac{iS_{cl}(x)}{\hbar}}$, denoting the classical action as $S_{cl}(x)$ and being $F(a,b)$ a constant depending on the time difference $t_b - t_a$. For the free particle and the harmonic oscillator as well as other physical problems [36, 37], this is an exact result.

For the BG statistics, we define the Quantropy functional:

$$\Phi_0 = -\int K(x)\ln K(x)dx - \alpha\int K(x)dx - \lambda\int (S_{cl}(x)K(x))dx. \quad (17)$$

The extrema condition $\frac{\delta\Phi_0}{\delta K(x)} = 0$ gives as solution the propagator dependence $K(x) = e^{-1-\alpha-\lambda S_{cl}(x)}$, with $\lambda = 1/i\hbar$, where the normalization constant α determines $F(a,b)$.

The integrated Quantropy functional for the new S_+ statistic is studied in [1, 6–10] is given by the following:

$$\Phi_+ = \int \left(1 - K(x)^{K(x)}\right)dx - \alpha\int K(x)dx - \lambda\int (S_{cl}(x))K(x)^{K(x)+1}dx. \quad (18)$$

The extrema condition $\frac{\delta\Phi_+}{\delta K(y)} = 0$ with functional derivatives leads to the equation:

$$0 = \int \left(-\delta(x-y)K(x)^{K(x)}\right)(\ln K(x)+1)dx - \alpha\int \delta(x-y)dx \\ -\lambda\int S_{cl}(x)K(x)^{K(x)+1}\delta(x-y)\left(\ln K(x) + \frac{K(x)+1}{K(x)}\right)dx, \quad (19)$$

and by changing the variable notation y to x, this can be written as

$$\lambda S_{cl}(x) = \frac{-1 - \ln K(x) - \alpha K(x)^{-K(x)}}{1 + K(x) + \ln K(x)}. \quad (20)$$

Using our knowledge to solve this type of equation from the statistical physics case, presented in **Section 2**, this gives for the modified propagator the following series solution:

$$K_+(x) = N_+e^{-\lambda S_{cl}}\Big(1 - e^{-\lambda S_{cl}}(\lambda S_{cl})^2 \\ + e^{-2\lambda S_{cl}}(\lambda S_{cl})^2\left(-1 - 2(\lambda S_{cl}) + 3(\lambda S_{cl})^2\right) \\ -\frac{1}{6}e^{-2S_{cl}\lambda}\times\left(-3S_{cl}^2\lambda^2 - 4S_{cl}^3\lambda^3 + 6S_{cl}^4\lambda^4 + 24S_{cl}^5\lambda^5 \\ -16S_{cl}^6\lambda^6\right) + \ldots\Big). \quad (21)$$

where N_+ is the normalization constant. This is obtained by taking the normalization $\alpha = -1$. A different normalization would change the coefficients in the expansion (21).

The maximization constraint for the new S_- statistic is given by

$$\Phi_- = \int \left(K(x)^{-K(x)} - 1\right)dx - \alpha\int K(x)dx \\ -\lambda\int (S_{cl})K(x)^{-K(x)+1}dx. \quad (22)$$

The extrema condition $\frac{\delta\Phi_-}{\delta K(x)} = 0$ gives the equation:

$$\lambda S_{cl}(x) = \frac{1 + \ln K(x) + \alpha K(x)^{K(x)}}{1 - K(x) - \ln K(x)}. \quad (23)$$

Using our knowledge of this type of equation from the statistical physics case, we obtain for the modified propagator the series solution:

$$K_-(x) = N_-e^{-\lambda S_{cl}}\Big(1 + e^{(-\lambda S_{cl})}(\lambda S_{cl})^2 \\ -e^{(-\lambda S_{cl})}(\lambda S_{cl})^2(-1 - 2(\lambda S_{cl}) + 3(\lambda S_{cl})^2 \\ +\frac{1}{6}e^{-2S_{cl}\lambda}\left(-3S_{cl}^2\lambda^2 - 4S_{cl}^3\lambda^3 + 6S_{cl}^4\lambda^4 + 24S_{cl}^5\lambda^5 - 16S_{cl}^6\lambda^6\right) + \ldots\Big), \quad (24)$$

N_- is the normalization constant. In the case of Tsallis statistics, the functional is given by the following:

$$\Phi_q = \int \frac{(1 - K(x)^q)}{(q-1)}dx - \alpha\int K(x)dx - \lambda\int (S_{cl})K(x)^q dx, \quad (25)$$

and the solution to $\frac{\delta\Phi_q}{\delta K(x)} = 0$ is

$$K_q(x) = N_q \exp_q(-\lambda S_{cl}(x)), \\ = N_q(1 - (1-q)\lambda S_{cl})^{\frac{1}{1-q}}. \quad (26)$$

We have still to discuss the normalization of the different Kernels. This q-propagator is related to the q-wave function for the free particle nonlinear quantum mechanics of [3]. We explore this case which has been studied by other means in the literature [3, 13, 24].

5 FREE PARTICLE PROPAGATORS

In this section, we write a modified propagator up to third order for the free particle in the case of the statistics S_+, S_-, and S_q for $q = 1 - \delta$ and $q = 1 + \delta$ with $\delta > 0$. The values of q less or equal than one are considered in order to compare the different propagators. We determine when the corrections to the usual propagator play an important role which turns to be in the quantum regime characterized by $S_{cl} \approx \hbar$. First, we describe the procedure; then, we describe the normalization; and in the last subsection, we summarize our results.

5.1 Superposition of Kernels

Now, we proceed to describe a generalized Kernel. The generalized complex probability distribution given by expansion (21) can be regarded as a superposition of

Kernels. Furthermore, the superposition will carry to the wave functions. In order to normalize the superposition, we consider that the total Kernel expansion integration is the same as the usual (1 for the free particle), as is explicit in the Quantropy functional (**4**). We show that this coincides with the result for the normalization obtained from propagating the wave function [34]. For the free particle, the unnormalized Kernel is as follows:

$$K_0(x,t;0,0)=\left(\frac{2\hbar\epsilon i\pi}{m}\right)^{(n-1)/2}\left(\frac{1}{n}\right)^{1/2}\exp\left(\frac{imx^2}{2\hbar t}\right),$$

where n is the number of divisions of the time interval and ϵ is an infinitesimal time parameter that satisfies $t=\epsilon n$. This expression arises from computing the path integral to get the following:

$$\begin{aligned}K_0(x,t;0,0)&=\int e^{iS/\hbar}Dx\\&=\int\exp\left(\frac{im}{2\hbar\epsilon}\sum_n(x_n-x_{n-1})^2\right)d^nx\\&=\left(\frac{i\pi}{2A}\right)^{\frac{1}{2}}\left(\frac{2i\pi}{3A}\right)^{\frac{1}{2}}\left(\frac{3i\pi}{4A}\right)^{\frac{1}{2}}\times\ldots\\&\times\left(\frac{(n-1)i\pi}{nA}\right)^{\frac{1}{2}}\exp\left(\frac{iA(x_0-x_n)^2}{n}\right),\end{aligned}\tag{27}$$

with $A=\frac{m}{2\epsilon\hbar}$. This calculation is performed dividing the integral in multiple Gaussian integrals (see [34]). The normalization constant is given by $N=\left(\frac{2\pi i\hbar\epsilon}{m}\right)^{-\frac{n}{2}}$; hence, the normalized propagator reads

$$K_1(x,t;0,0)=\left(\frac{2\hbar ti\pi}{m}\right)^{-1/2}\exp\left(\frac{imx^2}{2\hbar t}\right).\tag{28}$$

We define the unnormalized Kernel for the free particle as

$$k(x,t;1)=\exp\left(\frac{imx^2}{2\hbar t}\right)=e^{-\lambda A},\tag{29}$$

and the first two corrections in K_+ are given by

$$\begin{aligned}k(x,t;2)&=\left(\frac{mx^2}{\hbar t}\right)^2\exp\left(\frac{imx^2}{\hbar t}\right)=(\lambda A)^2\frac{\partial^2}{\partial\lambda^2}e^{-\lambda A},\\k(x,t;3)&=-\left(\frac{m^2x^4}{8\hbar^2t^2}+\frac{m^3x^6}{8i\hbar^3t^3}+3\frac{m^4x^8}{32\hbar^4t^4}\right)\times\exp\left(\frac{3imx^2}{2\hbar t}\right),\\&=\frac{1}{2}\left(-\frac{\lambda^2}{3^2}\frac{\partial^2}{\partial\lambda^2}+2\frac{\lambda^3}{3^3}\frac{\partial^3}{\partial\lambda}+3\frac{\lambda^4}{3^4}\frac{\partial^4}{\partial\lambda^4}\right)e^{-3\lambda A}.\end{aligned}$$

Thus, the generalized Kernel associated with S_+ entropy is given by

$$K_+(x,t)=N_+(k(x,t;1)+k(x,t;2)+k(x,t;3)+\ldots).$$

The normalization constant is determined by the requirement $\int_{-\infty}^{\infty}K_+(x,t)dx=1$, and up to the first corrections is given by $N_+=\frac{1}{1+3/(16\sqrt{2})}=0.883\ldots$. The reason for this normalization is also understood by an argument presented in the following, motivated by Feynmann and Hibbs procedure [34].

Let us also discuss the normalization of the modified propagator with respect to the usual one, as shown in the standard case [34]. We start considering the original unnormalized Kernel for the free particle computed from the path integral:

$$K_{1,0}(x,t;0,0)=2^{\frac{N-1}{2}}\left(\frac{\pi i\hbar t}{mN}\right)^{\frac{N-1}{2}}(N)^{-1/2}\exp\left(\frac{imx^2}{2\hbar t}\right).$$

To determine the normalization constant in the Feynman and Hibbs method, we can apply formulas (2–34) and (4-3) on their book [34] to write the new infinitesimal Kernel between position x_i and x_{i+1}, with $\Delta x_i=x_{i+1}-x_i$, in a time ϵ as follows:

$$\begin{aligned}&K_+(i_{i+1},i)=\frac{1}{A}\exp\left(\frac{i\epsilon}{\hbar}L\left(\frac{\Delta x_i}{\epsilon},\frac{x_{i+1}+x_i}{2},\frac{t_{i+1}+t_i}{2}\right)\right),\\&\left(1+(i\epsilon/\hbar)^2L\left(\frac{\Delta x_i}{\epsilon},\frac{x_{i+1}+x_i}{2},\frac{t_{i+1}+t_i}{2}\right)^2\exp\left(i\epsilon L(v,\bar{x},\bar{t})/\hbar\right)+\ldots\right).\end{aligned}\tag{30}$$

The method consists in writing the wave function at a position x at a time $t+\epsilon$ in terms of the wave function at position $y=x+\eta$ at a time t, explicitly

$$\begin{aligned}\psi(x,t+\epsilon)&=\int_{-\infty}^{\infty}K_+(x,y,\epsilon)\psi(y,t)dy,\\&=\int_{-\infty}^{\infty}\frac{1}{A}\exp\left(\frac{i\epsilon}{\hbar}L\left(\frac{x-y}{\epsilon},\frac{x+y}{2},t\right)\right)\times(1+\ldots)\psi(y,t)dy,\\&=\int_{-\infty}^{\infty}\frac{1}{A}\exp\left(\frac{i\epsilon}{\hbar}L\left(-\frac{\eta}{\epsilon},x+\frac{\eta}{2},t\right)\right)\times(1+\ldots)\psi(x+\eta,t)d\eta,\\&=\int_{-\infty}^{\infty}\frac{1}{A}\exp\left(\frac{im\eta^2}{2\hbar\epsilon}\right)\exp\left(-\frac{i\epsilon V\left(x+\frac{\eta}{2},t\right)}{\hbar}\right)\times(1+\ldots)\psi(x+\eta,t)d\eta.\end{aligned}\tag{31}$$

In the quantum standard theory, the normalization constant can be determined by expanding the LHS of (**31**) $\psi(x,t+\epsilon)=\psi(x,t)+\epsilon\partial_t\psi$ and the RHS $\psi(x+\eta)=\psi(x,t)+\eta\partial_x\psi+\frac{\eta^2}{2}\partial_x^2\psi$ and $\exp(-i\epsilon V/\hbar)=1-\frac{i\epsilon V}{\hbar}+\ldots$; then, we compare the leading term ϵ^0. This implies that

$$\frac{1}{A_0}\int_{-\infty}^{\infty}\exp\left(\frac{im\eta^2}{2\hbar\epsilon}\right)d\eta=1.\tag{32}$$

In a similarly fashion, one gets for the first correction to K_+ written in (**30**):

$$\frac{1}{A}\int_{-\infty}^{\infty}\exp\left(\frac{im\eta^2}{2\hbar\epsilon}\right)\left(1-\left(\frac{m\eta^2}{2\hbar\epsilon}\right)^2e^{\frac{im\eta^2}{2\hbar\epsilon}}+\ldots\right)d\eta=1.$$

It is worth to mention that the more important contribution to (31) is given for small η's, as well as in our generalized case. It is necessary to check this argument; in order to verify, let us consider the following integrals:

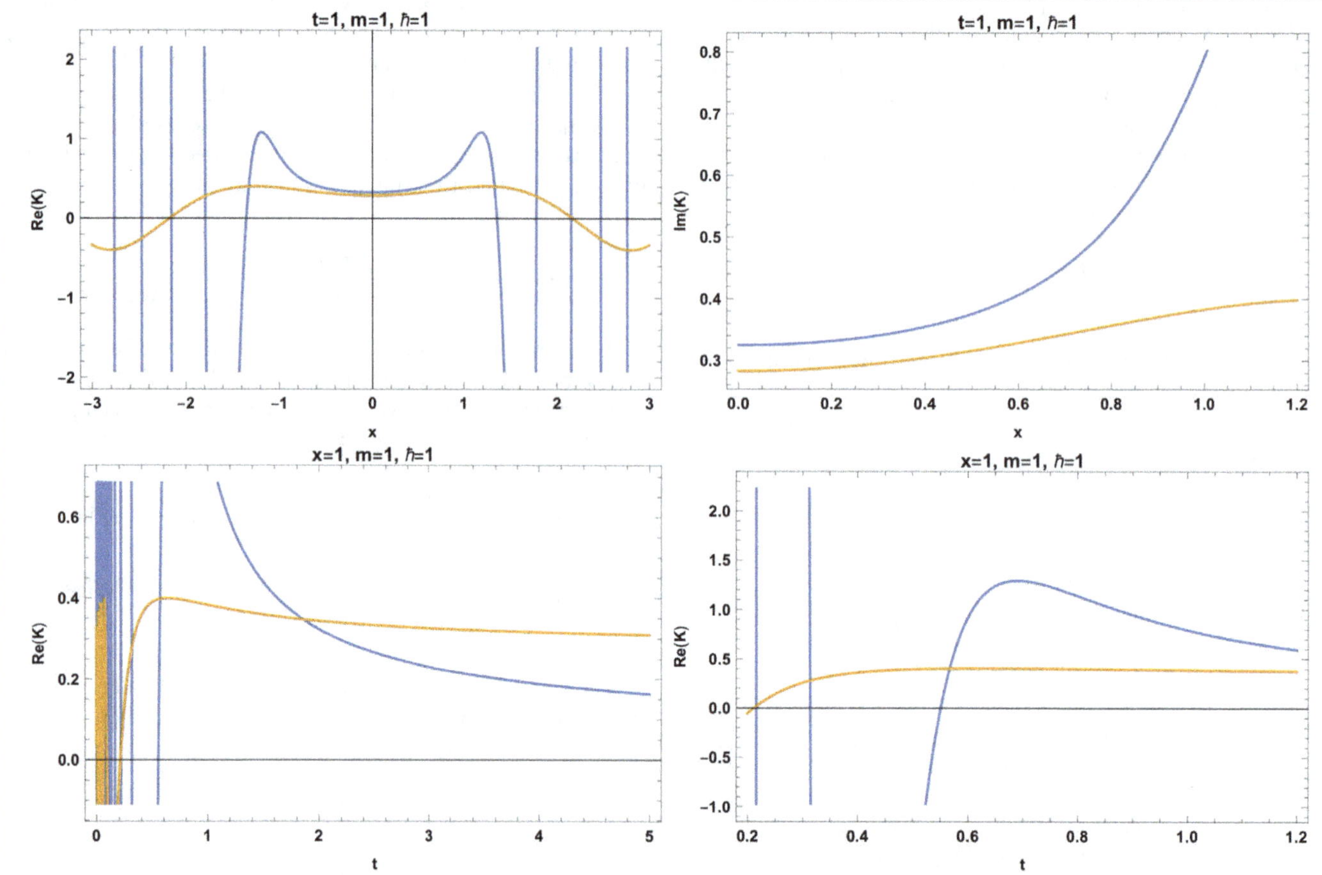

FIGURE 3 | Real parts of the modified propagator (blue line) vs. standard propagator (yellow line), for the free particle for the modified statistics S_-. We set the mass and the Planck constant to unity. Imposing $S_{cl} \lesssim \hbar$ for fixed $x = 1$ translates in $t \gtrsim 1/2$, and for fixed $t = 1$ translates in $x^2 \lesssim 2$.

$$\int e^{iCw^2} dw = \sqrt{i\pi/C}, \quad \int e^{iCw^2} w^4 dw = \frac{3\sqrt{\pi}}{4(-iC)^{5/2}},$$

$$\int e^{iCw^2} w^{2n+1} dw = 0, \quad n \in N.$$

The first correction gives the following relation:

$$A = A_0\left(1 + \frac{3}{16\sqrt{2}} + \ldots\right). \tag{33}$$

The previous normalization factor is a general feature to apply to any potential $V(x,t)$, in particular is valid for the cases discussed here. A similar expression holds for the normalization of K_- normalization, and this will be calculated in the next section.

5.2 Analysis of the Propagators

Here, we summarize the propagators obtained with the normalization methods described in previous subsections. The results for K_+ can be extrapolated to K_- and K_q because the method applied to obtain all of the propagators is the same.

Recall the standard propagator of the free particle from the space-time point $(0,0)$ to (x,t) is given by

$$K_1(x,t;0,0) = N_0 \exp\left(\frac{imx^2}{2\hbar t}\right). \tag{34}$$

The constant w.r.t. x is as follows: $N_0 = \sqrt{\frac{m}{2\pi i\hbar t}}$. For the case of S_+ and S_- statistics, the first two contributions to the modified propagators (21) and (24) read:

$$K_\pm = N_\pm \exp\left(\frac{imx^2}{2\hbar t}\right) \times \left(1 \mp \exp\left(\frac{imx^2}{2\hbar t}\right)\left(\frac{imx^2}{2\hbar t}\right)^2 + \ldots\right), \tag{35}$$

with $N_+ \sim N_0$. For the Tsallis statics, the associated propagator (26) is given by the expression:

$$K_q(x) = N_q\left(1 + (q-1)\left(\frac{mx^2}{2\hbar it}\right)\right)^{\frac{1}{1-q}}$$
$$= N_q \exp\left(\frac{imx^2}{2\hbar t}\right)\left(1 - (q-1)\left(\frac{m^2x^4}{2\hbar^2 t^2}\right) + \ldots\right). \tag{36}$$

We calculate the normalization constants for $K_\pm$ up to the first correction and exactly K_q to get the following:

$$N_+ = \sqrt{\frac{m}{2\pi i\hbar t}} \frac{1}{\left(1 + \frac{3}{16\sqrt{2}} + \ldots\right)}, \tag{37}$$

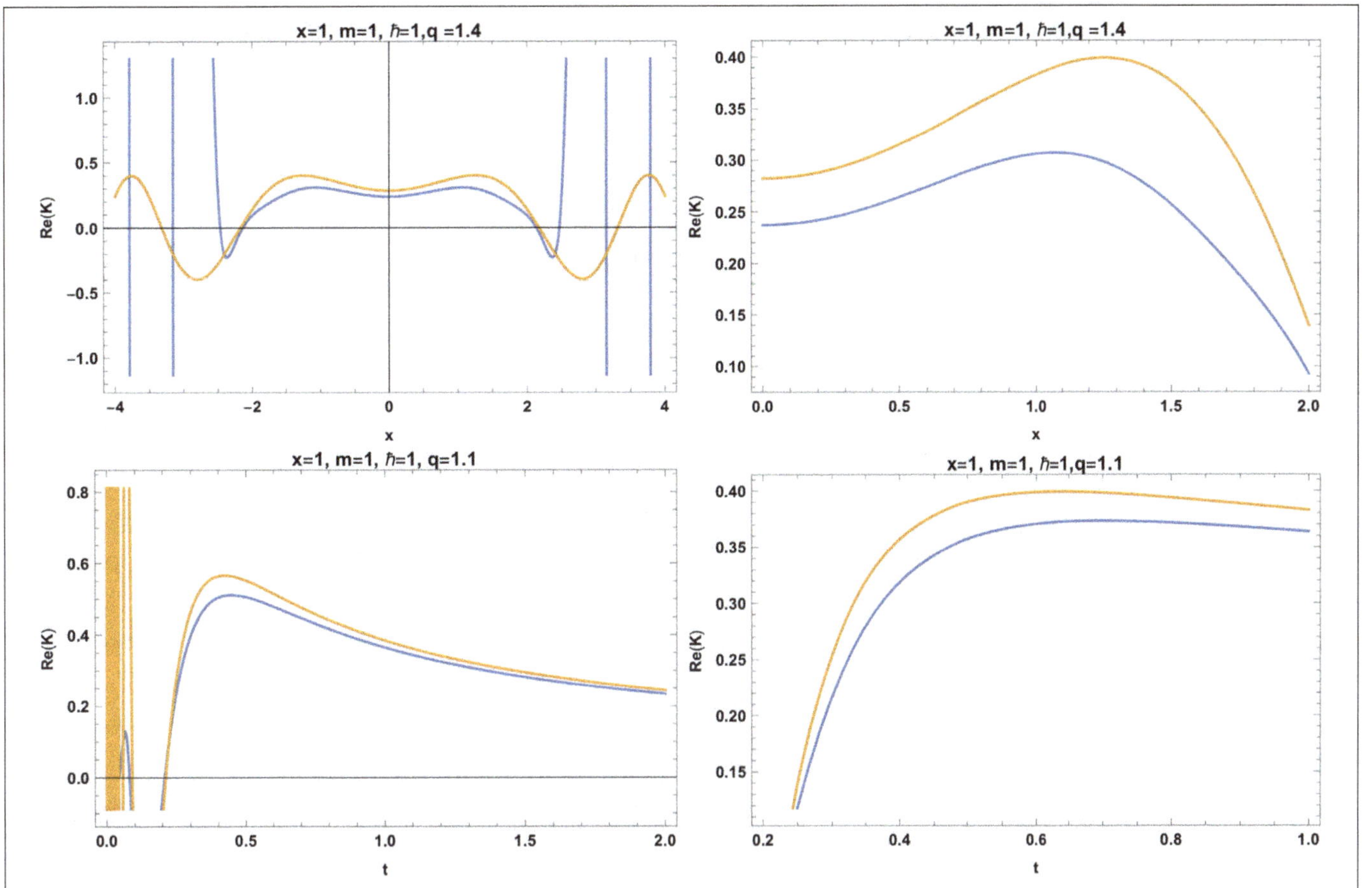

FIGURE 4 | Real parts of the modified propagator (blue line) vs. standard propagator (yellow line), for the free particle for the modified statistics of Tsallis for $q = 1.1$. We set the mass and the Planck constant to unity. Imposing $S_{cl} \lesssim \hbar$, it translates for fixed $x = 1$ in $t \gtrsim 1/2$ and for fixed $t = 1$ translates in $x^2 \lesssim 2$.

$$N_- = \sqrt{\frac{m}{2\pi i\hbar t}}\,\frac{1}{\left(1 - \frac{3}{16\sqrt{2}} + \ldots\right)}, \tag{38}$$

$$N_q = \sqrt{\frac{m}{2\pi i\hbar t}}\,\frac{\sqrt{(q-1)}\,\Gamma\left(\frac{1}{(q-1)}\right)}{\Gamma\left(\frac{1}{(q-1)} - \frac{1}{2}\right)}. \tag{39}$$

In the quantum regime $S_{cl} \approx \hbar$, the differences between the propagators K_+, K_-, K_q, and K_0 are shown in **Figures 2–4**. In **Figure 2** the K_- propagator is compared to the usual one. Also, **Figure 3** shows a comparision of the K_- propagator with the standard one. The last **Figure 4** shows a comparison between the propagators for the S_q statistics K_q for $q < 1$ and $q > 1$ with the usual one. The region of interest is the quantum regime with $S_{cl} \approx \hbar$. Furthermore, in the classical regime, the oscillations of the standard propagator grow averaging to zero as discussed in [34]. Thus, we are interested in comparing the corrections arising from different statistics in the quantum region of interest and, in the plots, the differences between the standard and the modified propagator can be observed. In the classical regime, there are oscillations that should sum up when computing more terms. When $S_{cl} \ll \hbar$, then both results coincide as shown in **Figure 5**. The modified propagators could be interpreted as describing a particle with an effective potential; this would give a spatially bounded wave function.

6 THE HARMONIC OSCILLATOR

In this section, we apply the formulation of our modified Quantropy of **Section 4** for the case of the harmonic oscillator. We compute the modified propagator constructed by a superposition as it was done previously. The extension of quantum systems employing the modified q-statistics has been made only for the case of the free particle [13] with different arguments. Our proposal allows to search the manifestation of nonextensive statistics in nonlinear quantum systems for generic potentials. We illustrate the procedure calculating only K_+, and the other propagators K_- and K_q could be similarly calculated.

For the harmonic oscillator with Lagrangian $\mathcal{L} = \frac{m}{2}\dot{x}^2 - \frac{m\omega^2}{2}x^2$, the path integral Kernel reads

$$K(a,b) = \left(\frac{m\omega}{2\pi i\hbar \sin\omega T}\right)^{1/2} \times \exp\left(\frac{im\omega}{2\hbar\sin\omega T}\left(\left(x_a^2 + x_b^2\right)\cos\omega T - 2x_a x_b\right)\right).$$

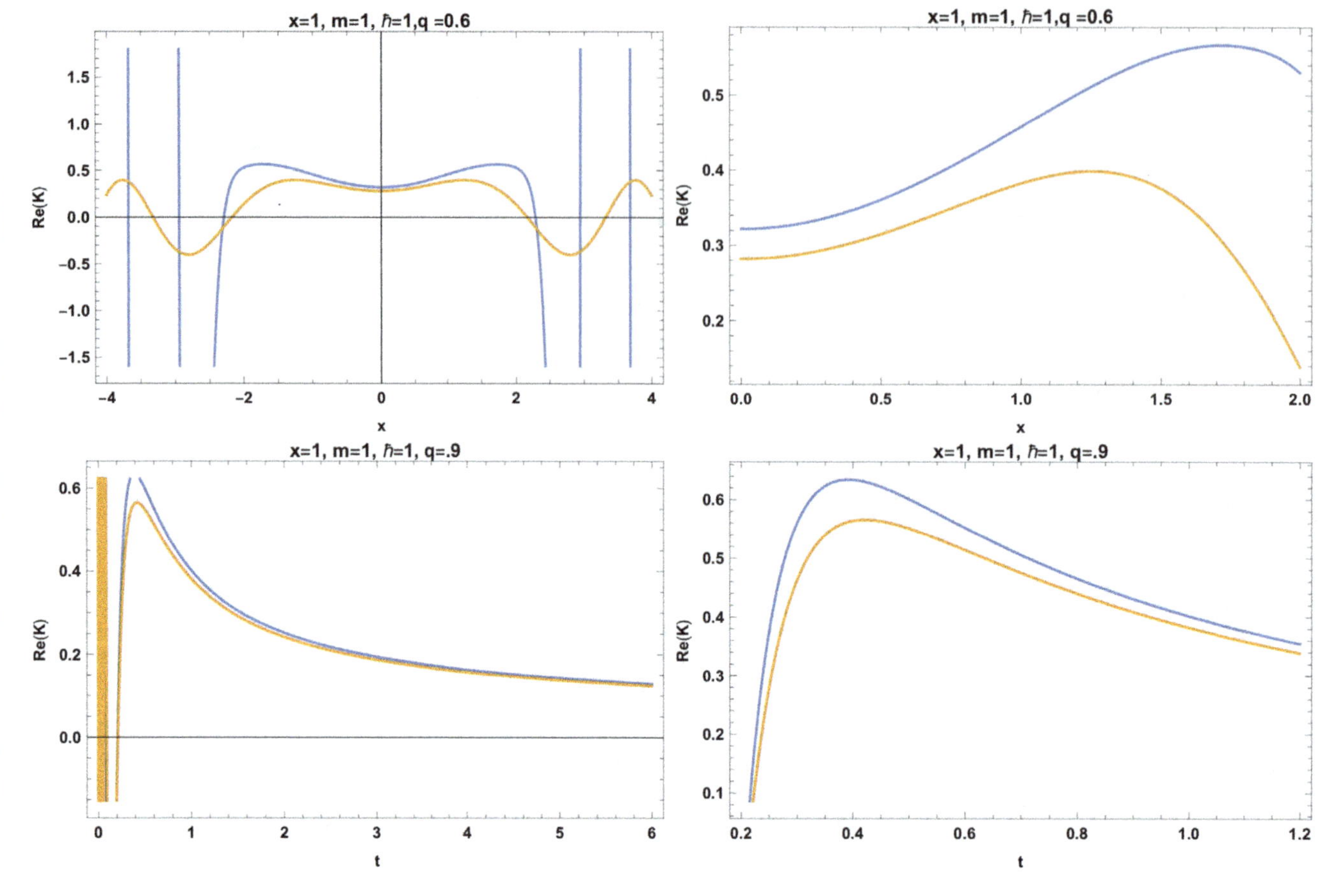

FIGURE 5 | Real parts of the modified propagator (blue line) vs. standard propagator (yellow line), for the free particle for the modified statistics of Tsallis for $q = 0.9$. We set the mass and the Planck constant to unity. Imposing $S_{cl} \lesssim \hbar$, it translates for fixed $x = 1$ in $t \gtrsim 1/2$ and for fixed $t = 1$ translates in $x^2 \lesssim 2$.

Next, following a similar procedure as for the free particle, we compute the generalized Kernel and normalize it. The unnormalized Kernel is given by the following:

$$K(x,t,1) = \exp\left(-\frac{1}{2}\lambda m\omega\cot(\omega t)x^2\right). \tag{40}$$

Now, we compute the next terms of the Kernel up to third order, which are as follows:

$$\begin{aligned} K(x,t,2) &= \left(\frac{\lambda}{2}\right)^2 \frac{\partial^2}{\partial\lambda^2}\exp(-2\lambda A) \\ &= \frac{1}{4}\lambda^2 m^2\omega^2 x^4 \exp\left(-\lambda m\omega\cot(\omega t)x^2\right), \end{aligned} \tag{41}$$

$$\begin{aligned} K(x,t,3) &= \frac{1}{2}\left(-\frac{\lambda^2}{3^2}\partial_\lambda^2 + 2\frac{\lambda^3}{3^3}\partial_\lambda^3 + 3\frac{\lambda^4}{3^4}\partial_\lambda^4\right)\exp(-3\lambda A) \\ &= \left(-\frac{1}{8}\lambda^2 m^2\omega^2\cot^2(\omega t)x^4 - \frac{1}{8}\lambda^3 m^3\omega^3\cot^3(\omega t)x^6 \right. \\ &\quad \left. + \frac{3}{32}\lambda^4 m^4\omega^4\cot^4(\omega t)x^8\right) \times \exp\left(\frac{-3}{2}\lambda m\omega x^2\cot(\omega t)\right). \end{aligned} \tag{42}$$

Thus, the total normalized propagator up to third order reads

$$\begin{aligned} K_+(x,t) = \frac{N_+}{\sqrt{2\pi}}\Bigg(&\left(\frac{m\omega\cot(t\omega)}{i\hbar}\right)^{\frac{1}{2}}\exp\left(\frac{-m\omega x^2\cot(t\omega)}{2i\hbar}\right) \\ &+ \frac{1}{4}\left(\frac{m\omega\cot(t\omega)}{i\hbar}\right)^{\frac{5}{2}}x^4\exp\left(\frac{-m\omega x^2\cot(t\omega)}{i\hbar}\right) - \frac{1}{8}\left[\left(\frac{m\omega\cot(t\omega)}{i\hbar}\right)^{\frac{5}{2}}x^4\right. \\ &\left. + \left(\frac{m\omega\cot(t\omega)}{i\hbar}\right)^{\frac{7}{2}}x^6 - \frac{3}{4}\left(\frac{m\omega\cot(t\omega)}{i\hbar}\right)^{\frac{9}{2}}x^8\right]\exp\left(\frac{-3m\omega x^2\cot(t\omega)}{2i\hbar}\right) + \dots\Bigg) \end{aligned} \tag{43}$$

where $N_+ = \frac{1}{\sqrt{\cos(\omega T)}}\frac{1}{\left(1+\frac{3}{16\sqrt{2}}+\frac{1}{96\sqrt{3}}\right)}$. The relative normalization of the modified propagator with respect to the normalization of the usual propagator is a result obtained in **Section 5** [see formulas (**31–33**)]. This result is universal, i.e., independent of the action. In **Figure 6**, we compare the propagator for the harmonic oscillator for the standard Quantropy and for the one based on S_+ and S_- statistics. We are interested in the quantum regime given by $S_{cl} \approx \hbar$. There are noticeable effects in that regime. Outside the quantum region, oscillations grow as in the usual case [34]. This behavior occurs in the classical region in which the modified Kernels will also not contribute.

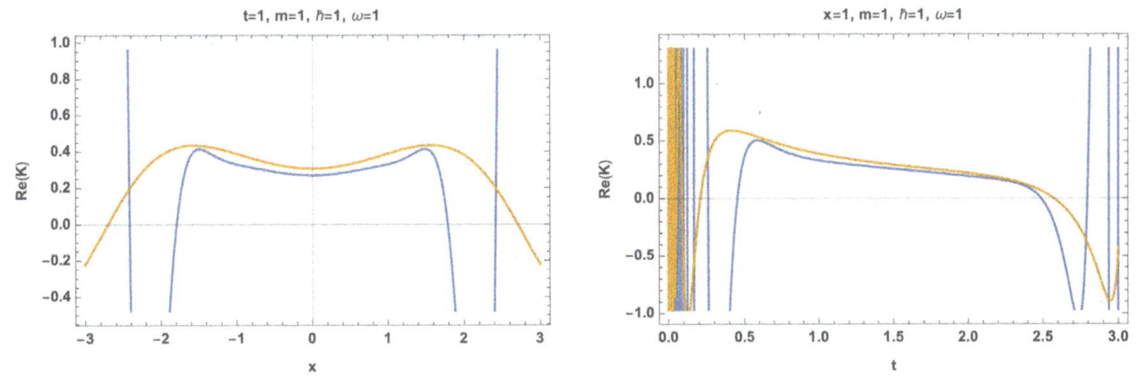

FIGURE 6 | We set the units $m = w = \hbar = 1$. The left image shows the real parts of the normalized propagators for a fixed time $t = 1$, and the region $S_{cl} \lesssim \hbar$ is given by $|x| \lesssim 1.765$. The blue line corresponds to the modified propagator K_+; meanwhile, the yellow one represents the usual propagator $K(x, t, 1)$. The right plot is the comparison between the propagators amplitude for $x = 1$ where the blue and the orange are the correction and the usual, respectively, and the quantum region is given by $t \gtrsim 0.464$.

7 POTENTIAL BARRIER

In this section, we apply the formulation of Quantropy developed previously to compute the propagators associated to a particle in an infinite potential barrier given as follows:

$$V(x) = \begin{cases} 0 & x > 0 \\ \infty & x \leq 0 \end{cases}.$$

The standard unnormalized propagator for this problem is given as follows [38]:

$$K(x, t; x_0, 0) = \exp\left[\frac{im}{2\pi\hbar t}(x - x_o)^2\right] - \exp\left[\frac{im}{2\pi\hbar t}(-x - x_o)^2\right], \tag{44}$$

where x_0 is the initial position. It is important to specify the initial position since the particle cannot be located at $x \leq 0$; thus, we set $x_0 = \epsilon$, where ϵ is a small positive non zero parameter. This allows to make a comparison with the free particle case. Notice that if we set $x_0 = 0$, the propagator vanishes since the x dependence is quadratic.

Now, we compute the nonlinear propagator associated to the statistics S_+: $K_+(x, t; \epsilon, 0)$ up to third order, which is given by expression (24). Since our theory is nonlinear, the superposition principle is not valid, i.e., we cannot consider the difference between the propagators of two free particles, like is done for the usual propagator $K(x, t; x_0, 0)$, rather in the semiclassical regime, we consider the logarithm:

$$S_{cl} = \frac{\hbar}{i}\ln(K(x, t; x_0, 0)), \tag{45}$$

This allows to substitute S_{cl} into the series expansion **(21)** $K_+(x)$:

$$\begin{aligned} K_+(x, t; \epsilon, 0) = {} & \left[e^{\frac{im(x-\epsilon)^2}{2\pi\hbar t}} - e^{\frac{im(-x-\epsilon)^2}{2\pi\hbar t}}\right] \times \Bigg[1 + \Big(e^{\frac{im(x-\epsilon)^2}{2\pi\hbar t}} \\ & - e^{\frac{im(-x-\epsilon)^2}{2\pi\hbar t}}\Big)\ln^2\left(e^{\frac{im(x-\epsilon)^2}{2\pi\hbar t}} - e^{\frac{im(-x-\epsilon)^2}{2\pi\hbar t}}\right) \\ & + \frac{1}{2}\left(e^{\frac{im(x-\epsilon)^2}{2\pi\hbar t}} - e^{\frac{im(-x-\epsilon)^2}{2\pi\hbar t}}\right)^2 \Big[3\ln^4\left(e^{\frac{im(x-\epsilon)^2}{2\pi\hbar t}} - e^{\frac{im(-x-\epsilon)^2}{2\pi\hbar t}}\right) \\ & + 2\ln^3\left(e^{\frac{im(x-\epsilon)^2}{2\pi\hbar t}} - e^{\frac{im(-x-\epsilon)^2}{2\pi\hbar t}}\right) - \ln^2\left(e^{\frac{im(x-\epsilon)^2}{2\pi\hbar t}} - e^{\frac{im(-x-\epsilon)^2}{2\pi\hbar t}}\right) \\ & \times\Big] + \frac{1}{6}\left(e^{\frac{im(x-\epsilon)^2}{2\pi\hbar t}} - e^{\frac{im(-x-\epsilon)^2}{2\pi\hbar t}}\right)^3 \Big[16\ln^6\left(e^{\frac{im(x-\epsilon)^2}{2\pi\hbar t}} - e^{\frac{im(-x-\epsilon)^2}{2\pi\hbar t}}\right) \\ & + 24\ln^5\left(e^{\frac{im(x-\epsilon)^2}{2\pi\hbar t}} - e^{\frac{im(-x-\epsilon)^2}{2\pi\hbar t}}\right) - 6\ln^4\left(e^{\frac{im(x-\epsilon)^2}{2\pi\hbar t}} - e^{\frac{im(-x-\epsilon)^2}{2\pi\hbar t}}\right) \\ & - 4\ln^3\left(e^{\frac{im(x-\epsilon)^2}{2\pi\hbar t}} - e^{\frac{im(-x-\epsilon)^2}{2\pi\hbar t}}\right) + 3\ln^2\left(e^{\frac{im(x-\epsilon)^2}{2\pi\hbar t}} - e^{\frac{im(-x-\epsilon)^2}{2\pi\hbar t}}\right)\Big]\Bigg]. \end{aligned} \tag{46}$$

Note that the propagator is ill-defined if ϵ is zero. This propagator oscillates faster than the usual propagator in x, and its amplitude is greater. However, the global behavior of both propagators is quite similar, and the oscillations and the amplitude grow as x increases. For the time dependence in both cases, the propagator tends to zero as t grows (see the set of graphics in **Figure 7**).

In the usual case, the propagator $K(x, t; x_0, 0)$ has the following interpretation, the first part corresponds to the classical path of the free particle from $(x_0, 0)$ to (x, t), while the second part corresponds to the classical path of a free particle from $(x_0, 0)$ bouncing off the wall and going to (x, t). The modified propagator expression suggests a similar interpretation, i.e., the whole propagator can be considered as the sum of both classical paths with the leading terms given by the standard free particle and nonlinear corrections which can be interpreted as an effective potential.

8 FINAL REMARKS

In this work, we explore the novel concept of Quantropy in Quantum Mechanics (Q.M.), which constitutes the analog of the entropy in Statistical Mechanics (S.M.). Mathematically, Quantropy can be regarded as an analytical continuation of

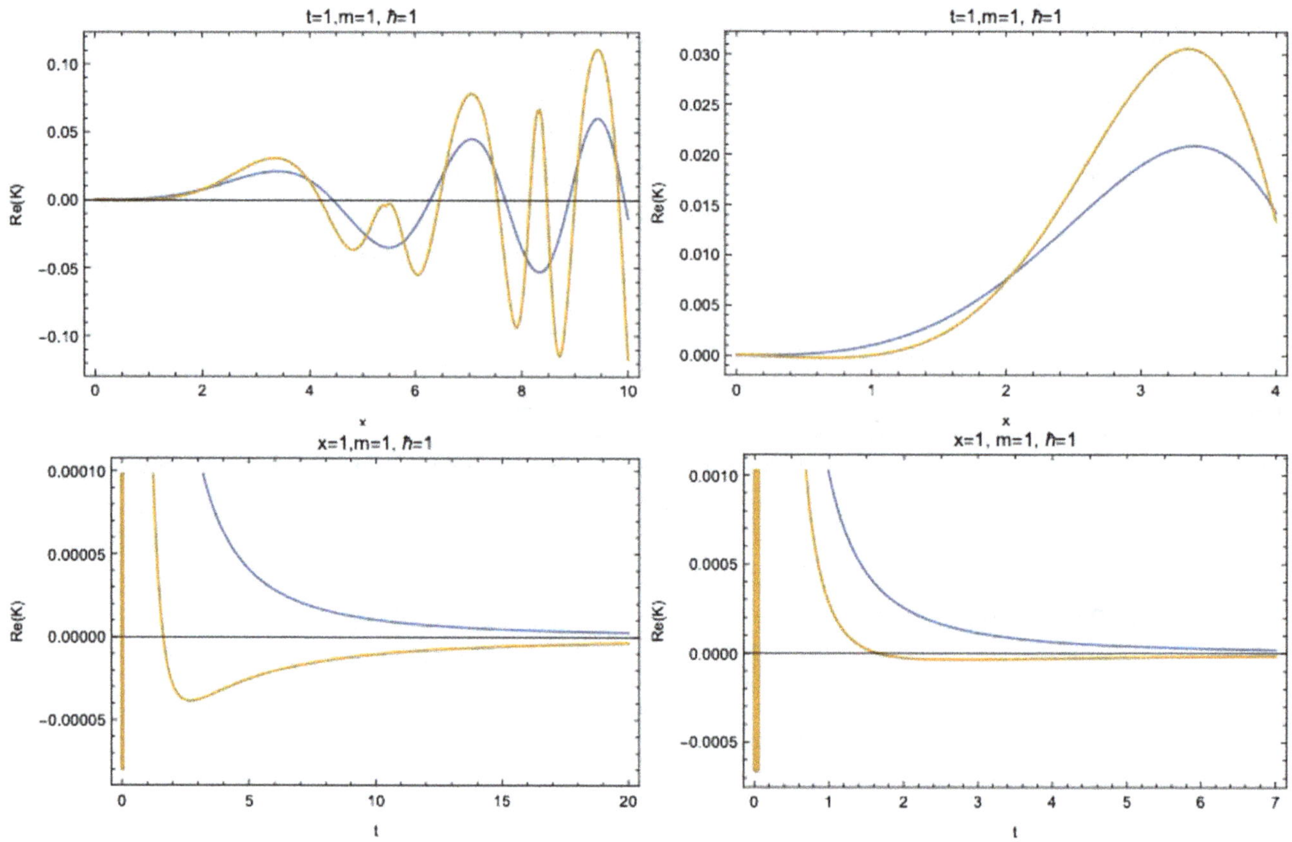

FIGURE 7 | We set the units $m = \hbar = 1$ and $\epsilon = 0.01$; the graphic shows the real parts of the modified propagator (blue line) vs. standard propagator (yellow line).

entropy, performed under the identification of the energy in S.M. to the action in Q.M., and the identification of the temperature to the Planck constant, the map reads: $E \rightarrow S$ and $T \rightarrow i\hbar$.

We establish a new definition of Quantropy, with the energy mapped to the classical action $E \rightarrow S_{cl}$, i.e., we consider that the main entity is the propagator $K(x)$ instead of the amplitude $a(x)$ of the path. Thus, we construct the propagator, a kind of integrated version of the Quantropy $Q_0 = -\int_X K(x) \ln K(x) dx$; in this way, the functional corresponding to the BG under the maximization procedure leads to $K_0 \sim \exp(iS_{cl}/\hbar)$.

We applied this concept to find the quantum mechanical implications of modified entropies S_+, S_-, and S_q, and the associated quantropies lead to generalized propagators, which imply a modified wave functional quantum mechanics (modified Schroedinger equation). Let us point out that the small probabilities limit of modified entropies S_+ and S_- leads to Boltzmann–Gibbs entropy. Similarly in the limit of small path amplitude, one recovers the original Quantropy from the modified functionals. This formalism could be a novel framework to study nonlinear quantum mechanics as those consider in [4, 12, 13]. We also provide an understating of modified propagators associated with Tsallis statistics, leading to wave functions corresponding to q-distributions.

Also, the result for the Tsallis statistics implies a propagator $K_q \sim \exp_q(iS_{cl}/\hbar)$, where $\exp_q(x)$ is the q-exponential; this is relevant since it makes contact with the Tsallis result of modified wave function for the free particle, whose solution is $\psi_q \sim \exp_q(i(kx - \omega t))$. The connection is due to the arguments of [34] in the discussion of the propagator for the free particle $K_0 \sim \exp(iS_{cl}/\hbar)$. They show that the propagator K_0 corresponds to the free particle wave function $\psi_0 \sim \exp(i(kx - \omega t))$. Thus, analogously K_q will lead to ψ_q. As a further work, we need to explore the relations in the case of the modified propagators K_+ and K_-. They will give rise to wave functions whose dependencies will be given by $\Psi_\pm = \exp_\pm(i(kx - \omega t))$, also for the free particle. In this case, we have a recurrent series solution but we do not have exact expressions for these generalized exponentials. As discussed, our proposal provides also generalized propagators K_+, K_-, and K_q for problems with interactions; we illustrated this by considering the K_+ associated with the harmonic oscillator and the infinite potential barrier.

There are hints from previous studies that the modified entropies considered here can be interpreted as linked with modified effective potentials. Therefore, these modifications to the free particle could be related to a usual quantum mechanics with an effective potential [9].

However, these effects could also lead to nonlinear quantum equations explored in the literature with modified wave functions [3, 12, 13, 26]. Furthermore, what we found here based on the concept of Quantropy could be linked to results for quantum systems in terms of usual entropy vs. the density matrix [10]. A system governed by a modified statistics (S_+, S_- or S_q) will lead to modified density-matrix distributions.

Moreover, the modified "propagators" K_+, K_-, and K_q actually are strictly no longer standard propagators because they lack the usual propagation property. This means that is not equivalent to propagate the particle from $(0,0)$ to (t_2, x_2), than to first propagate it from $(0,0)$ to (t_1, x_1) and then from (t_1, x_1) to (t_2, x_2). This occurs because, for example,

$$\int K_+(0,0;x_1,t_1)K_+(x_1,t_1;x_2,t_2)dx_1 \neq K_+(0,0;x_2,t_2).$$

This relates to the fact that in a quantum open systems where these generalized entropies are motivated, the nature of the processes is non-Markovian. Those systems in consideration are modeled with Master Equations (Stochastic) [39]. We consider that this formalism could be a natural framework to study nonlinear quantum mechanics.

We would like to explore further processes where the modified statistics in Quantropy play a central role. This could be done via modified wave functions, which could be interpreted as the usual quantum mechanics with an effective interaction [40] or from nonlinear quantum equations. The modified wave functions will correspond to the modified propagators obtained in this work. In this work, we obtained the modified propagators for the free particle, harmonic oscillator, and the infinite potential barrier associated to the different statistics K_+, K_-, and K_q, discussing the associated quantum behavior. We would also like to explore in future work the quantum mechanical evolution of other physical systems.

AUTHOR CONTRIBUTIONS

Here, NB certifies that every author in the present work contributed equally to the research and the writing of the manuscript on the following tasks: conceptualization; formal analysis; funding acquisition; investigation; project administration; resources; software; writing-original draft; and writing-review and editing. All authors contributed to the article and approved the submitted version.

ACKNOWLEDGMENTS

The authors thank Alejandro Cabo, Vishnu Jejjala, Oscar Loaiza-Brito, Miguel Sabido, Marco Ortega, Nelsón Flores-Gallegos, and Pablo López-Vázquez for useful discussions and comments. NB thanks PRODEP NPTC UGTO-515 Project, CIIC 181/2019 UGTO Project, CIIC 290/2020 UGTO Project, CIIC 2021 UGTO and CONACYT Project A1-S-37752. RS-S thanks CONACYT and PRODEP NPTC UDG-PTC-1368 Project and the MCFM PROINPEP-University of Guadalajara for supporting this work. OO thanks CONACYT Project 257919 and CIIC 188/2019 UGTO Project. NB, OO and CD thank the support of the University of Guanajuato.

REFERENCES

1. Obregón O. Superstatistics and gravitation. *Entropy* (2010) 12:2067–76. doi:10.3390/e12092067
2. Tsallis C. Possible generalization of Boltzmann-Gibbs statistics. *J Stat Phys* (1988) 52:479–87. doi:10.1007/BF01016429
3. Nobre FD, Rego-Monteiro MA, and Tsallis C. Nonlinear relativistic and quantum equations with a common type of solution. *Phys Rev Lett* (2011) 106:140601. doi:10.1103/PhysRevLett.106.140601
4. Sharma B, and Mital D. New nonadditive measures of inaccuracy. *J Math Sci* (1975) 10:122–33.
5. Beck C, and Cohen E. Physica A: statistical Mechanics and its applications. *Superstatistics* (2003) 322:267–275. doi:10.1016/S0378-4371(03)00019-0
6. Cabo Bizet N, Fuentes J, and Obregón O. Generalised asymptotic equivalence for extensive and non-extensive entropies. *EPL* (2019) 128:60004. doi:10.1209/0295-5075/128/60004
7. Obregón O, and Gil-Villegas A. Generalized information entropies depending only on the probability distribution. *Phys Rev E* (2013) 88:062146. doi:10.1103/PhysRevE.88.062146
8. Obregón O. Generalized information and entanglement entropy, gravitation and holography. *Int J Mod Phys A* (2015) 30:1530039. doi:10.1142/s0217751x15300392
9. Obregón O, Gil A, and Torres J. Computer simulation of effective potentials for generalized Boltzmann-gibbs statistics. *J Mol Liq* (2017) 248:364–9. doi:10.1016/j.molliq.2017.10.027
10. Cabo N, and Obregón O. Exploring the gauge/gravity duality of a generalized von neumann entropy. *Eur Phys J P* (2018) 133:55. doi:10.1140/epjp/i2018-11883-5
11. Baez J, and Pollard B. Quantropy. *Entropy* (2015) 17:772–89. doi:10.3390/e17020772
12. Nobre FD, Rego-Monteiro MA, and Tsallis C. A generalized nonlinear Schrödinger equation: classical field-theoretic approach. *Epl* (2012a) 97: 41001. doi:10.1209/0295-5075/97/41001
13. Nobre F, Rego-Monteiro M, and Tsallis C. Nonlinear q-generalizations of quantum equations: homogeneous and nonhomogeneous cases-an overview. *Entropy* (2017) 19:39. doi:10.3390/e19010039
14. Batle J, Plastino AR, Casas M, and Plastino A. Conditionalq-entropies and quantum separability: a numerical exploration. *J Phys A: Math Gen* (2002) 35: 10311–24. doi:10.1088/0305-4470/35/48/307
15. Nobre FD, Rego-Monteiro MA, and Tsallis C. A generalized nonlinear Schrödinger equation: classical field-theoretic approach. *Epl* (2012b) 97: 41001. doi:10.1209/0295-5075/97/41001
16. Plastino AR, and Tsallis C. Nonlinear Schroedinger equation in the presence of uniform acceleration. *J Math Phys* (2013) 54:041505. doi:10.1063/1.4798999
17. Curilef S, Plastino AR, and Plastino A. Tsallis' maximum entropy ansatz leading to exact analytical time dependent wave packet solutions of a nonlinear

Schrödinger equation. *Physica A: Stat Mech its Appl* (2013) 392:2631–42. doi:10.1016/j.physa.2012.12.041
18. Plastino A, and Rocca MC. From the hypergeometric differential equation to a non-linear Schrödinger one. *Phys Lett A* (2015) 379:2690–3. doi:10.1016/j.physleta.2015.08.015
19. Plastino AR, and Tsallis C. Dissipative effects in nonlinear klein-gordon dynamics. *Epl* (2016) 113:50005. doi:10.1209/0295-5075/113/50005
20. Plastino A, and Rocca MC. Hypergeometric foundations of fokker-planck like equations. *Phys Lett A* (2016) 380:1900–3. doi:10.1016/j.physleta.2016.03.047
21. Pennini F, Ferri G, and Plastino A. q-generalization of quantum phase-space representations. *Physica A: Stat Mech its Appl* (2015) 423:97–107. doi:10.1016/j.physa.2014.12.033
22. Majtey A, Plastino AR, and Plastino A. New features of quantum discord uncovered by q-entropies. *Physica A: Stat Mech its Appl* (2012) 391:2491–9. doi:10.1016/j.physa.2011.11.062
23. Nobre FD, and Plastino AR. A family of nonlinear Schrödinger equations admitting q -plane wave solutions. *Phys Lett A* (2017) 381:2457–62. doi:10.1016/j.physleta.2017.05.054
24. Plastino A, and Rocca MC. Quantum q-field theory: q-Schrödinger and q-Klein-Gordon fields. *EPL* (2017) 118:61004. doi:10.1209/0295-5075/118/61004
25. Plastino A, and Rocca MC. Tsallis' quantum q-fields. *Chin Phys C* (2018) 42:053102. doi:10.1088/1674-1137/42/5/053102
26. Chavanis P. Generalized euler, smoluchowski and Schrödinger equations admitting self-similar solutions with a tsallis invariant profile. *The Eur Phys J Plus* (2019) 134:353. doi:10.1140/epjp/i2019-12706-y
27. Khordad R, and Rastegar Sedehi HR. Magnetic susceptibility of graphene in non-commutative phase-space: extensive and non-extensive entropy. *Eur Phys J Plus* (2019) 134:133. doi:10.1140/epjp/i2019-12558-5
28. Sedehi HRR, and Khordad R. Entropy and specific heat of graphene at low and high temperatures under an external magnetic field. *Solid State Commun* (2020) 313:113911. doi:10.1016/j.ssc.2020.113911
29. Khordad R, Bahramiyan H, and Rastegar Sedehi HR. Effects of strain, magnetic field and temperature on entropy of a two dimensional gaas quantum dot under spin–orbit interaction. *Opt Quan Electronics* (2018) 50:294. doi:10.1007/s11082-018-1557-2
30. Servatkhah M, Khordad R, Firoozi A, Rastegar Sedehi HR, and Mohammadi A. Low temperature behavior of entropy and specific heat of a three dimensional quantum wire: shannon and tsallis entropies. *The Eur Phys J B* (2020) 93:111. doi:10.1140/epjb/e2020-10034-5
31. Abe S, and Okuyama S. Similarity between quantum mechanics and thermodynamics: entropy, temperature, and carnot cycle. *Phys Rev E* (2011) 83:021121. doi:10.1103/PhysRevE.83.021121
32. Yokoi Y, and Abe S. On quantum-mechanical origin of statistical mechanics. *J Phys Conf Ser* (2018) 1113:012012. doi:10.1088/1742-6596/1113/1/012012
33. Flego SP, Plastino A, and Plastino AR. Fisher information, the Hellmann-Feynman theorem, and the Jaynes reciprocity relations. *Ann Phys* (2011) 326:2533–43. doi:10.1016/j.aop.2011.07.009
34. Feynman R, and Hibbs A. *Quantum mechanics and path integrals international series in the earth and planetary sciences.* First Edition. New York, NY: McGraw-Hill College; Edición (1965).
35. Wentzel G. Eine Verallgemeinerung der Quantenbedingungen for die Zwecke der Wellenmechanik. *Z Physik* (1926) 38:518–29. doi:10.1007/bf01397171
36. Stack J. *Semiclassical approximation lecture notes in quantum mechanics.* Champaign, IL: University of Illinois Urbana-Champaign (2013).
37. Heller E. *The semiclassical way to dynamics and spectroscopy.* Princeton, NJ: Princeton University Press (2018).
38. Goodman M. Path integral solution to the infinite square well. *Am J Phys* (1981) 49:843–7. doi:10.1119/1.12720
39. de Vega I, and Alonso D. Dynamics of non-markovian open quantum systems. *Rev Mod Phys* (2017) 89:015001. doi:10.1103/RevModPhys.89.015001
40. Vignat C, Plastino A, Plastino AR, and Sanchez Dehesa J. Quantum potentials with q-Gaussian ground states. *Physica A: Stat Mech its Appl* (2012) 391:1068–73. doi:10.1016/j.physa.2011.09.031

2

On the Global Positivity Solutions of Non-Homogeneous Stochastic Differential Equations

Farai Julius Mhlanga* and Lazarus Rundora

Department of Mathematics and Applied Mathematics, University of Limpopo, Turfloop Campus, Sovenga, South Africa

Correspondence:
Farai Julius Mhlanga
farai.mhlanga@ul.ac.za

In this article, we treat the existence and uniqueness of strong solutions to the Cauchy problem of stochastic equations of the form $\mathrm{d}X_t = \alpha X_t \,\mathrm{d}t + \sigma X_t^\gamma \,\mathrm{d}B_t, X_0 = x > 0$. The construction does not require the drift and the diffusion coefficients to be Lipschitz continuous. Sufficient and necessary conditions for the existence of a global positive solution of non-homogeneous stochastic differential equations with a non-Lipschitzian diffusion coefficient are sought using probabilistic arguments. The special case $\gamma = 2$ and the general case, that is, $\gamma > 1$ are considered. A complete description of every possible behavior of the process X_t at the boundary points of the state interval is provided. For applications, the Cox-Ingersoll-Ross model is considered.

Keywords: geometric Brownian motion, Itô diffusion, Lipschitz continuous, scale function, speed measure

1. INTRODUCTION

The theory of stochastic differential equations was developed by [1]. Stochastic differential equations are valuable tools for modeling systems and processes with stochastic disturbances in many fields of science and engineering. For the general theory of stochastic differential equations, one can refer to [2–5]. Several authors have discussed results concerning the existence and uniqueness of solutions of stochastic differential equations [2, 6, 7]. Mishura and Posashkova [8] provided a sufficient condition on coefficients which ensures almost surely positivity of the trajectories of the solution of the stochastic differential equation with non-homogeneous coefficients and non-Lipschitz diffusion. Appleby et al. [9] investigated highly non-linear stochastic differential equations with delays and showed that properties on the coefficients of stochastic differential equations that guarantee stability also guarantee positivity of solutions as long as the initial value is non-zero. Xu et al. [10] investigated the global positive solution of a stochastic differential equation, where they generalized the mean-reverting constant elasticity of variance process by replacing the constant parameters with the parameters modulated by a continuous-time, finite-state, Markov chain. Zhang [11] treated the properties of solutions to stochastic differential equations with Sobolev diffusion coefficients and singular drifts. Bae et al. [12] proved the existence of and uniqueness of solution to stochastic differential equations under weakened and Hölder conditions and a weakened linear growth condition. Conditions for positivity of solutions of fractional stochastic differential equations with coefficients that do not satisfy the linear growth Lipschitz continuous conditions were obtained by [13].

The aim of this article is to prove the existence of a global positive solution of stochastic differential equations of the form

$$dX_t = \alpha X_t\, dt + \sigma X_t^{\gamma}\, dB_t, X_0 = x > 0, \tag{1}$$

where B_t is a standard Brownian motion, for different values of γ where α denotes the drift, σ denotes the volatility. $X = (X_t)_{t\geq 0}$ describes the underlying asset price. Such stochastic differential equations arise in modeling asset prices and interest rates on financial markets and it is crucial that X_t never becomes negative. Mao and Yuan [2] discussed the analytical properties when $\frac{1}{2} \leq \gamma \leq 1$ and showed that for a given initial value $X_0 = x > 0$, the solution of (1) remains positive with probability 1, namely, $X_t > 0$ for all $t \geq 0$ almost surely. The cases $\gamma = 0$ and $\gamma = 1$ give rise to the Ornstein-Uhlenbeck process and the Geometric Brownian motion, respectively, and this has been dealt with in the literature, see [4, 5, 14, 15]. When $\gamma > 1$, the diffusion coefficient of Equation (1) does not satisfy the linear growth condition, even though it is locally Lipschitz continuous. In view of this, it is not straightforward from the general theory of stochastic differential equations to obtain a unique global positive solution to Equation (1) that is defined for all $t \geq 0$. Nevertheless, there is a way to overcome such difficulties which we present in this article and we also provide detailed proofs that there is unique solutions to equations of the form (1). This article is an extension of the work done in [16] and [17] to non-homogeneous stochastic differential equations.

This article is structured as follows. In Section 2, we consider the existence of a positive global solution for non-homogeneous stochastic differential equations with non-Lipschitz coefficients. In particular, we treat the case $\gamma = 2$ and prove that if $\alpha \geq 0$ and $x \geq 0$ is arbitrary, then a unique strong solution of Equation (1) exists. Section 3 deals with the existence and uniqueness of a positive global solution to non-homogeneous stochastic differential equation (1). In particular, we consider the general case, that is, $\gamma > 1$. We provide a detailed proof of the existence of a unique solution to Equation (1). In Section 4, we investigate the behavior of the underlying process X_t at the boundaries of the state space $(0, \infty)$. The main tool used are simple probabilistic arguments. We only require the coefficients of our model to be continuous in the usual sense. In Section 5, we provide a brief conclusion.

2. EXISTENCE OF POSITIVE GLOBAL SOLUTIONS: $\gamma = 2$

We want to prove that a unique global positive solution to Equation (1) exists and investigate its properties. We notice that if $X_0 = x = 0$, then by strong uniqueness we have $X_t = 0$ for all $t \geq 0$. In addition, if a solution X_t exists for all $t < \tau(\omega) \leq \infty$ for some $x > 0$ and $X_T = 0$ for some $T = T(\omega) < \tau(\omega)$, then by the Strong Markov property we have$X_t = 0$ for all $t \in [T(\omega), \tau(\omega)]$. In particular, $x \geq 0$ implies that $X_t \geq 0$.

We call $(\Omega, \mathcal{F}, (\mathcal{F}_t), \mathbb{P})$ a stochastic basis if $(\Omega, \mathcal{F}, \mathbb{P})$ is a complete probability space and $(\mathcal{F}_t)$ is a right continuous filtration on Ω augmented by the $\mathbb{P}$-null sets. Let $B = (B_t)_{t\geq 0}$ be a standard Brownian motion defined on a stochastic basis $(\Omega, \mathcal{F}, (\mathcal{F}_t), \mathbb{P})$.

We consider a stochastic differential equation of the form

$$dX_t = \alpha(X_t)\, dt + \sigma(X_t)\, dB_t, \tag{2}$$

where the coefficients $\alpha : \mathbf{R} \to \mathbf{R}$ and $\sigma : \mathbf{R} \to \mathbf{R}$ are both Borel measurable functions. By the definition of stochastic differential, Equation (2) is equivalent to the stochastic integral equation:

$$X_t = x + \int_0^t \alpha(X_s)\, ds + \int_0^t \sigma(X_s)\, dB_s. \tag{3}$$

Definition 2.1. *[2, p. 48] An* **R***-valued stochastic process* $\{X_t\}_{t\in[0,T]}$ *is called a solution of Equation (2) if it has the following properties:*

1. $\{X_t\}$ *is continuous and* $\mathcal{F}_t$*-adapted.*
2. $\int_0^t \mid \alpha(X_s) \mid ds < \infty$ *and* $\int_0^t \sigma^2(X_s)ds < \infty$.
3. *Equation (3) holds for every* $t \in [0, T]$ *with probability* 1.

Definition 2.2. *[18, p. 167] A solution* (X, B) *of Equation (2) defined on* $(\Omega, \mathcal{F}, (\mathcal{F}_t), \mathbb{P})$ *is said to be a strong solution if* X *is adapted to the filtration* $\{\mathcal{F}_t^B\}$*, that is, the filtration of* $B = (B_t)_{t\geq 0}$ *completed with respect to* $\mathbb{P}$.

Definition 2.3. *[7, p. 300] A weak solution is a triple* $((\Omega, \mathcal{F}, \mathbb{P}), (B, X), (\mathcal{F}_t))$ *where* $(\Omega, \mathcal{F}, \mathbb{P})$ *is a probability space,* $\{\mathcal{F}_t\}$ *is a filtration of sub-*σ*-fields of* $\mathcal{F}$ *satisfying the usual conditions, X is a continuous, adapted* **R***-valued process, B is the standard Brownian motion such that Equation (3) is satisfied.*

Remark 2.1.

1. Definition 2.2 says that if the probability space $(\Omega, \mathcal{F}, \mathbb{P})$, the filtration $\{\mathcal{F}_t\}_{t\geq 0}$, the Brownian motion B_t and the coefficients $\alpha(x)$ and $\sigma(x)$ are all given in advance, and then the solution X_t is constructed, such a solution is called a strong solution.
2. Definition 2.3 says that if we are only given the coefficients $\alpha(x)$ and $\sigma(x)$, and we are allowed to construct a suitable probability space, a filtration and find a solution to the Equation (2), then such a solution is called a weak solution.

A solution $\{X_t\}_{t\in[0,T]}$ is said to be unique if any other solution $\{\overline{X}_t\}$ is indistinguishable from $\{X_t\}$, that is

$$\mathbb{P}(X_t = \overline{X}_t \quad \forall\ 0 \leq t \leq T) = 1.$$

Following [19], we impose the following hypothesis:

(H) The drift coefficient is globally Lipschitz, that is, for all $x, y \in \mathbf{R}$,

$$\mid \alpha(x) - \alpha(y) \mid\leq K \mid x - y \mid \tag{4}$$

where K is a fixed constant, while the diffusion coefficient is globally Hölder continuous, that is, for all $x, y \in \mathbf{R}$,

$$\mid \sigma(x) - \sigma(y) \mid\leq h(\mid x - y \mid) \tag{5}$$

where $h:[0,\infty) \to [0,\infty)$ is a strictly increasing function with $h(0)=0$ and the integral

$$\int_0^{\varepsilon} \frac{du}{h^2(u)} = \infty, \quad \forall \varepsilon > 0.$$

It is known [see [19]] that under the hypothesis (H), the strong uniqueness solution holds for the stochastic differential equation (2).

For the case $\gamma = 2$, Equation (1) takes the form

$$dX_t = \alpha X_t \, dt + \sigma X_t^2 \, dB_t. \tag{6}$$

If we let $Y_t = \ln | X_t |$ then an application of Itô's formula yields

$$dY_t = (\alpha - \frac{1}{2}\sigma^2 X_t^2) \, dt + \sigma X_t \, dB_t \tag{7}$$

which is equivalent to

$$\ln | X_t | = \ln | x | + \alpha t - \frac{1}{2}\sigma^2 \int_0^t X_s^2 \, ds + \sigma \int_0^t X_s \, dB_s. \tag{8}$$

This solution presents a challenge as the coefficients in Equation (8) do not satisfy the linear growth and Lipschitz conditions. However, there is a way to go around this. In the next result we prove the existence of a global solution to Equation (6) following arguments presented in [20].

Theorem 2.1. *Suppose $\alpha \geq 0$ and $x \geq 0$ is arbitrary, then the stochastic differential equation of the form (6) has a unique, strong solution X_t defined for all $t \geq 0$.*

Proof: The result is proved by a truncation procedure. For each $n \geq 1$, we set $\alpha = \alpha_n(x)$ and the truncation function

$$\sigma_n(x) = \begin{cases} \sigma x^2 & \text{if } | x | \leq n, \\ \sigma n^2 & \text{if } | x | > n \end{cases}.$$

Then, $\alpha_n(x)$ and $\sigma_n(x)$ satisfy the hypothesis (H). Hence, there is a unique solution $X_t = X_t^{(n)}$ defined for all t to the equation

$$X_t^{(n)} = x + \int_0^t \alpha_n(X_s^{(n)}) \, ds + \int_0^t \sigma_n(X_s^{(n)}) \, dB_s. \tag{9}$$

Define the stopping time

$$\tau_n = \inf\{t > 0; | X_t^{(n)} | \geq n\}, \quad n = 1, 2, ... \tag{10}$$

Then, by strong uniqueness we have

$$X_t^{(n)}(\omega) = X_t^{(n+1)}(\omega) \text{ for all } t \leq \tau_n \text{ a.s.} \tag{11}$$

Therefore,

$$\tau_n = \inf\{t > 0; | X_t^{(n)} | \geq n\} < \inf\{t > 0; | X_t^{(n+1)} | \geq n+1\} = \tau_{n+1}. \tag{12}$$

Hence, $\{\tau_n\}$ is an increasing sequence of stopping times. Put

$$\tau(\omega) = \lim_{n\to\infty} \tau_n(\omega) \leq \infty.$$

Then, for $t < \tau(\omega)$, a process X_t can be defined by setting

$$X_t(\omega) = X_t^{(n)}(\omega), \quad \text{if} \quad t < \tau_n(\omega). \tag{13}$$

It is clear that if $t < \tau(\omega)$ then one can easily show that $t < \tau_n(\omega)$ for some n. Therefore, by (11), this defines $X_t(\omega)$ uniquely. Hence, we have

$$X_t = x + \int_0^t \alpha X_s \, ds + \int_0^t \sigma X_s^2 \, dB_s \quad \text{for} \quad t < \tau(\omega). \tag{14}$$

3. EXISTENCE AND UNIQUENESS OF POSITIVE GLOBAL SOLUTION

In this section, we provide a detailed proof that there is a unique positive global solution to Equation (1). In particular, we focus on the case $\gamma > 1$. To establish the existence of a unique positive global solution, we need the following result.

Lemma 3.1. *[3, p. 57] The coefficients of Equation (1) satisfy the local Lipschitz condition for given initial condition $X_0 = x > 0$, that is, for every integer $k > 1$, there exists a positive constant L_k and $x, y \in [0, k]$ such that*

$$| \alpha x - \alpha y |^2 + | \sigma x^\gamma - \sigma y^\gamma |^2 \leq L_k | x - y |^2. \tag{15}$$

Therefore, there exists a unique local solution to Equation (1).

We now state our result in the following theorem.

Theorem 3.1. *For any given initial value $X_0 = x > 0$, α and $\sigma > 0$ there exists a unique positive global solution X_t to Equation (1) on $t \geq 0$ for $\gamma > 1$.*

Proof: It is clear that the coefficients of Equation (1) are locally Lipschitz continuous. Therefore, for any given initial value $X_0 = x > 0$, there is a unique local solution X_t, $t \in [0, \tau(\omega)]$ of Equation (1) where $\tau(\omega)$ is the explosion time. To prove that the solution is global, it suffices to show that $\tau(\omega) = \infty$ almost surely. We prove this by contradiction. If $\tau(\omega) \neq \infty$, then we can find a pair of positive constants ϵ and T such that

$$\mathbb{P}(\tau(\omega) \leq T) > \epsilon. \tag{16}$$

For each integer $n > 1$, we define a stopping time

$$\tau_n = \inf\{t \geq 0 : | X_t | \geq n\}. \tag{17}$$

Since $\tau_n \to \tau(\omega)$ almost surely, we can find a sufficiently large n_0 for which

$$\mathbb{P}(\tau_n \leq T) > \frac{\epsilon}{2}, \quad \text{for all } n \geq n_0. \tag{18}$$

For $\theta, \beta > 0$, we define a function $V \in C^2$ as

$$V(X) := \theta\sqrt{X} + \beta X^{-2}, \tag{19}$$

which is continuously twice differentiable in X. We observe that $V(X) \to +\infty$ as $X \to +\infty$ or $X \to 0$. For any $0 < t < T$, an application of Itô formula gives

$$dV(X_t) = \mathcal{L}V(X_t)\,dt + \sigma X^\gamma \left(\frac{1}{2}\theta X_t^{-\frac{1}{2}} - 2\beta X_t^{-3}\right) dB_t, \tag{20}$$

where

$$\begin{aligned}\mathcal{L}V(X_t) &= \alpha X_t \left(\frac{1}{2}\theta X_t^{-\frac{1}{2}} - 2\beta X_t^{-3}\right) \\ &\quad + \frac{1}{2}\sigma^2 X_t^{2\gamma}\left(-\frac{1}{4}\theta X_t^{-\frac{2}{3}} + 6\beta X_t^{-4}\right).\end{aligned} \tag{21}$$

By boundedness of polynomials, there exists a constant K such that

$$\alpha X_t \left(\frac{1}{2}\theta X_t^{-\frac{1}{2}} - 2\beta X_t^{-3}\right) + \frac{1}{2}\sigma^2 X_t^{2\gamma}\left(-\frac{1}{4}\theta X_t^{-\frac{2}{3}} + 6\beta X_t^{-4}\right) \le K. \tag{22}$$

Therefore, for any $t \in [0, T]$

$$\begin{aligned}\mathbf{E}[V(X_{t\wedge\tau_n})] &= V(x) + \mathbf{E}\left[\int_0^{t\wedge\tau_n} \mathcal{L}V(X_s)\,ds\right] \le V(x) + KT \\ &\quad + K\mathbf{E}\left[\int_0^t \mathbf{E}[V(X_{s\wedge\tau_n})]\,ds\right].\end{aligned} \tag{23}$$

The application of the Grownwall inequality yields

$$\mathbf{E}[V(X_{T\wedge\tau_n})] \le [V(x) + KT]e^{KT} \tag{24}$$

which is equivalent to

$$\mathbf{E}[V(X_{\tau_n})\mathbf{1}_{\{\tau_n \le T\}}] \le [V(x) + KT]e^{KT}. \tag{25}$$

On the other hand, we define

$$M_n = \inf\{V(X_t) \mid X_t > n,\ t \in [0, T]\}. \tag{26}$$

As $n \to +\infty$, $M_n \to +\infty$. It now follows from (18) and (26) that

$$[V(x) + KT]e^{KT} \ge M_n \mathbb{P}(\{\tau_n \le T\}) \ge \frac{1}{2}\epsilon M_n. \tag{27}$$

Letting $n \to +\infty$ yields a contradiction, so we must have $\tau(\omega) = \infty$ almost surely. Therefore, there exists a unique positive global solution X_t to Equation (1) for all $t \ge 0$.

4. ANALYSIS OF THE SOLUTION AT THE BOUNDARIES OF THE STATE SPACE

We now investigate the behavior of the underlying process X_t at the boundaries of the state space $(0, \infty)$ using probability arguments. X_t is the solution of the stochastic differential equation (1), where X_t is defined on the state space $(0, \infty)$, that is, the whole positive real line.

We first consider the Itô diffusion of the form

$$dX_t = \alpha(X_t)\,dt + \sigma(X_t)\,dB_t, \quad X_0 = x, \tag{28}$$

where $\alpha : \mathbf{R} \to \mathbf{R}$ and $\sigma : \mathbf{R} \to \mathbf{R}$ are functions satisfying the hypothesis (H). Note that here we do not have the time argument. We assume that the state space of X_t is a finite or infinite interval. Such a process is a continuous Markov process and under weak regularity conditions the drift coefficient $\alpha(x)$ and the diffusion coefficient $\sigma(x)$ are characterized, respectively, by

$$\alpha(x) = \lim_{h\downarrow 0} h^{-1}\mathbf{E}[X_h - x] \tag{29}$$

and

$$\sigma^2(x) = \lim_{h\downarrow 0} h^{-1}\mathbf{E}[(X_h - x)^2] = \lim_{h\downarrow 0} h^{-1}\mathrm{Var}(X_h). \tag{30}$$

For details about these conditions as well as the foregoing, see [21]. The above conditions can conveniently be weakened to give the following three conditions.

$$h^{-1}\mathbf{E}[(X_h - x)\mathbf{1}_{\{|X_h - x| \le 1\}}] \to \alpha(x), \tag{31}$$

$$h^{-1}\mathbf{E}[(X_h - x)^2\mathbf{1}_{\{|X_h - x| \le 1\}}] \to \sigma^2(x), \tag{32}$$

and

$$h^{-1}\mathbb{P}(|X_h - x| > \varepsilon) \to 0 \ \forall\ \varepsilon > 0, \tag{33}$$

where $\mathbf{1}_{\{\cdot\}}$ is the indicator function. These conditions enable us to perform the analysis of (28) without assuming the Lipschitz conditions to the coefficients. We will, however, assume that $\alpha(x)$ and $\sigma(x)$ are continuous.

Fix $q \in \mathbf{R}$ and define the scale function u by

$$u(x) = \int_q^x \exp\left(-\int_q^t \frac{2\alpha(z)}{\sigma^2(z)}\,dz\right) dt, \quad u(q) = 0. \tag{34}$$

The function u has a continuous, strictly positive derivative and u'' exists almost everywhere and satisfies

$$u''(x) = -\frac{2\alpha(x)}{\sigma^2(x)}u'(x). \tag{35}$$

We also introduce the speed measure

$$m(dx) = \frac{2}{u'(x)\sigma^2(x)}\,dx. \tag{36}$$

Now, let $p(t, x, y)$ be the transition density of X_t. Then, the Kolmogorov backward equation is given by

$$\frac{\partial p}{\partial t} = \frac{1}{2}\sigma^2(x)\frac{\partial^2 p}{\partial x^2} + \alpha(x)\frac{\partial p}{\partial x}. \tag{37}$$

At $t = 0$, $p(0, x, y) = \delta(x - y)$, where $\delta(\cdot)$ is Dirac's delta function.

Let $[a, b]$ be a fixed interval and start the process at $X_0 = x \in (a, b)$. We want to find the probability $p_+(x)$ that the process X_t hits b before it hits a. By the Markov property, we have

$$p_+(x) = \mathbf{E}[p_+(X_s)] + O\left(\mathbb{P}(|X_s - x| > \varepsilon)\right).$$

It follows from Equation (33) that

$$s^{-1}\left(\mathbb{P}(|X_s - x| > \varepsilon)\right) \to 0,$$

when $s \downarrow 0$ if $a + \varepsilon < x < b - \varepsilon$. Using the Itô's formula and Equation (37), we can show that $p_+(x)$ satisfies the Kolmogorov's backward equation

$$\frac{1}{2}\sigma^2(x)p''_+(x) + \alpha(x)p'_+(x) = 0, \tag{38}$$

for $x \in (a, b)$ with the boundary conditions $p_+(a) = 0$ and $p_+(b) = 1$. The explicit solution to Equation (38) is

$$p_+(x) = A\int_q^x \exp\left(-\int_q^t \frac{2\alpha(z)}{\sigma^2(z)}\,\mathrm{d}z\right)\,\mathrm{d}t + B. \tag{39}$$

We can write Equation (39) in the form

$$p_+(x) = Au(x) + B, \tag{40}$$

where $u(x)$ is of the form Equation (34) for a fixed $q \in (a, b)$, with A and B constants. Now, an application of boundary conditions $p_+(a) = 0$ and $p_+(b) = 1$ gives:

$$A = \frac{1}{u(b) - u(a)} \quad \text{and} \quad B = \frac{-u(a)}{u(b) - u(a)},$$

so that

$$p_+(x) = \frac{u(x) - u(a)}{u(b) - u(a)}. \tag{41}$$

Equations (34) and (41) will be important when applied to our specific problem.

Following similar arguments, we define

$$e(x) = \mathbf{E}[T_{ab}], \tag{42}$$

where $T_{ab} = \inf\{t > 0 : X_t \notin (a, b)\}$. An application of the Markov property gives

$$e(x) = s + \mathbf{E}[e(X_s)] + O(\mathbb{P}(|X_s - x| > \varepsilon)).$$

Dividing by s and letting s tend to 0 and an application of the Itô formula gives

$$\frac{1}{2}\sigma^2(x)e''(x) + \alpha(x)e'(x) = -1. \tag{43}$$

This equation can be solved by the standard Green function techniques as follows. The corresponding homogeneous equation is

$$e''(x) + \frac{2\alpha(x)}{\sigma^2(x)}e'(x) = 0$$

and its solution is

$$e(x) = \frac{u(x) - u(a)}{u(b) - u(a)}, \tag{44}$$

with boundary conditions $e(a) = 0$ and $e(b) = 1$ where $u(x)$ is defined in Equation (34). The Green function, $G_{(a,b)}(x, y)$, is calculated as

$$G_{(a,b)}(x, y) = \begin{cases} \frac{1}{W}\cdot e_1(x)e_2(y) & \text{if } x \le y, \\ \frac{1}{W}\cdot e_1(y)e_2(x) & \text{if } x \ge y, \end{cases} \tag{45}$$

where e_1 and e_2 take the form of Equation (44) and W is the Wronskian given by

$$W = \frac{u'(x)}{u(b) - u(a)}.$$

Therefore, the solution to Equation (43) is given by

$$e(x) = \int_a^b G_{(a,b)}(x, y)m(\mathrm{d}y),$$

where $G_{(a,b)}(x, y)$ is given by

$$G_{(a,b)}(x, y) = \begin{cases} \frac{2(u(x)-u(a))(u(b)-u(y))}{u(b)-u(a)} & \text{if} \quad x \le y, \\ \frac{2(u(y)-u(a))(u(b)-u(x))}{u(b)-u(a)} & \text{if} \quad x > y, \end{cases} \tag{46}$$

and $m(dy)$ is given by Equation (36).

We now consider Equation (1). We note that the diffusion coefficient $\sigma(x) = \sigma x^\gamma$ in Equation (1) is defined only on $(0, \infty)$, that is, the state space of the process is made up of the positive reals. The process X_t in Equation (1) is a diffusion process, and the coefficients σ and α are continuous on $(0, \infty)$. Following the arguments in [17], we investigate the behavior of Equation (1) at the boundaries of the state space. It is of interest whether or not the boundary points 0 and/or ∞ can be reached by the process X_t in a finite time.

A boundary point is said to be accessible if it can be reached in finite time with positive probability. Otherwise it is inaccessible [17]. The accessible boundary points are of two different types, namely, the exit and regular boundary points. For the exit boundary, the process is absorbed after the boundary is reached while the regular boundary point is imposed on a standard Brownian motion and could either be absorbed or reflected once the boundary is reached. The inaccessible boundaries are also of two types, namely, the entrance and natural boundary points. The boundary is said to be of entrance type if it is possible to start the process at infinity and then reach the interior of the state interval, otherwise it is called natural.

Let $[a, b]$ be a fixed interval and the process X_t starts in $X_0 = x \in (a, b)$. Let α and σ be continuous on a state interval whose interior is (c, d). We note that we may have $c = -\infty$ and/or $d = +\infty$. It is assumed that $\sigma^2(x) > 0$ on (c, d). Further, let $(\tilde{c}, \tilde{d}) = (u(c), u(d))$, where u is the scale function given by Equation (34).

Definition 4.1. *A natural upper boundary point d is said to be attracting if there is a positive probability that X_t shall converge to d as $t \to \infty$.*

The following classification theorem, taken from [17], will be the framework of the analysis of Equation (1).

Theorem 4.1. *Let u be the scale function given by Equation (34) and m(dy) be the speed measure given by Equation (36). Let b be a point in the interior of the state space (c, d). Then, the following statements hold.*

1. *A necessary and sufficient condition for d to be accessible is that* $u(d) < \infty$ *and* $\int_b^d (u(d) - u(y))m(\mathrm{d}y) < \infty$.
2. *An accessible boundary point d is regular if and only if* $\int_b^d m(\mathrm{d}y) < \infty$. *Otherwise it is exit boundary.*
3. *An inaccessible boundary point d is natural if and only if* $\int_b^d u(y)m(\mathrm{d}y) = \infty$.
4. *A natural boundary point d is attracting if and only if* $u(d) < \infty$ *and at the same time* $\int_b^d m(\mathrm{d}y) = \infty$.

We are now in a position of analyzing the non-homogeneous stochastic differential equation (1), repeated here for convenience,

$$\mathrm{d}X_t = \alpha X_t\,\mathrm{d}t + \sigma X_t^{\gamma}\,\mathrm{d}B_t,\ \ \gamma > 1,\ \ X_0 = x > 0. \tag{47}$$

This is a diffusion process with $\alpha(x) = \alpha x$, $\sigma(x) = \sigma x^{\gamma}$ and natural state interval $c = 0$ to $d = \infty$. Let b be a point in the interior of this state interval. From Equation (34) and (36), we calculate the scale function and speed measure, respectively, corresponding to Equation (47) to be

$$u(x) = \frac{\sigma^2}{2\alpha}\left(b^{2\gamma-1} - x^{2\gamma-1}\exp\left(\frac{\alpha}{\sigma^2(1-\gamma)}(b^{2-2\gamma} - x^{2-2\gamma})\right)\right),$$

and

$$m(dy) = \frac{1}{\sigma^2 y^{2\gamma}}\exp\left(-\frac{\alpha}{\sigma^2(1-\gamma)}(b^{2-2\gamma} - y^{2-2\gamma})\right)\mathrm{d}y. \tag{48}$$

It remains only to classify our boundary points on the basis of these results. We note that

$$u(d) = \frac{\sigma^2}{2\alpha}\left(b^{2\gamma-1} - d^{2\gamma-1}\exp\left(\frac{\alpha}{\sigma^2(1-\gamma)}(b^{2-2\gamma} - d^{2-2\gamma})\right)\right).$$

Since $d = \infty$ in our state space $(0, \infty)$, we use a limiting argument:

$$\lim_{d\to\infty} u(d) = \frac{\sigma^2 b^{2\gamma-1}}{2\alpha},$$

provided $0 < \gamma < \frac{1}{2}$. Now, since b is a finite fixed point in $(0, \infty)$, the limit is finite. We also need to investigate the integral

$$\int_b^d (u(d) - u(y))m(\mathrm{d}y) = \int_b^d \frac{1}{2\alpha y}\,\mathrm{d}y - \frac{1}{2\alpha}\exp\left(\frac{-\alpha}{\sigma^2(1-\gamma)}d^{2-2\gamma}\right)d^{2\gamma-1} \times \int_b^d \frac{1}{y^{2\gamma}}\exp\left(\frac{\alpha}{\sigma^2(1-\gamma)}y^{2-2\gamma}\right)\mathrm{d}y. \tag{49}$$

We observe that as $d \to \infty$ the term $\exp\left(-\frac{\alpha}{\sigma^2(1-\gamma)}d^{2-2\gamma}\right)$ approaches 0 provided $0 < \gamma < 1$. So in this case we remain with

$$\int_b^d (u(d) - u(y))m(\mathrm{d}y) \approx \int_b^d \frac{1}{2\alpha y}\,\mathrm{d}y \longrightarrow \infty \ \ as\ \ d \to \infty.$$

We therefore, according to Theorem 4.1, conclude that the upper boundary point $d = \infty$ is inaccessible if $0 < \gamma < 1$. Now for the lower boundary point 0 we have

$$u(0) = \frac{\sigma^2 b^{2\gamma-1}}{2\alpha} < \infty,$$

since b is a finite fixed point in the state space $(0, \infty)$. We also have

$$\int_0^b (u(0) - u(y))m(\mathrm{d}y) = \int_0^b \frac{1}{2\alpha y}\,\mathrm{d}y \to \infty,\ \text{as}\ y \to 0.$$

This shows that the boundary point 0 is inaccessible for all $\gamma \neq 1$. It remains to establish whether our boundary points are natural or not. Theorem 4.1 says the boundary point d is natural if and only if

$$\int_b^d u(y)m(\mathrm{d}y) = \infty.$$

Now,

$$\int_b^d u(y)m(\mathrm{d}y) = \int_b^d \left(\frac{b^{2\gamma-1}}{2\alpha y^{2\gamma}}\exp\left(-\frac{\alpha}{\sigma^2(1-\gamma)}(b^{2-2\gamma} - y^{2-2\gamma})\right) - \frac{1}{2\alpha y}\right)\mathrm{d}y$$
$$= \frac{b^{2\gamma-1}\exp\left(\frac{-\alpha}{\sigma^2(1-\gamma)}b^{2-2\gamma}\right)}{2\alpha}\int_b^d \frac{1}{y^{2\gamma}}\exp\left(\frac{\alpha}{\sigma^2(1-\gamma)}y^{2-2\gamma}\right)\mathrm{d}y - \int_b^d \frac{1}{2\alpha y}\,\mathrm{d}y.$$

We observe that for $0 < \gamma < 1$ the integral

$$\int_b^d \frac{1}{y^{2\gamma}}\exp\left(\frac{\alpha}{\sigma^2(1-\gamma)}y^{2-2\gamma}\right)\mathrm{d}y$$

explodes to infinity very fast as $d \to \infty$. Although the second integral,

$$\int_b^d \frac{1}{2\alpha y}\, dy,$$

also tends to infinity as $d \to \infty$, the whole integral $\int_b^d u(y)m(dy)$ tends to infinity as $d \to \infty$ because the second integral goes to infinity very slowly as compared to the first one. Hence,

$$\int_b^d u(y)m(dy) = \infty,$$

provided $0 < \gamma < 1$. This tells us that the boundary point $d = \infty$ is natural if $0 < \gamma < 1$.
Using similar arguments, we can show that

$$\int_0^b u(y)m(dy) = \infty.$$

Therefore, for $0 < \gamma < 1$, the boundary point 0 is natural. Next, we investigate if our natural boundary points are attracting. According to Theorem 4.1 the boundary point d is attracting if and only if $u(d) < \infty$ and at the same time $\int_b^d m(dy) = \infty$. Now we have already seen that $u(d) < \infty$ if $d = 0$ and/or $d = \infty$ provided $0 < \gamma < 1$. Further

$$\int_b^d m(dy) = \frac{1}{\sigma^2} \exp\left(-\frac{\alpha}{\sigma^2(1-\gamma)} b^{2-2\gamma}\right) \times \int_b^d \frac{1}{y^{2\gamma}} \exp\left(\frac{\alpha}{\sigma^2(1-\gamma)} y^{2-2\gamma}\right) dy,$$

which, for the reason given before, explodes to infinity as $d \to \infty$ for all $0 < \gamma < 1$. Therefore,

$$\int_b^\infty m(dy) = \infty,$$

for $0 < \gamma < 1$. Hence, the upper boundary point $d = \infty$ is attracting for $0 < \gamma < 1$. Similarly the lower boundary point 0 is shown to be attracting.

Now, we have established that both boundary points are attracting when $0 < \gamma < 1$. In this case we will show that, by Equation (41), our process will converge to ∞ with probability $p_+(x)$, where $x = X_0 \in (0, \infty)$. It turns out that

$$p_+(x) = \frac{\int_0^x \exp\left(-\frac{\alpha}{\sigma^2(1-\gamma)} y^{2-2\gamma}\right) dy}{\int_0^\infty \exp\left(-\frac{\alpha}{\sigma^2(1-\gamma)} y^{2-2\gamma}\right) dy}, \quad 0 < \gamma < 1. \quad (50)$$

Evaluating the integrals yields

$$p_+(x) = \lim_{y \to \infty} \left(\frac{x}{y}\right)^{2\gamma-1} \exp\left(\frac{\alpha}{\sigma^2(1-\gamma)}(y^{2-2\gamma} - x^{2-2\gamma})\right) = 0,$$

for $\frac{1}{2} < \gamma < 1$. This shows that although the upper boundary $d = \infty$ is attracting for $0 < \gamma < 1$, the process X_t will not converge to ∞ if $\frac{1}{2} < \gamma < 1$. Furthermore, the process X_t converges to 0 with probability $1 - p_+(x)$ which turns out to be 1 in this case. Thus it is certain that X_t will converge to 0 when $\frac{1}{2} < \gamma < 1$. We observe that if $0 < \gamma < \frac{1}{2}$ we have a problem since, in this case, it is not possible to proceed using a probabilistic argument. Our analysis is not complete if we do not consider the case $\gamma > 1$. We now proceed to make this analysis. As seen earlier

$$u(d) = \frac{\sigma^2}{2\alpha}\left(b^{2\gamma-1} - d^{2\gamma-1} \exp\left(\frac{\alpha}{\sigma^2(1-\gamma)}(b^{2-2\gamma} - d^{2-2\gamma})\right)\right).$$

If $\gamma > 1$, for example, if $\gamma = 2$, we have

$$u(d) \longrightarrow -\infty \quad as \quad d \to \infty,$$

since α, σ and b are fixed positive numbers. Therefore, we have

$$\lim_{d \to \infty} u(d) < \infty \quad \forall \ \gamma > 1.$$

Observe also that for such γ we have that

$$u(0) = \frac{\sigma^2 b^{2\gamma-1}}{2\alpha} < \infty,$$

effectively. Now, we consider again Equation (49). If $\gamma > 1$, for instance, $\gamma = 2$, we have

$$\int_b^d (u(d) - u(y))m(dy) = \int_b^d \frac{1}{2\alpha y}\, dy - \frac{d^3}{2\alpha} e^{\frac{\alpha}{\sigma^2 d^2}} \int_b^d \frac{1}{y^4} e^{-\frac{\alpha}{\sigma^2 y^2}}\, dy.$$

We immediately observe that as $d \to \infty$,
$\int_b^d (u(d) - u(y))m(dy) \to -\infty$ since the integral $\int_b^d \frac{1}{2\alpha y}\, dy$ goes to ∞ very slowly. Therefore, effectively we have

$$\int_b^d (u(d) - u(y))m(dy) < \infty,$$

for the boundary point $d = \infty$ and whenever $\gamma > 1$. This, together with $u(d) < \infty$ for $d = \infty$ and $\gamma > 1$, shows that the upper boundary point $d = \infty$ is always accessible whenever $\gamma > 1$. However, it is clear that if $d = 0$,

$$\int_0^b (u(0) - u(y))m(dy) = \int_0^b \frac{1}{2\alpha y}\, dy = \infty,$$

which shows that the lower boundary point 0 is always not accessible for $\gamma > 1$. In fact, the boundary point 0 is always inaccessible for all $\gamma \neq 1$ as also shown earlier. So from definition we have seen that the upper boundary point ∞ can be reached in finite time with positive probability provided $\gamma > 1$.

Finally we want to classify the accessible boundary point ∞, that is. is it regular or exit? From Theorem 4.1 it is regular if and only if $\int_b^d m(\mathrm{d}y) < \infty$. Now, as obtained earlier on

$$\int_b^d m(\mathrm{d}y) = \frac{1}{\sigma^2}\exp\left(-\frac{\alpha}{\sigma^2(1-\gamma)}b^{2-2\gamma}\right) \times \int_b^d \frac{1}{y^{2\gamma}}\exp\left(\frac{\alpha}{\sigma^2(1-\gamma)}y^{2-2\gamma}\right)\mathrm{d}y.$$

If $\gamma > 1$, the integral is always less than ∞ because of the negative exponent since α, σ are fixed positive numbers. So we have

$$\int_b^d m(\mathrm{d}y) < \infty \quad \forall \gamma > 1$$

since in this case the exponent is always negative. Hence for $\gamma > 1$, the accessible upper boundary point ∞ is always regular, i.e., apart from absorption and reflection there are also other possibilities after the boundary point is reached.
We, therefore, have the following result.

Theorem 4.2. *Let $x \in (0, \infty)$ with $\alpha \in \boldsymbol{R}$ arbitrary. Then, the stochastic differential equation (1) has a unique strong global solution $X_t : t \geq 0$. The solution has the following properties:*

1. *$x = 0$ implies that $X_t = 0$ for all $t > 0$ and $x \geq 0$ implies $X_t > 0$ for all $t \geq 0$.*
2. *If $\frac{1}{2} < \gamma < 1$, then $\lim_{t\to\infty} X_t = 0$ with probability $1 - p_+(x)$ where $p_+(x)$ is given by Equation (50).*
3. *If $\gamma > 1$, then $\lim_{t\to\infty} X_t = \infty$ with positive probability.*
4. *If $\gamma = 1$, we have the usual Geometric Brownian motion whereas if $\gamma = 0$, we have the Ornstein-Uhlenbeck process.*

In mathematical finance, our result is of particular interest for the Cox-Ingersoll-Roll (CIR) model which describes the stochastic evolution of interest rates $(r_t)_{t\geq 0}$ by the stochastic differential equation

$$\mathrm{d}r_t = \alpha(\mu - r)\,\mathrm{d}t + \sigma\sqrt{r_t}\,\mathrm{d}B_t, \quad t \geq 0,$$

with $r_0 \geq 0$ and $\alpha\mu \geq \frac{1}{2}\sigma^2$ where α, μ and σ denote real constants.

5. CONCLUDING REMARKS

In this article, we proved the existence of global positive solutions to non-homogeneous stochastic differential equations whose diffusion coefficient is non-Lispchitz. We relied on both the classical sense and probabilistic arguments. We provided detailed proofs in both cases. The probability arguments save as an alternative method of dealing with non-homogeneous stochastic differential equations where classical methods cannot be applied. Using the scale function and the speed of measure, we provided a complete classification of boundary types and boundary behavior of Equation (1). The results of this article can be applied to Cox-Ingersoll-Ross model. In addition, the positivity of solutions is important to other non-linear models that arise in sciences and engineering.

AUTHOR CONTRIBUTIONS

FM contributed to the conceptualization of the idea for research, development of the methodology, analysis of the model and writing up of the final article. LR conducted the primary research, identified the problem, and discussed the results. Both authors proof read the manuscript and approved the submitted version.

FUNDING

This work was supported by a Newton Fund's Operational Development Assistance grant, ID 32, under the SA-UK partnership. The grant is funded by the UK Department for Business, Energy and Industrial Strategy and Department of Higher Education and Training (DHET) and delivered by the British Council. For further information, please visit www.newtonfund.ac.uk.

ACKNOWLEDGMENTS

The authors are grateful to the referees for constructive comments and suggestions that have improved this article.

REFERENCES

1. Itô K. *On Stochastic Differential Equations.* New York, NY: American Mathematical Society (1951).
2. Mao X, Yuan C. *Stochastic Differential Equations with Markovian Switching.* London: Imperial College Press (2006). doi: 10.1142/p473
3. Mao X. *Stochastic Differential Equations and Applications.* London: Elsevier (2007). doi: 10.1533/9780857099402
4. Øksendal B. *Stochastic Differential Equations: An Introduction With Applications.* New York, NY: Springer Science & Business Media (2013).
5. Ikeda N, Watanabe S. *Stochastic Differential Equations and Diffusion Processes.* North-Holland: Elsevier (2014).
6. Protter PE. Stochastic differential equations. In: Rozavskii B, Yor M, editors. *Stochastic Integration and Differential Equations.* New York, NY: Springer (2005). p. 249–361. doi: 10.1007/978-3-662-10061-5_6
7. Karatzas I, Shreve S. *Brownian Motion and Stochastic Calculus.* vol. 113. New York, NY: Springer Science & Business Media (2012).
8. Mishura Y, Posashkova S. Positivity of solution of nonhomogeneous stochastic differential equation with non-lipschitz diffusion. *Theory Stochast Process.* (2008) 14:77–88. Available online at: http://tsp.imath.kiev.ua/files/172/1434_7.pdf
9. Appleby JA, Kelly C, Mao X, Rodkina A. Positivity and stabilisation for nonlinear stochastic delay differential equations. *Stochastics Int J Probabil Stochast Process.* (2009) 81:29–54. doi: 10.1080/17442500802214097

10. Xu R, Wu D, Yi R. Existence theorem for mean-reverting CEV process with regime switching. In: *2015 International Conference on Mechatronics, Electronic, Industrial and Control Engineering (MEIC-15)*. Atlantis Press (2015). p. 1560–3. doi: 10.2991/meic-15.2015.357
11. Zhang X. Stochastic differential equations with Sobolev diffusion and singular drift and applications. *Ann Appl Probabil.* (2016) 26:2697–732. doi: 10.1214/15-AAP1159
12. Bae MJ, Park CH, Kim YH. An existence and uniqueness theorem of stochastic differential equations and the properties of their solution. *J Appl Math Inform.* (2019) 37:491–506. doi: 10.14317/jami.2019.491
13. Kubilius K, Medziunas A. Positive solutions of the fractional SDEs with non-Lipschitz diffusion coefficient. *Mathematics*. MDPI (2021) 9:1–14. doi: 10.3390/math9010018
14. Duffie D. *Dynamic Asset Pricing Theory*. Princeton, NJ: Princeton University Press (2010).
15. Glasserman P. Monte carlo methods in financial engineering. In: *Stochastic Modelling and Applied Probability*. New York, NY: Springer-Verlag (2003). doi: 10.1007/978-0-387-21617-1
16. Rundora L. *Extension(s) of the Geometric Brownian Motion Model for Pricing of Assets*. Masters Dissertation, Harare: University of Zimbabwe (1997).
17. Helland I. *One-Dimensional Diffusion Processes and Their Boundaries*. (1996).
18. Revuz D, Yor M. *Continuous Martingales and Brownian Motion*. vol. 293. New York, NY: Springer Science & Business Media (2013).
19. Yamada T, Watanabe S. On the uniqueness of solutions of stochastic differential equations. *J Math Kyoto Univ.* (1971) 11:155–67. doi: 10.1215/kjm/1250523691
20. Lungu E, Øksendal B. Optimal harvesting from a population in a stochastic crowded environment. *Math Biosci.* (1997) 145:47–75. doi: 10.1016/S0025-5564(97)00029-1
21. Gualtierotti A. Statistics for stochastic processes: applications to engineering and finance. In: Dodge Y, editor. *Statistical Data Analysis and Inference*. North-Holland: Elsevier (1989). p. 543–55. doi: 10.1016/B978-0-444-88029-1.50054-X

3

Gibbs-Slice Sampling Algorithm for Estimating the Four-Parameter Logistic Model

Jiwei Zhang[1], Jing Lu[2], Hang Du[2,3] and Zhaoyuan Zhang[2,4]*

[1] Key Lab of Statistical Modeling and Data Analysis of Yunnan Province, School of Mathematics and Statistics, Yunnan University, Kunming, China, [2] Key Laboratory of Applied Statistics of MOE, School of Mathematics and Statistics, Northeast Normal University, Changchun, China, [3] School Affiliated to Longhua Institute of Educational Science, Shenzhen, China, [4] School of Mathematics and Statistics, Yili Normal University, Yili, China

***Correspondence:**
Jing Lu
luj282@nenu.edu.cn*

The four-parameter logistic (4PL) model has recently attracted much interest in educational testing and psychological measurement. This paper develops a new Gibbs-slice sampling algorithm for estimating the 4PL model parameters in a fully Bayesian framework. Here, the Gibbs algorithm is employed to improve the sampling efficiency by using the conjugate prior distributions in updating asymptote parameters. A slice sampling algorithm is used to update the 2PL model parameters, which overcomes the dependence of the Metropolis–Hastings algorithm on the proposal distribution (tuning parameters). In fact, the Gibbs-slice sampling algorithm not only improves the accuracy of parameter estimation, but also enhances sampling efficiency. Simulation studies are conducted to show the good performance of the proposed Gibbs-slice sampling algorithm and to investigate the impact of different choices of prior distribution on the accuracy of parameter estimation. Based on Markov chain Monte Carlo samples from the posterior distributions, the deviance information criterion and the logarithm of the pseudomarginal likelihood are considered to assess the model fittings. Moreover, a detailed analysis of PISA data is carried out to illustrate the proposed methodology.

Keywords: Bayesian inference, four-parameter logistic model, item response theory, model assessment, potential scale reduction factor, slice sampling algorithm

1. INTRODUCTION

Over the past four decades, item response theory (IRT) models have been extensively used in educational testing and psychological measurement (Lord and Novick, 1968; Van der Linden and Hambleton, 1997; Embretson and Reise, 2000; Baker and Kim, 2004). These are latent variable modeling techniques, in which the response probability is used to construct the interaction between an individual's "ability" and item level stimuli (difficulty, guessing, etc.), where the focus is on the pattern of responses rather than on composite or total score variables and linear regression theory. Specifically, IRT attempts to model individual ability using question-level performance instead of aggregating test-level performance, and it focuses more on the information provided by an individual on each question. In social sciences, IRT has been applied to attachment (Fraley et al., 2000), personality (Ferrando, 1994; Steinberg and Thissen, 1995; Gray-Little et al., 1997; Rouse et al., 1999), psychopathology (Reise and Waller, 2003; Loken and Rulison, 2010; Waller and Reise, 2010; Waller and Feuerstahler, 2017), attention deficit hyperactivity disorder (Lanza et al., 2005), and delinquency (Osgood et al., 2002), among others.

To explore these applications, it is necessary to establish how the appropriate IRT models should be built and what valuable educational psychological phenomena can be examined to guide practice. In the field of dichotomous IRT models, the one-parameter logistic (1PL) model and the Rasch model (Rasch, 1960), as well as their extensions, the two-parameter logistic model (2PL) (Birnbaum, 1957) and the three-parameter logistic model (3PL) (Birnbaum, 1968), have attracted increasing attention in recent years because of their attractive mathematical properties. However, compared with the widely used 1PL, 2PL, and 3PL models, the four-parameter logistic (4PL) model has languished in obscurity for nearly 30 years (Barton and Lord, 1981), although its importance has gradually been realized by many researchers over the past decade (Hessen, 2005; Loken and Rulison, 2010; Waller and Reise, 2010; Green, 2011; Liao et al., 2012; Yen et al., 2012; Magis, 2013; Waller and Feuerstahler, 2017). This growing interest can be attributed to the need to deal with a number of problems encountered in educational psychology, which can be explained well and indeed solved using the 4PL model. For example, in computerized adaptive testing (CAT), high-ability examinees might on occasion miss items that they should be able to answer correctly, owing to a number of reasons, including anxiety, carelessness, unfamiliarity with the computer environment, distraction by poor testing conditions, or even misreading of the question (Hockemeyer, 2002; Rulison and Loken, 2009). Chang and Ying (2008) demonstrated that the ability determined using the traditional 2PL model is underestimated when the examinee mistakenly answers several items at the beginning of the CAT. In addition, Rulison and Loken (2009) found that using the 3PL model could severely penalize a high-ability examinee who makes a careless error on an easy item (Barton and Lord, 1981; Rulison and Loken, 2009). In psychopathology studies, researchers found that subjects with severe psychopathological disorders may be reluctant to self-report their true attitudes, behaviors, and experiences, so it is obviously inappropriate to use the traditional 3PL model with lower asymptotic parameter to explain such behaviors (Reise and Waller, 2003; Waller and Reise, 2010). Descriptions of the applications of the 4PL model in other areas can be found in Osgood et al. (2002) and Tavares et al. (2004). In addition to the development of the 4PL model in terms of its applications, its theoretical properties have been investigated in some depth. For example, Ogasawara (2012) discussed the asymptotic distribution of the ability, and Magis (2013) systematically studied the properties of the information function and proposed a method for determining its maximum point.

The main reason why the 4PL model has not been more widely used is that an upper asymptotic parameter is added to the 3PL model, which makes parameter estimation more difficult. However, with the rapid development of computer technology in recent years, the estimation problem for complex models has been solved. At the same time, the development of statistical software makes it easier for psychometricians to study complex models such as the 4PL model. Several researchers have used existing software to estimate the 4PL model. For example, Waller and Feuerstahler (2017) investigated 4PL model item and person parameter estimations using marginal maximum likelihood (MML) with the *mirt* (Chalmers, 2012) package, which uses MML via the expectation-maximization (EM) algorithm to estimate simple item response theory models. This is a different approach to that adopted here, where we use a Gibbs-slice sampling algorithm based on augmented data (auxiliary variables). Our Gibbs-slice sampling algorithm is in a fully Bayesian framework, and the posterior samples are drawn from the full conditional posterior distribution, whereas the MML–EM algorithm used in the *mirt* package is in a frequentist framework. Parameter estimates are obtained by an integral operation in the process of implementing the EM algorithm. Loken and Rulison (2010) used WinBUGS (Spiegelhalter et al., 2003) to estimate the 4PL model parameters in a Bayesian framework. However, convergence of parameter estimation is not completely achieved in the case of some non-informative prior distributions for WinBUGS. The reason for this may be that WinBUGS does not explicitly impose the monotonicity restriction $c < d$ on the 4PL model, i.e., it does not assume that the lower asymptote parameter c is smaller than the upper asymptote parameter d. (The introduction of parameters in the 4PL model will be described in section 2, and further discussion of these two parameters can be found in Culpepper, 2016 and Junker and Sijtsma, 2001). Thus, the prior Gibbs samplers do not strictly enforce an identification condition, and this leads to estimator non-convergence. More specifically, the prior distributions of the upper and lower asymptote parameters are given by the following informative priors (Loken and Rulison, 2010, p. 513):

$$c_j \sim N(0.22, 0.05), \qquad d_j \sim N(0.84, 0.05).$$

If we choose the non-informative prior distributions

$$c_j \sim N(0.22, 10^5), \qquad d_j \sim N(0.84, 10^5),$$

then, from the value ranges of the upper and lower asymptote parameters, we find that the lower asymptote parameter can be larger than the upper asymptote parameter, $d_j < c_j$, which violates the model identification condition $c_j < d_j$ (this condition will be introduced in detail in section 2). In this case, using WinBUGS to infer the model parameters may lead to biased estimates when the sample size (the number of examinees) is small and the prior distributions then play an important role. To solve the above problems in using WinBUGS, Loken and Rulison (2010) employed strong informative prior distributions to obtain good recovery (Culpepper, 2016, p. 1,143). However, Culpepper (2016, p. 1,161) pointed out that the use of informative prior distribution may lead to serious deviations if it happens to be centered at the wrong values. Therefore, he proposed that recovery should also be dealt with by using some non-informative priors.

In the present study, a novel and highly effective Gibbs-slice sampling algorithm in the Bayesian framework is proposed to estimate the 4PL model. The Gibbs-slice sampling algorithm overcomes the defects of WinBUGS that affect the convergence of parameter estimation based on the monotonicity restriction. Moreover, the algorithm can obtain good recovery results by using various types of prior distribution. In the following sections, we will introduce the theoretical foundation of the slice sampling algorithm in detail, and we will then analyze the advantages of the slice sampling algorithm over two traditional Bayesian algorithms.

The rest of this paper is organized as follows. Section 2 contains a short introduction to the 4PL model, its reparameterized form, and model identification restrictions. In section 3, the theoretical foundation of the slice sampling algorithm is presented and its advantages compared with traditional Bayesian algorithms are analyzed. In section 4, three simulation studies focus respectively on the performance of parameter recovery, an analysis of the flexibility and sensitivity of different prior distributions for the slice sampling algorithm, and an assessment of model fittings using two Bayesian model selection criteria. In section 5, the quality of the Gibbs-slice sampling algorithm is investigated using an empirical example. We conclude the article with a brief discussion in section 6.

2. MODELS AND MODEL IDENTIFICATIONS

The 1PL and 2PL models have been widely used to fit binary item response data. Birnbaum (1968) modified the 2PL model to give the now well-known 3PL model, which includes a lower asymptote parameter to represent the contribution of guessing to the probability of correct response. To characterize the failure of high-ability examinees to answer easy items, Barton and Lord (1981) introduced an upper asymptote parameter into the 3PL model, giving the 4PL model:

$$\begin{aligned} P_{ij} &= P(y_{ij} = 1 \mid a_j, b_j, c_j, d_j, \theta_i) \\ &= c_j + (d_j - c_j)\frac{\exp[1.7a_j(\theta_i - b_j)]}{1 + \exp[1.7a_j(\theta_i - b_j)]} \end{aligned} \tag{1}$$

for $i = 1, \ldots, N$ and $j = 1, \ldots, J$, where N is the total number of examinees participating in the test and J is the test length. Here, y_{ij} is the binary response of the ith examinee with latent ability level θ_i to answer the jth item and is coded as 1 for a correct response and 0 for an incorrect response, P_{ij} is the corresponding probability of correct response, a_j is the item discrimination parameter, b_j is the item difficulty parameter, c_j is the item lower asymptote (pseudo-guessing) parameter, and d_j is the item upper asymptote parameter. The 4PL model reduces to the other models as special cases: $d_j = 1$ gives the 3PL model, $c_j = 0$ gives the 2PL model, and $a_j = 1$ gives the 1PL model. Following Culpepper (2016), we reparameterize the traditional 4PL model to construct a new 4PL model by defining a slipping parameter similar to that in cognitive diagnostic tests:

$$\begin{aligned} P_{ij} &= P(y_{ij} = 1 \mid a_j, b_j, c_j, \gamma_j, \theta_i) \\ &= c_j + (1 - \gamma_j - c_j)\frac{\exp[1.7a_j(\theta_i - b_j)]}{1 + \exp[1.7a_j(\theta_i - b_j)]}, \end{aligned} \tag{2}$$

where $\gamma_j = 1 - d_j$.

One identification restriction is that the upper asymptote must exceed the lower asymptote: $d_j > c_j$. Equivalently, the restriction $0 < c_j + \gamma_j < 1$ must be satisfied for the reparameterized 4PL model, Meanwhile, either the scale of latent abilities or the scale of item parameters must be restricted to identify the two0parameter IRT models. Three methods are widely used to identify two-parameter IRT models.

1. Fx the mean population level of ability to zero and the variance population level of ability to one (Lord and Novick, 1968; Bock and Aitkin, 1981; Fox and Glas, 2001; Fox, 2010), i.e., $\theta \sim N(0, 1)$.
2. Restrict the sum of item difficulty parameters to zero and the product of item discrimination parameters to one (Fox, 2001; Fox, 2005, 2010), i.e., $\sum_{j=1}^{J} b_j = 0$ and $\prod_{j=1}^{J} a_j = 1$.
3. Fix the item difficulty parameter at a specific value, most often zero, and restrict the discrimination parameter to a specific value, most often one (Fox, 2001; Fox, 2010), i.e., $b_1 = 0$ and $a_1 = 1$. The basic idea here is to identify the two-parameter logistic model by anchoring an item discrimination parameter to an arbitrary constant, typically $a_1 = 1$, for a given item. Meantime, a location identification constraint is imposed by restricting a difficulty parameter, typically $b_1 = 0$, for a given item. Based on the fixed anchoring values of the item parameters, other parameters are estimated on the same scale. The estimated difficulty or discrimination values of item parameters are interpreted based on their positions relative to the corresponding anchoring values. For details, see Fox (2010, p. 87).

In the present study, the main aim is to evaluate the accuracy of parameter estimation obtained by the slice sampling algorithm for different types of prior distributions. Therefore, the first of the above methods is used to eliminate the trade-offs between ability θ and the difficulty parameter b in location, and between ability θ (difficulty parameter b) and the discrimination parameter a in scale.

3. THEORETICAL FOUNDATION AND ANALYSIS OF THE ADVANTAGES OF THE SLICE SAMPLING ALGORITHM

3.1. Theoretical Foundation of the Slice Sampling Algorithm

The motivation for the slice sampling algorithm (Damien et al., 1999; Neal, 2003; Bishop, 2006; Lu et al., 2018) is that we can use the auxiliary variable approach to sample from posterior distributions arising from Bayesian non-conjugate models. The theoretical basis for this algorithm is as follows.

Suppose that the simulated values are generated from a target density function $t(x)$ given by $t(x) \propto \phi(x)\prod_{i=1}^{N} l_i(x)$ that cannot be sampled directly, where $\phi(x)$ is a known density from which samples can be easily drawn and $l_i(x)$ are non-negative invertible functions, which do not have to be density functions. We introduce the auxiliary variables represented by the vector $\boldsymbol{\delta} = (\delta_1, \ldots, \delta_N)'$, each element of which is from $(0, +\infty)$ and where $\delta_1, \ldots, \delta_N$ are mutually independent. The inequalities $\delta_i < l_i(x)$ are established, and the joint density can be written as

$$t(x, \delta_1, \ldots, \delta_N) \propto \phi(x) \prod_{i=1}^{N} \mathrm{I}\{\delta_i < l_i(x)\}, \tag{3}$$

where the indicator function $\mathrm{I}(A)$ takes the value 1 if A is true and the value 0 if A is false. If the auxiliary variables are integrated out, the marginal distribution $t(x)$ is obtained as

$$t(x) = \int_0^{l_1(x)} \cdots \int_0^{l_N(x)} t(x, \delta_1, \ldots, \delta_N)\, d\delta_N \cdots d\delta_1,$$
$$\propto \phi(x) \int_0^{l_1(x)} \cdots \int_0^{l_N(x)} 1\, d\delta_N \cdots d\delta_1 = \phi(x) \prod_{i=1}^{N} l_i(x). \quad (4)$$

Using the invertibility of the function $l_i(x)$, we can then obtain the set $\Lambda_{\delta_i} = \{x \mid \delta_i < l_i(x)\}$. The simulated values are generated from the Gibbs sampler based on the auxiliary variables by repeatedly sampling from the full conditional distributions, proceeding as follows at iteration r:

- Sample $\delta_i^{(r)} \sim \text{Uniform}\left(0, l_i(x^{(r-1)})\right)$, $i = 1, \ldots, N$.
- Sample $x^{(r)} \sim \Lambda_{\delta_i} = \{x \mid \delta_i^{(r)} < l_i(x)\}$.

We thereby derive a horizontal "slice" under the density function. Thus, a Markov chain based on the new Gibbs sampler can be constructed by sampling points alternately from the uniform distribution under the density curve and only concerning the horizontal "slice" defined by the current sample points.

3.2. Advantages of the Slice Sampling Algorithm Compared With the Metropolis–Hastings Algorithm

In the Bayesian framework, we first consider the benefits of the slice sampling algorithm compared with the traditional Metropolis–Hastings (MH) algorithm (Metropolis et al., 1953; Hastings, 1970; Tierney, 1994; Chib and Greenberg, 1995; Chen et al., 2000). It is known that the MH algorithm relies heavily on the tuning parameters of the proposal distribution for different data sets. In addition, the MH algorithm is sensitive to step size. If the step size is too small, the chain will take longer to traverse the target density. If the step size is too large, there will be inefficiencies due to high rejection rate. More specifically, researchers should ensure that each parameter candidate is no more than 50% accepted by adjusting the tuning parameters of the MH algorithm. Further, for example, when we draw two-dimensional item parameters at the same time in the 2PL model, the probability of acceptance will be reduced to around 25% (Patz and Junker, 1999, p. 163). Thus, the sampling efficiency of the MH algorithm is greatly reduced. However, the slice sampling algorithm avoids the retrospective tuning that is needed in the MH algorithm if we do not know how to choose a proper tuning parameter or if no value for the tuning parameter is appropriate. It always keeps the drawn samples accepted, thus increasing the sampling efficiency. Next, we show that the slice sampling algorithm is more efficient than a particular independent MH chain.

Let us use the MH algorithm to obtain samples from the posterior distribution $t(x)$ given by $t(x) \propto \phi(x)l(x)$, where $\phi(x)$ is selected as a special proposal distribution. Let x^* be a candidate value from the proposal distribution $\phi(x)$ and let $x^{(r)}$ be the current point. The probability of the new candidate being accepted, $\min\{1, l(x^*)/l(x^{(r)})\}$, is determined by a random number u from Uniform(0, 1). Essentially, if $u < l(x^*)/l(x^{(r)})$, then $x^{(r+1)} = x^*$; otherwise, $x^{(r+1)} = x^{(r)}$. The process is to draw the candidate first and then determine whether or not to "move" or "stay" by using the random number u. The "stay" process will lead to a reduction in the sampling efficiency of the MH algorithm. By contrast, suppose we consider the inverse process of the above sampling to draw the random number u first. To achieve the purpose of moving, we need to draw the candidate x^* from $\phi(x)$ such that $u < l(x^*)/l(x^{(r)})$. Therefore, x^* can be regarded as a sample from $\phi(x)$ restricted to the set $\Theta_u(r) = \{x \mid l(x) > ul(x^{(r)})\}$. In this case, the chain will always be moved, thus improving the sampling efficiency.

In addition, with the MH algorithm, it is relatively difficult to sample parameters with monotonicity or truncated interval restrictions. Instead, it is possible to improve the accuracy of parameter estimation by employing strong informative prior distributions to avoid violating the restriction conditions (Culpepper, 2016). For example, the prior distributions of the lower asymptote and upper asymptote parameters used in Loken and Rulison (2010) are, respectively Beta(5, 17) and Beta(17, 5), and these two parameters are fairly concentrated in the range of 0.227–0.773. However, the advantage of the slice sampling algorithm is that it can easily draw the posterior samples from any prior distribution as long as these distributions have a reasonable value range of parameters. See the following sections for details.

3.3. Advantages of the Slice Sampling Algorithm Compared With the Gibbs Algorithm

The idea of the slice sampling algorithm is to draw the posterior samples from a truncated prior distribution by introducing auxiliary variables, where the truncated interval is deduced from the likelihood function. This differs from the approach of the Gibbs algorithm (Geman and Geman, 1984; Gelfand and Smith, 1990), which is to generate posterior samples by sweeping through each variable to sample from its conditional distribution, with the remaining variables fixed at their current values. However, slice sampling algorithm can be conceived of as extensions of the Gibbs algorithm. In particular, when the parameters in which we are interested are represented by a multidimensional vector $\boldsymbol{X}$, we cannot use the slice sampling algorithm directly to obtain the multivariate set $\boldsymbol{\Theta}_u = (\Theta_u^1, \ldots, \Theta_u^k, \ldots, \Theta_u^p)$, where p is the dimension of $\boldsymbol{X}$. Therefore, a Gibbs sampler is employed to draw the samples from the full conditional distribution $l(x_k \mid \boldsymbol{x}_{(-k)}, u)$ for $k = 1, \ldots, p$, which is a realization of $t(\boldsymbol{X})$. This involves sampling from $\phi(x_k \mid \boldsymbol{x}_{(-k)})$ restricted to the set $\Theta_u^k = \{x_k \mid l(x_k, \boldsymbol{x}_{(-k)}) > u\}$, where the premise must be satisfied that $l(x_k, \boldsymbol{x}_{(-k)})$ is invertible for all k given $\boldsymbol{x}_{(-k)}$.

It is well-known that the Gibbs algorithm can quickly and effectively draw samples from the posterior distribution owing to the fact that the full conditional posterior distribution is easy to sample using the conjugate prior distribution. However, the Gibbs algorithm is not valid for Bayesian non-conjugate models such as the 2PL model. By comparison, the slice sampling algorithm for estimating the 2PL model has the advantage of a flexible prior distribution being introduced to obtain samples from the full conditional posterior distributions rather than being restricted to using the conjugate distributions, which is required in Gibbs sampling and is limited to the use of the normal ogive framework (Tanner and Wong, 1987; Albert, 1992; Béguin and Glas, 2001; Fox and Glas, 2001; Fox, 2010; Culpepper, 2016). The slice sampling algorithm allows the use of informative prior distributions and non-informative prior distributions, and even if an inappropriate prior distribution is adopted, it can still obtain satisfactory results. That is, any prior distribution can be used as long as the values sampled from it are in a reasonable range of the parameter support set. For example, for the discrimination parameter, the following prior distributions can be considered: the informative prior $\log N(0,1)$, the non-informative priors $N(0,1000)I(a>0)$, and the inappropriate priors Exp(1) and Gamma(2, 3).

4. BAYESIAN INFERENCE

4.1. Bayesian Estimation

In the present study, an efficient Gibbs-slice sampling algorithm in a fully Bayesian framework is used to estimate the following 4PL model. The sampling process of Gibbs-slice sampling algorithm consists of two parts. One part is the Gibbs sampling algorithm, which is used to update the guessing and slipping parameters from the truncated Beta distributions by introducing auxiliary variables (Béguin and Glas, 2001; Fox, 2010; Culpepper, 2016). The efficiency of Gibbs sampling is greatly improved by the use of conjugate prior distributions (Tanner and Wong, 1987; Albert, 1992). The other part is the slice sampling algorithm, which samples the 2PL model from the truncated full conditional posterior distributions by introducing different auxiliary variables.

Next, the specific sampling process of the Gibbs-slice sampling algorithm is described.

Gibbs Steps

First, following Béguin and Glas (2001), we introduce an auxiliary variable η_{ij}, where $\eta_{ij}=1$ indicates that examinee i has the ability to answer item j correctly and $\eta_{ij}=0$ otherwise. The purpose of introducing this auxiliary variable is to separate the guessing and slipping parameters from the 4PL model and make it easier to implement Gibbs sampling for the guessing and slipping parameters through the conjugate Beta distributions. Letting $\Delta=(\theta_i,a_j,b_j,c_j,\gamma_j)$, we can obtain the full conditional distribution of η_{ij} based on Bayes' theorem:

$$\begin{aligned}
P(\eta_{ij}=1\mid y_{ij}=1,\Delta)&=\frac{P(\eta_{ij}=1,y_{ij}=1,\Delta)}{P(y_{ij}=1\mid\Delta)}\\
&=\frac{(1-\gamma_j)P^*_{ij}}{c_j+(1-\gamma_j-c_j)P^*_{ij}},\\
P(\eta_{ij}=0\mid y_{ij}=1,\Delta)&=\frac{P(\eta_{ij}=0,y_{ij}=1,\Delta)}{P(y_{ij}=1\mid\Delta)}\\
&=\frac{c_j(1-P^*_{ij})}{c_j+(1-\gamma_j-c_j)P^*_{ij}},\\
P(\eta_{ij}=1\mid y_{ij}=0,\Delta)&=\frac{P(\eta_{ij}=1,y_{ij}=0,\Delta)}{P(y_{ij}=0\mid\Delta)}\\
&=\frac{\gamma_jP^*_{ij}}{1-c_j-(1-\gamma_j-c_j)P^*_{ij}},\\
P(\eta_{ij}=0\mid y_{ij}=0,\Delta)&=\frac{P(\eta_{ij}=0,y_{ij}=0,\Delta)}{P(y_{ij}=0\mid\Delta)}\\
&=\frac{(1-c_j)(1-P^*_{ij})}{1-c_j-(1-\gamma_j-c_j)P^*_{ij}}.
\end{aligned}\tag{5}$$

where

$$P^*_{ij}=\frac{\exp[1.7a_j(\theta_i-b_j)]}{1+\exp[1.7a_j(\theta_i-b_j)]}.$$

The priors of the guessing and slipping parameters follow the Beta distributions, i.e., $c_j\sim\text{Beta}(\nu_0,u_0),\gamma_j\sim\text{Beta}(\nu_1,u_1)$. However, the guessing and slipping parameters themselves satisfy the following truncated restrictions owing to model identification (Junker and Sijtsma, 2001; Culpepper, 2016):

$$\Xi=\{(c_j,\gamma_j)\mid 0\le c_j<1,0\le\gamma_j<1,0\le c_j<1-\gamma_j\}.\tag{6}$$

The joint posterior distribution of the guessing and slipping parameters can be written as

$$\begin{aligned}
p(c_j,\gamma_j\mid \boldsymbol{y}_j,\boldsymbol{\eta}_j)\propto\prod_{i=1}^{N}\left[(1-\gamma_j)^{\eta_{ij}}c_j^{(1-\eta_{ij})}\right]^{y_{ij}}\left[\gamma_j^{\eta_{ij}}(1-c_j)^{(1-\eta_{ij})}\right]^{(1-y_{ij})}\\
p(c_j,\gamma_j)I((c_j,\gamma_j)\in\Xi)\propto c_j^{\widehat{\kappa}_{00}+\nu_0-1}(1-c_j)^{\widehat{\kappa}_{01}+u_0-1}\\
\gamma_j^{\widehat{\kappa}_{10}+\nu_1-1}(1-\gamma_j)^{\widehat{\kappa}_{11}+u_1-1}I((c_j,\gamma_j)\in\Xi).
\end{aligned}\tag{7}$$

Let $\boldsymbol{y}_j'=(y_{1j},\ldots,y_{Nj})$, $\boldsymbol{\eta}_j'=(\eta_{1j},\ldots,\eta_{Nj})$, and

$$\widehat{\kappa}_{00}=(\boldsymbol{1}_N-\boldsymbol{\eta}_j)'\boldsymbol{y}_j,\qquad \widehat{\kappa}_{01}=(\boldsymbol{1}_N-\boldsymbol{\eta}_j)'(\boldsymbol{1}_N-\boldsymbol{y}_j),$$
$$\widehat{\kappa}_{10}=\boldsymbol{\eta}_j'(\boldsymbol{1}_N-\boldsymbol{y}_j),\qquad \widehat{\kappa}_{11}=\boldsymbol{\eta}_j'\boldsymbol{y}_j.$$

The full conditional posterior distributions of (c_j,γ_j) can be written as

$$\begin{aligned}
c_j^{(r)}\mid\gamma_j^{(r-1)}&\sim\text{Beta}(\widehat{\kappa}_{00}+\nu_0,\widehat{\kappa}_{01}+u_0)I(0\le c_j^{(r)}<1-\gamma_j^{(r-1)}),\\
\gamma_j^{(r)}\mid c_j^{(r)}&\sim\text{Beta}(\widehat{\kappa}_{10}+\nu_1,\widehat{\kappa}_{11}+u_1)I(0\le\gamma_j^{(r)}<1-c_j^{(r)}).
\end{aligned}\tag{8}$$

Slice Steps

Supposing that the guessing and slipping parameters have been updated by the Gibbs algorithm, we update the parameters in the 2PL model using the slice sampling algorithm. Two additional independent auxiliary variables λ_{ij} and φ_{ij}, defined on the intervals

$$\left(0, \frac{P_{ij}^{(r)} - c_j^{(r)}}{1 - \gamma_j^{(r)} - c_j^{(r)}}\right) \quad \text{and} \quad \left(0, \frac{1 - \gamma_j^{(r)} - P_{ij}^{(r)}}{1 - \gamma_j^{(r)} - c_j^{(r)}}\right),$$

are introduced to facilitate sampling, where r is the number of iterations. In fact, $(P_{ij} - c_j)/(1 - \gamma_j - c_j)$ is the correct response probability of the 2PL model, while $(1 - \gamma_j - P_{ij})/(1 - \gamma_j - c_j)$ is the corresponding incorrect response probability. Therefore, the joint likelihood of a, b, c, γ, θ based on the auxiliary variables λ and φ can be written as

$$\begin{aligned} p(\boldsymbol{y} \mid \boldsymbol{a}, \boldsymbol{b}, \boldsymbol{\theta}, \boldsymbol{c}, \boldsymbol{\gamma}, \boldsymbol{\lambda}, \boldsymbol{\varphi}) & \\ \propto \prod_{i=1}^{N}\prod_{j=1}^{J} & \left[\mathrm{I}(y_{ij} = 1)\mathrm{I}\left(0 < \lambda_{ij} \le \frac{P_{ij} - c_j}{1 - \gamma_j - c_j}\right)\right. \\ & \left. + \mathrm{I}(y_{ij} = 0)\mathrm{I}\left(0 < \varphi_{ij} \le \frac{1 - \gamma_j - P_{ij}}{1 - \gamma_j - c_j}\right)\right]. \end{aligned} \tag{9}$$

Equivalently,

$$\begin{aligned} p(\boldsymbol{y} \mid \boldsymbol{a}, \boldsymbol{b}, \boldsymbol{\theta}, \boldsymbol{c}, \boldsymbol{\gamma}, \boldsymbol{\lambda}, \boldsymbol{\varphi}) \propto \textstyle\prod_{i=1}^{N}\prod_{j=1}^{J} \mathrm{I}(y_{ij} = 1)\mathrm{I}(0 < \lambda_{ij} \le P_{ij}^*) \\ + \mathrm{I}(y_{ij} = 0)\mathrm{I}(0 < \varphi_{ij} \le Q_{ij}^*), \end{aligned} \tag{10}$$

where

$$P_{ij}^* = 1 - Q_{ij}^* = \frac{\exp[1.7a_j(\theta_i - b_j)]}{1 + \exp[1.7a_j(\theta_i - b_j)]} = \frac{P_{ij} - c_j}{1 - \gamma_j - c_j},$$
$$Q_{ij}^* = \frac{1}{1 + \exp[1.7a_j(\theta_i - b_j)]} = \frac{1 - \gamma_j - P_{ij}}{1 - \gamma_j - c_j}.$$

Integrating out the two random variables λ and φ in (10), the joint likelihood based on responses can be obtained:

$$\begin{aligned} p(\boldsymbol{y} \mid \boldsymbol{a}, \boldsymbol{b}, \boldsymbol{\theta}, \boldsymbol{c}, \boldsymbol{\gamma}, \boldsymbol{\lambda}, \boldsymbol{\varphi}) &\propto \prod_{i=1}^{N}\prod_{j=1}^{J} \mathrm{I}(y_{ij} = 1)E_\lambda[\mathrm{I}(0 < \lambda_{ij} \le P_{ij}^*)] \\ &\quad + \mathrm{I}(y_{ij} = 0)E_\varphi[\mathrm{I}(0 < \varphi_{ij} \le Q_{ij}^*)] \\ &\propto \prod_{i=1}^{N}\prod_{j=1}^{J} (P_{ij}^*)^{(y_{ij}=1)}(Q_{ij}^*)^{(y_{ij}=0)}, \end{aligned} \tag{11}$$

where E_λ is an expectation operation for the random variable λ. We know that $\boldsymbol{\eta}$, $\boldsymbol{\lambda}$, and $\boldsymbol{\varphi}$ are independent of each other. Therefore, the joint posterior distribution based on the auxiliary variables can be written as

$$\begin{aligned} p(\boldsymbol{\eta}, \boldsymbol{\theta}, \boldsymbol{a}, \boldsymbol{b}, \boldsymbol{c}, \boldsymbol{\gamma}, \boldsymbol{\lambda}, \boldsymbol{\varphi} \mid \boldsymbol{y}) \propto p(\boldsymbol{\eta} \mid \boldsymbol{a}, \boldsymbol{b}, \boldsymbol{\theta}, \boldsymbol{c}, \boldsymbol{\gamma}, \boldsymbol{y})p(\boldsymbol{\lambda}, \boldsymbol{\varphi} \mid \boldsymbol{a}, \boldsymbol{b}, \boldsymbol{\theta}, \boldsymbol{c}, \boldsymbol{\gamma}, \boldsymbol{y}) \\ \times p(\boldsymbol{\theta})p(\boldsymbol{a})p(\boldsymbol{b})p(\boldsymbol{c}, \boldsymbol{\gamma})\mathrm{I}((\boldsymbol{c}, \boldsymbol{\gamma}) \in \Xi). \end{aligned} \tag{12}$$

The specific form can be represented as

$$\begin{aligned} p(\boldsymbol{\eta}, \boldsymbol{a}, \boldsymbol{b}, \boldsymbol{\theta}, \boldsymbol{c}, \boldsymbol{\gamma}, \boldsymbol{\lambda}, \boldsymbol{\varphi} \mid \boldsymbol{y}) \propto \prod_{i=1}^{N}\prod_{j=1}^{J} &\left[(1 - \gamma_j)^{\eta_{ij}} c_j^{(1-\eta_{ij})}\right]^{y_{ij}} \\ &\left[\gamma_j^{\eta_{ij}}(1 - c_j)^{(1-\eta_{ij})}\right]^{(1-y_{ij})} \\ &\times \left[\mathrm{I}(y_{ij} = 1)\mathrm{I}(0 < \lambda_{ij} \le P_{ij}^*)\right. \\ &\left. + \mathrm{I}(y_{ij} = 0)\mathrm{I}(0 < \varphi_{ij} \le Q_{ij}^*)\right] \\ &\times \prod_{j=1}^{J} p(a_j)p(b_j)p(c_j, \gamma_j)\mathrm{I}((c_j, \gamma_j) \in \Xi) \\ &\prod_{i=1}^{N} p(\theta_i). \end{aligned} \tag{13}$$

The detailed slice sampling algorithm is given below.

First, we update the auxiliary variables λ_{ij} and φ_{ij} when given θ_i, a_j, b_j, c_j, γ_j, and y_{ij}. According to (13), the auxiliary variables λ_{ij} and φ_{ij} have the following interval constraints:

$$0 < \lambda_{ij} \le P_{ij}^* = \frac{P_{ij} - c_j}{1 - \gamma_j - c_j} \qquad \text{when} \quad y_{ij} = 1,$$
$$0 < \varphi_{ij} \le Q_{ij}^* = \frac{1 - \gamma_j - P_{ij}}{1 - \gamma_j - c_j} \qquad \text{when} \quad y_{ij} = 0.$$

Therefore, the full conditional posterior distributions of λ_{ij} and φ_{ij} can be written as

$$\lambda_{ij} \mid \theta_i, a_j, b_j, c_j, \gamma_j, y_{ij} \sim \text{Uniform}\left(0, \frac{P_{ij} - c_j}{1 - \gamma_j - c_j}\right) \quad \text{when} \quad y_{ij} = 1, \tag{14}$$

$$\varphi_{ij} \mid \theta_i, a_j, b_j, c_j, \gamma_j, y_{ij} \sim \text{Uniform}\left(0, \frac{1 - \gamma_j - P_{ij}}{1 - \gamma_j - c_j}\right) \quad \text{when} \quad y_{ij} = 0. \tag{15}$$

Next, we update the difficulty parameter b_j. The prior of the difficulty parameter is assumed to follow a normal distribution with mean μ_b and variance σ_b^2. According to (10), $\forall i$, when $y_{ij} = 1$, we have $0 < \lambda_{ij} \le P_{ij}^*$, and the following inequality can be established:

$$a_j(\theta_i - b_j) \ge \frac{1}{1.7}\log\left(\frac{\lambda_{ij}}{1 - \lambda_{ij}}\right), \text{ or equivalently}$$
$$b_j \le \theta_i - \frac{1}{1.7a_j}\log\left(\frac{\lambda_{ij}}{1 - \lambda_{ij}}\right).$$

In fact, this inequality is obtained through the following calculation process:

$$0 < \lambda_{ij} \le P_{ij}^*, \text{ or equivalently } 0 < \lambda_{ij} \le \frac{\exp[1.7a_j(\theta_i - b_j)]}{1 + \exp[1.7a_j(\theta_i - b_j)]},$$

from which

$$\lambda_{ij} + \lambda_{ij}\exp[1.7a_j(\theta_i - b_j)] \le \exp[1.7a_j(\theta_i - b_j)], \text{ or equivalently}$$
$$\frac{\lambda_{ij}}{1 - \lambda_{ij}} \le \exp[1.7a_j(\theta_i - b_j)].$$

Therefore, we have

$$\log\left(\frac{\lambda_{ij}}{1-\lambda_{ij}}\right) \le [1.7a_j(\theta_i - b_j)], \text{ or equivalently}$$

$$a_j(\theta_i - b_j) \ge \frac{1}{1.7}\log\left(\frac{\lambda_{ij}}{1-\lambda_{ij}}\right).$$

Finally, we obtain the following inequality:

$$b_j \le \theta_i - \frac{1}{1.7a_j}\log\left(\frac{\lambda_{ij}}{1-\lambda_{ij}}\right).$$

In the same way, $\forall i$, when $y_{ij} = 0$, we have $0 < \varphi_{ij} \le Q_{ij}^*$. Therefore, the following inequality can be obtained:

$$a_j(\theta_i - b_j) \le \frac{1}{1.7}\log\left(\frac{1-\varphi_{ij}}{\varphi_{ij}}\right), \text{ or equivalently}$$

$$b_j \ge \theta_i - \frac{1}{1.7a_j}\log\left(\frac{1-\varphi_{ij}}{\varphi_{ij}}\right).$$

Using the above inequalities $0 < \lambda_{ij} \le P_{ij}^*$ and $0 < \varphi_{ij} \le Q_{ij}^*$), we can obtain a truncated interval about the difficulty parameter b_j:

$$\theta_i - \frac{1}{1.7a_j}\log\left(\frac{1-\varphi_{ij}}{\varphi_{ij}}\right) \le b_j \le \theta_i - \frac{1}{1.7a_j}\log\left(\frac{\lambda_{ij}}{1-\lambda_{ij}}\right).$$

If this truncated interval is narrow, the sampling efficiency is improved and the parameter can converge fast. Therefore, we need to limit the upper and lower bounds of the truncated interval. In fact, we can obtain a maximum of $\theta_i - (1/1.7a_j)\log[(1-\varphi_{ij})/\varphi_{ij}]$ among all the examinees who correctly answer the jth item. Similarly, we can obtain a minimum of $\theta_i - (1/1.7a_j)\log[\lambda_{ij}/(1-\lambda_{ij})]$ among all the examinees who mistakenly answer the jth item. Finally, the full conditional posterior distribution of b_j can be obtained as a truncated prior distribution, with the truncated interval between maximum and minimum. The specific mathematical expressions are as follows.

Let $D_j = \{i \mid y_{ij} = 1, 0 < \lambda_{ij} \le P_{ij}^*\}$ and $F_j = \{i \mid y_{ij} = 0, 0 < \varphi_{ij} \le Q_{ij}^*\}$. Then, given $a_j, c_j, \gamma_j, \boldsymbol{\theta}, \boldsymbol{\lambda}, \boldsymbol{\varphi}$, and $\boldsymbol{y}$, the full conditional posterior distribution of b_j is

$$b_j \mid a_j, c_j, \gamma_j, \boldsymbol{\theta}, \boldsymbol{\lambda}, \boldsymbol{\varphi}, \boldsymbol{y} \sim N(\mu_b, \sigma_b^2)\mathrm{I}(b_j^L \le b_j \le b_j^U), \quad (16)$$

where

$$b_j^L = \max_{i\in F_j}\left\{\theta_i - \frac{1}{1.7a_j}\log\left(\frac{1-\varphi_{ij}}{\varphi_{ij}}\right)\right\} \quad \text{and}$$

$$b_j^U = \min_{i\in D_j}\left\{\theta_i - \frac{1}{1.7a_j}\log\left(\frac{\lambda_{ij}}{1-\lambda_{ij}}\right)\right\}.$$

Subsequently, we update the discrimination parameter a_j. To ensure that this parameter is greater than zero, we use a truncated normal distribution with mean μ_a and variance σ_a^2 as a prior distribution, $N(\mu_a, \sigma_a^2)\mathrm{I}(a_j > 0)$. Under the condition $y_{ij} = 1, \forall i$, $\theta_i - b_j > 0$, we have $0 < \lambda_{ij} \le P_{ij}^*$, while under the condition $y_{ij} = 0, \forall i, \theta_i - b_j < 0$, we have $0 < \varphi_{ij} \le Q_{ij}^*$. The following inequalities concerning the discrimination parameter a_j can be established using a procedure similar to that used above to derive the truncated interval for the difficulty parameter b_j:

$$a_j \ge \frac{1}{1.7(\theta_i - b_j)}\log\left(\frac{\lambda_{ij}}{1-\lambda_{ij}}\right),$$

$$a_j \ge \frac{1}{1.7(\theta_i - b_j)}\log\left(\frac{1-\varphi_{ij}}{\varphi_{ij}}\right).$$

Similarly, when $y_{ij} = 1, \forall i, \theta_i - b_j < 0$, we have $0 < \lambda_{ij} \le P_{ij}^*$, and when $y_{ij} = 0, \forall i, \theta_i - b_j > 0$, we have $0 < \varphi_{ij} \le Q_{ij}^*$, from which we obtain

$$a_j \le \frac{1}{1.7(\theta_i - b_j)}\log\left(\frac{\lambda_{ij}}{1-\lambda_{ij}}\right)$$

$$a_j \le \frac{1}{1.7(\theta_i - b_j)}\log\left(\frac{1-\varphi_{ij}}{\varphi_{ij}}\right).$$

Let

$$\Delta_j = \{i \mid y_{ij} = 1, \theta_i - b_j > 0, 0 < \lambda_{ij} \le P_{ij}^*\},$$
$$H_j = \{i \mid y_{ij} = 0, \theta_i - b_j < 0, 0 < \varphi_{ij} \le Q_{ij}^*\},$$
$$\nabla_j = \{i \mid y_{ij} = 1, \theta_i - b_j < 0, 0 < \lambda_{ij} \le P_{ij}^*\},$$
$$\Lambda_j = \{i \mid y_{ij} = 0, \theta_i - b_j > 0, 0 < \varphi_{ij} \le Q_{ij}^*\}.$$

Given b_j, c_j, γ_j, $\boldsymbol{\lambda}$, $\boldsymbol{\varphi}, \boldsymbol{\theta}$, and $\boldsymbol{y}$, the full conditional posterior distribution of a_j is given by

$$a_j \mid b_j, c_j, \gamma_j, \boldsymbol{\lambda}, \boldsymbol{\varphi}, \boldsymbol{\theta}, \boldsymbol{y} \sim N(\mu_a, \sigma_a^2)\mathrm{I}(0 < a_j^L \le a_j \le a_j^U), \quad (17)$$

where

$$a_j^L = \max\left\{0, \max_{i\in\Delta_j}\left\{\frac{1}{1.7(\theta_i - b_j)}\log\left(\frac{\lambda_{ij}}{1-\lambda_{ij}}\right)\right\}, \max_{i\in H_j}\left\{\frac{1}{1.7(\theta_i - b_j)}\log\left(\frac{1-\varphi_{ij}}{\varphi_{ij}}\right)\right\}\right\},$$

$$a_j^U = \min\left\{\min_{i\in\nabla_j}\left\{\frac{1}{1.7(\theta_i - b_j)}\log\left(\frac{\lambda_{ij}}{1-\lambda_{ij}}\right)\right\}, \min_{i\in\Lambda_j}\left\{\frac{1}{1.7(\theta_i - b_j)}\log\left(\frac{1-\varphi_{ij}}{\varphi_{ij}}\right)\right\}\right\}.$$

In fact, the discrimination parameter is set to be greater than zero in the item response theory. Therefore, the prior distribution for the discrimination parameter is assumed to be a normal distribution truncated at 0. Based on the likelihood information, we can obtain the truncation interval of the discrimination parameter. However, the left endpoint of the truncation interval may be <0. In this case, we need to add 0 to the truncation interval to restrict the left endpoint in 17.

Finally, we update the latent ability θ_i. The prior of θ_i is assumed to follow a normal distribution, $\theta_i \sim N(\mu_\theta, \sigma_\theta^2)$. The latent ability θ_i is sampled from the following normal distribution with truncated interval between θ_i^L and θ_i^U:

$$\theta_i \mid \boldsymbol{\lambda}, \boldsymbol{\varphi}, \boldsymbol{a}, \boldsymbol{b}, \boldsymbol{c}, \boldsymbol{\gamma}, \boldsymbol{y} \sim N(\mu_\theta, \sigma_\theta^2)\mathrm{I}(\theta_i^L \le \theta_i \le \theta_i^U), \quad (18)$$

where

$$\theta_i^L = \max_{j \in D_i} \left\{ \frac{1}{1.7a_j} \log\left(\frac{\lambda_{ij}}{1-\lambda_{ij}} \right) + b_j \right\},$$

$$\theta_i^U = \min_{j \in F_i} \left\{ \frac{1}{1.7a_j} \log\left(\frac{1-\varphi_{ij}}{\varphi_{ij}} \right) + b_j \right\}.$$

4.2. Bayesian Model Assessment

In this paper, two Bayesian model assessment methods are considered to fit three different models (the 2PL, 3PL, and 4PL models), namely, the deviance information criterion (DIC; Spiegelhalter et al., 2002) and the logarithm of the pseudomarginal likelihood (LPML; Geisser and Eddy, 1979; Ibrahim et al., 2001). These two criteria are based on the log-likelihood functions evaluated at the posterior samples of the model parameters. Therefore, the DIC and LPML of the 4PL model can be easily computed. Write $\mathbf{\Omega} = (\mathbf{\Omega}_{ij}, i = 1, \ldots, N, j = 1, \ldots, J)$, where $\mathbf{\Omega}_{ij} = (\theta_i, a_j, b_j, c_j, \gamma_j)'$. Let $\{\mathbf{\Omega}^{(1)}, \ldots, \mathbf{\Omega}^{(R)}\}$ denote an MCMC sample from the full conditional posterior distribution in (8) and (16)–(18), where $\mathbf{\Omega}^{(r)} = (\mathbf{\Omega}_{ij}^{(r)}, i = 1, \ldots, N, j = 1, \ldots, J)$ and $\mathbf{\Omega}_{ij}^{(r)} = (\theta_i^{(r)}, a_j^{(r)}, b_j^{(r)}, c_j^{(r)}, \gamma_j^{(r)})'$ for $i = 1, \ldots, N, j = 1, \ldots, J$, and $r = 1, \ldots, R$. The joint likelihood function of the responses can be written as

$$L(\boldsymbol{Y} \mid \mathbf{\Omega}) = \prod_{i=1}^{N} \prod_{j=1}^{J} f(y_{ij} \mid \theta_i, a_j, b_j, c_j, \gamma_j), \quad (19)$$

where $f(y_{ij} \mid \theta_i, a_j, b_j, c_j, \gamma_j)$ is the probability of response. The logarithm of the joint likelihood function in (19) evaluated at $\mathbf{\Omega}^{(r)}$ is given by

$$\log L(\boldsymbol{Y} \mid \mathbf{\Omega}^{(r)}) = \sum_{i=1}^{N} \sum_{j=1}^{J} \log f(y_{ij} \mid \theta_i^{(r)}, a_j^{(r)}, b_j^{(r)}, c_j^{(r)}, \gamma_j^{(r)}). \quad (20)$$

Since the joint log-likelihoods for the responses, $\log f(y_{ij} \mid \theta_i^{(r)}, a_j^{(r)}, b_j^{(r)}, c_j^{(r)}, \gamma_j^{(r)})$, $i = 1, \ldots, N$ and $j = 1, \ldots, J$, are readily available from MCMC sampling outputs, $\log f(y_{ij} \mid \theta_i^{(r)}, a_j^{(r)}, b_j^{(r)}, c_j^{(r)}, \gamma_j^{(r)})$ in (20) is easy to compute. We now calculate DIC as follows:

$$\text{DIC} = \widehat{\text{Dev}(\mathbf{\Omega})} + 2P_D = \widehat{\text{Dev}(\mathbf{\Omega})} + 2\left[\overline{\text{Dev}(\mathbf{\Omega})} - \widehat{\text{Dev}(\mathbf{\Omega})}\right], \quad (21)$$

where

$$\overline{\text{Dev}(\mathbf{\Omega})} = -\frac{2}{R} \sum_{r=1}^{R} \log L(\boldsymbol{Y} \mid \mathbf{\Omega}^{(r)}) \quad \text{and}$$

$$\widehat{\text{Dev}(\mathbf{\Omega})} = -2 \max_{1 \le r \le R} \log L(\boldsymbol{Y} \mid \mathbf{\Omega}^{(r)}).$$

In (21), $\overline{\text{Dev}(\mathbf{\Omega})}$ is a Monte Carlo estimate of the posterior expectation of the deviance function $\text{Dev}(\mathbf{\Omega}) = -2 \log L(\boldsymbol{Y} \mid \mathbf{\Omega})$, $\widehat{\text{Dev}(\mathbf{\Omega})}$ is an approximation of $\text{Dev}(\widehat{\mathbf{\Omega}})$, where $\widehat{\mathbf{\Omega}}$ is the posterior mode, when the prior is relatively non-informative, and $P_D = \overline{\text{Dev}(\mathbf{\Omega})} - \widehat{\text{Dev}(\mathbf{\Omega})}$ is the effective number of parameters. Based on our construction, both DIC and P_D given in (21) are always non-negative. The model with a smaller DIC value fits the data better.

Letting $U_{ij,\max} = \max 1 \le r \le R\{-\log f(y_{ij} \mid \theta_i^{(r)}, a_j^{(r)}, b_j^{(r)}, c_j^{(r)}, \gamma_j^{(r)})\}$, we obtain a Monte Carlo estimate of the conditional predictive ordinate (CPO; Gelfand et al., 1992; Chen et al., 2000) as

$$\log(\widehat{\text{CPO}_{ij}}) = -U_{ij,\max} - \log\left\{ \frac{1}{R} \sum_{r=1}^{R} \exp[- \log f(y_{ij} \mid \theta_i^{(r)}, a_j^{(r)}, b_j^{(r)}, c_j^{(r)}, \gamma_j^{(r)}) - U_{ij,\max}] \right\}. \quad (22)$$

Note that the maximum value adjustment used in $\log(\widehat{\text{CPO}_{ij}})$ plays an important role in numerical stabilization in computing $\exp[-\log f(y_{ij} \mid \theta_i^{(r)}, a_j^{(r)}, b_j^{(r)}, c_j^{(r)}, \gamma_j^{(r)}) - U_{ij,\max}]$ in (22). A summary statistic of the $\widehat{\text{CPO}_{ij}}$ is the sum of their logarithms, which is called the LPML and given by

$$\text{LPML} = \sum_{i=1}^{N} \sum_{j=1}^{J} \log(\widehat{\text{CPO}_{ij}}). \quad (23)$$

The model with a larger LPML has a better fit to the data.

5. SIMULATION STUDIES

5.1. Simulation Study 1

This simulation study is conducted to evaluate the recovery performance of the Gibbs-slice sampling algorithm based on different simulation conditions.

5.1.1. Simulation Design

The following manipulated conditions are considered: (a) test length $J = 20$ or 40 and (b) number of examinees $N = 500$, $1,000$, or $2,000$. Fully crossing different levels of these two factors yield six conditions (two test lengths × three sample sizes). Next, the true values of the parameters are given. True values of the item discrimination parameters a_j are generated from a uniform distribution, i.e., $a_j \sim U(0.5, 2.5)$, $j = 1, 2, \ldots, J$. The item difficulty parameters b_j are generated from a standardized normal distribution. The item guessing and slipping parameters (c_j, γ_j) are generated from $c_j \sim U(0, 0.25)$ and $\gamma_j \sim U(0, 0.25)I(\gamma_j < 1 - c_j)$. The ability parameters of examinees θ_i are also generated from a standardized normal distribution. In addition, we adopt non-informative prior distributions for the item parameters, i.e., $a_j \sim N(0, 10^5)I(0, +\infty)$, $b_j \sim N(0, 10^5)$, $g_j \sim \text{Beta}(1, 1)$, and $\gamma_j \sim \text{Beta}(1, 1)$, $j = 1, 2, \ldots, J$. The prior for the ability parameters is assumed to follow a standardized normal distribution owing to the model identification restrictions. One hundred replications are considered for each simulation condition.

5.1.2. Convergence Diagnostics

To evaluate the convergence of parameter estimation, we only consider convergence in the case of minimum sample sizes owing

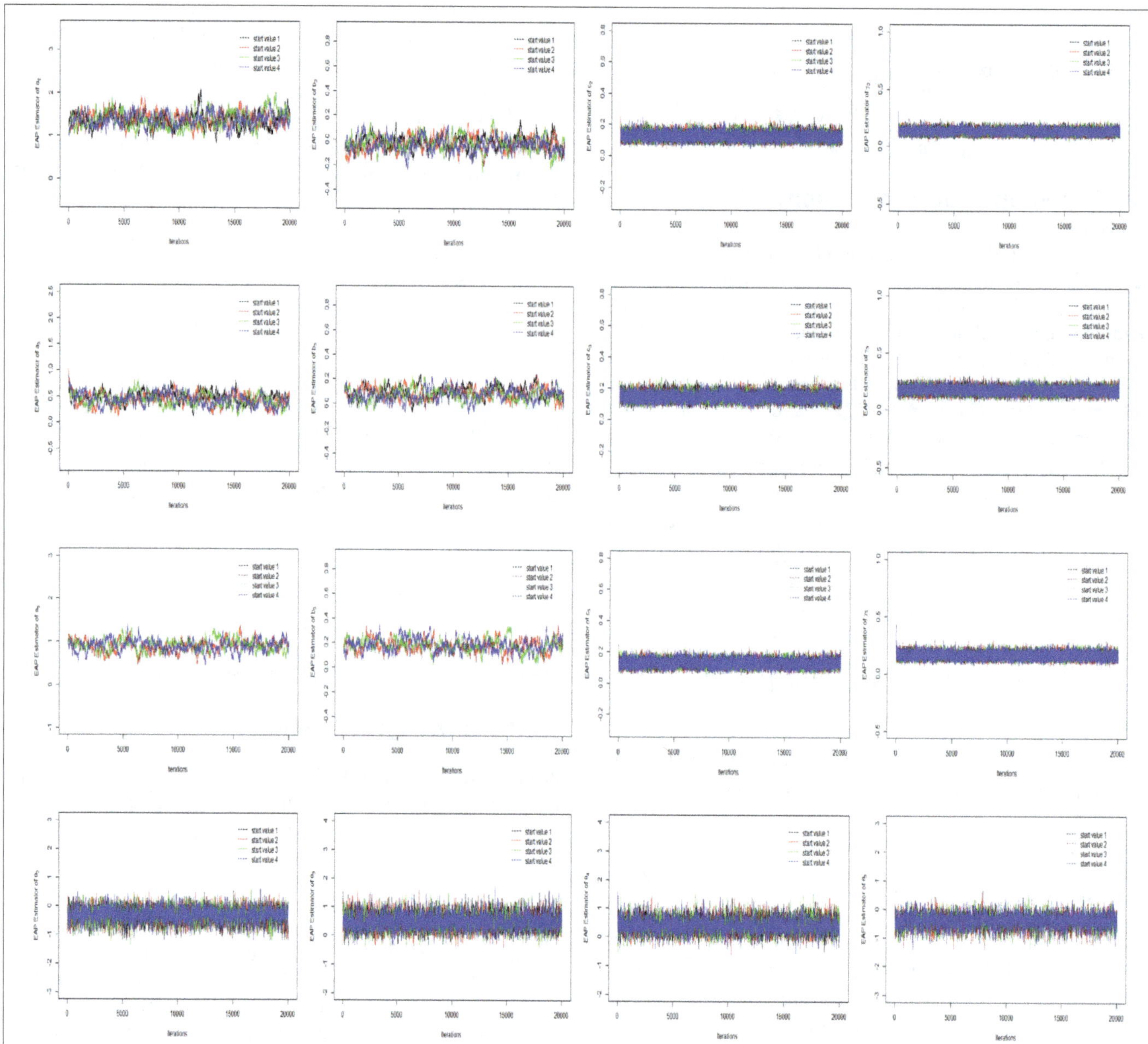

FIGURE 1 | The trace plots of three randomly selected items and persons for the Simulation Study 1.

to space limitations. That is, the test length is fixed at 20, and the number of examinees is 500. Two methods are used to check the convergence of our algorithm: the "eyeball" method to monitor convergence by visually inspecting the history plots of the generated sequences (Zhang et al., 2007), and the Gelman–Rubin method (Gelman and Rubin, 1992; Brooks and Gelman, 1998) to check the convergence of the parameters.

The convergence of the Gibbs-slice sampling algorithm is checked by monitoring the trace plots of the parameters for consecutive sequences of 20,000 iterations. The first 10,000 iterations are set as the burn-in period. As an illustration, four chains started at overdispersed starting values are run for each replication. The trace plots of three randomly selected items and persons are shown in **Figure 1**. In addition, the potential scale reduction factor (PSRF) ($\widehat{R}$; Brooks and Gelman, 1998) values of all item and person parameters are shown in **Figure 2**. We find that the PSRF values of all parameters are <1.2, which ensures that all chains converge as expected.

5.1.3. Item Parameter Recovery

The accuracy of the parameter estimates is measured by four evaluation criteria, namely, the Bias, mean squared error (MSE), standard deviation (SD), and coverage probability (CP) of the 95% highest probability density interval (HPDI) statistics. Let η

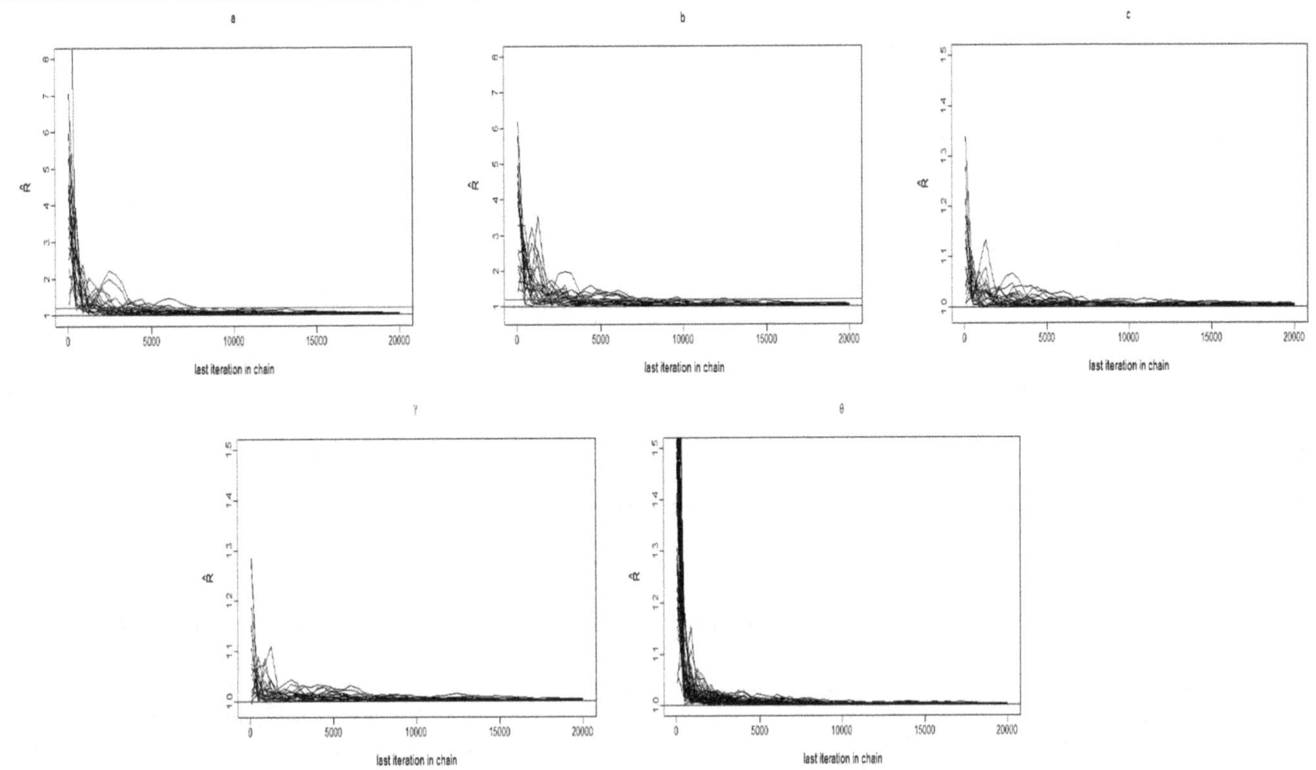

FIGURE 2 | The trace plots of $\widehat{R}$ for the Simulation Study 1.

be the parameter of interest. Assume that $M = 100$ data sets are generated. Also, let $\widehat{\eta}^{(m)}$ and $\mathrm{SD}^{(m)}(\eta)$ denote the posterior mean and the posterior standard deviation of η obtained from the mth simulated data set for $m = 1, \ldots, M$. The Bias for the parameter η is defined as

$$\mathrm{Bias}(\eta) = \frac{1}{M}\sum_{m=1}^{M}(\widehat{\eta}^{(m)} - \eta), \tag{24}$$

the MSE for η is defined as

$$\mathrm{MSE}(\eta) = \frac{1}{M}\sum_{m=1}^{M}(\widehat{\eta}^{(m)} - \eta)^2, \tag{25}$$

and the average of the posterior standard deviation is defined as

$$\mathrm{SD}(\eta) = \frac{1}{M}\sum_{m=1}^{M}\mathrm{SD}^{(m)}(\eta). \tag{26}$$

Bias and MSE are important criteria used to evaluate the accuracy of parameter estimation in a simulation study. These criteria are used to investigate the relative distance between the parameter estimator and the true value. The greater the distance between the parameter estimator and the true value, the lower is the accuracy of parameter estimation and the poorer is the performance of the algorithm. However, for real data analysis, it is impossible to calculate Bias and MSE. The SD, on the other hand, can be calculated from the posterior samples of a Markov chain in simulation studies and real data analysis. In our simulation study, we calculate the average SD through repeated experiments to eliminate the error caused by randomness in a single simulation experiment.

The coverage probability is defined as

$$\mathrm{CP}(\eta) = \frac{\#\text{ of 95\% HPDIs containing } \eta \text{ in } M \text{ simulated data sets}}{M}. \tag{27}$$

The average Bias, MSE, SD, and CP for item parameters based on six different simulation conditions are shown in **Table 1**. The following conclusions can be drawn.

1. Given the total test length, when the number of individuals increases from 500 to 2,000, the average MSE and SD for discrimination, difficulty guessing, and slipping parameters show a decreasing trend. For example, let us consider a total test length of 20 items. When the number of examinees increases from 500 to 2,000, the average MSE and the average SD of all discrimination parameters decrease from 0.0625 to 0.0474 and from 0.1460 to 0.0759, respectively. The average MSE and the average SD of all difficulty parameters decrease from 0.0505 to 0.0263 and

TABLE 1 | Evaluating the accuracy of item parameters based on six different simulated conditions in Simulation Study 1.

	No. of examinees 500				No. of examinees 1,000				No. of examinees 2,000			
Item parameter	**Bias**	**MSE**	**SD**	**CP**	**Bias**	**MSE**	**SD**	**CP**	**Bias**	**MSE**	**SD**	**CP**
NO. OF ITEMS = 20												
Discrimination[a]	−0.0087	0.0625	0.1460	0.9514	−0.0217	0.0513	0.1037	0.9504	0.0005	0.0474	0.0759	0.9486
Difficulty[b]	−0.0000	0.0505	0.0559	0.9385	0.0000	0.0389	0.0390	0.9412	−0.0000	0.0263	0.0260	0.9285
Guessing[c]	−0.0215	0.0092	0.0247	0.9325	−0.0453	0.0045	0.0193	0.9378	−0.0830	0.0023	0.0156	0.9515
Slipping[γ]	0.0132	0.0060	0.0260	0.9342	−0.0176	0.0038	0.0217	0.9628	−0.0558	0.0025	0.0166	0.9548
NO. OF ITEMS = 40												
Discrimination[a]	−0.0029	0.0842	0.1482	0.9546	−0.0035	0.0705	0.0962	0.9390	−0.0129	0.0594	0.0638	0.9781
Difficulty[b]	−0.0000	0.0443	0.0561	0.9543	−0.0000	0.0325	0.0389	0.9495	0.0000	0.0224	0.0267	0.9652
Guessing[c]	−0.0238	0.0075	0.0250	0.9385	−0.0625	0.0059	0.0201	0.9322	−0.0677	0.0033	0.0154	0.9418
Slipping[γ]	−0.0061	0.0035	0.0264	0.9310	−0.0169	0.0025	0.0209	0.9438	−0.0407	0.0024	0.0152	0.9422

Note that the Bias, MSE, SD, and CP denote the average Bias, MSE, SD, and CP for the parameters. [a]Discrimination parameters, [b]Difficulty parameters, [c]Guessing parameters, [γ]Slipping parameters.

from 0.0559 to 0.0260, respectively. The average MSE and the average SD of all guessing parameters decrease from 0.0092 to 0.0023 and from 0.0247 to 0.0156, respectively. The average MSE and the average SD of all slipping parameters decrease from 0.0060 to 0.0025 and from 0.0260 to 0.0166, respectively.

2. Under the six simulated conditions, the average CPs of the discrimination, difficulty guessing, and slipping parameters are about 0.9500.
3. When the number of examinees is fixed at 500, 1,000, or 2,000, and the number of items is fixed at 40, the average MSE and SD show that the recovery results of the discrimination, difficulty, guessing, and slipping parameters are close to those in the case where the total test length is 20, which indicates that the Gibbs-slice sampling algorithm is stable and there is no reduction in accuracy owing to an increase in the number of items.

In summary, the Gibbs-slice sampling algorithm provides accurate estimates of the item parameters in term of various numbers of examinees and items. Next, we will explain why the Bias criterion is useful, and why it seems irrelevant in the simulation study.

If we want to determine whether our algorithm estimates the parameter accurately, we need more information to infer the parameter, which requires a large sample size. Here, Bias is an important criterion to evaluate the accuracy of parameter estimation. Let us give an example to illustrate the role of Bias. In Simulation Study 1, suppose that we investigate the accuracy of the algorithm in estimating a discrimination parameter. When the number of examinees increases from 500 to 2,000, the Bias of the discrimination parameter should show a decreasing trend. The result of Bias reduction further verifies that a greater number of samples are needed to improve the accuracy of parameter estimation.

In Simulation Study 1, we cannot enumerate the Bias of each item parameter one by one because there are too many simulation conditions and we are subject to space limitations. Therefore, we choose to calculate the average Bias of the parameter of interest. Next, we take the discrimination parameters as an example to further explain why Bias seems irrelevant in Simulation Study 1. Suppose that we have obtained 40 Biases of discrimination parameters, that the Bias values of these 40 discrimination parameters are either positive or negative, and that the average Bias of all 40 items is close to 0. However, the near-zero value of the average Bias does not show whether the parameter estimation is accurate or the result is caused by the positive and negative superposition of the 40 Biases. In fact, we find that the Bias for each item discrimination parameter show a decreasing trend with increasing number of examinees. To sum up, we do not analyze the results of the average Bias in the simulation studies, but Bias is indeed an important criterion to evaluate the accuracy of each parameter estimation.

5.1.4. Ability Parameter Recovery

Next, we evaluate the recovery of the latent ability using four accuracy evaluation criteria. The following conclusions can be obtained from **Table 2**.

1. Given a fixed number of examinees (500, 1,000, or 2,000), when the number of items increases from 20 to 40, the average MSE and SD for the ability parameters also show a decreasing trend.
2. Under the six simulated conditions, the average CP of the ability is also about 0.9500.
3. Given a fixed number of examinees (500, 1,000, or 2,000), when the number of items increases from 20 to 40, the correlation between the estimates and the true values tends to increase. For example, for 500 examinees, when the number of items increases from 20 to 40, the correlation between the estimates and the true values increases from 0.8631 to 0.9102.
4. Given a fixed number of items (20 or 40), when the number of examinees increases from 500 to 2,000, the correlation

TABLE 2 | Evaluating the accuracy of person parameters based on six different simulated conditions in Simulation Study 1.

No. of items	No. of examinees	Bias	MSE	SD	CP	Correlation with true value
20	500	0.0545	0.2783	0.2523	0.9428	0.8631
	1,000	0.0149	0.2923	0.2636	0.9675	0.8764
	2,000	0.0052	0.3341	0.2961	0.9322	0.8599
40	500	0.0315	0.2346	0.2180	0.9274	0.9102
	1,000	0.0764	0.2553	0.2343	0.9626	0.9182
	2,000	0.0439	0.3042	0.2866	0.9542	0.9225

Note that the Bias, MSE, SD, and CP denote the average Bias, MSE, SD, and CP for the ability parameters.

between the estimates and the true values remains basically the same.

In summary, it is shown again that the Gibbs-slice sampling algorithm is effective and that the estimated results are accurate under various simulation conditions.

5.2. Simulation Study 2

Culpepper (2016) conducted an additional simulation study to confirm that the guessing and slipping parameters could give good recovery results in the process of Gibbs sampling regardless of whether informative or non-informative priors were used. Therefore, in this simulation study, we also adopt non-informative prior distributions for the guessing and slipping parameters in the Gibbs step to eliminate biased estimation of parameters due to wrong choices of the prior distributions, i.e., $c \sim \text{Beta}(1, 1)$ and $\gamma \sim \text{Beta}(1, 1)$, and we focus on the influence of different prior distributions on the accuracy of parameter estimation in the process of implementing the slice sampling algorithm. Note that in this simulation study, we do not focus on the accuracy of the guessing and slipping parameters, since Culpepper (2016) has already verified the accuracy of these two parameters in the case of the Gibbs algorithm under different types of prior distributions.

This simulation study is designed to show that the slice sampling algorithm is sufficiently flexible to recover various prior distributions of the item (discrimination and difficulty) and person parameters, and to address the sensitivity of the algorithm with different priors. Three types of prior distributions are examined: informative priors, non-informative priors, and inappropriate priors.

5.2.1. Simulation Design

The number of the examinees $N = 1,000$, and the test length $J = 20$. The true values for the items and persons are the same as in Simulation Study 1. One hundred replications are considered for each simulation condition. The following three kinds of prior distributions are considered in implementing the slice sampling algorithm:

(i) informative prior: $a \sim \log N(0, 1)$, $b \sim N(0, 1)$, and $\theta \sim N(0, 1)$;
(ii) non-informative prior: $a \sim N(0, 1000)\mathrm{I}(0, +\infty)$, $b \sim \text{Uniform}(-1000, 1000)$, and $\theta \sim N(0, 1000)$;
(iii) inappropriate prior: (1) $a \sim \text{Exp}(1)$, $b \sim t(1)$, and $\theta \sim t(1)$; (2) $a \sim \text{Gamma}(3, 2)$, $b \sim \text{Cauchy}(1, 3)$, and $\theta \sim \text{Cauchy}(1, 3)$.

The Gibbs-slice sampling algorithm is iterated 20,000 times. The first 10,000 iterations are discarded as burn-in. The PSRF values of all parameters are <1.2. The Bias, MSE, and SD of $\boldsymbol{a}$ and $\boldsymbol{b}$ based on the three kinds of prior distribution are shown in **Figure 3**.

5.2.2. Item Parameter Recovery

From **Figure 3**, we can see that the Bias, MSE, and SD of $\boldsymbol{a}$ and $\boldsymbol{b}$ are almost the same under different prior distributions. This shows that accuracy of parameter estimation can be guaranteed by the slice sampling algorithm, no matter what prior distribution is chosen, as long as the values sampled from this distribution belong to a reasonable parameter support set. In addition, the Bias, MSE and SD of $\boldsymbol{a}$ and $\boldsymbol{b}$ fluctuate around 0, which shows that the slice sampling algorithm is accurate and effective in estimating the item parameters.

5.2.3. Ability Parameter Recovery

Next, we evaluate the recovery of the latent ability based on different prior distributions in **Table 3**. We find that the MSE of ability parameters is between 0.2676 and 0.3014, and the corresponding SD is between 0.2436 and 0.3026 for all three kinds of prior distribution, which indicates that the choice of prior distribution has little impact on the accuracy of the ability parameters. In summary, the slice sampling algorithm is accurate and effective in estimating the person parameters. It is not sensitive to the specification of priors.

5.3. Simulation Study 3

In this simulation study, we use two Bayesian model assessment criteria to evaluate the model fittings. Two issues warrant further study. The first is whether the two criteria can accurately identify the true models under different design conditions. The second is that we study the phenomena of over-fitting and under-fitting between the true model and the fitting models.

5.3.1. Simulation Design

In this simulation, a number of individuals $N = 1,000$ is considered and the test length is fixed at 40. Three item response models are considered: the 2PL, 3PL, and 4PL models. Thus, we evaluate model fitting in the following three cases:

- Case 1: 2PL model (true model) vs. 2PL model, 3PL model, or 4PL model (fitted model).
- Case 2: 3PL model (true model) vs. 2PL model, 3PL model, or 4PL model (fitted model).

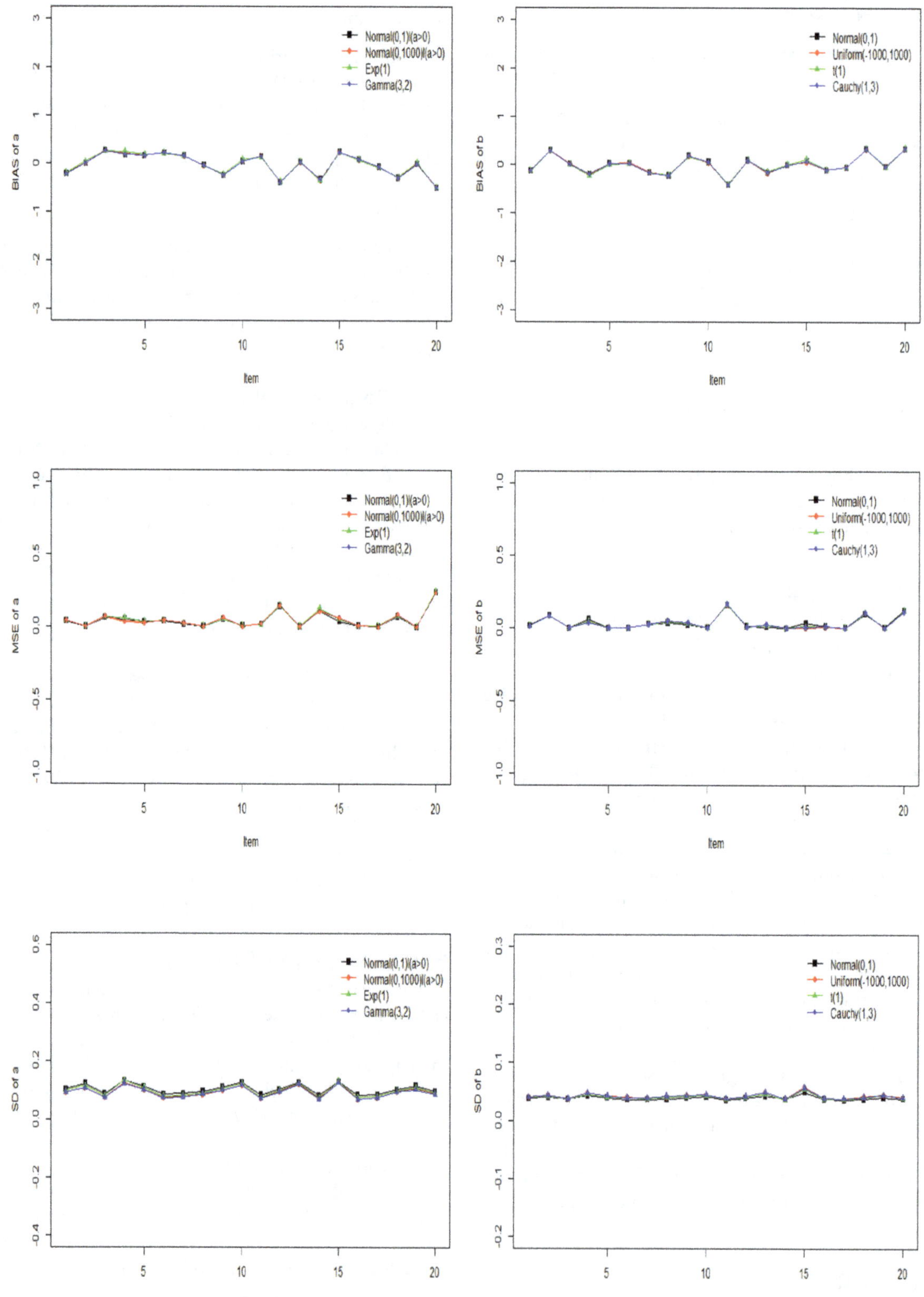

FIGURE 3 | The Bias, MSE, and SD of discrimination and difficulty parameters based on different priors.

TABLE 3 | Evaluating the accuracy of person parameters based on different prior distributions in Simulation Study 2.

Parameter	Accuracy evaluation index	Prior distribution			
		$N(0,1)$	$N(0,1,000)$	$t(1)$	$Cauchy(1,3)$
θ	Bias	0.0064	0.0149	0.0087	0.0238
	MSE	0.2676	0.2923	0.3014	0.2713
	SD	0.2436	0.3026	0.2810	0.2983

Note that the Bias, MSE, and SD denote the average Bias, MSE, and SD for the ability parameters.

- Case 3: 4PL model (true model) vs. 2PL model, 3PL model, or 4PL model (fitted model).

The true values and prior distributions for the parameters are specified in the same way as in Simulation Study 1. To implement the MCMC sampling algorithm, chains of length 20,000 with an initial burn-in period of 10,000 are chosen. There are 100 replications for each simulation condition. The potential scale reduction factor (PSRF; Brooks and Gelman, 1998) values of all item and person parameters for each simulation condition are <1.2. The results of Bayesian model assessment based on the 100 replications are shown in **Table 4**.

From **Table 4**, we find that when the 2PL model is the true model, the 2PL model is chosen as the best-fitting model according to the results of DIC and LPML, which is what we expect to see. The medians of DIC and LPML are, respectively 29,333.1917 and −14,881.2617. The second best-fitting model is the 3PL model. The differences between the 2PL and 3PL models in the medians of DIC and LPML are −1234.1551 and 650.9820, respectively. The 4PL model is the worst model to fit the data. This is because the data are generated from a simple 2PL model, and the complex 4PL model is used to fit this data, which leads to over-fitting. The differences between the 2PL and 4PL models in the medians of DIC and LPML are −5369.4761 and 2805.5087, respectively. When the 3PL model is the true model, the DIC and LPML consistently choose the 3PL model as the best-fitting model, with the corresponding median values being 24,866.9338 and −12,523.6985, respectively. The second best-fitting model is the 2PL model. The differences between the 3PL and 4PL models in the medians of DIC and LPML are −7786.6968 and 3934.9003, respectively, while the corresponding differences between the 3PL and 2PL models are −7569.1249 and 3886.7071. This shows that when the data are generated from the 3PL model, the simple 2PL model is more appropriate to fit the data compared with the complex 4PL model. When the 4PL model is the true model, the two criteria consistently select the 4PL model as the best-fitting model. The other two models suffer from serious under-fitting. The differences between the 4PL and 2PL models in the medians of DIC and LPML are −7807.8880 and 4339.4735, respectively, while the corresponding differences between the 4PL and 3PL models are −1104.4156 and 634.0753. The failure to select the 2PL (3PL) model is attributed to the under-fitting caused by a few parameters. That is, the guessing and slipping parameters in the 4PL model play an important role in adjusting the probability of

TABLE 4 | The results of Bayesian model assessment in the Simulation Study 3.

				True model 2PL	3PL	4PL
Fitted model	2PL	DIC	Q_1	**29319.0702**	30539.6070	34676.5622
			Median	**29333.1917**	30567.3468	34702.6678
			Q_3	**29341.0284**	30591.9937	34722.2367
			IQR	21.9582	52.3867	45.6745
		LPML	Q_1	**−14888.3688**	−15543.2057	−17701.0943
			Median	**−14881.2617**	−15532.2437	−17686.7704
			Q_3	**−14875.4347**	−15515.0444	−17670.9324
			IQR	12.9341	28.1613	30.1319
	3PL	DIC	Q_1	32431.0873	**24857.3160**	32648.0788
			Median	32436.0587	**24866.9338**	32653.6306
			Q_3	32442.8955	**24878.2528**	32660.3940
			IQR	11.8082	20.9368	12.3152
		LPML	Q_1	−16413.9390	**−12528.9444**	−16462.3200
			Median	−16410.4056	**−12523.6985**	−16458.5988
			Q_3	−16406.8835	**−12517.8991**	−16453.9725
			IQR	7.0555	11.0453	8.3427
	4PL	DIC	Q_1	35560.7897	28870.1192	**27768.0166**
			Median	35583.7535	28880.2811	**27775.8655**
			Q_3	35611.7761	28890.8003	**27780.0024**
			IQR	50.9863	20.6810	11.9857
		LPML	Q_1	−18320.2375	−14603.6126	**−13965.3888**
			Median	−18302.6986	−14597.3004	**−13963.2251**
			Q_3	−18288.7386	−14593.5979	**−13958.0409**
			IQR	31.4988	10.0147	7.3479

Note that the boldface values indicate that the corresponding model is the best fitted model with the smallest DIC and largest LPML values.

the tail of the item characteristic curve. In summary, the Bayesian assessment criteria are effective for identifying the true models and can be used in the following empirical example.

6. EMPIRICAL EXAMPLE

In this example, the 2015 computer-based PISA (Program for International Student Assessment) science data are used. Among the many countries that have participated in this computer-based assessment of sciences, we choose students from the USA as the object of analysis. The original sample size of students is 658, and 110 students with Not Reached (original code 6) or Not Response (original code 9) are removed, with Not Reached and Not Response (omitted) being treated as missing data. The final 548 students answer 16 items. All 16 items are scored using a dichotomous scale. The descriptive statistics for these PISA data are shown in **Table 5**. We find that three items, DR442Q05C, DR442Q06C, and CR442Q07S, have lower correct rates than the other items, with the corresponding values being 25.7, 23.2, and 28.5%, respectively. The correct rate represents the proportion at which all examinees answer each item correctly. Moreover, the four items with the highest correct rates are

TABLE 5 | The descriptive statistics for PISA 2015 released computer-based sciences items.

Item	Item code	Correct rate (%)	Item	Item code	Correct rate (%)
1	CR083Q01S	54.2	9	CR442Q07S	28.5
2	CR083Q02S	83.6	10	CR245Q01S	53.8
3	CR083Q03S	75.2	11	CR245Q02S	60.0
4	CR083Q04S	66.6	12	CR101Q01S	43.6
5	DR442Q02C	80.1	13	CR101Q02S	87.6
6	DR442Q03C	76.5	14	CR101Q03S	57.7
7	DR442Q05C	25.7	15	CR101Q04S	80.1
8	DR442Q06C	23.1	16	CR101Q05S	48.7

Note that the correct rate represents the percentage of all examinees who correctly answer each item.

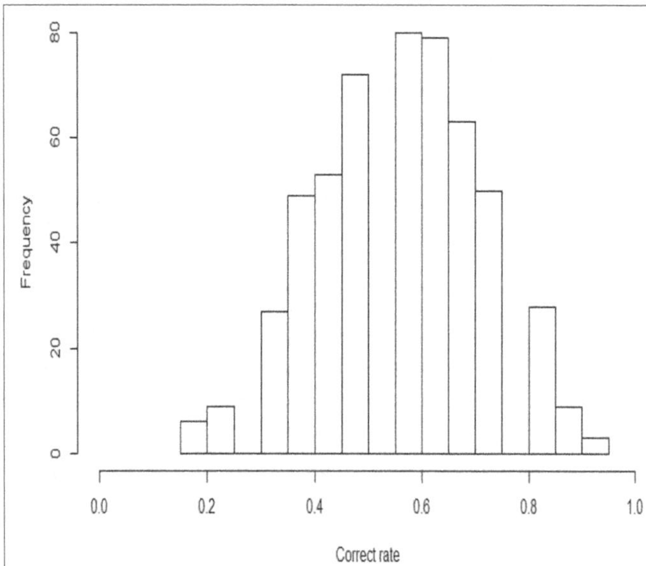

FIGURE 4 | Frequency histograms of the correct rates for 548 examinees.

CR101Q02S (87.6%), CR083Q02S (83.6%) DR442Q02C (80.1%), and CR101Q04S (80.1%). The frequency histogram of the correct rates for the 548 examinees is shown in **Figure 4**.

6.1. Bayesian Model Assessment

We consider three models to fit the PISA data: the 2PL, 3PL, and 4PL models. In the estimation procedure, the same non-informative priors as in Simulation Study 1 are utilized for the unknown parameters. In all of the Bayesian computations, we use 20,000 MCMC samples after a burn-in of 10,000 iterations for each model to compute all posterior estimates. The convergence of the chains is checked by PSRF. The PSRF values of all item and ability parameters for each model are <1.2. On this basis, the results of Bayesian model assessment for the PISA data are shown in **Table 6**.

According to DIC and LPML in **Table 6**, we find that the 4PL model is the best-fitting model compared with the 2PL and 3PL models. The values of DIC and LPML are 10,854.2075 and −5494.4088, respectively. The second best-fitting model is the 3PL model. The differences between the 4PL and 3PL models

TABLE 6 | The results of Bayesian model assessment for the PISA data.

Model	DIC	LPML
2PL model	14206.9508	−7290.7545
3PL model	12230.3819	−6168.9428
4PL model	**10854.2075**	**− 5494.4088**

Note that the boldface values indicate that the corresponding model is the best fitted model with the smallest DIC and largest LPML values.

in DIC and LPML are −1376.1744 and 674.5340, respectively. This shows that the introduction of slipping parameters in the 3PL model is sufficient to fit these PISA data. The worst-fitting model is the 2PL model. This is attributed to the relatively simple structure of this model, which makes it unable to describe changes in probability at the end of the item characteristic curve caused by guessing or slipping. The differences between the 4PL and 2PL models in DIC and LPML are −3353.7433 and 1796.3457, respectively.

Next, we will use the 4PL model to analyze the PISA data in detail based on the results of the model assessment.

6.2. Analysis of Item Parameters

The estimated results for the item parameters are shown in **Table 7**, from which we find that the expected a posteriori (EAP) estimations of the 11 item discrimination parameters are greater than one. This indicates that these items can distinguish the differences between abilities well. The five items with the lowest discrimination are items 16 (CR101Q05S), 10 (CR245Q01S), 12 (CR101Q01S), 2 (CR083Q02S), and 5 (DR442Q02C) in turn. The EAP estimates of the discrimination parameters for these five items are 0.6681, 0.6792, 0.7348, 0.8083, and 0.8901. In addition, the EAP estimates of seven of the difficulty parameters are less than zero, which indicates that these seven items are easier than the other nine items. The five most difficult items are items 8 (DR442Q06C), 7 (DR442Q05C), 9 (CR442Q07S), 12 (CR101Q01S), and 16 (CR101Q05S) in turn. The EAP estimates of the difficulty parameters for these five items are 1.2528, 1.2203, 1.0804, 0.4521, and 0.3102. The corresponding correct rates in **Table 5** for these five items are 23.1, 25.7, 28.5, 43.63 and 48.7%, respectively. The most difficult five items have low correct rates, which is consistent with our intuition. The EAP estimates of the guessing parameters for the 16 items range from 0.0737 to 0.1840. The five items with the highest guessing parameters are items 2 (CR083Q02S), 5 (DR442Q02C), 13 (CR101Q02S), 15 (CR101Q04S), and 3 (CR083Q03S) in turn. The EAP estimates of the guessing parameters for these five items are 0.1840, 0.1791, 0.1790, 0.1673, and 0.3102. We find that the five items with high guessing parameters also have high correct rates. The corresponding correct rates for these five items are 83.6, 80.1, 87.6, 80.1, and 75.2%. This shows that these five items are more likely to be guessed correctly than the other 11 items. In addition, the five easiest slipping items are items 8 (DR442Q06C), 7 (DR442Q05C), 9 (CR442Q07S), 12 (CR101Q01S), and 16 (CR101Q05S) in turn. The EAP estimates of the slipping parameters for these five items are 1.785, 1.619, 1.581, 0.1481, and 0.1431. We find that the more difficult an

TABLE 7 | The estimation results of item parameter for the PISA data.

PARM	EAP	SD	HPDI	PARM	EAP	SD	HPDI
a_1	1.0416	0.1227	[0.8215, 1.2856]	b_1	0.1939	0.0615	[0.0861, 0.3222]
a_2	0.8083	0.1316	[0.5715, 1.0665]	b_2	−0.8496	0.0815	[−0.9936, −0.6793]
a_3	1.1171	0.1513	[0.8327, 1.4101]	b_3	−0.5071	0.0625	[−0.6214, −0.3699]
a_4	1.1119	0.1308	[0.8813, 1.3996]	b_4	−0.1947	0.0563	[−0.3030, −0.0876]
a_5	0.8901	0.1034	[0.6847, 1.0933]	b_5	−0.6969	0.0623	[−0.8230, −0.5741]
a_6	1.2772	0.1719	[0.9642, 1.6355]	b_6	−0.5966	0.0700	[−0.7525, −0.4675]
a_7	1.3404	0.1348	[1.0800, 1.5839]	b_7	1.2203	0.0778	[1.0635, 1.3738]
a_8	1.1202	0.1608	[0.7827, 1.4713]	b_8	1.2528	0.0966	[1.0313, 1.4246]
a_9	1.2377	0.1475	[0.9338, 1.5149]	b_9	1.0804	0.0819	[0.9117, 1.2155]
a_{10}	0.6792	0.1125	[0.4780, 0.9079]	b_{10}	0.1669	0.0640	[0.0423, 0.2832]
a_{11}	1.0720	0.1214	[0.8432, 1.3184]	b_{11}	0.0258	0.0512	[−0.0617, 0.1330]
a_{12}	0.7348	0.0897	[0.5528, 0.9035]	b_{12}	0.4521	0.0548	[0.3448, 0.5506]
a_{13}	1.1994	0.1706	[0.8682, 1.5513]	b_{13}	−1.1843	0.0841	[−1.3510, −1.0305]
a_{14}	1.0083	0.1219	[0.7666, 1.2336]	b_{14}	0.0985	0.0525	[0.0029, 0.2053]
a_{15}	1.2047	0.1707	[0.8618, 1.5329]	b_{15}	−0.7719	0.0667	[−0.9095, −0.6543]
a_{16}	0.6681	0.0924	[0.4999, 0.8482]	b_{16}	0.3102	0.0584	[0.2012, 0.4321]
c_1	0.1344	0.0254	[0.0870, 0.1853]	γ_1	0.1170	0.0225	[0.0738, 0.1616]
c_2	0.1840	0.0363	[0.1137, 0.2545]	γ_2	0.0736	0.0142	[0.0466, 0.1023]
c_3	0.1650	0.0315	[0.1065, 0.2285]	γ_3	0.0804	0.0155	[0.0506, 0.1106]
c_4	0.1532	0.0292	[0.1006, 0.2137]	γ_4	0.0950	0.0182	[0.0605, 0.1306]
c_5	0.1791	0.0343	[0.1131, 0.2461]	γ_5	0.0781	0.0149	[0.0495, 0.1077]
c_6	0.1607	0.0309	[0.1014, 0.2210]	γ_6	0.0749	0.0148	[0.0458, 0.1032]
c_7	0.0737	0.0147	[0.0459, 0.1023]	γ_7	0.1619	0.0314	[0.1034, 0.2261]
c_8	0.0805	0.0152	[0.0507, 0.1096]	γ_8	0.1785	0.0339	[0.1145, 0.2470]
c_9	0.0842	0.0159	[0.0549, 0.1165]	γ_9	0.1581	0.0307	[0.0983, 0.2178]
c_{10}	0.1561	0.0279	[0.1024, 0.2115]	γ_{10}	0.1313	0.0248	[0.0832, 0.1786]
c_{11}	0.1485	0.0268	[0.0996, 0.2035]	γ_{11}	0.1028	0.0197	[0.0646, 0.1408]
c_{12}	0.1361	0.0243	[0.0897, 0.1842]	γ_{12}	0.1481	0.0275	[0.0967, 0.2040]
c_{13}	0.1790	0.0354	[0.1118, 0.2484]	γ_{13}	0.0607	0.0118	[0.0368, 0.0827]
c_{14}	0.1469	0.0268	[0.0952, 0.1991]	γ_{14}	0.1100	0.0211	[0.0697, 0.1523]
c_{15}	0.1673	0.0322	[0.1057, 0.2299]	γ_{15}	0.0716	0.0143	[0.0444, 0.1006]
c_{16}	0.1505	0.0266	[0.0991, 0.2028]	γ_{16}	0.1431	0.0268	[0.0931, 0.1960]

PARM denotes parameter, EAP is the expected a posteriori estimation, SD denotes the standard deviation, and HPDI denotes the highest probability density interval.

item is, the more likely is the examinee to slip in answering it, which leads to a reduction in the correct rate. The SDs of the discrimination parameters range from 0.0897 to 0.1719, those of the difficulty parameters from 0.0512 to 0.0966, those of the guessing parameters from 0.0147 to 0.0363, and those of the slipping parameters from 0.0118 to 0.0339.

6.3. Analysis of Person Parameters

The histograms of the posterior estimates of the ability parameters are shown in **Figure 5**. Most of the estimated abilities of the examinees are near zero. The number of examinees with high ability (the estimates are between 0 and 1.2) is more than the number with low ability (the estimates are between −1.2 and 0). The ability parameter posterior histogram is consistent with the frequency histogram of the correct rate (**Figure 4**). That is, the trend of change in the correct rate in the histogram is same as that in the ability posterior histogram. The number of examinees with high correct rate is more than the number with low correct rate. It is once again verified that the results of the estimation are accurate.

7. DISCUSSION

In this paper, an efficient Gibbs-slice sampling algorithm in a fully Bayesian framework has been proposed to estimate the 4PL model. This algorithm, as its name suggests, can be conceived of as an extension of the Gibbs algorithm. The sampling process consists of two parts. One part is the Gibbs algorithm, which is used to update the guessing and slipping parameters when non-informative uniform priors are employed for cases that are prototypical of educational and psychopathology items. This part implements sampling by using a conjugate prior and greatly increases efficiency. The other part is the slice sampling algorithm, which samples the 2PL IRT model from the truncated

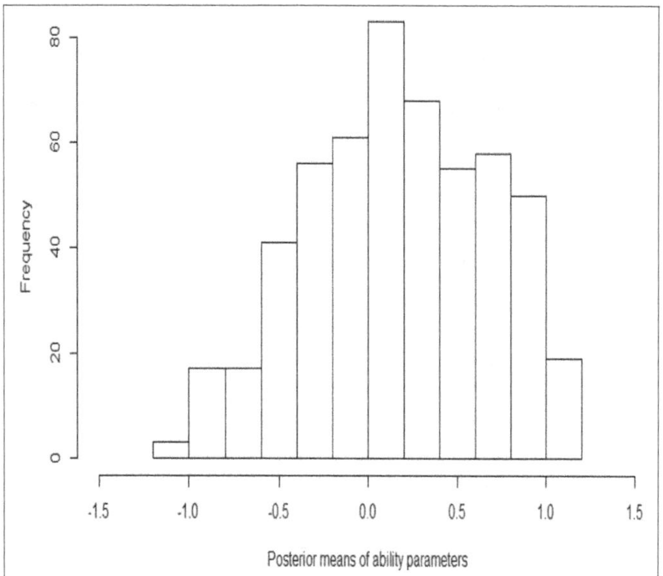

FIGURE 5 | The histograms of the posterior estimates of ability parameters.

full conditional posterior distribution by introducing auxiliary variables. The motivations for the slice sampling algorithm are manifold. First, this algorithm has the advantage of flexibility in the choice of prior distribution to obtain samples from the full conditional posterior distributions, rather than being restricted to using the conjugate distributions as in the Gibbs sampling process, which is also limited to the normal ogive framework. This allows the use of informative priors, non-informative priors, and inappropriate priors for the item parameters. Second, the Metropolis–Hastings algorithm depends on the proposal distributions and variances (tuning parameters) and is sensitive to step size. If the step size is too small, the chain will take longer to traverse the target density. If the step size is too large, there will be inefficiencies due to a high rejection rate. However, the slice sampling algorithm can automatically tune the step size to match the local shape of the target density and draw samples with acceptance probability equal to one. Thus, it is easier and more efficient to implement.

However, the computational burden of the Gibbs-slice sampling algorithm becomes intensive, especially when a large numbers of examinees or items are considered, or a large MCMC sample size is used. Therefore, it is desirable to develop a standalone R package associated with C++ or Fortran software for more a extensive large-scale assessment program. In fact, the new algorithm based on auxiliary variables can be extended to estimate some more complex item response and response time models, for example, the graded response model or the Weibull response time model. Only DIC and LPML have been considered in this study, but other Bayesian model selection criteria such as marginal likelihoods may also be potentially useful to compare different IRT models. These extensions are beyond the scope of this paper but are currently under further investigation.

AUTHOR CONTRIBUTIONS

JZ completed the writing of the article and provided article revisions. JL provided original thoughts. JZ, JL, HD, and ZZ provided key technical support. All authors contributed to the article and approved the submitted version.

FUNDING

This work was supported by the Fundamental Research Funds for the Central Universities of china (Grant No. 2412020QD025), and the "Youth Development Project" of School of Mathematics and Statistics of Northeast Normal University and the Foundation for Postdoctoral of Yunnan University (Grant No. CI76220200).

REFERENCES

1. Albert, J. H. (1992). Bayesian estimation of normal ogive item response curves using Gibb ssampling. *J. Educ. Stat.* 17, 251–269.
2. Baker, F. B., and Kim, S. H. (2004). *Item Response Theory: Parameter Estimation Techniques*. New York, NY: Marcel Dekker.
3. Barton, M. A., and Lord, F. M. (1981). *An Upper Asymptote for the Three-Parameter Logistic Item Response Model*. Princeton, NJ: Educational Testing Service.
4. Béguin, A. A., and Glas, C. A. W. (2001). MCMC estimation of multidimensional IRT models. *Psychometrika* 66, 541–561. doi: 10.1007/BF02296195
5. Birnbaum, A. (1957). *Efficient Design and Use of Tests of a Mental Ability for Various Decision-Making Problems. Series Report No. 58-16*. Randolph Air Force Base, TX: USAF School of Aviation Medicine.
6. Birnbaum, A. (1968). "Some latent trait models and their use in inferring an examinee's ability," in *Statistical Theories of Mental Test Scores*, eds F. M. Lord and M. R. Novick (Reading, MA: MIT Press), 397–479.
7. Bishop, C. M. (2006). "Slice sampling," in *Pattern Recognition and Machine Learning*, eds M. Jordan, J. Kleinberg, B. Schölkopf (New York, NY: Springer), 523–558.
8. Bock, R. D., and Aitkin, M. (1981). Marginal maximum likelihood estimation of item parameters: application of an EM algorithm. *Psychometrika* 46, 443–459. doi: 10.1007/BF02293801
9. Brooks, S. P., and Gelman, A. (1998). Alternative methods for monitoring convergence of iterative simulations. *J. Comput. Graph. Stat.* 7, 434–455. doi: 10.1080/10618600.1998.10474787

Chalmers, R. P. (2012). mirt: A multidimensional item response theory package for the Renvironment. *J. Stat. Softw.* 48, 1–29. doi: 10.18637/jss.v048.i06

10. Chang, H.-H., and Ying, Z. (2008). To weight or not to weight? Balancing influence of initial items in adaptive testing. *Psychometrika* 73, 441–450. doi:10.1007/s11336-007-9047-7
11. Chen, M.-H., Shao, Q.-M., and Ibrahim, J. G. (2000). *Monte Carlo Methods in Bayesian Computation*. New York, NY: Springer.
12. Chib, S., and Greenberg, E. (1995). Understanding the Metropolis- Hastings algorithm. *Am. Stat.* 49, 327–335. doi: 10.1080/00031305.1995.10476177
13. Culpepper, S. A. (2016). Revisiting the 4-parameter item response model: Bayesian estimation and application. *Psychometrika* 81, 1142–1163. doi: 10.1007/s11336-015-9477-6

14. Damien, P., Wakefield, J., and Walker, S. (1999). Gibbs sampling for Bayesian non-conjugate and hierarchical models by auxiliary variables. *J. R. Stat. Soc. Ser. B* 61, 331–344. doi: 10.1111/1467-9868.00179
15. Embretson, S. E., and Reise, S. P. (2000). *Item Response Theory for Psychologists.* Mahwah, NJ: Erlbaum.
16. Ferrando, P. J. (1994). Fitting item response models to the EPI-A impulsivity subscale. *Educ. Psychol. Measure.* 54, 118–127. doi: 10.1177/0013164494054001016
17. Fox, J.-P. (2010). *Bayesian Item Response Modeling: Theory and Applications.* New York, NY: Springer.
18. Fox, J. P. (2005). Multilevel IRT using dichotomous and polytomous items. *Br. J. Math. Stat. Psychol.* 58, 145–172. doi: 10.1348/000711005X38951
19. Fox, J. P., and Glas, C. A. W. (2001). Bayesian estimation of a multilevel IRT model using Gibbs sampling. *Psychometrika* 66, 271–288. doi: 10.1007/BF02294839
20. Fraley, R. C., Waller, N. G., and Brennan, K. A. (2000). An item response theory analysis of self-report measures of adult attachment. *J. Pers. Soc. Psychol.* 78, 350–365. doi: 10.1037/0022-3514.78.2.350
21. Geisser, S., and Eddy, W. F. (1979). A predictive approach to model selection. *J. Am. Stat. Assoc.* 74, 153–160. doi: 10.1080/01621459.1979.10481632
22. Gelfand, A. E., Dey, D. K., and Chang, H. (1992). "Model determination using predictive distributions with implementation via sampling-based methods (with discussion)," in *Bayesian Statistics, Vol. 4*, eds J. M. Bernardo, J. O. Berger, A. P. Dawid, and A. F. M. Smith (Oxford: Oxford University Press), 147–167.
23. Gelfand, A. E., and Smith, A. F. M. (1990). Sampling-based approaches to calculating marginal densities. *J. Am. Stat. Assoc.* 85, 398–409. doi: 10.1080/01621459.1990.10476213
24. Gelman, A., and Rubin, D. B. (1992). Inference from iterative simulation using multiple sequences. *Stat. Sci.* 7, 457–472. doi: 10.1214/ss/1177011136
25. Geman, S., and Geman, D. (1984). Stochastic relaxation, Gibbs distributions, and the Bayesian restoration of images. *IEEE Trans. Pattern Anal. Machine Intell.* 6, 721–741. doi: 10.1109/TPAMI.1984.4767596
26. Gray-Little, B., Williams, V. S. L., and Hancock, T. D. (1997). An item response theory analysis of the Rosenberg Self-Esteem Scale. *Pers. Soc. Psychol. Bull.* 23, 443–451. doi: 10.1177/0146167297235001
27. Green, B. F. (2011). A comment on early student blunders on computer-based adaptive tests. *Appl. Psychol. Measure.* 35, 165–174. doi: 10.1177/0146621610377080
28. Hastings, W. K. (1970). Monte Carlo sampling methods using Markov chains and their applications. *Biometrika* 57, 97–109. doi: 10.1093/biomet/57.1.97
29. Hessen, D. J. (2005). Constant latent odds- ratios models and the Mantel-Haenszel null hypothesis. *Psychometrika* 70, 497–516. doi: 10.1007/s11336-002-1040-6
30. Hockemeyer, C. (2002). A comparison of non-deterministic procedures for the adaptive assessment of knowledge. *Psychol. Beiträge* 44, 495–503.
31. Ibrahim, J. G., Chen, M.-H., and Sinha, D. (2001). *Bayesian Survival Analysis.* New York, NY: Springer.
32. Junker, B. W., and Sijtsma, K. (2001). Cognitive assessment models with few assumptions, and connections with nonparametric item response theory. *Appl. Psychol. Measure.* 25, 258–272. doi: 10.1177/01466210122032064
33. Lanza, S. T., Foster, M., Taylor, T. K., and Burns, L. (2005). Assessing the impact of measurement specificity in a behavior problems checklist: An IRT analysis. Technical Report 05-75. The Pennsylvania State University; The Methodology Center, University Park, PA.
34. Liao, W.-W., Ho, R.-G., Yen, Y.-C., and Cheng, H.-C. (2012). The four-parameter logistic item response theory model as a robust method of estimating ability despite aberrant responses. *Soc. Behav. Pers.* 40, 1679–1694. doi: 10.2224/sbp.2012.40.10.1679
35. Loken, E., and Rulison, K. (2010). Estimation of a four-parameter item response theory model. *Br. J. Math. Stat. Psychol.* 63, 509–525. doi: 10.1348/000711009X474502
36. Lord, F. M., and Novick, M. R. (1968). *Statistical Theories of Mental Test Scores.* Reading, MA: Addison-Wesley.
37. Lu, J., Zhang, J. W., and Tao, J. (2018). Slice-Gibbs sampling algorithm for estimating the parameters of a multilevel item response model. *J. Math. Psychol.* 82, 12–25. doi: 10.1016/j.jmp.2017.10.005
38. Magis, D. (2013). A note on the item information function of the four-parameter logistic model. Appl. Psychol. Measure. 37, 304–315. doi: 10.1177/0146621613475471
39. Metropolis, N., Rosenbluth, A. W., Rosenbluth, M. N., Teller, A. H., and Teller, E. (1953). Equation of state calculations by fast computing machines. *J. Chem. Phys.* 21, 1087–1092. doi: 10.1063/1.1699114
40. Neal, R. (2003). Slice sampling. *Ann. Stat.* 31, 705–767. doi: 10.1214/aos/1056562461
41. Ogasawara, H. (2012). Asymptotic expansions for the ability estimator in item response theory. *Comput. Stat.* 27, 661–683. doi: 10.1007/s00180-011-0282-0 Osgood, D. W., McMorris, B. J., and Potenza, M. T. (2002). Analyzing multipleitem measures of crime and deviance I: item response theory scaling. *J. Quant. Criminol.* 18, 267–296. doi: 10.1023/A:1016008004010
42. Patz, R. J., and Junker, B. W. (1999). A straight forward approach to Markov chain Monte Carlo methods for item response models. *J. Educ. Behav. Stat.* 24, 146–178. doi: 10.3102/10769986024002146
43. Rasch, G. (1960). *Probabilistic Models for Some Intelligence and Attainment Tests.* Copenhagen: Danish Institute for Educational Research.
44. Reise, S. P., and Waller, N. G. (2003). How many IRT parameters does it take to model psychopathology items? *Psychol. Methods* 8, 164–184. doi: 10.1037/1082-989X.8.2.164
45. Rouse, S. V., Finger, M. S., and Butcher, J. N. (1999). Advances in clinical personality measurement: an item response theory analysis of the MMPI-2 PSY-5 scales. *J. Pers. Assess.* 72, 282–307. doi: 10.1207/S15327752JP720212
46. Rulison, K. L., and Loken, E. (2009). I've fallen and I can't get up: Can high-ability students recover from early mistakes in CAT? *Appl. Psychol. Measure.* 33, 83–101. doi: 10.1177/0146621608324023
47. Spiegelhalter, D. J, Thomas, A., Best, N. G., and Lunn, D. (2003). *WinBUGS Version 1.4 User Manual.* Cambridge: MRC Biostatistics Unit. Available online at: http://www.mrc-bsu.cam.ac.uk/wp-content/uploads/manual14.pdf
48. Spiegelhalter, D. J., Best, N. G., Carlin, B. P., and Van Der Linde, A. (2002). Bayesian measures of model complexity and fit. *J. R. Stat. Soc. Ser. B* 64, 583–639. doi: 10.1111/1467-9868.00353
49. Steinberg, L., and Thissen, D. (1995). "Item response theory in personality research," in *Personality Research, Methods, and Theory: A Festschrift Honoring Donald W. Fiske*, eds P. E. Shrout and S. T. Fiske (Hillsdale, NJ: Erlbaum), 161–181.
50. Tanner, M. A., and Wong, W. H. (1987). The calculation of posterior distributions by data augmentation. *J. Am. Stat. Assoc.* 82, 528–550. doi: 10.1080/01621459.1987.10478458
51. Tavares, H. R., de Andrade, D. F., and Pereira, C. A. (2004). Detection of determinant genes and diagnostic via item response theory. *Genet. Mol. Biol.* 27, 679–685. doi: 10.1590/S1415-47572004000400033
52. Tierney, L. (1994). Markov chains for exploring posterior distributions (with discussions). *Ann. Stat.* 22, 1701–1762. doi: 10.1214/aos/1176325750
53. Van der Linden, W. J., and Hambleton, R. K. (eds.). (1997). *Handbook of Modern Item Response Theory.* New York, NY: Springer.
54. Waller, N. G., and Feuerstahler, L. M. (2017). Bayesian modal estimation of the four-parameter item response model in real, realistic, and idealized data sets. *Multivar. Behav. Res.* 52, 350–370. doi: 10.1080/00273171.2017.12 92893
55. Waller, N. G., and Reise, S. P. (2010). "Measuring psychopathology with non-standard IRT models: fitting the four-parameter model to the MMPI," in *Measuring Psychological Constructs: Advances in Modelbased Approaches*, eds S. Embretson and J. S. Roberts (Washington, DC: American Psychological Association), 147–173.
56. Yen, Y.-C., Ho, R.-G., Laio, W.-W., Chen, L.-J., and Kuo, C.-C. (2012). Anempirical evaluation of the slip correction in the four parameter logistic models with computerized adaptive testing. *Appl. Psychol. Measure.* 36, 75–78. doi: 10.1177/0146621611432862
57. Zhang, Z., Hamagami, F., Wang, L., Grimm, K. J., and Nesselroade, J. R. (2007). Bayesian analysis of longitudinal data using growth curve models. *Int. J. Behav. Dev.* 31, 374–383. doi: 10.1177/0165025407 077764

4

A Knowledge-Based Smart Trainer System for Transferring Knowledge from Coaches to Long Jump Students

Teerawat Kamnardsiri[1], Pattaraporn Khuwuthyakorn[2], Sirinun Boripuntakul[3] and Worawit Janchai[2]**

[1]Department of Digital Game, College of Arts, Media and Technology, Chiang Mai University, Chiang Mai, Thailand, [2]College of Arts, Media and Technology, Chiang Mai University, Chiang Mai, Thailand, [3]Department of Physical Therapy, Faculty of Associated Medical Sciences, Chiang Mai University, Chiang Mai, Thailand

***Correspondence:**
Worawit Janchai
worawit.j@cmu.ac.th

Background: The constructive and specific feedback in guiding long jump athletes to improve their performance in each phase is part of the critical process for achieving desired long jump distance. However, to date, the potential approach for assisting a coach in capturing long jump movement and transferring their knowledge to long jump students is not well-established.

Objectives: To investigate the performance of long jump students and evaluate transferring knowledge from coaches to long jump students using a Knowledge-Based Smart Trainer (*KBST*) System.

Methods: Twenty-two participants (fifteen males, mean age = 15.33 ± 1.95 years; seven females, mean age = 14.57 ± 2.07 years) participated in the study. All participants were recruited from eleven sports schools in Thailand. Each participant was instructed to perform the long jump movement, including running, take-off, and landing, for three attempts (*Test 1*, *Test 2,* and *Test 3*). *Test 1* was the conventional approach (coaches provided the feedback based on their experience). *Test 2* and *Test 3* were the KBST system approach (coaches provided the feedback based on the results from KBST system). Two cameras were used to record the participant movement from the starting position to the landing position. The capture data were analyzed by KBST system program. The outcome measures were starting position, maximum velocity, maximum velocity position, and take-off angle. Repeated-Measures ANOVA was conducted to compare the long jump performance across the three trials. The statistical significance was set at p-value < 0.05.

Results: There was a statistically significant difference between *Test 1* and *Test 3* for long jump distance (mean difference = 0.292; Std. Error = 0.129; Sig. = 0.34). However, the mean of take-off angle was similar across the three trials (*Test 1* = 12.16°, *Test 2* = 12.71°, and *Test 3* = 12.95°, respectively).

Conclusion: The *KBST* system was effective in improving long jump students' performance and also transferring knowledge from the coach to long jump students.

Keywords: knowledge-based smart trainer, knowledge transfer, long jump, image processing, computer vision, smart system

INTRODUCTION

The long jump is one part of track and field sports in school athletic events. The long jump was contained in the first Olympics in ancient Greece, circa 708 BC. Long jump biomechanics is comprised of four phases, including the approach-run phase, take-off phase, flight phase, and landing phase (Linthorne, 2008; Kamnardsiri et al., 2015a). The desired long jump distance requires a proper strategic movement in each phase, following through with the judge's criteria for evaluation (Santos and Shannon, 1989; Alexander, 1990; Hay, 1993). Specifically, the optimal action in each phase has an effect on the flight distance; thus, the athlete's performance in each phase has to be concerned.

Commonly, one of the key factors that influence the long jumpers' performance is an experienced expert in the long jump biomechanics who is able to provide suggestions and feedback to the long jumper. However, the availability of qualified coaches in this field has still been quite limited recently. Almost all of the long jump coaches only passed the basic level (Level I-II course) of the IAAF Coaches Education and Certification System (Worldathletics, 2005). In this circumstance, a Human Visual System (HVS) error would be inevitable due to constructive feedback that is specific for athletes to upgrade their skills relying more on their coaches to be proficiently skilled when providing advice (Level III to Level V). Consequently, there might be some negative effects on the long jumper's performance, whereas incorrect suggestions are unintentionally provided by the coaches. To overcome this shortcoming, coaches need some accessibility approaches to assist them in capturing long jump movement and then giving appropriate feedback to the long jumper (Kamnardsiri et al., 2019a).

Motion capture systems and timing gates were employed as gold standard approaches for measuring speed and take-off angle (Graham-Smith and Lees, 2005; van der Kruk and Reijne, 2018). Nevertheless, these approaches are costly and time-consuming in the configuration system, which might be a practical barrier in that working field. Besides, some systems have a cost limitation for procurement.

Computer technology has been developed in supporting coaches for measuring and analyzing data from the athlete, for example, video analysis (Bryan, 2004), game analysis (O'Donoghue, 2013; O'Donoghue, 2015), sports tracking (Gade and Moeslund, 2018), human motion tracking (Krzeszowski et al., 2017; Shah et al., 2017), and especially computer vision techniques that are an essential role in the development of tools for analyzing the movement, recreation and sports activities, and athletic skills of participants (Leo et al., 2017). In the previous works, several researchers studied computer sciences in sports; for instance, Theodorou et al. (2017) investigated the relative influence of step length and step frequency on step velocity during the approach-run of high-level long jumpers using a high-speed video camera. Scott et al. (1997) studied the association between non-long jumpers and experienced long jumpers in footfall placement between trials using the manually panned system capturing the video with a camera record. Hsu et al. (2006) developed automatic motion detection during a standing long jump. Video sequence images were segmented by using the background subtraction technique. The features of all frames were extracted by the Pose Estimation and the Hue-Saturation-Value (HSV) space. They employed the genetic algorithm-based (GA-based) approach to learn the joints of the person. The results showed that the joints of the person could generate the required movements for studying the performance. Dubois et al. (2012) used an image processing technique for measuring the freestyle swimming biomechanics. Video footage was analyzed using an image processing technique. Autodetection for arms and legs data was performed, whereas data were analyzed for improving the movement of each athlete. Kamnardsiri et al. (2019a) investigated the requirement of the Knowledge-Based System for training long jump from coaches (sports schools) and designed the system from the long jump experts (passed the Level III course from IAAF). The results indicated that the students acquired "*Correct after jump from only coach suggestion*" as 100% and also "*Need the expert system*" is 100% to help coaches improve the capabilities of the long jumper. The coaches employed the video camera to capture long jump movement at 71.43%. Therefore, the Knowledge-Based Smart Trainer (*KBST*) system is one of the considerations in systems procedure for transferring the coach's knowledge to the long jump athlete. This study aims to evaluate the performance of the long jump students and transferring of knowledge from the coach to the long jump students using the *KBST* system.

The present study aims to investigate the performance of long jump students with and without the *KBST* system. We hypothesized that after applying the *KBST* approach, the performance of long jump students would demonstrate a significant improvement. This information would address an effective and practical approach for transferring experts' knowledge to coaches and long jump students.

KNOWLEDGE-BASED SMART TRAINER SYSTEM

According to the Knowledge-Based System for long jump training, the approach-run phase was designed by long jump experts and developed by using the Matlab R2015a in combination with the Computer Vision System Toolbox and the Image Processing Toolbox. The system design is comprised of the 1) design Knowledge-Based System framework (Kamnardsiri et al., 2015a; Kamnardsiri et al., 2015b) and 2) knowledge-based condition for training the long jump (Kamnardsiri et al., 2019a). Moreover, the Knowledge-Based System for Training Long Jump Athletes was developed and then validated for knowledge transferring (Kamnardsiri et al., 2018; Kamnardsiri et al., 2019b).

In this study, we applied the system design and development of Kamnardsiri et al. (2019a) and then modified the suggestion phase of both the approach-run and take-off phases. Also, we added the take-off phase into the system and then redefined the name as the Knowledge-Based Smart Trainer (*KBST*) system, which is comprised of four sections (**Figure 1**).

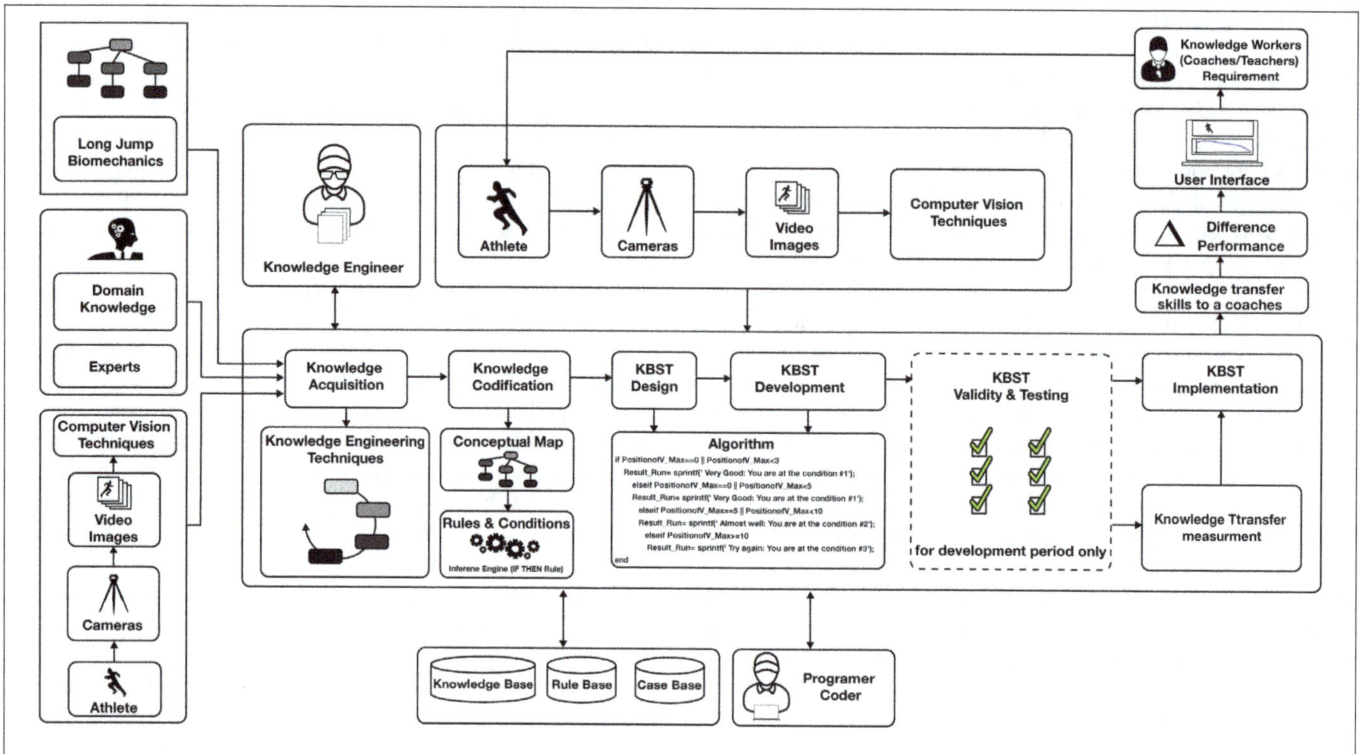

FIGURE 1 | The Knowledge-Based Smart Trainer system framework.

Domain Knowledge

The domain knowledge involves a combination of knowledge from professionals' experiences (long jump experts) at long jump biomechanics. All experts need to pass at least the Level III course of the IAAF (Worldathletics, 2005). Moreover, computer vision techniques are considered for collecting video data.

KBST Process

To develop the system, the *KBST* system employed the process of Knowledge-Based System (KBS) development process (Akerkar and Sajja, 2009; Sajja and Akerkar, 2010). The main key to this process is a Knowledge Engineer (KE) (Studer et al., 1998; Laudon, 2003; Laudon and Laudon, 2011). KBS applies technology into the organization that is managed with the organization's assets, for instance, expert systems, rule-based systems, case-based systems, and database management systems (Becerra-Fernandez and Sabherwal, 2014). The *KBST* process consists of the following four parts:

- Experts' knowledge is captured using a knowledge acquisition process with knowledge engineering techniques and codified into the conceptual map, rules, and conditions (inference engine).
- Design and development of the *KBST* system are performed to convert experts' knowledge into the algorithm to produce the knowledge engine.
- Validity and testing are operated only in the earlier implementation development period of the *KBST* system. The importance of this process is the programmer or coder.
- The indirect knowledge transfer process is measured from the results of the athlete to check the knowledge transfer from the *KBST* system to coaches after testing.

Computer Vision

Computer vision techniques (Weinland et al., 2011) are used for capturing input data from standard video cameras, and then the data is loaded into the computer. Besides, all captured data are processed by computer vision techniques to be compared with the inference engine.

Knowledge Workers

Knowledge workers in this study make the knowledge suggestion and the results are computed with Algorithm 1 and Algorithm 2 of the *KBST* system. The knowledge is transferred to coaches (knowledge workers). Therefore, the *KBST* system is able to assist coaches in monitoring and planning programs to improve the individual skills of the long jump student.

MATERIALS AND METHODS

Participants

Twenty-two long jump students aged between 15 and 20 years were recruited from six sports schools in Thailand: 1) Lampang province (five students), 2) Suphanburi province (three students), 3) Nakhon Sawan province (two students), 4) Khonkaen province (four students), 5) Trang province (four students), and 6) Nakhon Si Thammarat province (four students) were included in participation event.

Algorithm 1 Approach-run suggestion algorithm.

Input: ***V_Max***, ***PositionofV_Max*****Output: Result_Run, Result_suggesstion**

Begin
if *V_Max* < 5 then
Result_Run = **"*Sorry your maximum velocity* (*V_{max}*): *V_MAX* m.s^{-1}. that is not suitable for long jump. The maximum speed should be more than 5 m.s^{-1}."**
Result_suggesstion = "Suggestion:

You should try again with more speed-up.
You should practise your speed performance."
elseif *PositionofV_Max* == 0 || ***PositionofV_Max*** <=5 then
Result_Run = **"*Very Good: You are at the Rule #1***
Maximum velocity (*V_{max}*): *V_MAX* m.s^{-1}.
Maximum velocity position (*$V_{max\ position}$*): *PositionofV_Max* m."
Result_suggesstion = "Suggestion:
This is good running but you have to control the velocity to take-off.
Your maximum velocity is this position, so you should control the speed for taking-off.
Please do not reduce your speed and keep this speed level or increase your speed-up before the next jump.
You should move the starting-point equivalent lose space to the take-off board from the last starting-position but the speed is still the same with each running attempt."
else if *PositionofV_Max* > 5 || ***PositionofV_Max*** <= 10 then
Result_Run = ***"Almost well: You are at the Rule #2***
***Maximum velocity* (*V_{max}*): *V_MAX* m.s^{-1}.**
Maximum velocity position* (*$V_{max\ position}$*): *PositionofV_Max m."
Result_suggesstion = "Suggestion:
Before reaching to the take-off board, you should have two check makes.
You have to shift in from start point to take-off board.
Good, but you should keep this maximum speed to take-off board. The controlled maximum speed should be less than 5 meters from the take-off board.
In the real competition situation, you will adjust the new marker using around 14-17 soles (2 strides of running) to come take-off board.
You should move the starting-point equivalent lose space to the take-off board from the last starting-position but the speed is still the same with each running attempt.
Do not take more long stride or short stride of a leg before take-off"
elseif *PositionofV_Max* > 10
Result_Run = **"*Try again: You are at the Rule #3***
Maximum velocity (*V_{max}*): *V_MAX* m.s^{-1}.
Maximum velocity position (*$v_{max\ position}$*): *PositionofV_Max* m."
Result_suggesstion = "Suggestion:
Number of stride ratio of all distance from the start point to the first check mark should be more than the last period before reaching to the take-off board.
You should slightly increase to the take-off board. It is unable to use this speed for taking-off. You should practise the approach-run again.
You may start with over fast running that is why Vmax appears in this period. It affects the athlete who is unable to preserve and control the speed for taking-off.
You should start with around 10-20 m. with long stride from the take-off board, then each stride adjusts your speed at 5-10 m. and the maximum speed should be around 0-5 m. from the Take-off board.
You should gradually increase speed from starting position to check mark point, then perform the fastest speed around 0-5 m from take-off board.
Increasing speed, the jumper gradually straightens up and run to become more upright and go along with a smooth stride pattern.
Rhythm in each approach increases your pace at the same rate and takes care to keep the degree of muscle contraction uniform.
One checkmark at a distance which is the midpoint of the run is usually sufficient."
end if
end

Participants took part in the evaluation of performance improvement from the system. In addition, participants were recruited based on the following inclusion criteria: those who were Thai nationals, in healthy condition, identified as long jump athletes between the ages of 15 and 20 years, able to run a fast 25–30 m dash and jump into the sandpit, and able to comprehend instructions and willing to participate. Moreover, the exclusion criteria were the presence of physical conditions such as a musculoskeletal injury that preclude participants from completing the testing protocol, taking alcohol 6 h before testing, or using drug regimens that had a negative effect on the running performance. All methods have been followed in agreement with the relevant guidelines and with the principles outlined in the Declaration of Helsinki and were approved by the Chiang Mai University, Research Ethics Committee, Chiang Mai, Thailand (*number CMUREC 61/044*). All of the participants provided their informed consent from their parents prior to participation. The acceptable participants were informed of the study's purposes before signing the informed consent.

Protocol

Participants wore a black sport long-sleeved shirt and black sport jogger pants. Each participant performed the long jump movement (running, take-off, and landing) for three tests (*Test 1*, *Test 2*, and *Test 3*). Prior to testing, all participants were engaged in warm-up protocol, including stretching, dynamic exercises, and jogging. After that, they performed the long jump under the supervision of the coach until achieving their regular maximal performance (*Test 1*). Therefore, *Test 1* was the traditional approach or baseline trial in which participants obtained usual coach feedback without *KBST* results (e.g., "*Try Again*," "*Almost well*" or "*Good*"). Following the

Algorithm 2 Take-off suggestion algorithm.

Input: ***Take_off_angle***
Output: Result_Jump, Result_suggesstion
Begin
if ***Take_off_angle*** <= 5 then
Result_Jump = "Sorry you did not take-off! Your take-off angle is: *Take_off_angle* that is not suitable for the take-off phase."
Result_suggesstion = "Suggestion:
The take-off angle should be more than 5.
You should try again with more take-off angle.
You should practice your take-off performance."
elseif ***Take_off_angle*** >= 18 then
Result_Jump = "Very good: Your take-off angle is: *Take_off_angle*."
Result_suggesstion = "Suggestion:
The swinging leg movement must begin from the hip that is shifted forwards and upwards.
The forward drive must be braced, so that the driving force can be transmitted to the body.
During the swing, the leg remains well flexed at the knee joint, the thigh should be reached at the horizontal at the end of the take-off action.
A forward inclination of the line of force can be increased through an effective swing, hence creating favourable conditions for the backward rotation.
The transference of movement is from swinging the leg and the arms to the trunk, the arms moved up but were quickly braced. The elbows are slightly turned out, hence the shoulders can be raised more easily and balanced that is more easy to be maintained."
elseif ***Take_off_angle*** >= 45 then
Result_Jump = "Fair: Your take-off angle is: *Take_off_angle*."
Result_suggesstion = "Suggestion:
You should reduce the take-off angle.
Take-off angle could be around 18-22 degree as an appropriation for long jump.
When starting to jump the take-off leg side could be straight.
You must be careful to maintain the spring in your muscles.
You should increase your horizontal velocity of running before take-off.
You do not wait until the foot has touched the ground, but place the leg, with a quick grabbing movement, back and down."
else
Result_Jump = "Fair: Your take-off angle is: *Take_off_angle*."
Result_suggesstion = "Suggestion:
The swinging leg movement must begin from the hip that is shifted forwards and upwards. Your horizontal velocity must be transformed into a vertical velocity momentum, which is directed forward and upward, take-off foot should be set somewhat further forward.
The body during take-off could be straight and a bit lean toward.
Practicing neural correlation and muscle movement is very important; athletes could practice this skilled.
You must be careful to maintain the spring in your muscles.
You must assume a more upright position, since the forward lean used in sprints would prevent a good take-off lift.
You should take a lower approach of your CG in the penultimate stride and lift all parts of your body during take-off.
During the last strides, knees are picked up high in the front in preparation for the energetic forward drive of the free leg.
The last stride must, in the first place, bring the flight phase rapidly to an end, so the curve of the CG of the jumper has no descending branch but moves only upwards after lowering the penultimate stride.
Synchronize to relax your arm movements."
end if
end

protocol for testing of the *KBST* system, participants performed long jump for two attempts (*Test 2* and *Test 3*, respectively) with sufficient rest between each test. The coach acquired the results from the *KBST* and provided the participants with the suggestion after *Test 1* and *Test 2*. The starting position ($Start_{Position}$), maximum velocity (V_{max}), maximum velocity position ($V_{maxPosition}$), and take-off angle (θ) were assessed from a total of all long jumps on the synthetic running track, flat and dry. The protocol for testing of the *KBST* system consisted of four located markers that were *M1* at the starting point of the long jump sand pit, *M2* at the take-off position, *M3* at 5 m far from the take-off position, and *M4* at 10 m far from the take-off position. The first camera was posited about 10 m far from the take-off position (*M2*). The second camera was posited about 25 m far from the marker (*M4*). Both cameras were fixed on the three-way head tripod with a height about 1.5 m from the floor (**Figure 2**).

Materials for motion capture in this study are comprised of the following: 1) Camera 1, standard video camera (*Nikon 1 J1, Japan*) with 10–30 mm lens; 2) Camera 2, standard video camera (*Casio ZR-55, Japan*) with 25 mm lens; 3) two tripods and three-way heads; 4) floor markers *M1*, *M2*, *M3*, and *M4*; 5) black sport long-sleeved shirt; 6) black sport jogger pants.

Procedure

Evaluation of performance improvement using the *KBST* was conducted by comparing *Test 1*, *Test 2*, and *Test 3*. The phase consists of four substeps as follows: 1) capturing the long jump *Test 1*, giving suggestion from *KBST*, and correcting jump; 2) capturing the long jump *Test 2*, giving suggestion, from *KBST* and correcting jump; 3) capturing the long jump *Test 3* and giving suggestion from *KBST*; 4) analyzing the captured data.

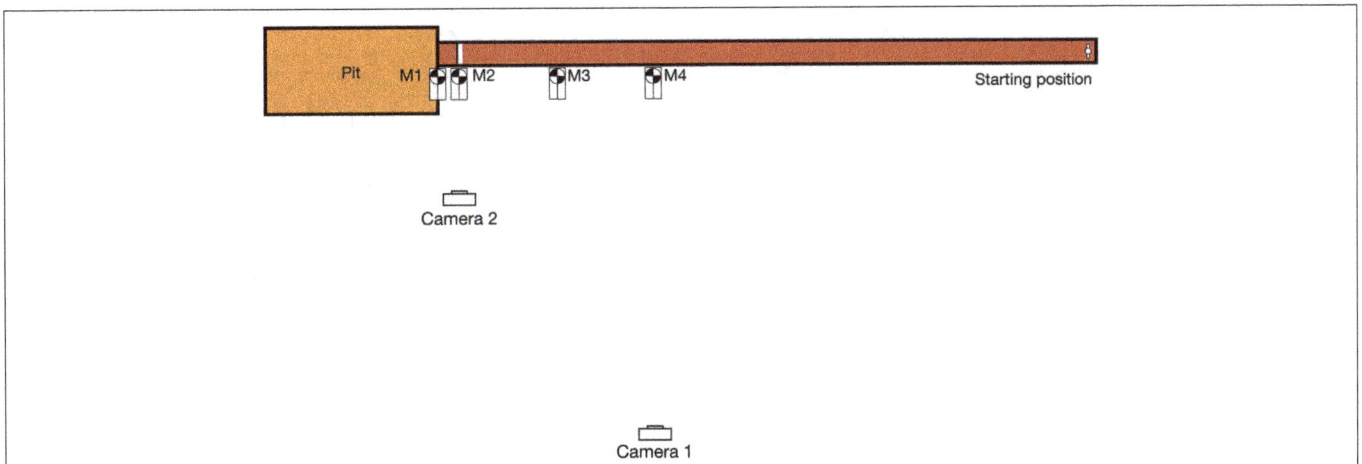

FIGURE 2 | The protocol of the Knowledge-Based Smart Trainer system for testing.

Data Acquisition and Analysis

The approach-run phase standard video camera (Camera 1) was set at the original size of 1,280 × 720 pixels and 60 Hz for capturing data. The camera was employed to capture the participant running from the starting position to the take-off position. The take-off and landing phase standard video camera (Camera 2) was set at the size of 640 × 480 pixels and 120 Hz for capturing data. All captured data were downloaded and analyzed using the *KBST* system that was utilized for calculating the starting position ($Start_{Position}$), maximum velocity (V_{max}), maximum velocity position ($V_{maxPosition}$), and take-off angle (θ). The task of collecting data was organized three times (*Test 1*, *Test 2*, and *Test 3*).

Statistical Analyses

All data were analyzed using IBM SPSS Statistics 21 for Windows (SPSS Inc., Chicago, IL, United States) with a statistically significant difference of $p < 0.05$. The descriptive statistics (means, SD, and variance) were employed for calculating demographic profiles and long jump variables of the participants. The One-Way Repeated-Measures ANOVA was employed for analyzing the significantly significant differences across the three tests.

Data Collection Location

This study was conducted at six sports schools in Thailand in Lampang province, Suphanburi province, Nakhon Sawan province, Khonkaen province, Trang province, and Nakhon Si Thammarat province. The time for testing was around from 9:00 am to 11:00 am.

RESULTS

Demographics Profile of Participants

Twenty-two participants (15 males: age 15.33 (1.95) years, weight 62.93 (8.97) kg, height 174.60 (9.09) cm, BMI 20.54 (1.27) kg m^{-2}; 7 females: age 14.57 (2.07) years, weight 50.57 (4.64) kg, height 161.14 (4.52) cm, BMI 19.45 (1.22) kg m^{-2}) successfully completed three tests without any incident. The participants' demographic characteristics indicated that the participants in this study are long jump students as a relatively healthy group (**Table 1**).

Long Jump Performance of Participants

The long jump variables contained the starting position ($Start_{Position}$), maximum velocity (V_{max}), and maximum velocity position ($V_{maxPosition}$), take-off angle (θ), and long jump distance. These variables were calculated by the *KBST* system. The variable of long jump distance was only evaluated with a tape measure. Twenty-two long jump students performed three long jumping tests. The results indicated that the majority of the long jump students improved the mean of long jump distance as *Test 1* = 4.88 m, *Test 2* = 5.09 m, and *Test 3* = 5.17 m, respectively. There was the significant difference at the 0.05 level between *Test 1* and *Test 3* (mean difference = 0.292*; Std. error = 0.129; Sig. = 0.34). However, the mean of take-off angle does not indicate a difference as *Test 1* = 12.16°, *Test 2* = 12.71°, and *Test 3* = 12.95°, thus illustrating that it has improved a little bit, whereas the mean of velocity was slightly declined from *Test 1* = 7.30 m s^{-1} as *Test 2* = 6.93 m s^{-1} and *Test 3* = 6.86 m s^{-1} (**Figure 3** and **Table 2**).

The Percentage of Suggestion From *KBST* System

For the percentage of the suggestion, the coach, acquiring the long jump biomechanics information of individual participant after

TABLE 1 | The participants' demographic characteristics.

Variable	Male (n = 15)	Female (n = 7)	All (N = 22)
Age (years)	15.33 ± 1.95	14.57 ± 2.07	15.09 ± 1.97
Weight (kg)	62.93 ± 8.97	50.57 ± 4.64	59.00 ± 9.72
Height (cm)	174.60 ± 9.09	161.14 ± 4.52	170.31 ± 10.09
BMI (kg m^{-2})	20.54 ± 1.27	19.45 ± 1.22	20.19 ± 1.33

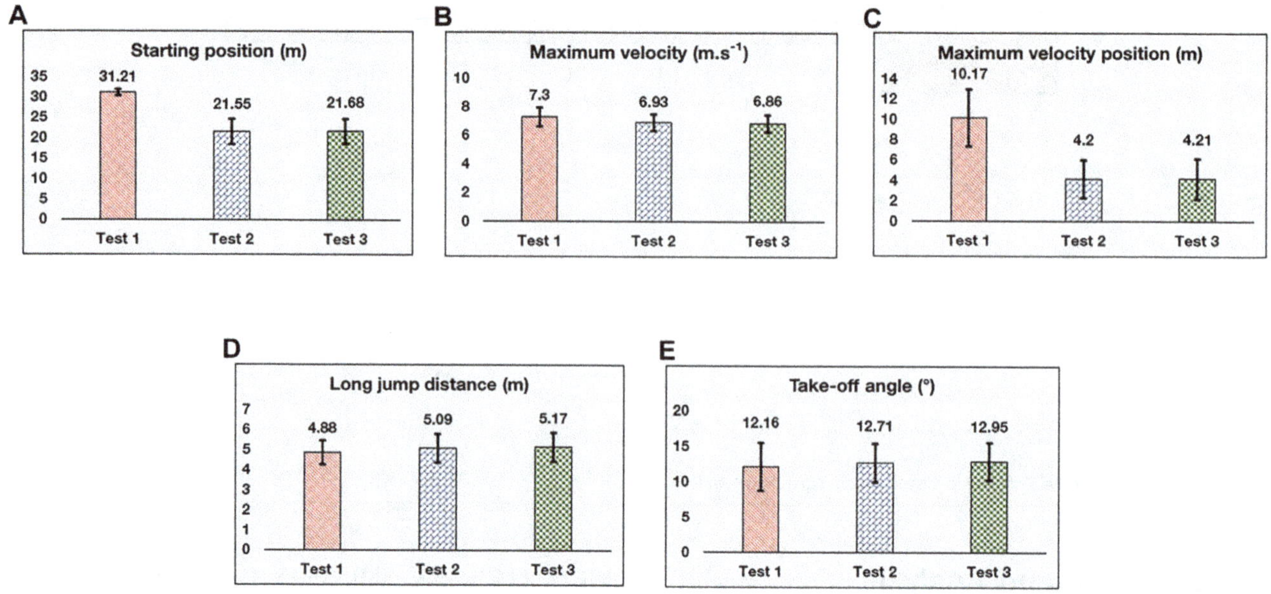

FIGURE 3 | Three long jump performance tests: **(A)** starting position, **(B)** maximum velocity, **(C)** maximum velocity position, **(D)** long jump distance, and **(E)** take-off angle.

the *KBST* system, had the data calculated from the video sequences of two cameras. The results demonstrated that the majority of the "*Try again*" suggestion was approximately 59% and the "*Almost well*" suggestion was about 36% in *Test 1*. After the KBST system provided the long jump suggestion contents to each teacher, participants practiced individually with some coaching in the long jump biomechanics. The results of the "*Very good*" suggestion in *Test 2* and *Test 3* were about 82 and 73%, respectively. However, in the take-off phase, the results revealed that *Test 1*, *Test 2*, and *Test 3* had the "*Fair*" suggestion from the system at about 95%. Nevertheless, the "*Very good*" suggestion was significantly 1% less in approximation (**Table 3**).

Evaluation of Long Jump Students' Performance Improvement

The performance of long jump students was evaluated using the statistical analysis as One-Way Repeated-Measures ANOVA for calculating the significantly significant difference among tests. The results presented that the mean difference between *Test 1* and *Test 3* was 0.292 at a significant level ($p < 0.05$). Nonetheless, the mean difference between *Test 1* and *Test 2* showed no difference (**Table 4**).

DISCUSSION

This study demonstrated that the *KBST* system can be used as a video-based assessment tool for transferring knowledge from coaching to the long jump students. Two standard cameras were employed in this study for capturing long jump data from each student and then Algorithm 1 and Algorithm 2 calculated the long jump performance of the students. Furthermore, the *KBST* can also provide suggestions from experts' knowledge that can be transferred to the coach and then each long jump student obtains that knowledge to enhance their performance. Finally, long jump students can be retrained and retested from the new suggestion of the coach.

This study found that the relationships between take-off distance, take-off angle, and run-up speed were in agreement

TABLE 2 | Descriptive statistics of the long jump tests (*N* = *22*).

Variable	Minimum	Maximum	Mean	SD	Variance
Starting point (m)					
Test 1	30.00	32.93	31.21	0.73	0.53
Test 2	15.51	28.81	21.55	3.10	9.63
Test 3	15.63	28.95	21.68	3.06	9.37
Maximum velocity (m s^{-1})					
Test 1	6.24	9.05	7.30	0.66	0.43
Test 2	5.96	7.92	6.93	0.58	0.34
Test 3	5.67	7.90	6.86	0.60	0.37
Maximum velocity position (m)					
Test 1	2.48	14.02	10.17	2.80	7.86
Test 2	1.47	9.11	4.20	1.85	3.43
Test 3	1.53	8.82	4.21	1.99	3.95
Take-off angle (°)					
Test 1	6.90	18.19	12.16	3.40	11.54
Test 2	8.38	19.43	12.71	2.71	7.33
Test 3	8.91	18.60	12.95	2.64	6.96
Long jump distance (m) with the tape measure					
Test 1	4.01	6.01	4.88	0.60	0.36
Test 2	3.97	6.58	5.09	0.71	0.51
Test 3	3.83	6.32	5.17	0.71	0.51

TABLE 3 | Percentage of the suggestion by the *KBST* system (n = 22).

Long jump phase	*Test 1* (%)	*Test 2* (%)	*Test 3* (%)
Approach-run			
Very good	1 (4.50)	18 (81.80)	16 (72.70)
Almost well	8 (36.40)	4 (18.20)	6 (27.30)
Try again	13 (59.10)	0 (0.00)	0 (0.00)
Take-off			
Very good	1 (4.50)	2 (9.10)	1 (4.50)
Fair	21 (95.50)	20 (90.90)	21 (95.50)

with the previous studies (Alexander, 1990; Hay, 1993; Seyfarth et al., 2000; Linthorne et al., 2005; Bridgett and Linthorne, 2006). Unfortunately, the take-off angle in this study was lower than the optimum take-off angle for long jump athletes that is observed from the values of world-class long jumpers, about 15°–27° (Hay, 1986; Lees et al., 1993; Lees et al., 1994; Arampatzis et al., 1999; Jaitner et al., 2001; Bridgett and Linthorne, 2006; Linthorne, 2008).

When transferring knowledge, the *KBST* suggested that the teacher provides the feedback in the simplest form to help the students focus on areas to improve their long jumps. The results found that *Test 3* shows more improvement in the approach-run phase as a "*Very good*" suggestion from the *KBST* of about 72%; nevertheless, the take-off phase is still "*Fair*" according to the suggestion of three tests. There are several pieces of evidence of transferring knowledge from coaching to improve the athletes' skill from the knowledge management theory (Ray and Ray, 2008; Ennis, 2015; Predoiu et al., 2016; Kendellen and Camiré, 2017).

To support the main contribution of this study, we examined the *KBST* system and method with the ability to capture kinematic data from the long jump student and then processed and analyzed the captured data. In addition, the *KBST* system can be used for transferring experts' knowledge to coaches and long jump students. This method is aimed to overwhelm the difficulties in the traditional coaching technique. Additionally, this method concentrates on creating a low-cost system that measures and analyzes the data from individual long jump students and then gives suggestions to them. The *KBST* system is also easy to set up for calibrating and application.

The study encountered some noise data in the computing methods, such as environmental changes, brightness changes, low-light situation, uncertain moving objects, tree leaves moving, color tracking of the body clothes, and water reflection. Those data could produce much more feature extraction techniques and would also lead to a state extension in the process of analyzing. However, this study's achievement is still in the working process of collecting data from the long jump students, which still needs study cooperation with the Thai National Team long jump athletes for measuring and monitoring performances of each long jump athlete.

There were a few limitations in the study of the *KBST* system. The first is on capturing the volume of the system that depended on the camera positioning; for instance, if the volume of capturing the athlete running around was 50 m, the camera has to be placed at approximately 35–40 m from the athlete's body. So, we have to consider the area for using the *KBST* system. The second is the color of the athlete's suit; according to the real environment, the system is set for detection of just lowering the value of color such as dark gray value and black value. Therefore, the appropriate athlete's suit is dark gray or black color, whereas other colors are restricted to use. Third, the *KBST* system is unable to be captured under a low-light environment situation. It works at normal illumination. In addition, the *KBST* system has to employ assistance capturers for capturing both the running and jumping phases, so the *KBST* simultaneous automatic capturing for two cameras is needed. Finally, the present study used a pre-post design, in which *Test 1* was set as a pre-*KBST* intervention (baseline or control trial) and *Test 2* and *Test 3* were the first and second trials of post-*KBST* intervention, respectively. However, the control group comparison is needed to confirm our findings.

Future studies need to develop an autonomous system for capturing the Thai National Team long jump athletes and for applying the *KBST* system to other kinds of sports.

In conclusion, the *KBST* system is easy to use, cost-effective, and markerless. It offers an alternative approach to capture and analyze the long jump performance of an individual athlete. In particular, the *KBST* system provides suggestion knowledge from experts and is shown to improve all long jump performance parameters from the three long jump tests at six sports schools.

TABLE 4 | The significant difference among the tests of long jump distance performance.

Long jump test		Mean difference	Std. error	Sig.[a]	95% confidence interval for difference[a]	
A	B	(A and B)			Lower bound	Upper bound
Test 1	*Test 2*	−0.207	0.140	0.153	−0.498	0.084
	Test 3	−0.292[b]	0.129	0.034	−0.560	−0.024
Test 2	*Test 1*	0.207	0.140	0.153	−0.084	0.498
	Test 3	−0.085	0.072	0.256	−0.235	0.066
Test 3	*Test 1*	0.292[b]	0.129	0.034	0.024	0.560
	Test 2	0.085	0.072	0.256	−0.066	0.235

Note: Based on estimated marginal means.

[a]*Adjustment for multiple comparisons: least significant difference (equivalent to no adjustments).*

[b]*The mean difference is significant at the 0.05 level.*

ETHICS STATEMENT

The studies involving human participants were reviewed and approved by CMUREC 61/044. Written informed consent to participate in this study was provided by the participants' legal guardian/next of kin.

AUTHOR CONTRIBUTIONS

TK was involved in conceptualization, methodology, system development, evaluation, formal analysis, investigation, data curation, writing of the original draft, and project administration. PK was involved in conceptualization, methodology, investigation, and evaluation. SB was involved in conceptualization, methodology, formal analysis, validation, investigation, data curation, and writing—reviewing and editing. WJ was involved in conceptualization, methodology, formal analysis, investigation, supervision, project administration, and writing—reviewing and editing.

ACKNOWLEDGMENTS

The authors would like to acknowledge the Motion Capture and Motion Analysis Laboratory, College of Arts, Media and Technology, Chiang Mai University, Chiang Mai, Thailand. Furthermore, the authors would like to thank Wacharee Rittiwat, Permsak Suriyachan, and also Kittisak Sukon (the long jump coach of Thailand National Team) for supporting knowledge of the long jump techniques.

REFERENCES

1. Akerkar, R., and Sajja, P. (2009). *Knowledge-based systems.* Sudbury, MA: Jones and Bartlett Publishers.
2. Alexander, R. M. (1990). Optimum take-off techniques for high and long jumps. *Philos. Trans. R. Soc. Lond. B Biol. Sci.* 329, 3–10. doi:10.1098/rstb.1990.0144
3. Arampatzis, A., Brüggemann, G. P., and Metzler, V. (1999). The effect of speed on leg stiffness and joint kinetics in human running. *J. Biomech.* 32, 1349–1353. doi:10.1016/S0021-9290(99)00133-5
4. Becerra-Fernandez, I., and Sabherwal, R. (2014). *Knowledge management: systems and processes.* 2nd Edn. London, United Kingdom: Routledge.
5. Bridgett, L. A., and Linthorne, N. P. (2006). Changes in long jump take-off technique with increasing run-up speed. *J. Sports Sci.* 24, 889–897. doi:10.1080/02640410500298040
6. Bryan, J. (2004). Video analysis software and the investigation of the conservation of mechanical energy. *Contemp. Issues Technol. Teach. Educ.* 4, 284–298. Available at: https://citejournal.org/volume-4/issue-3-04/science/video-analysis-software-and-the- investigation-of-the-conservation-of-mechanical-energy (Accessed September 18, 2020).
7. Dubois, R. P., Thiel, D. V., and James, D. A. (2012). "Using image processing for biomechanics measures in swimming," in Proceedings of the ninth conference of the international sports engineering association, Lowell, MA, July 9–13, 2012 (Amsterdam, Netherlands: Elsevier Ltd.), 807–812. doi:10.1016/j.proeng.2012. 04.138
8. Ennis, C. D. (2015). Knowledge, transfer, and innovation in physical literacy curricula. *J. Sport Health Sci.* 4, 119–124. doi:10.1016/j.jshs.2015.03.001
9. Gade, R., and Moeslund, T. B. (2018). Constrained multi-target tracking for team sports activities. *IPSJ Trans. Comput. Vis. Appl.* 10, 1–11. doi:10.1186/s41074- 017-4570038-z
10. Graham-Smith, P., and Lees, A. (2005). A three-dimensional kinematic analysis of the long jump take-off. *J. Sports Sci.* 23 (9), 891–903. doi:10.1080/02640410400022169 Hay, J. G. (1986). The biomechanics of the long jump. *Exerc. Sport Sci. Rev.* 14, 401–446.
11. Hay, J. G. (1993). Citius, altius, longius (faster, higher, longer): the biomechanics of jumping for distance. *J. Biomech.* 26, 7–21. doi:10.1016/0021-9290(93)90076-Q
12. Hsu, H. H., Hsieh, S. W., Chen, W. C., Chen, C. J., and Yang, C. Y. (2006). "Motion analysis for the standing long jump," in Proceedings of the 26th IEEE international conference distributed computing system's workshops, Lisboa, Portugal, July 4–7, 2006 (New Jersey: IEEE), 1–47. doi:10.1109/ICDCSW. 2006.74
13. Jaitner, T., Mendoza, L., and Schöllhorn, W. (2001). Analysis of the long jump technique in the transition from approach to takeoff based on time-continuous kinematic data. *Eur. J. Sport Sci.* 1, 1–12. doi:10.1080/17461390100071506
14. Kamnardsiri, T., Janchai, W., Khuwuthyakorn, P., and Rittiwat, W. (2018). On the validity of the knowledge transferring assessment of the knowledge-based system for training long jump athletes: the approach-run phase. *Eurasian J. Anal. Chem.* 13, 594–602.
15. Kamnardsiri, T., Janchai, W., Khuwuthyakorn, P., Suriyachan, P., and Rittiwat, W. (2019a). Requirement analysis to design the knowledge-based system for long jump coaching. *Int. J. Knowl. Manage. Stud.* 10, 118–137. doi:10.1504/IJKMS. 2019.099120
16. Kamnardsiri, T., Janchai, W., Khuwuthyakorn, P., and Rittiwat, W. (2019b). Implementation and validity of the long jump knowledge-based system: case of the approach run phase. *PeerJ Prepr.* 7, e27524v1. doi:10.7287/peerj. preprints.27524v1
17. Kamnardsiria, T., Janchaia, W., Khuwuthyakorna, P., Suwansrikhama, P., Klaphajoneb, J., and Suriyachanc, P. (2015a). Knowledge-based system framework for training long jump athletes using action recognition. *J. Adv. Inf. Technol.* 6, 182–193. doi:10.12720/jait.6.4.182-193
18. Kamnardsiri, T., Janc hai, W., Khuwuthyakorn, P., Suwansrikham, P., Klaphajone, J., and Suriyachan, P. (2015b). "Framework of knowledge-based system for practising long jumpers using movement recognition," in Proceedings of the 12th international conference on intellectual capital and knowledge management and organisational learning, Bangkok, Thailand, November, 2015 (Academic Conferences International Limited), 372.
19. Kendellen, K., and Camiré, M. (2017). Examining the life skill development and transfer experiences of former high school athletes. *Int. J. Sport Exerc. Psychol.* 15, 395–408. doi:10.1080/1612197X.2015.1114502
20. Krzeszowski, T., Przednowek, K., Wiktorowicz, K., and Iskra, J. (2017). Estimation of hurdle clearance parameters using a monocular human motion tracking method. *Comput. Methods Biomech. Biomed. Eng.* 19, 1319–1329. 478 doi:10. 1080/10255842.2016.1139092
21. Laudon, K. C. (2003). *Essentials of management information systems: managing the digital firm.* Upper Saddle River, NJ: Pearson.
22. Laudon, K. C., and Laudon, J. P. (2011). *Essentials of management information systems.* Upper Saddle River, NJ: Pearson.
23. Lees, A., Fowler, N., and Derby, D. (1993). A biomechanical analysis of the last stride, touch-down and take-off characteristics of the women's long jump. *J. Sports Sci.* 11, 303–314. doi:10.1080/02640419308730000
24. Lees, A., Graham-Smith, P., and Fowler, N. (1994). A biomechanical analysis of the last stride, touchdown, and takeoff characteristics of the men's long jump. *J. Appl. Biomech.* 10, 61–78. doi:10.1123/jab.10.1.61

25. Leo, M., Medioni, G., Trivedi, M., Kanade, T., and Farinella, G. M. (2017). Computer vision for assistive technologies. *Comput. Vis. Image Understand.* 154, 1–15. doi:10.1016/j.cviu.2016.09.001
26. Linthorne, N. P., Guzman, M. S., and Bridgett, L. A. (2005). Optimum take-off angle in the long jump. *J. Sports Sci.* 23, 703–712. doi:10.1080/02640410400022011
27. Linthorne, N. P. (2008). *Routledge handbook of biomechanics and human movement science: biomechanics of the long jump*. London, United Kingdom: Routledge, 340–353.
28. O'Donoghue, P. (2015). "Game analysis," in *The computer science in sport: research and practice* (New York, NY: Routledge), 154–186.
29. O'Donoghue, P. (2013). "Match analysis for coaches," in *An introduction to sports coaching: connecting theory to practice*. 2nd Edn (New York, NY: Routledge), 161–176.
30. Predoiu, R., Ramsey, P., and Arsenescu, F. (2016). "Intellectual abilities in top male junior tennis players," in The European proceedings of social and behavioural sciences EpSBS, Bucharest, Romania, June 10–13, 2015 (London, United Kingdom: Future Academy), Vol. 11, 490–495. doi:10.15405/epsbs.2016.06.68 Ray, J. M., and Ray, R. D. (2008). Train-to-Code: an adaptive expert system for training systematic observation and coding skills. *Behav. Res. Methods* 40, 673–693. doi:10.3758/brm.40.3.673
31. Sajja, P. S., and Akerkar, R. (2010). "Knowledge-based systems for development." in *Advanced knowledge based systems: model, applications and research*, Editors P. S. Sajja and R. Akerkar (Kolhapur, India: TMRF), Vol. 1, 1–11. Available at: http://www.tmrfindia.org/eseries/ebookV1-C1.pdf (Accessed September 20, 2020).
32. Santos, J., and Shannon, K. (1989). *Track: the field events*. New York, NY: Sports illustrated.
33. Scott, M. A., Li, F. X., and Davids, K. (1997). Expertise and the regulation of gait in the approach phase of the long jump. *J. Sports Sci.* 15, 597–605. doi:10.1080/ 026404197367038
34. Seyfarth, A., Blickhan, R., and Van Leeuwen, J. L. (2000). Optimum take-off techniques and muscle design for long jump. *J. Exp. Biol.* 203 (4), 741–750.
35. Shah, J. H., Chen, Z., Sharif, M., Yasmin, M., and Fernandes, S. L. (2017). A novel biomechanics-based approach for person re-identification by generating dense color sift salience features. *J. Mech. Med.* 17 (07), 174001. doi:10.1142/ S0219519417400115
36. Studer, R., Benjamins, V. R., and Fensel, D. (1998). Knowledge engineering: principles and methods. *Data Knowl. Eng.* 25, 161–197. doi:10.1016/S0169-023X(97)00056-6 Theodorou, A. S., Panoutsakopoulos, V., Exell, T. A., Argeitaki, P., Paradisis, G. P., and Smirniotou, A. (2017). Step characte ristic interaction and asymmetry during the approach phase in long jump. *J. Sports Sci.* 35 (4), 346–354. doi:10.1080/02640414.2016.1164884
37. van der Kruk, E., and Reijne, M. M. (2018). Accuracy of human motion capture systems for sport applications; state-of-the-art review. *Eur. J. Sport Sci.* 18 (6), 806–819. doi:10.1080/17461391.2018.1463397
38. Weinland, D., Ronfard, R., and Boyer, E. (2011). A survey of vision-based methods for action representation, segmentation and recognition. *Comput. Vis. Image Understand.* 115, 224–241. doi:10.1016/j.cviu.2010.10.002
39. Worldathletics (2005). The IAAF coaches education and certification system. Available at: https://www.worldathletics.org/news/news/the-iaaf-coa ches- education-and-certification (Accessed September 18, 2020).

5

A Numerical Analysis of the Leakage Characteristics of an Embankment Dam Slope with Internal Erosion

Yu Jia[1], *Yu Ding*[1]*, *Xuan Wang*[1,2], *Jiasheng Zhang*[1,2] *and Xiaobin Chen*[1,2]

[1]*School of Civil Engineering, Central South University, Changsha, China,* [2]*National Engineering Research Center of High-Speed Railway Construction Technology, Central South University, Changsha, China*

Correspondence:
Yu Ding
ding_yu@csu.edu.cn

Leakage is a common defect of embankment dam slopes, and is usually accompanied by internal erosion. This study establishes a mathematical two-phase seepage coupling model that explicitly considers the effects of internal erosion. With the help of finite element software COMSOL Multiphysics®, we use this model to study the characteristics of fluid seepage, the fraction of leakage volume that is made up of migrated fine particles (migrated fine particles volume fraction), porosity, permeability, and displacement in the leakage process of a three-dimensional embankment dam slope, as well as to study the influence of the water level and leakage outlet size on the aforementioned features. Our results show that water seepage velocity gradually increases with time, especially at the downstream leakage outlet. Therefore, the erosion and migration of fine particles occur primarily at the downstream leakage outlet, resulting in a significant increase in the migrated fine particles volume fraction, porosity, and permeability. In addition, the maximum migrated fine particles volume fraction, the maximum porosity, and the maximum permeability in the slope increase nonlinearly with time. Curves for maximum displacement with a strength reduction factor can be divided into three stages: the early stable stage, the midterm nonlinear growth stage (creeping state), and the later rapid growth stage (instability state). We also find that the rise of the water level promotes the erosion and migration of fine particles on the slope and that the maximum migrated fine particles volume fraction and the maximum porosity increase with the water level. The values of the strength reduction factor for the creeping state decrease as the water level rises. Furthermore, the larger the size of the downstream leakage outlet, the larger the maximum migrated fine particles volume fraction and maximum porosity. However, the relationship between maximum displacement and strength reduction factor is nearly unaffected by the size of the leakage outlet.

Keywords: embankment dam slope, internal erosion, two-phase seepage characteristics, deformation characteristics, porosity, permeability, migrated fine particles

1 INTRODUCTION

The embankment dam slope is a key component of flood control and disaster mitigation systems, and their seepage characteristics are important research topics in geotechnical and hydraulic engineering (Foster et al., 2000a; Cesali and Federico, 2019; Yang et al., 2019; Jiang et al., 2020). More specifically, leakage is a common type of damage to embankment dam slopes (Foster et al.,

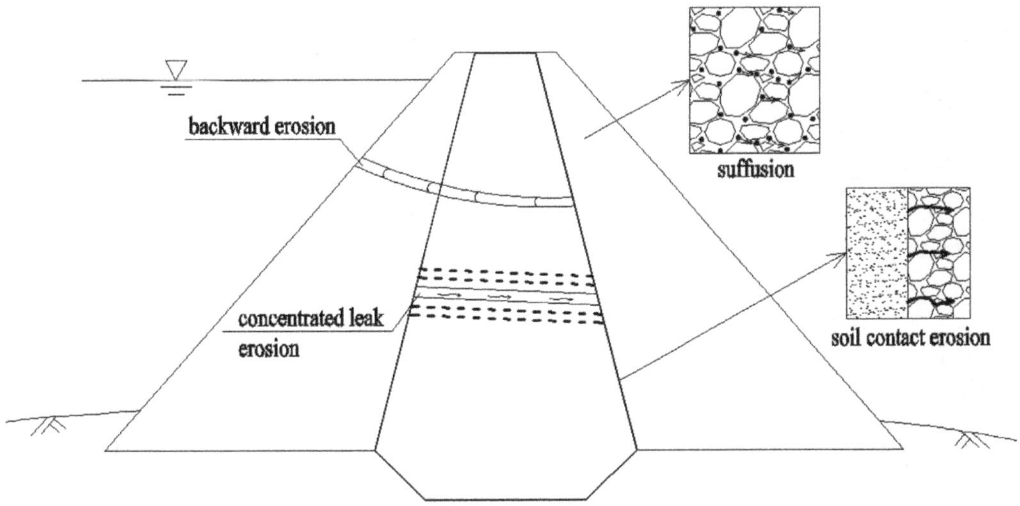

FIGURE 1 | Diagram of internal erosion in a dam slope: concentrated leak erosion, backward erosion, soil contact erosion, and suffusion (Fell and Fry, 2007).

2000b; Chaney et al., 2000; Razavi et al., 2020a) and is usually accompanied by internal erosion. The internal erosion occurs completely inside the slope or the foundation, and usually causes danger and potentially dangerous situations (Ni et al., 2018; Xie et al., 2019). During internal erosion, fluid drives fine particles to move along the pores of the soil skeleton, and this migration can result in a change in the mechanical properties of the soil such as porosity and permeability. Furthermore, changes in porosity and permeability can further change seepage characteristics. Internal erosion consists primarily of four different types: Concentrated leak erosion, backward erosion, soil contact erosion, and suffusion (Fell and Fry, 2007; Richards and Reddy, 2007). As shown in **Figure 1**, concentrated leak erosion occurs mostly at cracks in the slope, backward erosion occurs upward from and along the seepage path, contact erosion is more likely to occur through the slope on the contact surface of different fillers or on the contact surface between the slope and the structure, and suffusion usually occurs in gap-graded soil layers. This study focuses on suffusion, which corresponds to the detachment and migration of fine particles within the pores of the soil skeleton by means of seepage flow.

Sterpi (2003) tested the erosion characteristics of fine particles caused by seepage and established an erosion equation for fine particles on test results. By combining this with the mass conservation equation for fine particles, the erosion and migration process of fine particles during seepage can be simulated. Deng et al. (2020) conducted an experimental study on the characteristics of horizontal seepage-induced suffusion under controlled vertical stress, and their test results showed that after the initiation of suffusion, alternating decreases and increases in the coefficient of permeability occurred, possibly due to alternating clogging and dredging of soil pores by movable fine particles. Additionally, the authors observed an uneven distribution of local hydraulic gradients along the flow path and considered that this could be regarded as a sign of the initiation of suffusion. Ke and Takahashi (2012) and Ke and Takahashi (2014) studied the effect of internal erosion on soil strength and found that the internal erosion indicated by a loss of fine particles caused changes in the void ratio and a significant increase in hydraulic conductivity, resulting in a decrease in soil strength from its initial value. Another study, by Ma et al. (2021), studied the characteristics of non-Darcy hydraulic properties and deformation behaviors of granular gangues using laboratorial, theoretical, and *in situ* approaches, and their results showed that the porosity and permeability of granular gangues decreased gradually and that the crushing ratio of the sample increased gradually with an increase in original grain size grade and a decrease in stress rate, respectively. The authors then constructed an improved model to predict permeability evolution using the fractal dimension of granular gangues.

Other studies have focused on the numerical simulation of internal erosion characteristics. For example, Yang et al. (2019) and Yang et al. (2020) established a coupled mathematical model of the seepage field and stress field under the influence of internal erosion and used the model to simulate the seepage and deformation characteristics of a three-dimensional dam under the influence of internal erosion and studied the failure mode of the dam under different boundary conditions. Based on thermodynamic principles and porous media theory, Zhang et al. (2013) also created a mathematical model that considered the effects of internal erosion. The model was used to simulate the evolution of erosion inside the dam, the characteristics of the spatial and temporal distributions of porosity, and the concentration of moving fine particles was. Razavi et al. (2020b) investigated the settlement of the Esfarayen earth-filled dam and believed that the main reason for the excessive settlement was internal erosion. Here, the broken zone gradually formed a large hole that eventually caused the sudden settling of the surface of the dam. Subsequently, a testing program was developed that could determine the starting time of internal erosion. Based on the two-phase flow theory, and Ma et al. (2019), Ma et al. (2020), the authors also established a resistance model of water–sediment flow in fractures and quantitatively analyzed the

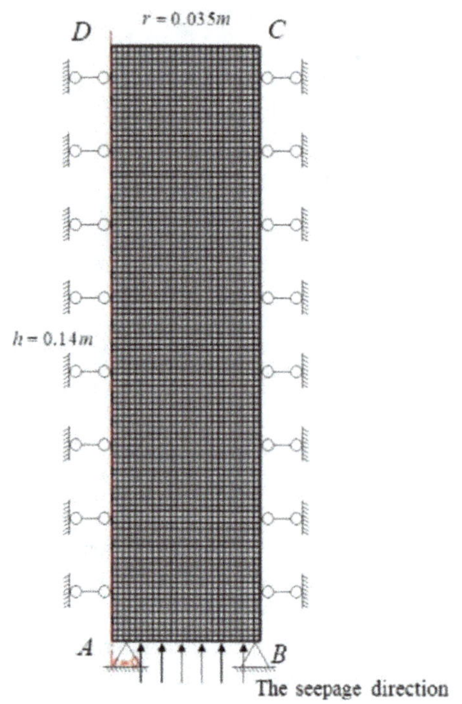

FIGURE 2 | Illustration of the test.

TABLE 1 | Physical parameters of the soil (Sterpi, 2003).

Parameter	Value
Water density ρ_w (kg/m^3)	1,000
Air density ρ_a (kg/m^3)	1.25
Soil density ρ_s (kg/m^3)	2,720
The dynamic viscosity coefficient of water μ_w (Pa·s)	1.0×10^{-3}
The dynamic viscosity coefficient of air μ_a (Pa·s)	1.8×10^{-5}
Initial volume fraction of fine particles C_0	0.001
Critical volume fraction of fine particles C_{cr}	0.23
The migration coefficient ζ (m^{-1})	1
Standard atmospheric pressure p_{atm} (Pa)	1.01×10^5
Elastic modulus E (GPa)	96.76
Poisson's ratio ν	0.13
Initial porosity ϕ_0	0.34
Maximum porosity ϕ_{max}	0.49
Initial permeability k_0 (m^2)	1.0×10^{-8}

characteristics of water–sediment flow in rock fractures using computational fluid dynamics (CFD). They then validated their calculations with a laboratory-scale test.

In this study, we construct a mathematical two-phase seepage coupling model to examine the effects of internal erosion on an embankment dam slope based on two-phase and porous media seepage, the constitutive model of fine particles migration, the Mohr–Coulomb criterion, and the principle of effective stress in porous media. We then implement a finite element model of the embankment dam slope with COMSOL Multiphysics® software in order to simulate the model's behavior and study the influence of internal erosion on the hydraulic characteristics (porosity and permeability) and on the deformation characteristics of the embankment dam slope itself.

2 THE MATHEMATICAL MODEL OF TWO-PHASE SEEPAGE

2.1 Assumptions

In this study, we consider the saturated soil to be a saturated porous medium composed of three phases: the solid phase (coarse particles and fine particles that constitute the skeleton), the fluid phase (water), and the migrated fine particles phase, where fine particles migrate with the fluid. To establish the mathematical model for two-phase seepage that considers the effects of internal erosion, we make the following three assumptions. First, we assume that the saturated soil is uniform and isotropic, second, that the seepage of water follows Darcy's law, and third, that the migration velocity of fine particles is equal to the water seepage velocity.

2.2 Seepage Field Equations

2.2.1 Mass Conservation Equations

The mass conservation equations for water-migrated fine particles and solids can be expressed as (Wang, 2006):

$$\frac{\partial}{\partial t}[\phi(1-C)\rho_w] + div[\phi(1-C)\rho_w v_{ws}] = 0 \quad (1)$$

$$\frac{\partial}{\partial t}(\phi C \rho_p) + div(\phi C \rho_p v_{ps}) = m_s \quad (2)$$

$$\frac{\partial}{\partial t}[(1-\phi)\rho_s] + div[(1-\phi)\rho_s v_s] = -m_s \quad (3)$$

where ϕ is the porosity; C is the migrated fine particles volume fraction, which is defined as the volume ratio of migrated fine particles to pores; ρ_w is the density of water (kg/m^3); ρ_p is the density of migrated fine particles (kg/m^3); ρ_s is the density of solid (kg/m^3); v_{ws} is the velocity vector of water relative to solid (m/s); v_{ps} is the velocity vector of migrated fine particles to solid ((m/s)); v_s is the velocity vector of solid (m/s); $v_s = 0$ in this study; and m_s is the mass rate of migrated fine particles (kg/(m^{3}s)).

2.2.2 Equations of Motion

The seepage of water conforms to Darcy's law, as follows (Darcy, 1856):

$$\phi(1-C)v_{ws} = \phi C v_{ps} = -\frac{k}{\mu_w}(\nabla p_w - \rho_w g) \quad (4)$$

where k is the permeability of porous media (m^2), p_w is the pore water pressure (Pa), μ_w is the dynamic viscosity coefficient of water (Pa · s), and g is the gravitational acceleration vector (N/kg).

We can substitute **Eq. 4** into **Eqs 1, 2** to obtain the governing equations of mass conservation as follows:

$$(1-C)\rho_w\frac{\partial \phi}{\partial t} - \phi\rho_w\frac{\partial C}{\partial t} + \nabla\left[-\rho_w\frac{kk_{rw}}{\mu_w}(\nabla p_w - \rho_w g)\right] = 0 \quad (5)$$

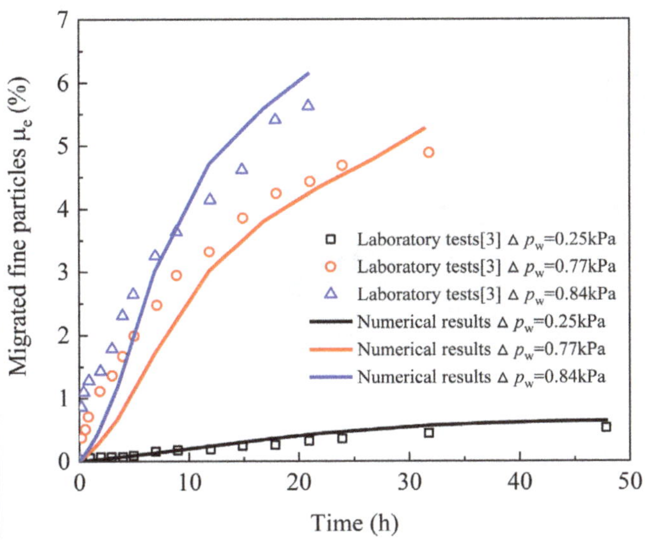

FIGURE 3 | Evolution curves for the mass percentage of migrated fine particles.

$$C\rho_p\frac{\partial\phi}{\partial t}+\phi\rho_p\frac{\partial C}{\partial t}+\nabla\left[-\rho_p\frac{kk_{rw}}{\mu_w}\left(\nabla p_w-\rho_w g\right)\right]=m_s \tag{6}$$

$$\frac{\partial\phi}{\partial t}=\frac{m_s}{\rho_s} \tag{7}$$

2.2.3 Constitutive Equation of Fine Particles Migration

With a gradual increase in fluid velocity, the rest of the fine particles in the porous medium gradually begin to roll, slide, suspend, and finally migrate with the fluid. This process of fine particles migration has an important influence on changes in pore structure and hydraulic characteristics. According to references (Sakthivadivel, 1966a; Sakthivadivel, 1966b; Vardoulakis et al., 1996; Stavropoulou et al., 1998), the mass fraction of fine particles migration is a comprehensive result of porosity, water seepage velocity, and other factors. Because the migrated fine particles cause changes in pore structure, and the limit of the range of the changes is the pore structure of the same material in the loose state. That is, the increase in porosity caused by fine particles migration cannot exceed the maximum porosity in principle. Then:

TABLE 2 | Physical parameters for soils (Deng, 2006; Ding et al., 2020).

Soil	Heavy silty loam
Water density ρ_w (kg/m^3)	1,000
Soil density ρ_s (kg/m^3)	1,680
The dynamic viscosity coefficient of water μ_w (Pa·s)	1.0×10^{-3}
Initial volume fraction of fine particles C_0	0.01
Critical volume fraction of fine particles C_{cr}	0.30
The migration coefficient ζ (m^{-1})	2
Standard atmospheric pressure p_{atm} (Pa)	1.01×10^5
Elastic modulus E (MPa)	45
Poisson's ratio ν	0.44
Initial porosity ϕ_0	0.40
Maximum porosity ϕ_{max}	0.63
Initial permeability k_0 (m^2)	4.2×10^{-13}
Cohesion c (kPa)	15
Frictional angle φ (°)	28
Frictional angle at saturation φ_{sat} (°)	17

$$\frac{m_s}{\rho_s}=\zeta\left(\phi_{\max}-\phi\right)\left(C-C^2/C_{cr}\right)|v_{ws}| \tag{8}$$

where $\phi_{\max}$ is the maximum porosity; C_{cr} is the critical value of the migrated fine particles volume fraction for which the two competing phenomena, erosion and deposition (clogging), balance each other; $|v_{ws}|$ is the velocity vector of water (m/s); and ζ is the migration coefficient of fine particles (m^{-1}).

2.2.4 Permeability Evolution Equation

According to the Kozeny–Carman equation, the relationship between the permeability and porosity of porous media is (Singh and Wallender, 2008):

$$k=k_0\left(\frac{\phi}{\phi_0}\right)^3\left(\frac{1-\phi_0}{1-\phi}\right)^2 \tag{9}$$

where k_0 is the initial absolute permeability of porous media and ϕ_0 is the initial porosity of porous media.

2.2.5 The Stress Field Equations for Saturated Soil

According to the principle of effective stress in saturated porous media:

$$\sigma_{ij}=\sigma'_{ij}+\phi p_w\delta_{ij}=\lambda\varepsilon_V\delta_{ij}+2\mu\varepsilon_{ij} \tag{10}$$

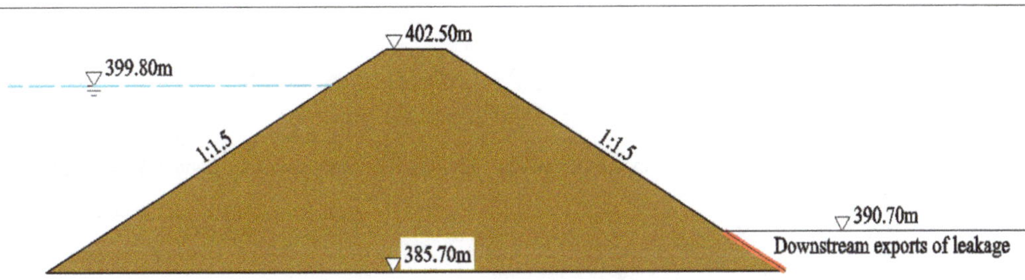

FIGURE 4 | Schematic diagram for leakage of the embankment dam slope (Ding et al., 2020).

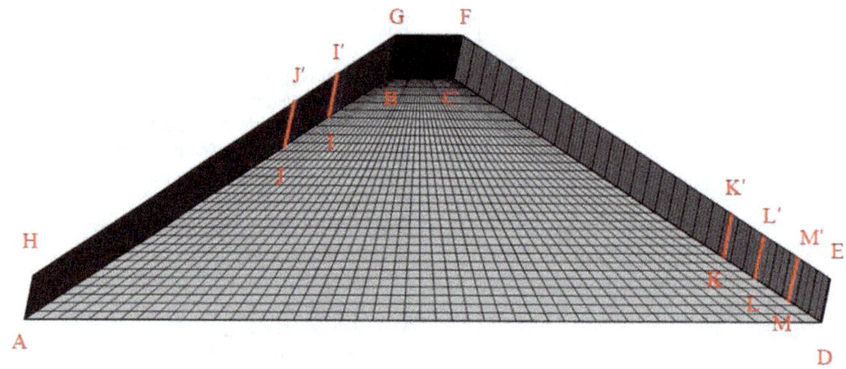

FIGURE 5 | Mesh of the numerical model.

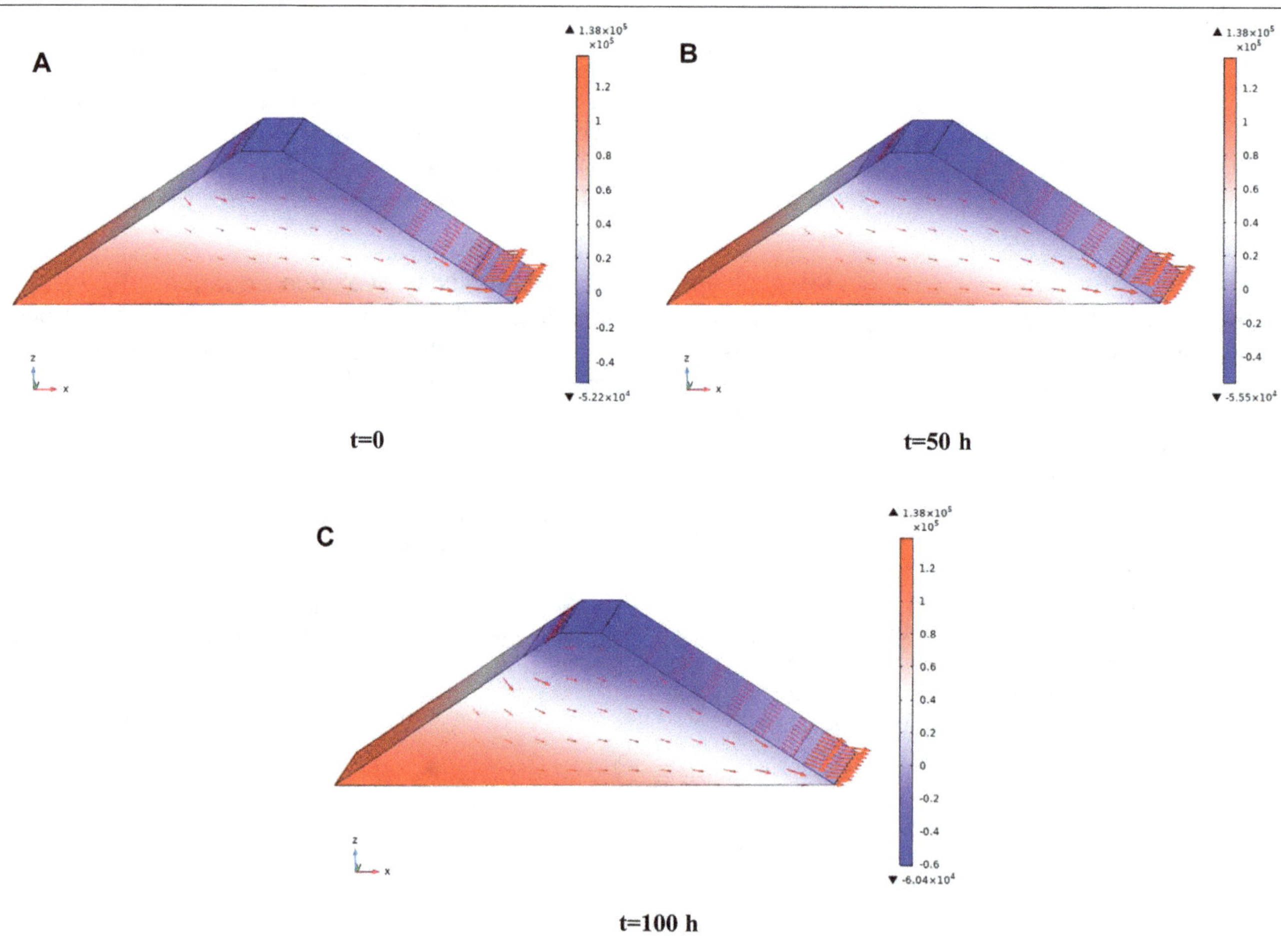

FIGURE 6 | Spatial distribution of the pore water pressure (unit: Pa) and water seepage velocity in the embankment dam slope. **(A)** $t = 0$, **(B)** $t = 50$ h, and **(C)** $t = 100$ h.

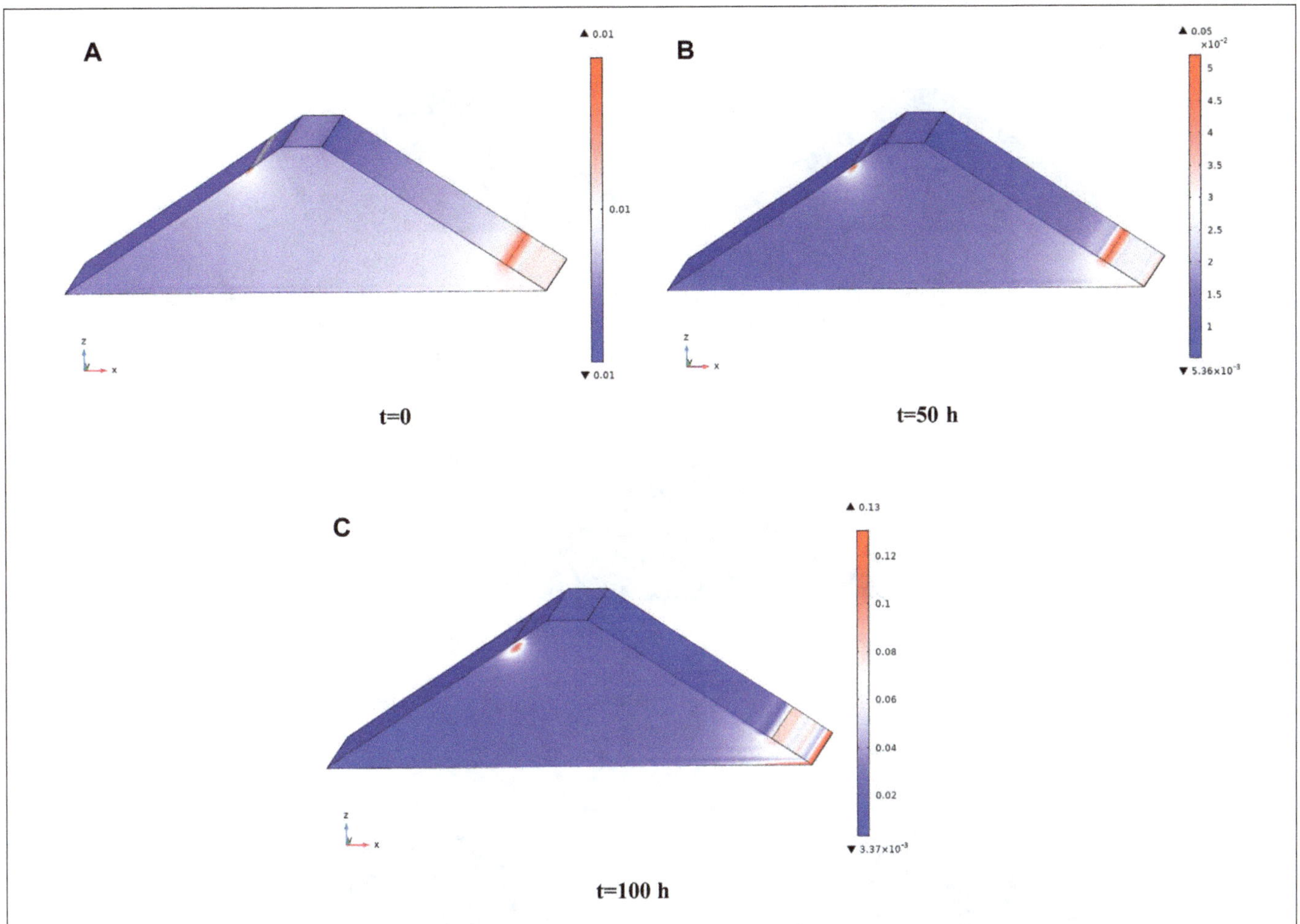

FIGURE 7 | Spatial distribution of the migrated fine particles volume fraction in the embankment dam slope. **(A)** t = 0, **(B)** t = 50 h, and **(C)** t = 100 h.

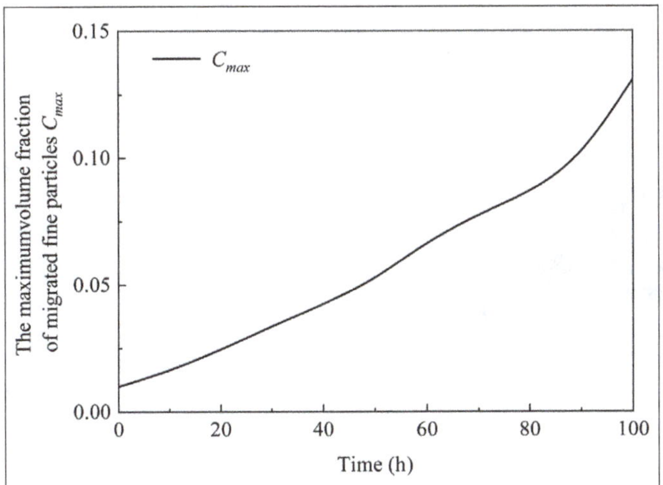

FIGURE 8 | Evolution curve of the maximum migrated fine particles volume fraction.

where σ_{ij} is the total stress tensor, σ'_{ij} is the effective stress tensor, δ_{ij} is the indicator symbol such that $\delta_{ij} = \begin{cases} 1, i = j \\ 0, i \neq j \end{cases}$; ε_V is the volumetric strain of the solid skeleton, ε_{ij} is the strain of the solid skeleton, and λ, μ is the constant of Lamé.

The stress balance equation of porous media can thus be expressed as:

$$\sigma_{ij,j} + \rho g = 0 \tag{11}$$

where ρ is the average density of the porous medium (kg/m^3).

According to the Mohr–Coulomb criterion and the principle of effective stress in saturated porous media:

$$\tau = c_e + \sigma' \tan\varphi_e = c_e + (\sigma - \phi p_w)\tan\varphi_e \tag{12}$$

$$c_e = c/F_s \tag{13}$$

$$\varphi_e = \arctan(\tan\varphi/F_s) \tag{14}$$

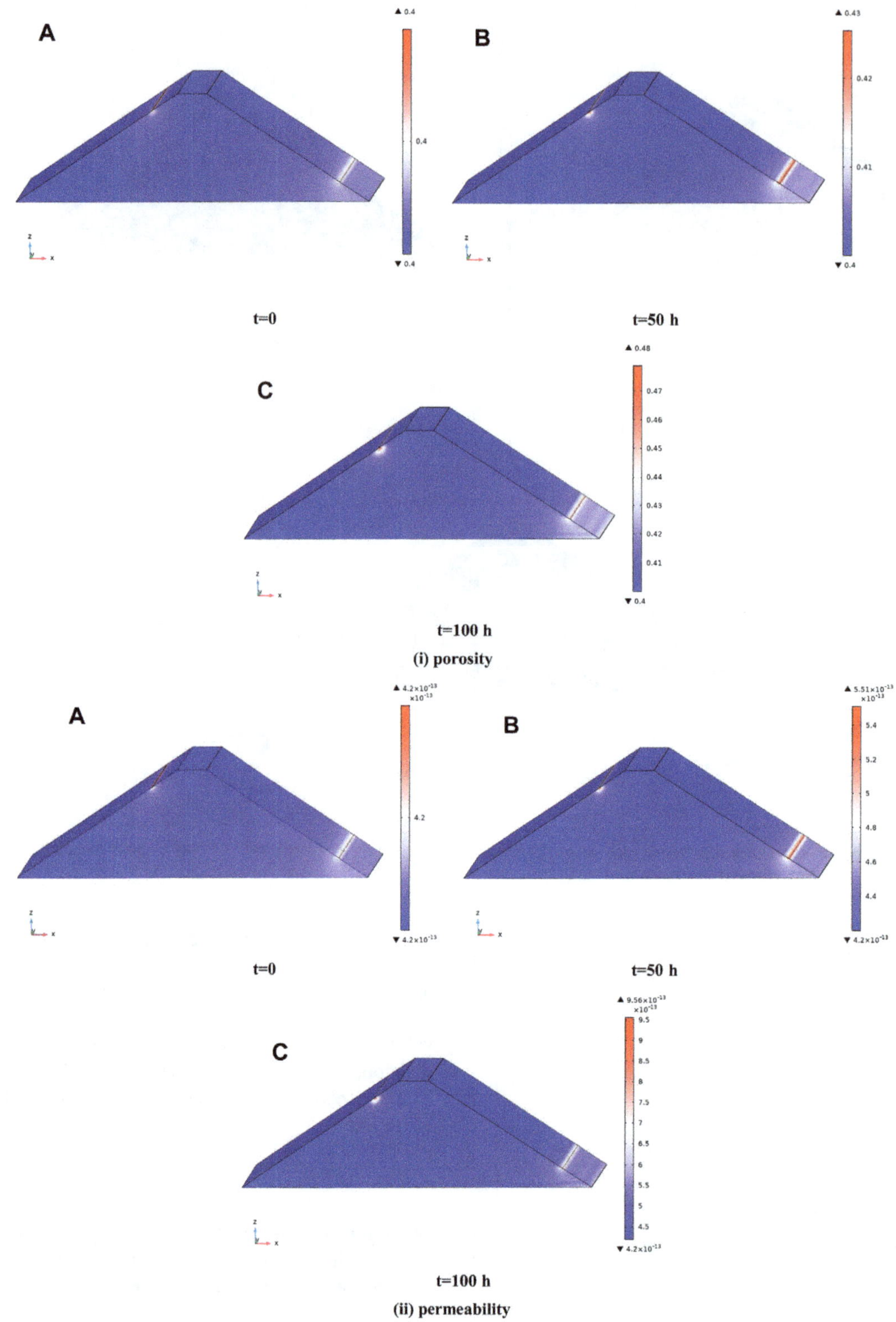

FIGURE 9 | Spatial distribution of porosity and permeability in the embankment dam slope. **(A)** $t = 0$, **(B)** $t = 50$ h, **(C)** $t = 100$ h, **(i)** porosity, and **(ii)** permeability.

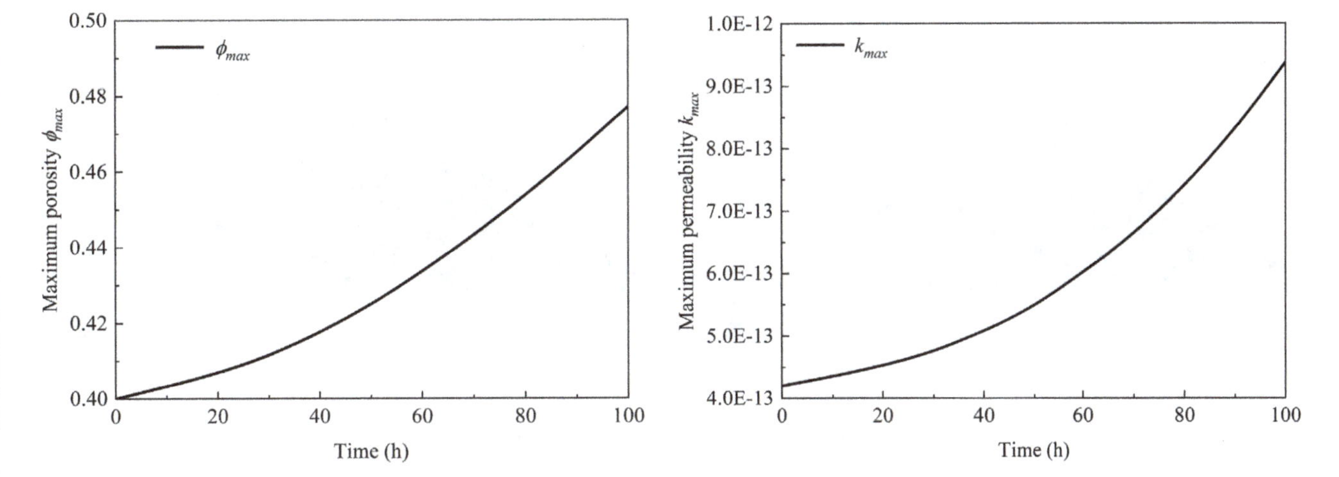

FIGURE 10 | Evolution curves of maximum porosity and maximum permeability.

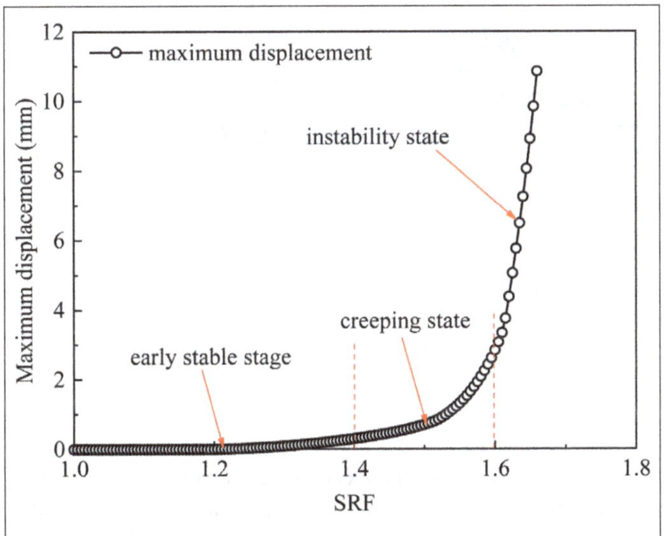

FIGURE 11 | Variation diagram of maximum displacement with the SRF.

where τ is shear strength, c is cohesive, c_e is cohesive at the ultimate state, F_s is the strength reduction factor (SRF), φ is internal friction angle, φ_e is the internal friction angle at the ultimate state, σ is the total stress, and σ' is the effective stress.

2.2.6 Boundary and Initial Conditions

We solve the mathematical model with the help of COMSOL Multiphysics®. The fundamental variables (p_w, C, and u_s) were discretized in space by the Lagrange method and discretized in time by the implicit Euler backward difference method. Other variables, such as $\dot{m}$ and k, can be calculated from **Eqs 8, 9**, respectively. To solve these highly nonlinear equations, boundary conditions and initial conditions must be given. There are two commonly used boundary conditions, namely, the Dirichlet boundary and the Neumann boundary. The Dirichlet boundary conditions are $p_w = \bar{p}_w$ on Γ_{p_w}, $C = \bar{C}$ on Γ_C, $u_s = \bar{u}_s$ on Γ_{u_s}, and the Neumann boundary conditions are $q_w = -v_{ws} \cdot n$ on Γ_{q_w}, $q_C = -Cv_{ps} \cdot n$ on Γ_C, and $\sigma \cdot n = \bar{t}$ on Γ_σ. The initial conditions are $p_w = p_{w0}$, $C = C_0$, and $u_s = u_{s0}$ at $t = 0$.

3 MODEL VALIDATION

Sterpi (2003) has previously conducted a seepage test for the erosion characteristics of fine particles in the soil. In this study, we used the model proposed in the previous section to conduct numerical calculations for this test and to compare the calculation results with the test results to determine the accuracy of the model. A two-dimensional axisymmetric model with a radius of 0.035 m and a height of 0.14 m is established in Sterpi's study, as shown in **Figure 2**. The soil tested was well-graded compacted sand and gravel with a size range of 0–2 mm, and a content of fine particles (with a particles size less than 0.075 mm) by mass was 23%. The physical parameters for the soil are shown in **Table 1** of her study (Sterpi, 2003). Our numerical model adopted a mapping mesh with 2,500 meshes and an average unit mass of 0.98. The mesh of the numerical model is shown in **Figure 2**.

The initial conditions for the calculation model were as follows: p_{w0} was set as the hydrostatic pressure, u_{s0} was 0, and C_0 was 0.001. The boundary conditions for the seepage field are, AB was an infiltration boundary, CD was the outflow boundary, and boundaries AD and BC were impermeable. The boundary conditions for the stress field were that AD was an axisymmetric boundary, the horizontal constraint was imposed on the boundary BC, fixed constraints were imposed on the boundary AB, and the remaining boundary was free.

Figure 3 shows the evolution curves for the mass percentage for migrated fine particles. We can see that the mass percentage for migrated fine particles increases with time, and the numerical results are close to those from the experimental tests. Although there is some error between the calculated results and the experimental results in the groups of $\Delta p_w = 0.77$ and 0.84, the maximum error is only 8.5%, which is in the acceptable range.

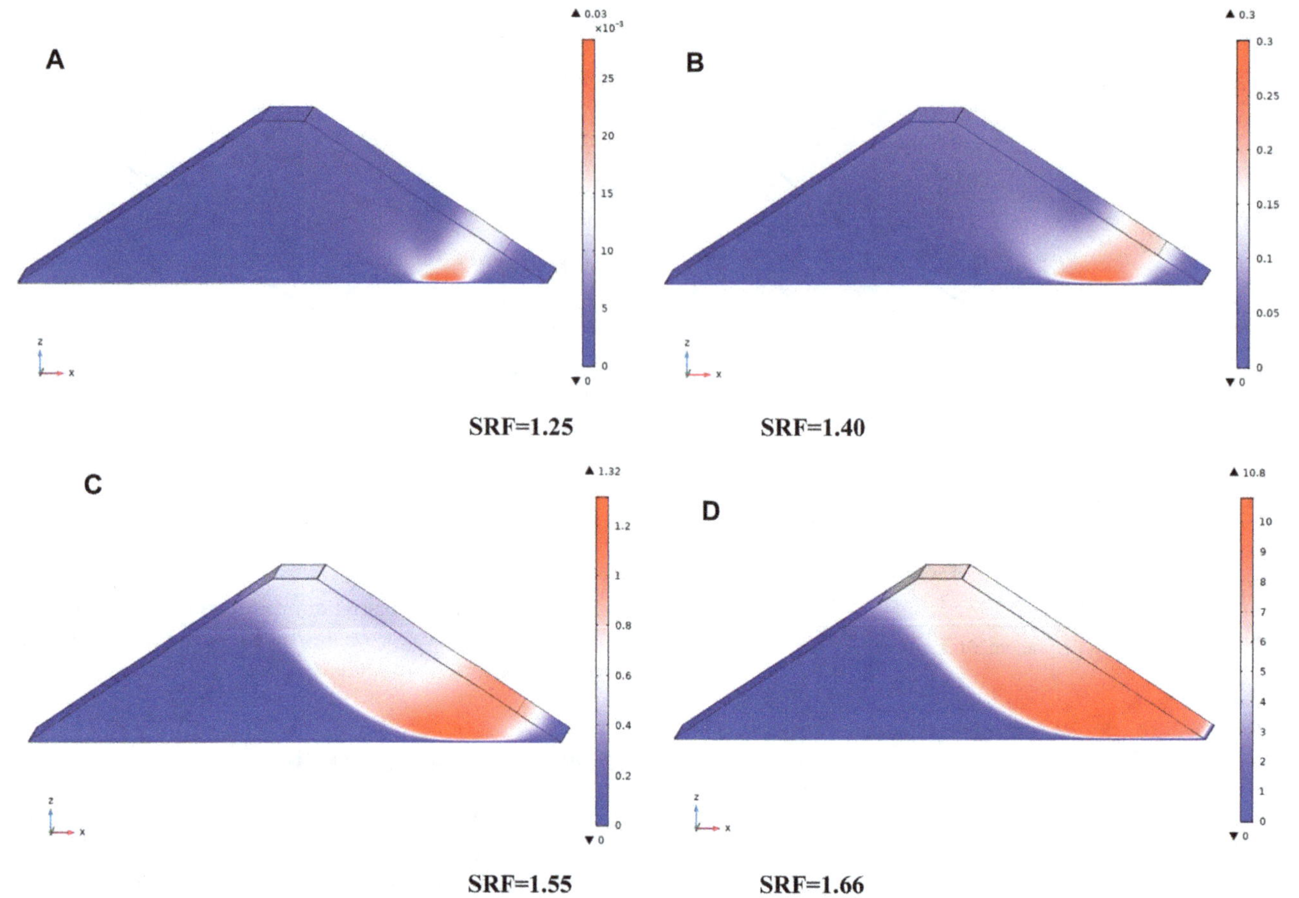

FIGURE 12 | Spatial distribution of displacement (in mm) in the dam slope under different SRFs. **(A)** SRF = 1.25, **(B)** SRF = 1.40, **(C)** SRF = 1.55, and **(D)** SRF = 1.66.

This error may be caused by the migration coefficient, which should be a variable related to testing conditions; the actual value of the migration coefficient may vary greatly when the hydraulic gradient is large. However, we used a fixed value in the numerical simulation process.

4 THE NUMERICAL STUDY ON THE SEEPAGE CHARACTERISTICS OF AN EMBANKMENT DAM SLOPE

4.1 Description of the Problem

The embankment dam featured in this study is located in Xinjiang Province, China, and a diagram of it is shown in **Figure 4** (Ding et al., 2020). The top width is 4.50 m, the bottom width is 64.50 m, and the height is 16.80 m. The gradient of the upstream and downstream dam slopes is 1:1.5, with a normal water level of 399.80 m. The dam slope is made of heavy silty loam, and the mechanical parameters of this soil are shown in **Table 2** (Deng, 2006; Ding et al., 2020). After more than 20 years of operation, leakage occurred at the foot of the downstream dam slope, seriously threatening the safety of the dam slope.

In the process of numerical calculation, we established a three-dimensional dam slope model as shown in **Figure 5**. In this model, we adopted a hexahedral mesh, and a total of 40,600 grid points were created, with an average grid mass of 0.82.

For the seepage field, AII'H is the hydrostatic pressure boundary, which is also the fine particles inlet boundary, and the hydrostatic pressure is given by $p_w = \rho_w g h$ (h is the vertical height of the corresponding boundary surface). The edge LDEL' (vertical height is 3.0 m) is the atmospheric pressure boundary, and this is also the outlet boundary for fine particles. The initial pore water pressure is calculated from the Darcy seepage field, as shown in **Figure 6A**. For the stress field, ADEH is the fixed boundary, AII'H is the hydrostatic pressure boundary (the hydrostatic pressure is equivalent to the external load), and the rest of the boundary is a free boundary.

4.2 Analysis of the Numerical Results

4.2.1 Variation in Pore Water Pressure and Water Seepage Velocity

Figure 6 shows the spatial distribution of the pore water pressure and the water seepage velocity inside the dam slope at different times. The surface where the pore water

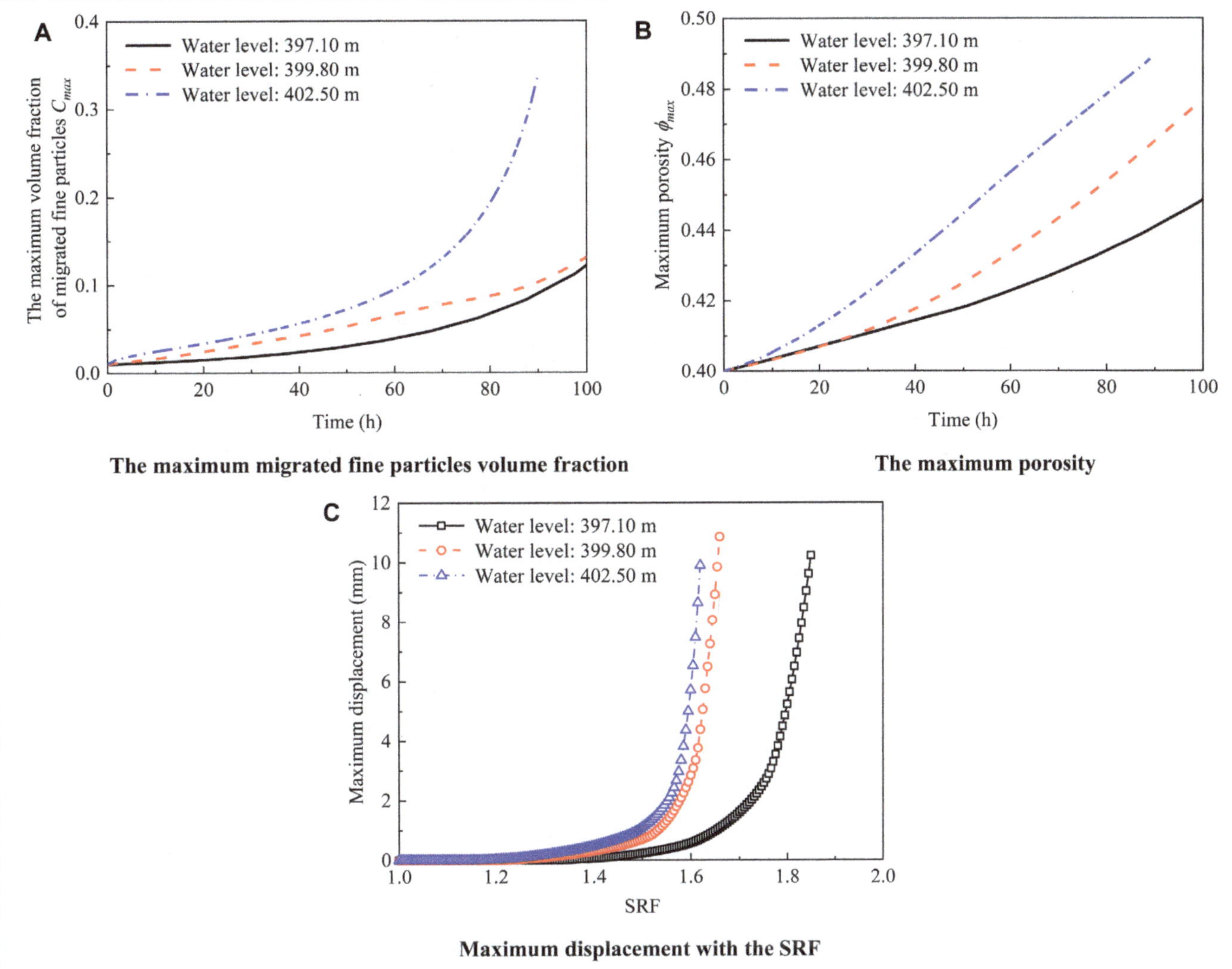

FIGURE 13 | The influence of the water level on the maximum migrated fine particles volume fraction, maximum porosity, and maximum displacement. **(A)** The maximum migrated fine particles volume fraction. **(B)** The maximum porosity. **(C)** Maximum displacement with the SRF.

pressure equals 0 is called the free surface. We can see from **Figure 6** that the water seepage velocity gradually increases with time and that the water seepage velocity at the downstream exports is greater than that at the interior of the dam slope. This is because the porosity and permeability gradually increase (**Figure 9** and **Figure 10** for details) as the fine particles gradually erode, and the increase in permeability is concentrated at the downstream outlet. Therefore, the water seepage velocity at the downstream exports increases at a relatively large rate.

4.2.2 Variation in Migrated Fine Particles Volume Fraction

The spatial distribution of the migrated fine particles volume fraction inside the embankment dam slope is shown in **Figure 7**. At $t = 0$, the migrated fine particles volume fraction inside the dam slope is relatively small and uniformly distributed. As time passes, however, the fluid gradually erodes the fine particles and the erosion of fine particles occurs primarily at the downstream exports of leakage. With an increase in distance from the exports of leakage, the migrated fine particles volume fraction decreases. The points with the highest increases in migrated fine particles volume fraction correlate to the points with the highest seepage velocity, as shown in **Figure 6**, and porosity, as shown in **Figure 9**. The reason for this may be inferred from the constitutive equation of fine particles migration (**Eq. 8**), which is a function of the seepage velocity and porosity of the soil.

The evolution curve of the maximum migrated fine particles volume fraction simulated in the dam slope is shown in **Figure 8**. We can see that the maximum migrated fine particles volume fraction increased with time. Specifically, when the time is less than 50 h, the maximum migrated fine particles volume fraction increases approximately linearly. After 50 h, the maximum migrated fine particles volume fraction increases nonlinearly, until 100 h when the

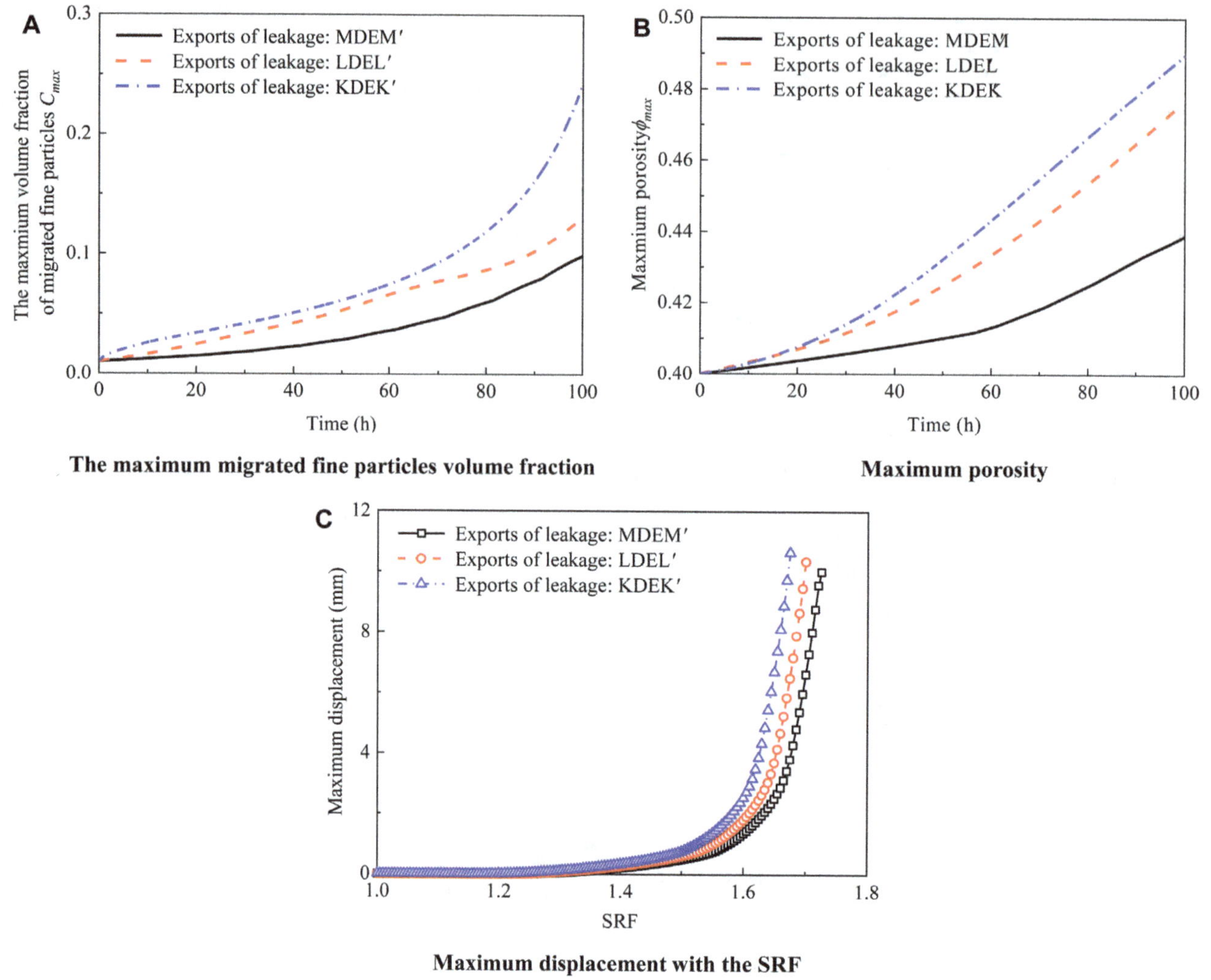

FIGURE 14 | The influence of downstream leakage outlets on the maximum migrated fine particles volume fraction, maximum porosity, and maximum displacement. **(A)** The maximum migrated fine particles volume fraction. **(B)** Maximum porosity. **(C)** Maximum displacement with the SRF.

maximum migrated fine particles volume fraction of 0.13 is reached.

4.2.3 Variation in Porosity and Permeability

The spatial distributions of porosity and permeability are shown in **Figure 9**, and the evolution curves of maximum porosity and maximum permeability are shown in **Figure 10**. Combining **Figure 9** and **Figure 10**, we can see that with the gradual erosion of fine particles, the porosity and permeability inside the dam slope gradually increase as well, and the porosity and permeability at the downstream exports of leakage increase particularly significantly. The maximum porosity and maximum permeability evolution curves show the same pattern as well. That is, the maximum porosity or maximum permeability increase with time. The porosity increases from the initial 0.40 to 0.48, an increase of 20%, and the permeability increases from the initial 4.2×10^{-13} m^2 to 9.4×10^{-13} m^2, an increase of 2.33 times.

4.2.4 Variation in Displacement

Figure 11 shows the variation diagram of the maximum deformation with the SRF. The maximum displacement can be divided into three stages with the change in the SRF, the early stable stage, the midterm nonlinear growth stage (creeping state), and the later rapid growth stage (instability state). When the SRF is less than 1.40, the maximum displacement of the dam slope hardly changes with the SRF, indicating that the dam slope is in a stable state. However, when the SRF is between 1.40 and 1.60, the maximum displacement increases nonlinearly with the SRF, indicating that the dam slope is in a creeping state. Finally, when the SRF is greater than 1.60, the maximum displacement increases approximately linearly and rapidly with the SRF, indicating that the dam slope is in a state of instability. **Figure 12** shows the spatial distribution of displacement in the dam slope under different SRFs. When the SRF is 1.25, the displacement is first generated inside the dam slope near the downstream exports of leakage, which is approximately 0.03 mm.

As the SRF continues to increase, the area with larger displacement gradually extends to the upstream and downstream exports. After this, when the SRF is at 1.55, a through-sliding surface has been formed inside the dam slope. Here, as the SRF continues to increase, the displacement generated inside the dam slope increases rapidly, and the maximum displacement is approximately 10.8 mm, which occurrs at the downstream exports of leakage. As mentioned earlier, the erosion of fine particles occurrs primarily at the downstream exports of leakage, and changes to the porosity and permeability at the downstream exports of leakage are the most significant, leading to a significant decrease in the strength parameters of the soil at the foot of the slope. Therefore, the displacement is also large.

4.3 The Influence of the Water Level

The water level of the dam often changes due to rainfall and other reasons, and it is therefore of practical engineering significance to study the influence of the water level on the seepage characteristics of the dam slope (Shen, 2016). To analyze the influence of water level conditions on seepage and deformation characteristics, two different water level conditions were selected for calculation: 402.50 and 397.10 m (Deng, 2006). **Figure 13** shows the influence of these water levels on the maximum migrated fine particles volume fraction, maximum porosity, and maximum displacement with the SRF. We conclude that the maximum migrated fine particles volume fraction at different water levels behave similarly. That is, it increases nonlinearly with time in both cases. In addition, the higher the water level, the larger the maximum migrated fine particles volume fraction, and the higher the water level, the greater the maximum porosity as well. The relationship between the maximum displacement and the SRF shows that the higher the water level, the smaller the SRF when the creeping state begins and the dam slope fails. With water levels of 397.10, 399.80, and 402.50 m, the SRFs corresponding to the onset of the creeping state are approximately 1.50, 1.40, and 1.30, respectively, and the corresponding SRFs for dam slope damage are 1.85, 1.66, and 1.62, respectively. This also shows that the higher the water level (the larger the head difference between the upstream and downstream of the dam slope), the easier it is for fine particles to erode, and for migration and leakage damage to occur inside the dam slope.

4.4 The Influence of Downstream Leakage Outlets

To analyze the influence of downstream leakage outlets on seepage and deformation characteristics, two different downstream leakage outlets were selected for calculation: MDEM' (vertical height of 2.0 m) and KDEK' (vertical height of 4.0 m). **Figure 14** shows the influence of downstream leakage outlets on the maximum migrated fine particles volume fraction, maximum porosity, and maximum displacement with the SRF. Here we see that the maximum migrated fine particles volume fraction and maximum porosity corresponding to the same time increases with the increase in leakage outlet size. However, the trend of the maximum displacement with the strength reduction factor change is seldom basically unaffected by the size of the leakage outlet.

4.5 A Comparative Analysis Between This Research and Existing Research

The seepage of a dam slope is often accompanied by the erosion of fine particles, and this migration can cause changes in hydraulic parameters (such as porosity and permeability), which in turn affect the seepage and deformation characteristics of the dam slope. The numerical calculation model established in this study can calculate the fluid–solid coupling characteristics, the characteristics of fine particles migration, and the evolution characteristics of hydraulic parameters. Compared with the numerical calculation results in the study (Deng, 2006) that did not consider fine particles erosion, the strength reduction factor obtained in our study is small. From this point of view, considering the erosion of fine particles and changes in hydraulic parameters in the calculation process is more in line with actual engineering conditions.

5 CONCLUSION

In this study, we constructed a mathematical model for two-phase seepage in an embankment dam slope that considered the effects of internal erosion. By using COMSOL Multiphysics®, we were able to study the characteristics of seepage, hydraulic parameters, and deformation of a three-dimensional embankment dam slope leakage process and have reached the following conclusions.

First, the water seepage, velocity gradually increases with time, especially at the leakage outlet. Therefore, the erosion and migration of fine particles primarily occur at the downstream leakage outlet, resulting in a significant increase in migrated fine particles volume fraction, porosity, and permeability. The maximum migrated fine particles volume fraction increases from 0.01 to 0.13 during the duration of our study, the maximum porosity increases by 20%, and the maximum permeability grew to 2.33 times higher than that at the initial moment.

Second, the maximum displacement of the dam slope can be divided into three stages with changes in the SRF: the early stable stage, the midterm nonlinear growth stage (creeping state), and the later rapid growth stage (instability state). In actual engineering, when the maximum displacement is in the midterm nonlinear growth stage, some reinforcement measures should be taken, such as building a dam seepage wall, building a pressure relief well, and strengthening the downstream dam foot.

Third, a rise in the water level promotes the erosion and migration of fine particles in the dam slope, and the maximum volume fraction of fine particles and maximum porosity in the dam slope also increase with the water level. Additionally, the higher the water level, the smaller the SRF when the creeping state begins and the dam slope fails.

Finally, the maximum migrated fine particles volume fraction and maximum porosity for a given time increase with leakage outlet size. However, the behavior of the maximum displacement with the SRF change is nearly unaffected by the size of the leakage outlet.

Our mathematical model can calculate the fluid–solid coupling and the fine particles migration characteristics in an embankment dam slope, but the numerical calculations in this study were performed under the assumption of homogeneous dam material. However, in nature, the seepage path and the mechanism of fine particles migration may become more complex than modeled here within a heterogeneous medium. Modeling heterogeneous dam materials is the focus of future research.

AUTHOR CONTRIBUTIONS

YJ: writing—original draft, methodology, data curation, formal analysis, and visualization. YD: investigation, data curation, and writing—review and editing. XW: writing—review and editing and supervision. JZ: conceptualization and supervision. XC: resources and project administration.

ACKNOWLEDGMENTS

The authors are grateful for the constructive comments and suggestions from the reviewers. The authors thank AiMi Academic Services (www.aimieditor.com) for the English language editing and review services.

REFERENCES

1. Cesali, C., and Federico, V. (2019). *Detection of Permeability Defects within Dams and Levees through Coupled Seepage and Heat Transport Analyses*. Cham, Switzerland: LNCE.
2. Chaney, R., Demars, K., Reddi, L., Lee, I.-M., and Bonala, M. (2000). Comparison of Internal and Surface Erosion Using Flow Pump Tests on a Sand-Kaolinite Mixture. *Geotech. Test. J.* 23 (1), 116–122. doi:10.1520/ gtj11129j
3. Darcy, H. (1856). *Les Fontaines Publiques De La Ville De Dijon*. Paris: Gallica.
4. Deng, G., Zhang, L.-L., Chen, R., Liu, L.-l., Shu, K.-X., and Zhou, Z.-L. (2020). Experimental Investigation on Suffusion Characteristics of Cohesionless Soils along H orizontal Seepage Flow under Controlled Vertical Stress. Front. Earth Sci. 8 (195), 195. doi:10.3389/feart.2020.00195
5. Deng, Y. Y. (2006). *Analysis the Seepage Destory Mechanisms of Disease Earth Dam*. Shihezi: Shihezi University.
6. Ding, Y., Zhang, J.-s., Jia, Y., Chen, X.-b., Wang, X., and Meng, F. (2020). Study on Two-phase Fluid-Solid Coupling Characteristics in Saturated Zone of Subgrade Considering the Effects of Fine Particles Migration. *Appl. Sci.* 10 (21), 7539. doi:10.3390/app10217539
7. Fell, R. F., and Fry, J. J. (2007). *The State of the Art of Assessing the Likelihood of Internal Erosion of Embankment Dams, Water Retaining Structures and Their Foundations*. London, United Kingdom: Taylor & Francis, 1–23.
8. Foster, M., Fell, R., and Spannagle, M. (2000). The Statistics of Embankment Dam Failures and Accidents. *Can. Geotech. J.* 37 (5), 1000–1024. doi:10.1139/ t00-030
9. Foster, M., Fell, R., and Spannagle, M. (2000). The Statistics of Embankment Dam Failures and Accidents. *Can. Geotech. J.* 37 (5), 1000–1024. doi:10.1139/ t00-030
10. Jiang, X., Zhanyuan, Z., Chen, H., Deng, M., Niu, Z., Deng, H., et al. (2020). Natura l Dam Failure in Slope Failure Mode Triggered by Seepage. *Geomatics, Nat. Hazards Risk* 11 (1), 698–723. doi:10.1080/19475705.2020.1746697
11. Ke, L., an d Takahashi, A. (2014). Experimental Investigations on Suffusion Characteristics and its Mechanical Consequences on Saturated Cohesionless Soil. *Soils Found.* 54 (4), 713–730. doi:10.1016/j.sandf.2014.06.024
12. Ke, L., and Takahashi, A. (2012). Strength Reduction of Cohesionless Soil Due to Internal Erosion Induced by One-Dimensional Upward Seepage Flow. *Soils Found.* 52 (4), 698–711. doi:10.1016/j.sandf.2012.07.010
13. Ma, D., Duan, H., Liu, W., Ma, X., and Tao, M. (2020). Water-Sediment Two- phase Flow Inrush Hazard in Rock Fractures of Overburden Strata during Coal Mining. *Mine Water Environ.* 39, 308–319. doi:10.1007/s10230-020- 00687-6
14. Ma, D., Wang, J., Cai, X., Ma, X., Zhang, J., Zhou, Z., et al. (2019). Effects of Height/Diameter Ratio on Failure and Damage Properties of Granite under Coupled Bending and Splitting Deformation. *Eng. Fract. Mech.* 220, 106640. doi:10. 1016/j.engfracmech.2019.106640
15. Ma, D., Zhang, J., Duan, H., Huang, Y., Li, M., Sun, Q., et al. (2021). Reutilization of Gangue Wastes in Underground Backfilling Mining: Overburden Aquifer Protection. *Chemosphere* 264, 128400. doi:10.1016/j.chemosphere.2020.128400
16. Ni, X. D., Zuo, X. Y., Sun, D. M., and Liu, Z. (2018). Study on the Piping Erosion in Levee Foundation by Sand Box Model Test. *Adv. Eng. Sci.* 50 (2), 24–31. doi:10. 15961/j.jsuese.201700849
17. Razavi, S. K., Bonab, M. H., and Dabaghian, A. (2020). Investigation into the Internal Erosion and Local Settlement of Esfarayen Earth-Fill Dam. *J. Geotech. Geoenviron.* 146 (4), 4020006–4020013. doi:10.1061/(asce) gt.1943-5606.0002216
18. Razavi, S. K., Hajialilue Bonab, M., and Dabaghian, A. (2020). Investigation into the Internal Erosion and Local Settlement of Esfarayen Earth-Fill Dam. *J. Geotech. Geoenviron. Eng.* 146 (4), 04020006. doi:10.1061/(ASCE) GT. 1943-5606.0002216
19. Richards, K. S., and Reddy, K. R. (2007). Critical Appraisal of Piping Phenomena in Earth Dams. *Bull. Eng. Geol. Environ.* 66 (4), 381–402. doi:10.1007/s10064-007- 0095-0
20. Sakthivadivel, R. (1966). *Theory and Mechanism of Filtration of Non-colloidal Fines through A Porous Medium*. Berkekey: University of California.
21. Sakthivadivel, S. I. R. (1966). *A Review of Filtration Theories*. Berkekey: University of California.
22. Shen, H. (2016). *Study on Fluid-Solid Coupling Theroy of Gravelly Soil Influenced by Internal Erosion and its Engineering Application*. Shanghai: Shanghai Jiao Tong University.
23. Singh, P. N., and Wallender, W. W. (2008). Effects of Adsorbed Water Layer in Predicting Saturated Hydraulic Conductivity for Clays with Kozeny-Carman Equation. *J. Geotech. Geoenviron. Eng.* 134 (6), 829–836. doi:10.1061/(asce)1090-0241(2008)134:6(829)
24. Stavropoulou, M., Papanastasiou, P., and Vardoulakis, I. (1998). Coupled Wellbore Erosion and Stability Analysis. *Int. J. Numer. Anal. Meth. Geomech.* 22 (9), 749–769. doi:10.1002/(sici)1096-9853(199809)22:9<749::aid-nag944>3.0.co;2-k
25. Sterpi, D. (2003). Effects of the Erosion and Transport of Fine Particles Due to Seepage Flow. *Int. J. Geomech.* 3 (1), 111–122. doi:10.1061/(asce)1532-3641(2003)3:1(111)
26. Vardoulakis, I., Stavropoulou, M., and Papanastasiou, P. (1996). Hydro-Mechanical Aspects of the Sand Production Problem. *Transp. Porous Med.* 22 (2), 225–244. doi:10.1007/BF01143517
27. Wang, X. D. (2006). *Fundamentals of Seepage Mechanics*. Beijing: Petroleum Industry Press.
28. Xie, Q., Liu, J., Han, B., Li, H., Li, Y., and Jiang, Z. (2019). Experimental Investigation of Interfacial Erosion on Culvert-Soil Interface in Earth Dams. *Soils Found.* 59 (3), 671–686. doi:10.1016/j.sandf.2019.02.004
29. Yang, J., Yin, Z.-Y., Laouafa, F., and Hicher, P.-Y. (2019). Internal Erosion

in Dike- on-Foundation Modeled by a Coupled Hydromechanical Approach. *Int. J. Numer. Anal. Methods Geomech.* 43 (3), 663–683. doi:10.1002/nag.2877

30. Yang, J., Yin, Z. Y., Laouafa, F., and Hicher, P. Y. (2020). Three-dimensional Hydromechanical Modeling of Internal Erosion in Dike-on-foundation. *Int. J. Numer. Anal. Methods Geomech.* 44 (8), 1200–1218. doi:10.1002/nag.3057
31. Zhang, X. S., Wong, H., Leo, C. J., Bui, T. A., Wang, J. X., Sun, W. H., et al. (2013). A Thermodynamics-Based Model on the Internal Erosion of Earth Structures.
32. *Geotech. Geol. Eng.* 31 (2), 479–492. doi:10.1007/s10706-012-9600-8

6

On the (k,s)-Hilfer-Prabhakar Fractional Derivative with Applications to Mathematical Physics

*Muhammad Samraiz [1], Zahida Perveen [1], Gauhar Rahman [2], Kottakkaran Sooppy Nisar [3] and Devendra Kumar [4]**

[1] Department of Mathematics, University of Sargodha, Sargodha, Pakistan, [2] Department of Mathematics, Shaheed Benazir Bhutto University, Upper Dir, Pakistan, [3] Department of Mathematics, College of Arts and Sciences, Prince Sattam Bin Abdulaziz University, Wadi Aldawaser, Saudi Arabia, [4] Department of Mathematics, University of Rajasthan, Jaipur, India

***Correspondence:**
Devendra Kumar
devendra.maths@gmail.com

In this paper we introduce the (k, s)-Hilfer-Prabhakar fractional derivative and discuss its properties. We find the generalized Laplace transform of this newly proposed operator. As an application, we develop the generalized fractional model of the free-electron laser equation, the generalized time-fractional heat equation, and the generalized fractional kinetic equation using the (k, s)-Hilfer-Prabhakar derivative.

Keywords: modified (k, s) fractional integral operator, (k, s)-Prabhakar fractional derivative, (k, s)-Hilfer-Prabhakar fractional derivative, fractional heat equation, fractional kinetic equation

1. INTRODUCTION

Fractional calculus is the area of mathematical analysis that deals with the study and application of integrals and derivatives of arbitrary order. In recent decades, fractional calculus has become of increasing significance due to its applications in many fields of science and engineering [1–5]. The first application of fractional calculus was given by Abel [6] and includes the solution to the tautocrone problem. Fractional calculus also has applications in biophysics, wave theory, polymers, quantum mechanics, continuum mechanics, field theory, Lie theory, group theory, spectroscopy, and other scientific areas [7–9]. Although this calculus has a long history, over the past few decades it has attracted greater attention because of the fascinating results obtained when it is used to model certain real-world problems [10–13]. What makes fractional calculus special is that there are numerous types of fractional operators, so any scientist modeling real-world phenomena can choose the operator that fits their purposes the best. Each classical fractional derivative is usually defined in terms of a specific integral. Among the most well-known concepts of fractional derivatives are the Riemann-Liouville, Caputo, Grünwald-Letnikov, and Hadamard derivatives [10, 14, 15], whose formulations involve single-kernel integrals and which are used to investigate, for example, memory effect problems [16].

The Riemann-Liouville fractional derivative is remarkable, but it has some drawbacks when used to model physical phenomena because of its improper physical conditions. Caputo's great contribution was to develop a concept of fractional derivative appropriate for physical conditions [17]. A number of other families of fractional operators have been established, such as the Liouville, Erdlyi-Kober, Hadamard, Grünwald-Letnikov, Hilfer, Hilfer-Prabhakar, and k-Hilfer-Prabhakar operators, to mention just a few [10, 18–20]. Because there are so many concepts of fractional operator, it has become necessary to define generic fractional operators, of which the classical ones are particular cases. One class of extensions of Riemann-Liouville fractional operators comprises the so-called k-Prabhakar integral operators, which can be found in [21]. Inspired by the definitions of k-Prabhakar integral operators and k-Hilfer-Prabhakar derivatives [20], the authors introduced the (k,s)-Hilfer fractional derivative, which unifies a large class of fractional operators [19, 20]; [Samraiz et al., accepted].

In recent years, the generalization of integral and differential operators has become an important subject of research in fractional calculus [9, 20, 22–28]. Different special functions, including the Gauss hypergeometric function, Mittag-Leffler-style functions, the Wright function, Meijer's G function, and Fox's H function, appear in the kernels of several generalizations of the integral operators. R. Hilfer introduced the Hilfer fractional derivative in [9], which is a generalization of the Riemann-Liouville and Caputo fractional derivatives.The Prabhakar integral and derivative operators are obtained from the Riemann-Liouville integral operator by extending its kernel to involve the three-parameter Mittag-Leffler function [19].

This paper is motivated by the rich applications of fractional differential equations (FDEs) in physics, economics, engineering, and many other branches of science [8, 10, 13, 17]. Since no general method exists that can be used to analytically solve every FDE, one of the most pressing and challenging tasks is to develop suitable methods for finding analytical solutions to certain classes of FDEs [29–31]. Researchers have become interested in fractional interpretations of the classical integral transforms, i.e., Laplace and Fourier transforms [32–34], in the past few years. It can be shown that integral transformations such as the Laplace, Fourier, generalized Laplace, and ρ-Laplace transforms are useful methods for obtaining analytical solutions to some classes of FDEs. In this framework, we use a generalized Laplace transform to obtain analytical solutions to certain classes of FDEs that contain (k,s)-Hilfer-Prabhakar fractional derivatives. Given the wide range of fractional operators available in the literature, it can be difficult to choose the most suitable approach for a given problem. It is therefore essential to consider generalizations of classical fractional operators to aid in choosing an appropriate operator.

Diaz et al. [35] defined k-gamma and k-beta functions as follows.

Definition 1.1. *The k-gamma function is a generalization of the classical Γ function given by*

$$\Gamma_k(\theta) = \lim_{n\to\infty} \frac{n!k^n (nk)^{\frac{\theta}{k}-1}}{(\theta)_{n,k}}, \quad k>0,\ \mathrm{Re}(\theta)>0,$$

where $(\theta)_{n,k} = \theta(\theta+k)(\theta+2k)\cdots(\theta+(n-1)k)$ for $n \geq 1$ is called the Pochhammer k symbol. The integral representation is

$$\Gamma_k(\theta) = \int_0^\infty x^{\theta-1} e^{\frac{-x^k}{k}}\, dx, \quad \mathrm{Re}(\theta)>0.$$

Clearly, $\Gamma(\theta) = \lim_{k\to 1}\Gamma_k(\theta)$ and $\Gamma_k(\theta) = k^{\frac{\theta}{k}-1}\Gamma(\frac{\theta}{k})$.

Definition 1.2. *For $\mathrm{Re}(\theta) > 0$, $k > 0$, and $\mathrm{Re}(\zeta) > 0$, the k-beta function is given by*

$$B_k(\theta,\zeta) = \frac{1}{k}\int_0^1 \tau^{\frac{\theta}{k}-1}(1-\tau)^{\frac{\zeta}{k}-1}\, d\tau.$$

The functions Γ_k and B_k are related by an identity

$$B_k(\theta,\zeta) = \frac{\Gamma_k(\theta)\Gamma_k(\zeta)}{\Gamma_k(\theta+\zeta)}.$$

The k-Mittag-Leffler function given in [36] is defined as follows.

Definition 1.3. *Let $n \in N$, $k \in \mathbb{R}^+$, $\mu,\rho,\gamma \in \mathbb{C}$, $\mathrm{Re}(\rho) > 0$, and $\mathrm{Re}(\mu) > 0$. Then the k-Mittag-Leffler function is defined by*

$$E^{\gamma}_{k,\rho,\mu}(\theta) = \sum_{n=0}^{\infty} \frac{(\gamma)_{n,k}\theta^n}{\Gamma_k(\rho n+\mu)n!}.$$

The modified (k,s)-fractional integral operator involving the k-Mittag-Leffler function given in [Samraiz et al., accepted] is defined as follows.

Definition 1.4. *Let $s \in \mathbb{R}\backslash\{-1\}$, $k \in \mathbb{R}^+$, $\mu,\rho,\omega,\gamma \in \mathbb{C}$, $\mathrm{Re}(\rho) > 0$, $\mathrm{Re}(\gamma) > 0$, $\mathrm{Re}(\mu) > 0$, and $\Phi \in L^1[0,\beta]$. Then the modified (k,s)-fractional integral operator involving the k-Mittag-Leffler function is given by*

$$({}^{s}_{k}\mathfrak{J}^{\omega,\gamma}_{0^+;\rho,\mu}\Phi)(\theta) = \frac{(s+1)^{1-\frac{\mu}{k}}}{k} \int_0^\theta (\theta^{s+1}-\zeta^{s+1})^{\frac{\mu}{k}-1}\zeta^s E^{\gamma}_{k,\rho,\mu}(\omega(\theta^{s+1}-\zeta^{s+1})^{\frac{\rho}{k}}) \Phi(\zeta)\, d\zeta. \tag{1.1}$$

Definition 1.5 ([Samraiz et al., accepted]). *Let $s \in \mathbb{R}\backslash\{-1\}$, $k \in \mathbb{R}^+$, $\mu,\rho,\omega,\gamma \in \mathbb{C}$, $\mathrm{Re}(\rho) > 0$, $\mathrm{Re}(\mu) > 0$, $n = \left[\frac{\mu}{k}\right]+1$, and $\Phi \in L^1[0,\beta]$. Then the (k,s)-Prabhakar fractional derivative operator with the k-Mittag-Leffler function as its kernel is given by*

$$({}^{s}_{k}\mathfrak{D}^{\omega,\gamma}_{0^+;\rho,\mu}\Phi)(\theta) = \left(\frac{1}{\theta^s}\frac{d}{d\theta}\right)^n k^n ({}^{s}_{k}\mathfrak{J}^{\omega,-\gamma}_{0^+;\rho,nk-\mu}\Phi)(\theta). \tag{1.2}$$

Definition 1.6 ([Samraiz et al., accepted]). *Let $s \in \mathbb{R}\backslash\{-1\}$, $k \in \mathbb{R}^+$, $\mu,\rho,\omega,\gamma \in \mathbb{C}$, $\mathrm{Re}(\rho) > 0$, $\mathrm{Re}(\mu) > 0$, $n = \left[\frac{\mu}{k}\right]+1$, and $\Phi \in C^n[0,\beta]$ with $0<\theta<\beta<\infty$. Then the regularized version of the (k,s)-Prabhakar derivative is*

$${}^{C}_{s,k}\mathfrak{D}^{\omega,\gamma}_{0^+;\rho,\mu}\Phi(\theta) = k^{ns}{}_{k}\mathfrak{J}^{\omega,-\gamma}_{0^+;\rho,nk-\mu}\left(\frac{1}{\theta^s}\frac{d}{d\theta}\right)^n \Phi(\theta). \tag{1.3}$$

Definition 1.7. *Let $g \in C^n[\alpha,\beta]$ such that $g'(\zeta) > 0$ on $[\alpha,\beta]$. Then*

$$AC^n_g[\alpha,\beta] = \left\{\Phi:[\alpha,\beta]\to\mathbb{C} \text{ with } \Phi^{[n-1]} \in AC[\alpha,\beta]\right\},$$

where $\Phi^{[n-1]} = \left(\frac{1}{g'(\zeta)}\frac{d}{d\zeta}\right)^{n-1}\Phi$.

The generalized Laplace transform introduced by Jarad et al. [34] is presented in the following definition.

Definition 1.8. *Let Φ and g be real-valued functions on $[\alpha,\infty)$ such that $g(\zeta)$ is continuous and $g'(\zeta) > 0$ on $[\alpha,\infty)$. The generalized Laplace transform of Φ is*

$$L_g\{\Phi(\theta)\}(u) = \int_\alpha^\infty e^{-u(g(\theta)-g(\alpha))}\Phi(\theta)g'(\theta)\,d\theta$$

for all values of u.

Definition 1.9 ([34]). *Let Φ and Ψ be two piecewise-continuous functions on each interval $[0,T]$ that are of exponential order. The generalized convolution of Φ and Ψ is given by*

$$(\Phi *_g \Psi)(\theta) = \int_\alpha^\theta \Phi(\zeta)\Psi\big(g^{-1}(g(\theta)+g(\alpha)-g(\zeta))\big)g'(\zeta)\,d\zeta.$$

Theorem 1.10 ([34]). *Let $\Phi \in C_g^{n-1}[\alpha,T]$ be such that $\Phi^{[1]}$ is of g-exponential order. Let $\Phi^{[1]}$ be a piecewise-continuous function on the interval $[\alpha,T]$. Then the generalized Laplace transform of $\Phi^{[1]}(\zeta)$ exists and*

$$L_g\{\Phi^{[1]}(\theta)\}(u) = sL_g\{\Phi(\theta)\}(u) - \Phi(\alpha).$$

Proposition 1.11. *Let $s \in \mathbb{R}\backslash\{-1\}$, $k \in \mathbb{R}^+$, $\mu,\rho,\omega,\gamma \in \mathbb{C}$, $\mathrm{Re}(\rho) > 0$, $\mathrm{Re}(\mu) > 0$, and $\beta > 0$. Then the integral operator ${}_k^s\mathfrak{J}_{0^+;\rho,\mu}^{\omega,\gamma}$ is bounded on $C[0,\beta]$, i.e.,*

$$|({}_k^s\mathfrak{J}_{0^+;\rho,\mu}^{\omega,\gamma}\Phi)(\theta)| \le G\|\Phi\|_{C[0,\beta]},$$

where

$$\|\Phi\|_{C[0,\beta]} = \max\{|\Phi| : 0 < x < \beta\}$$

and

$$G = \frac{(s+1)^{-\frac{\mu}{k}}(\beta^{s+1})^{\mathrm{Re}(\frac{\mu}{k})}}{k} \sum_{n=0}^{\infty} \frac{|(\gamma)_{n,k}\omega^n|}{|\Gamma_k(\rho n+\mu)|n!}\frac{(\beta^{s+1})^{\mathrm{Re}(\frac{\rho}{k})n}}{\big[n\mathrm{Re}(\frac{\rho}{k})+\mathrm{Re}(\frac{\mu}{k})\big]}. \tag{1.4}$$

Theorem 1.12. *Let $s \in \mathbb{R}\backslash\{-1\}$, $k \in \mathbb{R}^+$, $\mu,\rho,\omega,\gamma \in \mathbb{C}$, $\mathrm{Re}(\rho) > 0$, $\mathrm{Re}(\gamma) > 0$, and $\mathrm{Re}(\mu) > 0$. Let $\Phi \in L^1[0,\beta]$ be a piecewise-continuous function on each interval $[0,\theta]$ that is of $g(\theta)$-exponential order. Then*

$$\begin{aligned} &L_g\{({}_k^s\mathfrak{J}_{0^+;\rho,\mu}^{\omega,\gamma}\Phi)(\theta)\}(u) \\ &= \big((s+1)(ku)\big)^{-\frac{\mu}{k}}\big(1-k\omega(ku)^{-\frac{\rho}{k}}\big)^{-\frac{\gamma}{k}}L_g\{\Phi(\theta)\}(u). \end{aligned}$$

Theorem 1.13 ([Samraiz et al., accepted]). *Let $k \in \mathbb{R}^+$, $s \in [0,\infty)$, $\mu,\rho,\omega,\gamma \in \mathbb{C}$, $\mathrm{Re}(\rho) > 0$, $\mathrm{Re}(\gamma) > 0$, $\mathrm{Re}(\mu) > 0$, and $g(\theta) = \theta^{s+1}$. Let $\Phi \in AC_g^n[0,\beta]$ and ${}_k^s\mathfrak{J}_{0^+;\rho,nk-m-\mu}^{\omega,\gamma}\Phi$ for $m = 0,1,2,\ldots,n-1$ be of $g(\theta)$-exponential order. Then*

$$\begin{aligned} &L_g\{({}_k^s\mathfrak{D}_{0^+;\rho,\mu}^{\omega,\gamma}\Phi)(\theta)\}(u) \\ &= (s+1)^{-\frac{nk-\mu}{k}}(ku)^{\frac{\mu}{k}}\big(1-k\omega(ku)^{-\frac{\rho}{k}}\big)^{\frac{\gamma}{k}}L_g\{\Phi(t)\}(u) \\ &\quad - \sum_{m=0}^{n-1} k^{n-m}u^{n-m-1}\Big({}_k^s\mathfrak{D}_{0^+;\rho,\mu-(n-m)k}^{\omega,-\gamma}\Phi\Big)(0^+), \end{aligned}$$

with $|k\omega(ku)^{-\frac{\rho}{k}}| < 1$.

Theorem 1.14 ([Samraiz et al., accepted]). *The generalized Laplace transform of the regularized version of the (k,s)-Prabhakar fractional derivative is*

$$\begin{aligned} &L_g\{{}_{s,k}^{C}\mathfrak{D}_{0^+;\rho,\mu}^{\omega,\gamma}\Phi(\theta)\}(u) \\ &= (s+1)^{-\frac{nk-\mu}{k}}\Big[(ku)^{\frac{\mu}{k}}\big(1-k\omega(ku)^{-\frac{\rho}{k}}\big)^{\frac{\gamma}{k}}L_g\{\Phi(\theta)\}(u) \\ &\quad - \sum_{m=0}^{n-1} k^{m+1}(ku)^{\frac{\mu-(m+1)k}{k}}\big(1-k\omega(ku)^{-\frac{\rho}{k}}\big)^{\frac{\gamma}{k}}\big(\Phi^{[m]}\big)(0^+)\Big], \end{aligned}$$

with $|k\omega(ku)^{-\frac{\rho}{k}}| < 1$.

2. THE *(k,s)*-HILFER-PRABHAKAR FRACTIONAL DERIVATIVE AND GENERALIZED LAPLACE TRANSFORMS

In this section we introduce a new family of operators called the (k,s)-Hilfer-Prabhakar fractional derivative. The generalized Laplace transforms of these operators are also studied in this section.

Definition 2.1. *Let $\Phi \in C^1[0,\beta]$, $0 < \theta < \beta < \infty$, $s \in \mathbb{R}\backslash\{-1\}$, $k,\rho > 0$, $\omega,\gamma \in \mathbb{R}$, $\mu \in (0,1)$, $\nu \in [0,1]$, and $\big(\Phi * {}_k^s\mathfrak{J}_{0^+;\rho,(1-\nu)(k-\mu)}^{\omega,-\gamma(1-\nu)}\big)(\theta) \in AC^1[0,\beta]$. The (k,s)-Hilfer-Prabhakar derivative is defined as*

$${}_k^s\mathfrak{D}_{0^+;\rho,\omega}^{\gamma,\mu,\nu}\Phi(\theta) = k\bigg({}_k^s\mathfrak{J}_{0^+;\rho,\nu(k-\mu)}^{\omega,-\gamma\nu}\Big(\frac{1}{\theta^s}\frac{d}{d\theta}\Big){}_k^s\mathfrak{J}_{0^+;\rho,(1-\nu)(k-\mu)}^{\omega,-\gamma(1-\nu)}\Phi\bigg)(\theta).$$

Note that if we choose $\nu = 0$ in the above definition, we get (1.2) corresponding to $m = 1$; and if we take $\nu = 1$, we obtain (1.3) corresponding to $m = 1$.

Theorem 2.2. *For $s \in \mathbb{R}\backslash\{-1\}$, $k,\rho > 0$, $\omega,\gamma \in \mathbb{R}$, $\mu \in (0,1)$, $\nu \in [0,1]$, and $\Phi \in L^1[0,\beta]$, the operator ${}_k^s\mathfrak{D}_{0^+;\rho,\omega}^{\gamma,\mu,\nu}$ is bounded on $C[0,\beta]$, i.e.,*

$$\|{}_k^s\mathfrak{D}_{0^+;\rho,\omega}^{\gamma,\mu,\nu}\Phi(\theta)\| \le C_1C_2\|\Phi\|_{[0,\beta]},$$

where

$$C_1 = \frac{(s+1)^{-\frac{\nu(k-\mu)}{k}}(\beta^{s+1})^{\mathrm{Re}(\frac{\nu(k-\mu)}{k})}}{k} \sum_{n=0}^{\infty}\frac{|(-\gamma\nu)_{n,k}\omega^n|}{|\Gamma_k(\rho n+\nu(k-\mu))|n!}\frac{(\beta^{s+1})^{\mathrm{Re}(\frac{\rho}{k})n}}{\big[n\mathrm{Re}(\frac{\rho}{k})+\mathrm{Re}(\frac{\nu(k-\mu)}{k})\big]} \tag{2.1}$$

and

$$\begin{aligned} C_2 = &\frac{(s+1)^{-\frac{(1-\nu)(k-\mu)-k}{k}}(\beta^{s+1})^{\mathrm{Re}(\frac{(1-\nu)(k-\mu)-k}{k})}}{k} \sum_{m=0}^{\infty}\frac{|(\gamma(\nu-1))_{m,k}\omega^m|}{|\Gamma_k(\rho m+(1-\nu)(k-\mu))|m!} \\ &\times \frac{(\beta^{s+1})^{\mathrm{Re}(\frac{\rho}{k})m}}{\big[m\mathrm{Re}(\frac{\rho}{k})+\mathrm{Re}(\frac{(1-\nu)(k-\mu)}{k})\big]}. \end{aligned} \tag{2.2}$$

Proof Using the estimates in Proposition 1.11, we get

$$
\begin{aligned}
&\|{}^{s}_{k}\mathfrak{D}^{\gamma,\mu,\nu}_{0^+;\rho,\omega}\Phi(\theta)\|\\
&= \left\|k\Big({}^{s}_{k}\mathfrak{J}^{\omega,-\gamma\nu}_{0^+;\rho,\nu(k-\mu)}\Big(\frac{1}{\theta^s}\frac{d}{d\theta}\Big)\Big({}^{s}_{k}\mathfrak{J}^{\omega,-\gamma(1-\nu)}_{0^+;\rho,(1-\nu)(k-\mu)}\Phi\Big)(\theta)\right\|\\
&\le C_1\left\|\Big(\frac{1}{\theta^s}\frac{d}{d\theta}\Big)\Big({}^{s}_{k}\mathfrak{J}^{\omega,-\gamma(1-\nu)}_{0^+;\rho,(1-\nu)(k-\mu)}\Phi\Big)(\theta)\right\|\\
&= C_1\left\|\Big({}^{s}_{k}\mathfrak{J}^{\omega,-\gamma(1-\nu)}_{0^+;\rho,(1-\nu)(k-\mu)-k}\Phi\Big)(\theta)\right\|\\
&\le C_1C_2\|\Phi\|_{[0,\beta]},
\end{aligned}
$$

where C_1 and C_2 are the constants defined by (2.1) and (2.2).

Proposition 2.3. *Let $s\in\mathbb{R}\backslash\{-1\}$, $k,\rho,\lambda>0$, $\omega,\gamma,\sigma\in\mathbb{R}$, $\mu\in(0,1)$, $\nu\in[0,1]$, $\lambda>\mu+\nu k-\mu\nu$, and $\Phi\in L^1[0,\beta]$. Then*

$$
\Big({}^{s}_{k}\mathfrak{D}^{\gamma,\mu,\nu}_{0^+;\rho,\omega}\big({}^{s}_{k}\mathfrak{J}^{\omega,\sigma}_{0^+;\rho,\lambda}\Phi\big)\Big)(\theta)=\big({}^{s}_{k}\mathfrak{J}^{\omega,\sigma-\gamma}_{0^+;\rho,\lambda-\mu}\Phi\big)(\theta).
$$

In particular,

$$
\Big({}^{s}_{k}\mathfrak{D}^{\gamma,\mu,\nu}_{0^+;\rho,\omega}\big({}^{s}_{k}\mathfrak{J}^{\omega,\gamma}_{0^+;\rho,\mu}\Phi\big)\Big)(\theta)=\Phi(\theta).
$$

Proof. By using Definition 2.1 and the semigroup property of the modified (k,s)-fractional integral operator with the k-Mittag-Leffler function, we obtain

$$
\begin{aligned}
&\Big({}^{s}_{k}\mathfrak{D}^{\gamma,\mu,\nu}_{0^+;\rho,\omega}\big({}^{s}_{k}\mathfrak{J}^{\omega,\sigma}_{0^+;\rho,\lambda}\Phi\big)\Big)(\theta)\\
&= k\Big({}^{s}_{k}\mathfrak{J}^{\omega,-\gamma\nu}_{0^+;\rho,\nu(k-\mu)}\Big(\frac{1}{\theta^s}\frac{d}{d\theta}\Big){}^{s}_{k}\mathfrak{J}^{\omega,-\gamma(1-\nu)}_{0^+;\rho,(1-\nu)(k-\mu)}\big({}^{s}_{k}\mathfrak{J}^{\omega,\sigma}_{0^+;\rho,\lambda}\Phi\big)\Big)(\theta)\\
&= k\Big({}^{s}_{k}\mathfrak{J}^{\omega,-\gamma\nu}_{0^+;\rho,\nu(k-\mu)}\Big(\frac{1}{\theta^s}\frac{d}{d\theta}\Big)\big({}^{s}_{k}\mathfrak{J}^{\omega,-\gamma(1-\nu)+\sigma}_{0^+;\rho,(1-\nu)(k-\mu)+\lambda}\Phi\big)\Big)(\theta)\\
&= \Big({}^{s}_{k}\mathfrak{J}^{\omega,-\gamma\nu}_{0^+;\rho,\nu(k-\mu)}\big({}^{s}_{k}\mathfrak{J}^{\omega,-\gamma(1-\nu)+\sigma}_{0^+;\rho,(1-\nu)(k-\mu)+\lambda-k}\Phi\big)\Big)(\theta)\\
&= \big({}^{s}_{k}\mathfrak{J}^{\omega,\sigma-\gamma}_{0^+;\rho,\lambda-\mu}\Phi\big)(\theta).
\end{aligned}
$$

This completes the proof.

Theorem 2.4. *Let $s\in\mathbb{R}\backslash\{-1\}$, $k,\rho,\lambda>0$, $\omega,\gamma\in\mathbb{R}$, $\mu\in(0,1)$, $\nu\in[0,1]$, $\lambda>\mu+\nu k-\mu\nu$, and $\Phi\in L^1[0,\beta]$. Then*

$$
\Big({}^{s}_{k}J^{\lambda}_{0^+}\big({}^{s}_{k}\mathfrak{D}^{\gamma,\mu,\nu}_{0^+;\rho,\omega}\Phi\big)\Big)(\theta)=\big({}^{s}_{k}\mathfrak{J}^{\omega,-\gamma}_{0^+;\rho,\lambda-\mu}\Phi\big)(\theta).
$$

Proof. By using Definition 2.1 and Theorem 2 in [Samraiz et al., accepted], we get

$$
\begin{aligned}
&\Big({}^{s}_{k}J^{\lambda}_{0^+}\big({}^{s}_{k}\mathfrak{D}^{\gamma,\mu,\nu}_{0^+;\rho,\omega}\Phi\big)\Big)(\theta)\\
&= k\Big({}^{s}_{k}J^{\lambda}_{0^+}\,{}^{s}_{k}\mathfrak{J}^{\omega,-\gamma\nu}_{0^+;\rho,\nu(k-\mu)}\Big(\frac{1}{\theta^s}\frac{d}{d\theta}\Big){}^{s}_{k}\mathfrak{J}^{\omega,-\gamma(1-\nu)}_{0^+;\rho,(1-\nu)(k-\mu)}\Phi\Big)(\theta)\\
&= k\Big({}^{s}_{k}\mathfrak{J}^{\omega,-\gamma\nu}_{0^+;\rho,\nu(k-\mu)+\lambda}\Big(\frac{1}{\theta^s}\frac{d}{d\theta}\Big){}^{s}_{k}\mathfrak{J}^{\omega,-\gamma(1-\nu)}_{0^+;\rho,(1-\nu)(k-\mu)}\Phi\Big)(\theta)\\
&= \Big({}^{s}_{k}\mathfrak{J}^{\omega,-\gamma\nu}_{0^+;\rho,\nu(k-\mu)+\lambda}\big({}^{s}_{k}\mathfrak{J}^{\omega,-\gamma(1-\nu)}_{0^+;\rho,(1-\nu)(k-\mu)-k}\Phi\big)\Big)(\theta)\\
&= \big({}^{s}_{k}\mathfrak{J}^{\omega,-\gamma}_{0^+;\rho,\lambda-\mu}\Phi\big)(\theta),
\end{aligned}
$$

and thus the result is proved.

Theorem 2.5. *The Laplace transform of the (k,s)-Hilfer-Prabhakar fractional derivative is*

$$
\begin{aligned}
&L_g\{{}^{s}_{k}\mathfrak{D}^{\gamma,\mu,\nu}_{0^+;\rho,\omega}\Phi(\theta)\}(u)\\
&= (s+1)^{\frac{\mu-k}{k}}(ku)^{\frac{\mu}{k}}\big(1-k\omega(ku)^{-\frac{\rho}{k}}\big)^{\frac{\gamma}{k}}\\
&\quad\times L_g\{\Phi(\theta)\}(u)-k(s+1)^{-\frac{\nu(k-\mu)}{k}}(ku)^{\frac{-\nu(k-\mu)}{k}}\\
&\quad\times\big(1-k\omega(ku)^{-\frac{\rho}{k}}\big)^{\frac{\gamma\nu}{k}}\,{}^{s}_{k}\mathfrak{J}^{\omega,-\gamma(1-\nu)}_{0^+;\rho,(1-\nu)(k-\mu)}\Phi(0^+).
\end{aligned}
$$

Proof. By using Definition 2.1, Theorem 1.12, and Theorem 1.10, we obtain

$$
\begin{aligned}
&L_g\{{}^{s}_{k}\mathfrak{D}^{\gamma,\mu,\nu}_{0^+;\rho,\omega}\Phi(\theta)\}(u)\\
&= k(s+1)^{-\frac{\nu(k-\mu)}{k}}(ku)^{\frac{-\nu(k-\mu)}{k}}\big(1-k\omega(ku)^{-\frac{\rho}{k}}\big)^{\frac{\gamma\nu}{k}}\\
&\quad L_g\Big\{{}^{s}_{k}\mathfrak{J}^{\omega,-\gamma(1-\nu)}_{0^+;\rho,(1-\nu)(k-\mu)}\overset{[1]}{}\Phi(\theta)\Big\}(u)\\
&= k(s+1)^{-\frac{\nu(k-\mu)}{k}}(ku)^{\frac{-\nu(k-\mu)}{k}}\big(1-k\omega(ku)^{-\frac{\rho}{k}}\big)^{\frac{\gamma\nu}{k}}\\
&\quad\times\Big[uL_g\{{}^{s}_{k}\mathfrak{J}^{\omega,-\gamma(1-\nu)}_{0^+;\rho,(1-\nu)(k-\mu)}\Phi(\theta)\}(u)-{}^{s}_{k}\mathfrak{J}^{\omega,-\gamma(1-\nu)}_{0^+;\rho,(1-\nu)(k-\mu)}\Phi(0^+)\Big]\\
&= (ku)(s+1)^{-\frac{\nu(k-\mu)}{k}}(ku)^{\frac{-\nu(k-\mu)}{k}}\big(1-k\omega(ku)^{-\frac{\rho}{k}}\big)^{\frac{\gamma\nu}{k}}\\
&\quad L_g\{{}^{s}_{k}\mathfrak{J}^{\omega,-\gamma(1-\nu)}_{0^+;\rho,(1-\nu)(k-\mu)}\Phi(\theta)\}(u)\\
&\quad-k(s+1)^{-\frac{\nu(k-\mu)}{k}}(ku)^{\frac{-\nu(k-\mu)}{k}}\big(1-k\omega(ku)^{-\frac{\rho}{k}}\big)^{\frac{\gamma\nu}{k}}\,{}^{s}_{k}\mathfrak{J}^{\omega,-\gamma(1-\nu)}_{0^+;\rho,(1-\nu)(k-\mu)}\Phi(0^+)\\
&= (ku)(s+1)^{-\frac{\nu(k-\mu)}{k}}(ku)^{\frac{-\nu(k-\mu)}{k}}\big(1-k\omega(ku)^{-\frac{\rho}{k}}\big)^{\frac{\gamma\nu}{k}}\\
&\quad\times\Big[(s+1)^{-\frac{(1-\nu)(k-\mu)}{k}}(ku)^{\frac{-(1-\nu)(k-\mu)}{k}}\big(1-k\omega(ku)^{-\frac{\rho}{k}}\big)^{\frac{\gamma(1-\nu)}{k}}\\
&\quad L_g\{\Phi(\theta)\}(u)\Big]\\
&\quad-k(s+1)^{-\frac{\nu(k-\mu)}{k}}(ku)^{\frac{-\nu(k-\mu)}{k}}\big(1-k\omega(ku)^{-\frac{\rho}{k}}\big)^{\frac{\gamma\nu}{k}}\,{}^{s}_{k}\mathfrak{J}^{\omega,-\gamma(1-\nu)}_{0^+;\rho,(1-\nu)(k-\mu)}\Phi(0^+)\\
&= (s+1)^{\frac{\mu-k}{k}}(ku)^{\frac{\mu}{k}}\big(1-k\omega(ku)^{-\frac{\rho}{k}}\big)^{\frac{\gamma}{k}}L_g\{\Phi(\theta)\}(u)\\
&\quad-k(s+1)^{-\frac{\nu(k-\mu)}{k}}\\
&\quad\times(ku)^{\frac{-\nu(k-\mu)}{k}}\big(1-k\omega(ku)^{-\frac{\rho}{k}}\big)^{\frac{\gamma\nu}{k}}\,{}^{s}_{k}\mathfrak{J}^{\omega,-\gamma(1-\nu)}_{0^+;\rho,(1-\nu)(k-\mu)}\Phi(0^+),
\end{aligned}
$$

which proves the result.

3. GENERALIZATION OF THE FREE-ELECTRON LASER EQUATION

The integrodifferential free-electron laser equation describes the unsaturated behavior of the free-electron laser. Several attempts have been made to solve the generalized fractional integrodifferential free-electron laser equation in recent years. In this section, we develop a generalized fractional model of the free-electron laser equation that involves the novel (k,s)-Hilfer-Prabhakar derivative.

Theorem 3.1. *The solution of the Cauchy problem*

$$
{}^{s}_{k}\mathfrak{D}^{\gamma,\mu,\nu}_{0^+;\rho,\omega}\Phi(\theta)=\lambda\,{}^{s}_{k}\mathfrak{J}^{\sigma,\omega}_{0^+;\rho,\mu}\Phi(\theta)+f(\theta), \tag{3.1}
$$

$$
{}^{s}_{k}\mathfrak{J}^{\omega,-\gamma(1-\nu)}_{0^+;\rho,(1-\nu)(k-\mu)}\Phi(0^+)=C,\quad C\ge 0, \tag{3.2}
$$

where $\theta \in (0,\infty)$, $f \in L^1[0,\infty)$, $\mu \in (0,1)$, $\nu \in [0,1]$, $\omega,\lambda \in \mathbb{R}$, $\rho > 0$, *and* $\gamma,\sigma \geq 0$, *is given by*

$$\Phi(\theta) = C\sum_{m=0}^{\infty} \lambda^m (s+1)^{-\frac{\nu(k-\mu)+\mu-k}{k}} (\theta^{s+1})^{\frac{\nu(k-\mu)+\mu(1+2m)}{k}-1} \times E^{(\gamma+\sigma)m-\gamma(\nu-1)}_{k,\rho,\nu(k-\mu)+\mu(1+2m)}(\omega(\theta^{s+1})^{\frac{\rho}{k}}) + \sum_{m=0}^{\infty} \lambda^m (s+1)^{2m} ({}^s_k\mathfrak{I}^{\omega,(\gamma+\sigma)m+\gamma}_{k,\rho,\mu(1+2m)} f)(\theta).$$

PROOF. By applying the generalized Laplace transform to both sides of (3.1) and using Theorems 2.5 and 1.12, we get

$$L_g\{{}^s_k\mathfrak{D}^{\gamma,\mu,\nu}_{0^+;\rho,\omega}\Phi(\theta)\}(u) = \lambda L_g\{{}^s_k\mathfrak{I}^{\sigma,\omega}_{0^+;\rho,\mu}\Phi(\theta)\}(u) + L_g\{f(\theta)\}(u),$$

which can also be written as

$$L_g\{\Phi(\theta)\}(u) = \frac{Ck(s+1)^{-\frac{\nu(k-\mu)}{k}}(ku)^{\frac{-\nu(k-\mu)}{k}}\left(1-k\omega(ku)^{\frac{\rho}{k}}\right)^{\frac{\gamma\nu}{k}}}{(s+1)^{\frac{\mu-k}{k}}(ku)^{\frac{\mu}{k}}\left(1-k\omega(ku)^{\frac{-\rho}{k}}\right)^{\frac{\gamma}{k}}} + \frac{L_g\{f(\theta)\}(u)}{(s+1)^{\frac{\mu-k}{k}}(ku)^{\frac{\mu}{k}}\left(1-k\omega(ku)^{\frac{-\rho}{k}}\right)^{\frac{\gamma}{k}}} \left(1-\lambda(ku)^{-\frac{2\mu}{k}}\left(1-k\omega(ku)^{\frac{-\rho}{k}}\right)^{-\frac{\gamma+\sigma}{k}}\right)^{-1}.$$

Using the binomial expansion gives

$$L_g\{\Phi(\theta)\}(u) = \frac{Ck(s+1)^{-\frac{\nu(k-\mu)}{k}}(ku)^{\frac{-\nu(k-\mu)}{k}}\left(1-k\omega(ku)^{\frac{\rho}{k}}\right)^{\frac{\gamma\nu}{k}}}{(s+1)^{\frac{\mu-k}{k}}(ku)^{\frac{\mu}{k}}\left(1-k\omega(ku)^{\frac{-\rho}{k}}\right)^{\frac{\gamma}{k}}} + \frac{L_g\{f(\theta)\}(u)}{(s+1)^{\frac{\mu-k}{k}}(ku)^{\frac{\mu}{k}}\left(1-k\omega(ku)^{\frac{-\rho}{k}}\right)^{\frac{\gamma}{k}}} \sum_{m=0}^{\infty}\lambda^m (ku)^{-2\frac{\mu m}{k}}\left(1-k\omega(ku)^{\frac{-\rho}{k}}\right)^{-\frac{(\gamma+\sigma)}{k}m}$$

$$= Ck\sum_{m=0}^{\infty}\lambda^m(s+1)^{-\frac{\nu(k-\mu)+\mu-k}{k}}(ku)^{-\frac{\nu(k-\mu)+\mu(1+2m)}{k}} \left(1-k\omega(ku)^{\frac{-\rho}{k}}\right)^{-\frac{(\gamma+\sigma)m-\gamma(\nu-1)}{k}} + \sum_{m=0}^{\infty}\lambda^m(s+1)^{-\frac{\mu-k}{k}}(ku)^{-\frac{\mu(1+2m)}{k}} \left(1-k\omega(ku)^{\frac{-\rho}{k}}\right)^{-\frac{\gamma+m(\gamma+\sigma)}{k}} L_g\{f(\theta)\}(u).$$

Applying the inverse Laplace transform, we obtain

$$\Phi(\theta) = C\sum_{m=0}^{\infty} \lambda^m (s+1)^{-\frac{\nu(k-\mu)+\mu-k}{k}} (\theta^{s+1})^{\frac{\nu(k-\mu)+\mu(1+2m)}{k}-1} E^{(\gamma+\sigma)m-\gamma(\nu-1)}_{k,\rho,\nu(k-\mu)+\mu(1+2m)}(\omega(\theta^{s+1})^{\frac{\rho}{k}}) + \sum_{m=0}^{\infty} \lambda^m (s+1)^{2m} ({}^s_k\mathfrak{I}^{\omega,(\gamma+\sigma)m+\gamma}_{k,\rho,\mu(1+2m)} f)(\theta),$$

hence the result.

REMARK 3.2. *If* $s = 0$, $k = 1$, $\gamma = \nu = 0$, $\rho = \sigma = 1$, $\mu \to 1$, $f(\theta) = 0$, $\omega = ir$, *and* $\lambda = -i\Pi p$ *(with* $r,p \in \mathbb{R}$*), then above Cauchy problem reduces to the following free-electron laser equation:*

$$\frac{d}{d\theta}\Phi(\theta) = -ip\Pi\int_0^{\theta}(\theta-t)e^{ir(\theta-t)}\Phi(t)\,dt, \quad \Phi(0) = 1.$$

COROLLARY 3.3. *If we take* $s = 0$ *and* $k = 1$*, then we get the Cauchy problem given in [19]:*

$$\mathfrak{D}^{\gamma,\mu,\nu}_{0^+;\rho,\omega}\Phi(\theta) = \lambda\mathfrak{I}^{\sigma,\omega}_{0^+;\rho,\mu}\Phi(\theta) + f(\theta), \tag{3.3}$$

$$\mathfrak{I}^{\omega,-\gamma(1-\nu)}_{0^+;\rho,(1-\nu)(k-\mu)}\Phi(0^+) = C, \quad C \geq 0, \tag{3.4}$$

where $\theta \in (0,\infty)$, $f \in L^1[0,\infty)$, $\mu \in (0,1)$, $\nu \in [0,1]$, $\omega,\lambda \in \mathbb{R}$, $\rho > 0$, *and* $\gamma,\sigma \geq 0$, *and its solution is given by*

$$\Phi(\theta) = C\sum_{m=0}^{\infty}\lambda^m(\theta)^{\nu(1-\mu)+\mu(1+2m)-1}E^{(\gamma+\sigma)m-\gamma(\nu-1)}_{\rho,\nu(1-\mu)+\mu(1+2m)}(\omega(\theta)^{\rho}) + \sum_{m=0}^{\infty}\lambda^m(\mathfrak{I}^{\omega,(\gamma+\sigma)m+\gamma}_{0^+,\rho,\mu(1+2m)}f)(\theta).$$

4. THE TIME-FRACTIONAL HEAT EQUATION

Lately, numerous papers have been devoted to mathematical analysis of variations of the time-fractional heat equation and its applications in mathematical physics and probability theory [see, for example, [37, 38] and the references therein]. This section focuses on the generalized time-fractional heat equation involving the (k,s)-Hilfer-Prabhakar derivative.

THEOREM 4.1. *The solution of the Cauchy problem*

$${}^s_k\mathfrak{D}^{\gamma,\mu,\nu}_{0^+;\rho,\omega}V(\theta,\zeta) = G\frac{\partial^2}{\partial\zeta^2}V(\theta,\zeta), \quad \zeta > 0,\ \theta \in \mathbb{R}, \tag{4.1}$$

$$\left[{}^s_k\mathfrak{I}^{-\gamma(1-\nu),\omega}_{0^+;\rho,(1-\nu)(k-\mu)}V(\theta,\zeta)\right]_{\zeta=0^+} = h(\theta), \tag{4.2}$$

$$\lim_{\theta\to\infty}V(\theta,\zeta) = 0, \tag{4.3}$$

where $s \in [0,\infty)$, $\mu \in (0,1)$, $\nu \in [0,1]$, $\omega \in \mathbb{R}$, $G,k,\rho > 0$, *and* $\gamma \geq 0$, *is given by*

$$V(\theta,\zeta) = \int_{-\infty}^{+\infty}dp\, e^{-ip\theta}\hat{h}(p) \frac{1}{2\Pi}\sum_{m=0}^{\infty}(-G)^m(s+1)^{-\frac{(1-\nu)(k-\mu)-(\mu-k)m}{k}}(\zeta^{s+1})^{\frac{\mu(m+1)-\nu(\mu-k)}{k}-1} \times E^{\gamma(m+1-\nu)}_{k,\rho,\mu(m+1)-\nu(k-\mu)}(\omega(\zeta^{s+1})^{\frac{\rho}{k}})p^{2m}\hat{h}(p).$$

PROOF. Let $\hat{V}(p,t) = F(V)(p,\zeta)$ denote the Fourier transform with respect to the space variable θ. Taking the Fourier transform of (4.1) and using (4.3), we obtain

$${}^s_k\mathfrak{D}^{\gamma,\mu,\nu}_{0^+;\rho,\omega}\hat{V}(p,\zeta) = -Gp^2\hat{V}(p,\zeta).$$

Now, applying the generalized Laplace transform to both sides of above equation, we get

$$\begin{aligned}
&L_g\{{}^s_k\mathfrak{D}^{\gamma,\mu,\nu}_{0^+;\rho,\omega}\hat{V}(p,\zeta)\}\\
&= -Gp^2L_g\{\hat{V}(p,\zeta)\}(u)(s+1)^{\frac{\mu-k}{k}}(ku)^{\frac{\mu}{k}}\left(1-k\omega(ku)^{-\frac{\rho}{k}}\right)^{\frac{\gamma}{k}}\\
&L_g\{\hat{V}(p,\zeta)\}(u)\\
&\quad - k(s+1)^{-\frac{\nu(k-\mu)}{k}}(ku)^{\frac{-\nu(k-\mu)}{k}}\left(1-k\omega(ku)^{-\frac{\rho}{k}}\right)^{\frac{\gamma\nu}{k}}\\
&[{}^s_k\mathfrak{J}^{\omega,-\gamma(1-\nu)}_{0^+;\rho,(1-\nu)(k-\mu)}\hat{V}(p,\zeta)]_{\zeta=0^+}\\
&= -Gp^2L_g\{\hat{V}(p,\zeta)\}(u),
\end{aligned}$$

which can be written as

$$\begin{aligned}
&L_g\{\hat{V}(p,\zeta)\}(u)\\
&= \frac{k(s+1)^{-\frac{\nu(k-\mu)}{k}}(ku)^{\frac{-\nu(k-\mu)}{k}}\left(1-k\omega(ku)^{-\frac{\rho}{k}}\right)^{\frac{\gamma\nu}{k}}\hat{h}(p)}{(s+1)^{\frac{\mu-k}{k}}(ku)^{\frac{\mu}{k}}\left(1-k\omega(ku)^{-\frac{\rho}{k}}\right)^{\frac{\gamma}{k}}+Gp^2}L_g\{\hat{V}(p,\zeta)\}(u)\\
&= k(s+1)^{\frac{\nu(\mu-k)-(\mu-k)}{k}}(ku)^{\frac{\nu(\mu-k)-\mu}{k}}\left(1-k\omega(ku)^{-\frac{\rho}{k}}\right)^{\frac{\gamma(\nu-1)}{k}}\hat{h}(p)\\
&\quad\times\left(1+\frac{Gp^2}{(s+1)^{\frac{\mu-k}{k}}(ku)^{-\frac{\mu}{k}}\left(1-k\omega(ku)^{-\frac{\rho}{k}}\right)^{\frac{\gamma}{k}}}\right)^{-1}\\
&= k(s+1)^{\frac{\nu(\mu-k)-(\mu-k)}{k}}(ku)^{\frac{\nu(\mu-k)-\mu}{k}}\left(1-k\omega(ku)^{-\frac{\rho}{k}}\right)^{\frac{\gamma(\nu-1)}{k}}\hat{h}(p)\\
&\quad\times\sum_{m=0}^{\infty}(-G)^m(s+1)^{\frac{(\mu-k)m}{k}}(ku)^{\frac{\mu}{k}m}\left(1-k\omega(ku)^{-\frac{\rho}{k}}\right)^{\frac{\gamma}{k}m}\\
&= k\sum_{m=0}^{\infty}(-G)^m(s+1)^{\frac{(1-\nu)(k-\mu)+(k-\mu)m}{k}}(ku)^{\frac{\nu(\mu-k)-\mu(m+1)}{k}}\\
&\quad\times\left(1-k\omega(ku)^{-\frac{\rho}{k}}\right)^{-\frac{\gamma(m+1-\nu)}{k}}\hat{h}(p).
\end{aligned}$$

Applying the inverse Laplace transform, we get

$$\begin{aligned}
\hat{V}(p,\zeta) &= \sum_{m=0}^{\infty}(-G)^m(s+1)^{\frac{(1-\nu)(k-\mu)+(k-\mu)m}{k}}(\zeta^{s+1})^{\frac{\mu(m+1)-\nu(\mu-k)}{k}-1}\\
&\quad\times E^{\gamma(m+1-\nu)}_{k,\rho,\mu(m+1)-\nu(k-\mu)}(\omega(\zeta^{s+1})^{\frac{\rho}{k}})p^{2m}\hat{h}(p).
\end{aligned}$$

Now, applying the inverse Fourier transform yields

$$\begin{aligned}
V(\theta,\zeta) &= \int_{-\infty}^{+\infty}dp\, e^{-ip\theta}\hat{h}(p)\\
&\frac{1}{2\Pi}\sum_{m=0}^{\infty}(-G)^m(s+1)^{-\frac{(1-\nu)(k-\mu)-(\mu-k)m}{k}}(\zeta^{s+1})^{\frac{\mu(m+1)-\nu(\mu-k)}{k}-1}\\
&\times E^{\gamma(m+1-\nu)}_{k,\rho,\mu(m+1)-\nu(k-\mu)}(\omega(\zeta^{s+1})^{\frac{\rho}{k}})p^{2m}\hat{h}(p).
\end{aligned}$$

REMARK 4.2. *If $s = 0$, $k = 1$, $\gamma = 0$, and $\mu \to 1$, then the above Cauchy problem reduces to*

$$\frac{\partial}{\partial\theta}V(\theta,\zeta) = G\frac{\partial^2}{\partial\theta^2}V(\theta,\zeta)$$

$$\left[V(\theta,\zeta)\right]_{\zeta=0^+} = h(\theta), \quad \lim_{\theta\to\infty}V(\theta,\zeta) = 0,$$

which is the heat equation.

COROLLARY 4.3. *If we take $s = 0$ and $k = 1$, we get the following Cauchy problem given in [19]:*

$$\mathfrak{D}^{\gamma,\mu,\nu}_{0^+;\rho,\omega}V(\theta,\zeta) = G\frac{\partial^2}{\partial\theta^2}V(\theta,\zeta), \quad \zeta > 0,\ \theta\in\mathbb{R},$$

$$\left[\mathfrak{J}^{-\gamma(1-\nu),\omega}_{0^+;\rho,(1-\nu)(1-\mu)}V(\theta,\zeta)\right]_{\zeta=0^+} = h(\theta),$$

$$\lim_{\theta\to\infty}V(\theta,\zeta) = 0,$$

where $\mu\in(0,1)$, $\nu\in[0,1]$, $\omega\in\mathbb{R}$, $R,\rho>0$, and $\gamma\geq 0$, with solution given by

$$\begin{aligned}
V(\theta,\zeta) &= \int_{-\infty}^{+\infty}dp\, e^{-ip\theta}\hat{h}(p)\,\frac{1}{2\Pi}\sum_{m=0}^{\infty}(-G)^m\zeta^{\mu(m+1)-\nu(\mu-1)-1}\\
&\times E^{\gamma(m+1-\nu)}_{\rho,\mu(m+1)-\nu(\mu-1)}(\omega\zeta^{\rho})p^{2m}\hat{h}(p).
\end{aligned}$$

5. GENERALIZATION OF THE FRACTIONAL KINETIC DIFFERINTEGRAL EQUATION

Fractional differential equations are important tools for developing mathematical models of numerous phenomena in fields such as physics, dynamic systems, control systems, and engineering. In mathematical modeling, kinetic equations describe the continuity of the motion of a substance and are basic equations of mathematical physics and the natural sciences. In this section, we consider an equation that generalizes kinetic equations. For related literature, we refer the reader to [39–42].

THEOREM 5.1. *Consider the Cauchy problem*

$$a{}^s_k\mathfrak{D}^{\gamma,\mu,\nu}_{0^+;\rho,\omega}N(t) - N_0f(t) = b{}^s_k\mathfrak{J}^{\omega,\sigma}_{0^+;\rho,q}N(t), \quad f\in L^1[0,\infty), \tag{5.1}$$

$${}^s_k\mathfrak{J}^{\omega,-\gamma(1-\nu)}_{0^+;\rho,(1-\nu)(k-\mu)}N(0) = d, \quad d\geq 0, \tag{5.2}$$

where $s\in[0,\infty)$, $\nu\in[0,1]$, $\omega\in\mathbb{C}$, $a,b\in\mathbb{R}$ $(a\neq 0)$, $\mu,\rho,q,k>0$, and $\gamma,\sigma\geq 0$. The solution to the problem is

$$\begin{aligned}
N(t) &= d\sum_{n=0}^{\infty}\left(\frac{b}{a}\right)^n(s+1)^{-\frac{\nu(k-\mu)+(\mu-k)(n+1)+qn}{k}}(t^{s+1})^{\frac{\nu(k-\mu)+\mu+(q+\mu)n}{k}-1}\\
&E^{(\gamma+\sigma)n+\gamma(1-\nu)}_{k,\rho,\nu(k-\mu)+\mu+(q+\mu)n}(\omega(t^{s+1})^{\frac{\rho}{k}})\\
&\quad+\frac{N_0}{a}\sum_{n=0}^{\infty}\left(\frac{b}{a}\right)^n(s+1)^{n+1}{}^s_k\mathfrak{J}^{\omega,(\gamma+\sigma)n+\gamma}_{0^+;\rho,(q+\mu)n+\mu}f(t).
\end{aligned}$$

PROOF. Applying the generalized Laplace transform to both sides of (5.1), we get

$$aL_g\{{}^s_k\mathfrak{D}^{\gamma,\mu,\nu}_{0^+;\rho,\omega}N(t)\}(u) - N_0L_g\{f(t)\}(u) = bL_g\{{}^s_k\mathfrak{J}^{\omega,\sigma}_{0^+;\rho,q}N(t)\}(u).$$

Using Theorems 2.5 and 1.12, we get

$$a\Big[(s+1)^{\frac{\mu-k}{k}}(ku)^{\frac{\mu}{k}}\left(1-k\omega(ku)^{-\frac{\rho}{k}}\right)^{\frac{\gamma}{k}}L_g\{N(t)\}(u)$$
$$-k(s+1)^{-\frac{\nu(k-\mu)}{k}}(ku)^{\frac{-\nu(k-\mu)}{k}}$$
$$\times\left(1-k\omega(ku)^{-\frac{\rho}{k}}\right)^{\frac{\gamma\nu}{k}}{}_k^s\mathfrak{J}^{\omega,-\gamma(1-\nu)}_{0^+;\rho,(1-\nu)(k-\mu)}N(0^+)\Big]$$
$$-N_0L_g\{f(t)\}(u)$$
$$=b(s+1)^{-\frac{\mu}{k}}(ku)^{-\frac{\mu}{k}}\left(1-k\omega(ku)^{-\frac{\rho}{k}}\right)^{-\frac{\sigma}{k}}L_g\{N(t)\},$$

which can be written as

$$\left[\frac{a-b(s+1)^{-\frac{\mu-k+q}{k}}(ku)^{-\frac{\mu+q}{k}}\left(1-k\omega(ku)^{-\frac{\rho}{k}}\right)^{-\frac{\gamma+\sigma}{k}}}{(s+1)^{-\frac{\mu-k}{k}}(ku)^{-\frac{\mu}{k}}\left(1-k\omega(ku)^{-\frac{\rho}{k}}\right)^{-\frac{\gamma}{k}}}\right]$$
$$L_g\{N(t)\}(u)$$
$$=akd(s+1)^{-\frac{\nu(k-\mu)}{k}}(ku)^{-\frac{\nu(k-\mu)}{k}}\left(1-k\omega(ku)^{-\frac{\rho}{k}}\right)^{\frac{\gamma\nu}{k}}$$
$$+N_0L_g\{f(t)\}(u),$$

$$L_g\{N(t)\}(u)$$
$$=akd\left[\frac{(s+1)^{-\frac{\nu(k-\mu)+(\mu-k)}{k}}(ku)^{-\frac{\nu(k-\mu)+\mu}{k}}\left(1-k\omega(ku)^{-\frac{\rho}{k}}\right)^{\frac{\gamma(\nu-1)}{k}}}{a-b(s+1)^{-\frac{\mu-k+q}{k}}(ku)^{-\frac{\mu+q}{k}}\left(1-k\omega(ku)^{-\frac{\rho}{k}}\right)^{-\frac{\gamma+\sigma}{k}}}\right]$$
$$+\left[\frac{(s+1)^{-\frac{\mu-k}{k}}(ku)^{-\frac{\mu}{k}}\left(1-k\omega(ku)^{-\frac{\rho}{k}}\right)^{-\frac{\gamma}{k}}}{a-b(s+1)^{-\frac{\mu-k+q}{k}}(ku)^{-\frac{\mu+q}{k}}\left(1-k\omega(ku)^{-\frac{\rho}{k}}\right)^{-\frac{\gamma+\sigma}{k}}}\right]$$
$$N_0L_g\{f(t)\}(u).$$

Taking $\left|\frac{b}{a}(s+1)^{-\frac{\mu-k+q}{k}}(ku)^{-\frac{\mu+q}{k}}\left(1-k\omega(ku)^{-\frac{\rho}{k}}\right)^{-\frac{\gamma+\sigma}{k}}\right|<1$ gives

$$L_g\{N(t)\}(u)$$
$$=\Big[kd(s+1)^{-\frac{\nu(k-\mu)+(\mu-k)}{k}}(ku)^{-\frac{\nu(k-\mu)+\mu}{k}}\left(1-k\omega(ku)^{-\frac{\rho}{k}}\right)^{\frac{\gamma(\nu-1)}{k}}$$
$$+(s+1)^{-\frac{\mu-k}{k}}(ku)^{-\frac{\mu}{k}}\left(1-k\omega(ku)^{-\frac{\rho}{k}}\right)^{-\frac{\gamma}{k}}a^{-1}N_0L_g\{f(t)\}(u)\Big]$$
$$\times\sum_{n=0}^{\infty}\left(\frac{b}{a}\right)^n(s+1)^{-\frac{(\mu-k+q)n}{k}}(ku)^{-\frac{(\mu+q)n}{k}}$$
$$\left(1-k\omega(ku)^{-\frac{\rho}{k}}\right)^{-\frac{(\gamma+\sigma)n}{k}}$$
$$=dk\sum_{n=0}^{\infty}\left(\frac{b}{a}\right)^n(s+1)^{-\frac{\nu(k-\mu)+(\mu-k)(n+1)+qn}{k}}(ku)^{-\frac{\nu(k-\mu)+\mu+(\mu+q)n}{k}}$$
$$\left(1-k\omega(ku)^{-\frac{\rho}{k}}\right)^{-\frac{(\gamma+\sigma)n+\gamma(1-\nu)}{k}}$$
$$+\frac{N_0}{a}\sum_{n=0}^{\infty}\left(\frac{b}{a}\right)^n(s+1)^{-\frac{(\mu-k)(n+1)+qn}{k}}(ku)^{-\frac{\mu+(\mu+q)n}{k}}$$
$$\left(1-k\omega(ku)^{-\frac{\rho}{k}}\right)^{-\frac{(\gamma+\sigma)n+\gamma}{k}}.$$

Applying the inverse Laplace transform, we get

$$N(t)=d\sum_{n=0}^{\infty}\left(\frac{b}{a}\right)^n(s+1)^{-\frac{\nu(k-\mu)+(\mu-k)(n+1)+qn}{k}}$$
$$(t^{s+1})^{\frac{\nu(k-\mu)+\mu+(q+\mu)n}{k}-1}E^{(\gamma+\sigma)n+\gamma(1-\nu)}_{k,\rho,\nu(k-\mu)+\mu+(q+\mu)n}\left(\omega(t^{s+1})^{\frac{\rho}{k}}\right)$$
$$+\frac{N_0}{a}\sum_{n=0}^{\infty}\left(\frac{b}{a}\right)^n(s+1)^{n+1}{}_k^s\mathfrak{J}^{\omega,(\gamma+\sigma)n+\gamma}_{0^+;\rho,(q+\mu)n+\mu}f(t),$$

which is the required result.

REMARK 5.2. *If we take $s=0$, $k=1$, $\nu=\gamma=\sigma=0$, $\mu\to 0$, $a=1$, and $b=-c^p$, then we get the following fractional kinetic equation given in [39]:*

$$N(t)-N_0f(t)=-c^pD^p_{0+}N(t),\quad N(0)=d,\quad d\geq 0,$$

where D^p_{0+} is the Riemann-Liouville fractional integral operator, defined as

$$D^p_{0+}N(t)=\frac{1}{\Gamma(p)}\int_0^t(t-\tau)^{p-1}N(\tau)\,d\tau.$$

Here $N(t)$ denotes the number density of a given species at time t, with $N_0=N(0)$ being the number density of that species at time $t=0$, c is a constant, and $f\in L^1[0;\infty)$.

COROLLARY 5.3. *If we take $s=0$ and $\nu=0$, then we get the following Cauchy problem given in [42]:*

$$a_k\mathfrak{D}^{\gamma,\mu}_{0^+;\rho,\omega}N(t)-N_0f(t)=b_k\mathfrak{J}^{\omega,\sigma}_{0^+;\rho,q}N(t),\quad f\in L^1[0,\infty),$$
$${}_k\mathfrak{J}^{\omega,-\gamma}_{0^+;\rho,k-\mu}N(0)=d,\quad d\geq 0,$$

where $\omega\in\mathbb{C}$, $a,b\in\mathbb{R}(a\neq 0)$, $\mu,\rho,q,k>0$, and $\gamma,\sigma\geq 0$. The solution to the problem is

$$N(t)=d\sum_{n=0}^{\infty}\left(\frac{b}{a}\right)^n t^{\frac{\mu+(q+\mu)n}{k}-1}E^{(\gamma+\sigma)n+\gamma}_{k,\rho,\mu+(q+\mu)n}\left(\omega(t)^{\frac{\rho}{k}}\right)$$
$$+\frac{N_0}{a}\sum_{n=0}^{\infty}\left(\frac{b}{a}\right)^n{}_k\mathfrak{J}^{\omega,(\gamma+\sigma)n+\gamma}_{0^+;\rho,(q+\mu)n+\mu}f(t).$$

6. CONCLUSION

A new generalized fractional derivative operator, referred to as the (k,s)-Hilfer-Prabhakar fractional derivative, is developed in this article. The generalized Laplace transform of the proposed operator is also studied. Potential applications of the proposed operator are discussed, which concern fractional models of the free-electron laser equation, heat equation, and kinetic equation that involve the new operator. The results in this article suggest that this novel operator can be used to solve various types of problems arising in mathematical physics and other fields.

AUTHOR CONTRIBUTIONS

MS, KN, and DK: Conceptualization. MS, ZP, GR, and KN: Writing original draft. KN and DK: Methodology.

ZP, GR, and DK: Formal analysis. MS and GR: Validation. KN and DK: Revision and final check. All authors contributed to the article and approved the submitted version.

REFERENCES

1. Kumar D, Singh J, Tanwar K, Baleanu D. A new fractional exothermic reactions model having constant heat source in porous media with power, exponential and Mittag-Leffler Laws. *Int J Heat Mass Transfer.* (2019) **138**:1222–7. doi: 10.1016/j.ijheatmasstransfer.2019.04.094
2. Kumar D, Singh J, Qurashi MA, Baleanu D. A new fractional SIRS-SI malaria disease model with application of vaccines, anti-malarial drugs, and spraying. *Adv Differ Equat.* (2019) **2019**:278. doi: 10.1186/s13662-019-2199-9
3. Kumar, D, Singh J, Baleanu, D. A new numerical algorithm for fractional Fitzhugh-Nagumo Equation arising in transmission of nerve impulses. *Nonlinear Dyn.* (2018) **91**:307–17. doi: 10.1007/s11071-017-3870-x
4. Kumar D, Singh J, Baleanu D. Analysis of regularized long-wave equation associated with a new fractional operator with Mittag-Leffler type kernel. *Physica A.* (2018) **492**:155–67. doi: 10.1016/j.physa.2017.10.002
5. Singh J, Kumar D, Baleanu D. New aspects of fractional Biswas-Milovic model with Mittag-Leffler law. *Math Model Nat Phenomena.* (2019) **14**:303. doi: 10.1051/mmnp/2018068
6. Abel, N. Solution de quelques problemes l'aide d'integrales definies. *Mag Nat.* (1823) **1**:1–27.
7. Herrmann R. *Fractional Calculus: An Introduction for Physicists.* 3rd revised ed. Singapore: World Scientific Publishing (2018).
8. Hilfer R. *Applications of Fractional Calculus in Physics.* Singapore: World Scientific Publishing (2000).
9. Hilfer R. *Threefold Introduction to Fractional Derivatives.* Anomalous Transport: Foundations and Applications. Weinheim: Germany Publishing (2008).
10. Kilbas A, Srivastava HM, Trujillo JJ. *Theory and Application of Fractional Differential Equations.* Amsterdam: North Holland Mathematics Studies (2006). p. 204.
11. Lorenzoand CF, Hartley TT. Variable order and distributed order fractional operators. *Nonlinear Dynam.* (2002) **29**:57–98. doi: 10.1023/A:1016586905654
12. Magin RL. *Fractional Calculus in Bioengineering.* Redding, CA: House Publishers (2006).
13. Podlubny I. *Fractional Differential Equations.* San Diego CA: Academic Press (1999).
14. Oldham KB, Spanier J. *The Fractional Calculus.* New York, NY: Academic Press (1974).
15. Baleanu D, Diethelm K, Scalas E, Trujillo JJ. *Fractional Calculus: Models and Numerical Methods.* Series on Complexity, Nonlinearity and Chaos. New Jersey, NJ: World Scientific Publishing (2012).
16. Mainardi F. *Fractional Calculus and Waves in Linear Viscoelasticity: An Introduction to Mathematical Models.* Singapore; London; Hong Kong: World Scientific (2010).
17. Diethelm K. *The Analysis of Fractional Differential Equations: An Application-Oriented Exposition Using Differential Operators of Caputo Type.* Berlin; Heidelberg: Springer (2010).
18. Hilfer R, Luchko Y, Tomovski Z. Operational method for solution of the fractional differential equations with the generalized Riemann-Liouville fractional derivatives. *Fract Calc Appl Anal.* (2009) **12**:299–318. Available online at: https://www.researchgate.net/publication/228746820_Operational_method_for_the_solution_of_fractional_differential_equations_with_generalized_Riemann-Liouville_fractional_derivatives
19. Garra R, Goreno R, Polito F, Tomovski Z. Hilfer-Prabhakar derivative and some applications. *Appl Math Comput.* (2014) **242**:576–89. doi: 10.1016/j.amc.2014.05.129
20. Panchal SK, Pravinkumar VD, Khandagale AD. k-Hilfer-Prabhakar fractional derivatives and its applications. *Indian J Math.* (2017) **59**:367–83. Available online at: http://www.amsallahabad.org/pdf/ijm593.pdf
21. Dorrego GA. Generalized Riemann-Liouville fractional operators associated with a generalization of the Prabhakar integral operator. *Prog Fract Differ Appl.* (2016) **2**:131–40. doi: 10.18576/pfda/020206
22. Nisar KS, Rahman G, Baleanu D, Mubeen S, Arshad M. The (k, s)-fractional calculus of k-Mittag-Leffler function. *Adv Differ Equat.* (2017) 118:1–2. doi: 10.1186/s13662-017-1239-6
23. Gaboury S, Tremblay R, Fugere B. Some relations involving a generalized fractional derivative operator. *J Inequal Appl.* (2013) 167:1–9. doi: 10.1186/1029-242X-2013-167
24. Rahman G, Nisar KS, Mubeen S. A new extension of extended Caputo fractional derivative operator. *Math Eng Sci Aerospace.* (2020) **11**:265–79.
25. Bohner M, Rahman G, Mubeen S, Nisar KS. A further extension of the extended Riemann-Liouville fractional derivative operator. *Turk J Math.* (2018) **42**:2631–42. doi: 10.3906/mat-1805-139
26. Nisar KS, Rahman G, Tomovski Z. On a certain extension of the Riemann-Liouville fractional derivative operator. *Commun Korean Math Soc.* (2019) **34**:507–22. doi: 10.4134/CKMS.c180140
27. Rahman G, Mubeen S, Nisar KS, Choi J. Certain extended special functions and fractional integral and derivative operators via an extended beta functions. *Nonlinear Funct Anal Appl.* (2019) **24**:1–13. Available online at: http://nfaa.kyungnam.ac.kr/journal-nfaa/index.php/NFAA/article/view/1148/990
28. Rahman G, Mubeen S, Nisar KS. On generalized k-fractional derivative operator. *AIMS Math.* (2020) **5**:1936–45. doi: 10.3934/math.2020129
29. Modanl M. On the numerical solution for third order fractional partial differential equation by difference scheme method. *Int J Optim Control Theor Appl.* (2019) **9**:1–5. doi: 10.11121/ijocta.01.2019.00678
30. Vivek D, Kanagarajan K, Sivasundaram S. Dynamics and stability results for Hilfer fractional type thermistor problem. *Fractal Fract.* (2017) 1:5. doi: 10.3390/fractalfract1010005
31. Ozdemir N, Avci D, Iskender B. The numerical solutions of a two-dimensional space-time Riesz-Caputo fractional diffusion equation. *Int J Optim Control Theor Appl.* (2011) **1**:17–26. doi: 10.11121/ijocta.01.2011.0028
32. Jarad F, Abdeljawad T. A modified Laplace transform for certain generalized fractional operators. *Results Nonlinear Anal.* (2018) **2**:88–98. doi: 10.3934/dcdss.2020039
33. Kerr FH. Namias fractional Fourier-Transforms on L2 and applications to differential equations. *J Math Anal Appl.* (1988) **136**:404–18.
34. Jarad F, Abdeljawad T. Generalized fractional derivatives and laplace transform. *Discrete Contin Dyn Syst Ser S.* (2020) **13**:709–22.
35. Diaz R, Pariguan E. On hypergeometric functions and pochhammer k-symbol. *Divulg Math.* (2007) **15**:179–92. Available online at: https://citeseerx.ist.psu.edu/viewdoc/download?doi=10.1.1.930.1846&rep=rep1&type=pdf
36. Dorrego GA, Cerutti RA. The k-Mittag-Leffler function. *Int J Contemp Math Sci.* (2012) **7**:705–16. Available online at: http://www.m-hikari.com/ijcms/ijcms-2012/13-16-2012/ceruttiIJCMS13-16-2012-2.pdf
37. Luchko Y. Some uniqueness and existence results for the initial-boundary-value problems for the generalized time-fractional diffusion equation. *Comput Math Appl.* (2010) **59**:1766–72. doi: 10.1016/j.camwa.2009.08.015
38. Luchko Y. Initial-boundary-value problems for the generalized multi-term time-fractional diffusion equation. *J Math Anal Appl.* (2011) **374**:538–48. doi: 10.1016/j.jmaa.2010.08.048
39. Saxena RK, Kalla SL. On the solutions of certain fractional kinetic equations. *Appl Math Comput.* (2008) **199**:504–11. doi: 10.1016/j.amc.2007.10.005
40. Choi J, Kumar D. Solutions of generalized fractional kinetic equations involving Aleph functions. *Math. Commun.* (2015) **20**:113–23. Available online at: https://www.mathos.unios.hr/mc/index.php/mc/article/view/1125
41. Kumar D, Choi J. Generalized fractional kinetic equations associated with aleph function. *Proc Jangjeon Math Soc.* (2016) **19**:145–55. Available online at: http://www.jangjeon.or.kr/etc/view.html?id=1684

7

Waves that Appear from Nowhere: Complex Rogue Wave Structures and their Elementary Particles

Nail Akhmediev *

Department of Theoretical Physics, Research School of Physics, The Australian National University, Canberra, NSW, Australia

***Correspondence:**
Nail Akhmediev
nna124@rsphysse.anu.edu.au

The nonlinear Schrödinger equation has wide range of applications in physics with spatial scales that vary from microns to kilometres. Consequently, its solutions are also universal and can be applied to water waves, optics, plasma and Bose-Einstein condensate. The most remarkable solution presently known as the Peregrine solution describes waves that appear from nowhere. This solution describes unique events localized both in time and in space. Following the language of mariners they are called "rogue waves". As thorough mathematical analysis shows, these waves have properties that differ them from any other nonlinear waves known before. Peregrine waves can serve as 'elementary particles' in more complex structures that are also exact solutions of the nonlinear Schrödinger equation. These structures lead to specific patterns with various degrees of symmetry. Some of them resemble "atomic like structures". The number of particles in these structures is not arbitrary but satisfies strict rules. Similar structures may be observed in systems described by other equations of mathematical physics: Hirota equation, Davey-Stewartson equations, Sasa-Satsuma equation, generalized Landau-Lifshitz equation, complex KdV equation and even the coupled Higgs field equations describing nucleons interacting with neutral scalar mesons. This means that the ideas of rogue waves enter nearly all areas of physics including the field of elementary particles.

Keywords: nonlinear schrodinger equation, rogue waves, peregrine wave, water waves, optical fibers

1 INTRODUCTION

The nonlinear Schrödinger equation (NLSE) has wide range of applications in physics with spatial scales that vary from microns to kilometres and even light years. It describes nonlinear wave phenomena in optics [1, 2], oceanography [3, 4], plasmas [5, 6], atmosphere [7], Bose-Einstein condensate [8, 9] and cosmology [10]. Taking into account the lowest order nonlinearity and dispersion, this equation describes nonlinear wave phenomena at the fundamental level. NLSE serves as a basic tool for understanding modulation instability [11, 12], solitons [13], periodic waves [14] and extreme waves [15, 16]. The ideas born in the studies of NLSE solutions can be transferred to many other systems. Despite being studied for nearly 50 years, the NLSE solutions have a rich structure and provide surprises for researchers even today [17].

Being a practical introduction to a special issue, this article provides a basic review of mathematical results on NLSE that are important for understanding the nonlinear phenomena in general. It leaves aside the complexities of inverse scattering technique [13], Darboux transformation [18], theta functions [19] and other sophistications of modern mathematics [20].

Instead, it provides solutions in explicit form so that everyone can appreciate their clarity, simplicity of making useful plots and most importantly, the possibility of using them in applications.

The NLSE describes systems with an infinite number of degrees of freedom. Being integrable, it has an infinite number of solutions that can be presented in analytic form. Among them, there are fundamental ones such as soliton solutions [13], Akhmediev breathers [21–24], the Peregrine solution [26–28] and the general doubly-periodic solutions [17, 22]. These can be considered as fundamental modes of the nonlinear system each with a specific single eigenvalue of the inverse scattering transform [13]. More complicated solutions are nonlinear superpositions of these fundamental ones. Although mathematically, these superpositions may look highly complicated, conceptually, they can be understood as a combination of fundamental solutions. The corresponding spectrum of eigenvalues for complex solutions is a combined set of individual eigenvalues.

Presently, the most studied combinations are multi-soliton solutions [13, 29]. Next in complexity are multi-Akhmediev breathers [25, 31]. Rogue waves can also be superimposed resulting in multi-rogue wave solutions. In contrast to multi-solitons and multi-ABs, multi-rogue waves are degenerate solutions. The eigenvalues corresponding to the individual contributions are located at the same point of the complex plane. Despite this complication, multi-rogue waves are also well studied [32–34] but, perhaps, their physics is less understood [35–37] than the physics of any other solutions of the NLSE. The main reason is the unusual set of rules that control their superposition [35]. On the other hand, the superpositions of elementary doubly-periodic solutions studied in [17, 22] still need to be constructed. This task is mathematically challenging and has not been addressed so far. Existence of "explicit" solutions in terms of theta functions [19] does not provide any clue for solving this highly involved task.

Building higher-order superpositions consisting of the same type of fundamental solutions, say, multi-soliton solutions is relatively simple task [38]. This may be done using techniques such as Darboux transformation [30]. Mixing different types of fundamental solutions is more difficult. However, this has also been done in a few recent works. Mixing them is the way to address problems such as rogue waves on top of a periodic background [39, 40]. The number of possibilities is literally infinite.

After what is said above, it may seem strange that the NLSE model is the simplest one among the existing nonlinear evolution equations. However, this is indeed the case. The NLSE provides the conceptual background for further developments in the science of rogue waves. We should keep in mind that the first known integrable equation which is the real KdV equation [41] does not have rogue wave solutions. It cannot be used as a mathematical platform for rogue wave research. On the contrary, the ideas developed in the studies of NLSE solutions can be further extended to many other systems. These include Hirota equation [42], Sasa-Satsuma equation [43], Davey-Stewartson equations [44–46], Sine-Gordon equation [47], Landau-Lifschitz equation [48] and many others. The basic concepts are valid not only for integrable equations but can be expanded to non-integrable cases [49, 50] and, to some extent, to dissipative systems [51, 52].

The generality of the concept of rogue waves can be further expanded to extreme events in nature. Clearly, the equations that describe natural phenomena are more complex than the NLSE [53]. Nevertheless, these are also evolution equations that can be solved if not analytically, then numerically [54]. Rogue waves must be part of complex evolution of the system with either regular or chaotic initial conditions. These rogue waves may take more complicated forms than a simple Peregrine wave. They can take the form of tornadoes or hurricanes. These are also formations that "appear from nowhere". Thus, they do belong to the class of rogue waves or "extreme events".

Our task here is well defined by the subject of the special issue. Therefore, we will concentrate on the Peregrine wave, its analogs and its higher-order combinations. This is a very small subset of the whole set of multi-parameter families of solutions of the NLSE. Nevertheless, this subset plays an important role in explaining extreme events in many physical situations. Understanding variety of complex phenomena starts with the studies of simple examples. These simple examples are listed in the present rendition.

2 NLSE

The relative simplicity of the nonlinear Schrödinger equation, its integrability [13] and its applicability to many weakly nonlinear dispersive systems made it a universal model for wave propagation. The most common applications include deep ocean water waves [3, 4] and waves in an optical fiber [1, 2]. Universality means that this equation can be written in a standard form that is applicable to all major physical settings. Variables in this form are dimensionless and there are no free parameters. Namely,

$$i\frac{\partial\psi}{\partial x}+\frac{1}{2}\frac{\partial^2\psi}{\partial t^2}+|\psi|^2\psi=0 \tag{1}$$

Here, we consider x as the propagation distance and t as the retarded time in a reference frame moving with the group velocity. The function ψ means the envelope of the wave packet. Being a complex function, it defines both, the amplitude of the envelope and the phase shift of the carrier wave.

For a given central frequency of narrow banded waves, the group velocity of the waves is well defined by the dispersion relation. In this case, the propagation distance and time in the moving frame are linearly related. This means that the time and the distance can be easily exchanged in **Eq. 1** [4]. This replacement creates an alternative form of the equation that has been used in the earlier descriptions of water wave propagation [4]. Evolution in time is also convenient in problems related to Bose-Einstein condensate. For experiments in water tanks and in optical fibers, it is more convenient to stick to the notations taken in **Eq. 1**. Then waves are evolving along the tank or along the fiber and the shape of the wave envelope can be observed moving along with the wave packet.

All solutions below will be given in dimensionless form that directly satisfy **Eq. 1**. For practical applications, solutions must be rescaled to dimensional variables, i.e. they must be expressed in meters (kilometres) along the tank or along the fiber, in seconds for the transverse variable and in the amplitude units for ψ. In the case of an optical fiber, this rescaling is given by:

$$X = L_{NL} \times x, \quad T = \sqrt{|\beta_2| L_{NL}} \times t, \quad \Psi = \sqrt{P_P} \times \psi, \tag{2}$$

where $L_{NL} = (\gamma P_P)^{-1}$ is the nonlinear length, γ is the nonlinear coefficient defined by the material of the fiber, β_2 is the group velocity dispersion defined by the fiber design, and P_P is the pump power. Coefficients in the rescaling can be transferred to the equation or used only with the transformation (2). The latter choice is more convenient as it allows us to keep the NLSE to be a universal model.

In the case of waves in deep water the rescaling takes the form:

$$X = \frac{x}{k\xi^2}, \quad T = \sqrt{2}\frac{t}{\omega\xi}, \quad \Psi = \frac{\psi\xi}{k}, \tag{3}$$

where X is dimensional distance along the tank, T is dimensional time in the frame moving with the group velocity $c_g = \frac{\omega}{2k}$, Ψ is the envelope of the water wave elevation, ω is the carrier frequency and k is carrier wavenumber that satisfies the dispersion relation $\omega = \sqrt{gk}$ with $g = 9.81$ m/s^2 being the gravitational acceleration. Dimensionless parameter ξ is the wave steepness defined as the product of the wave amplitude a and the wavenumber k.

Further adjustment of solutions can be done with the use of scaling transformation

$$\psi'(x,t) = \alpha\psi(\alpha x, \alpha^2 t) \tag{4}$$

Namely, if $\psi(x,t)$ is a solution of the NLSE (1), then ψ' is also a solution of the same equation. **Eq. 4** provides an additional tool for adjustment of initial conditions to the required levels in optical and hydrodynamic experiments.

3 PEREGRINE WAVE

One of the simplest non-singular rational solutions of the NLSE 1) is given by [15, 16, 26]:

$$\psi(x,t) = \left[4\,\frac{1+2ix}{1+4x^2+4t^2} - 1\right]e^{ix}, \tag{5}$$

It is known as the Peregrine solution or Peregrine soliton [27] or Peregrine breather [28]. The modulus of this complex solution is shown in **Figure 1**. The main feature of this solution is the localization of its central peak both in time and in space. The constant background represents a plane wave with an infinite source of energy. This solution has all features of rogue waves in the ocean [15, 16]. It represents an unexpected wave event on an otherwise flat background.

Peregrine solution has been observed in various experiments. The most notable ones are in water waves [55] and in optical

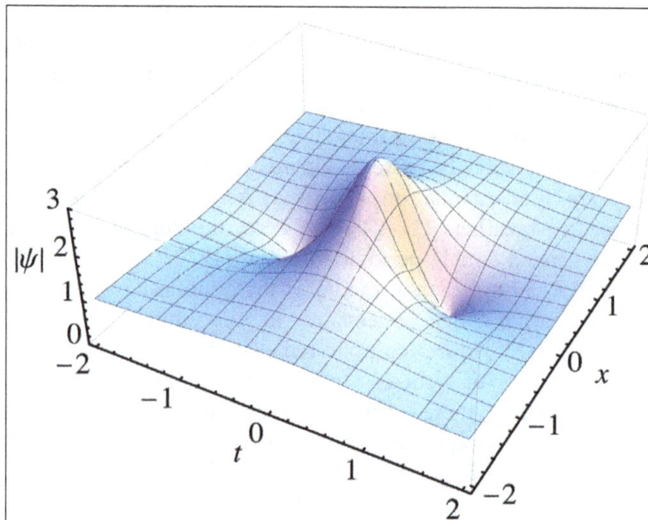

FIGURE 1 | A single Peregrine wave of the NLSE [15]. The maximum amplitude at the origin is three while the background amplitude is 1. The solution has two zeros along the t axis at each side of the maximum. This simple wave structure appears in more complicated rogue wave patterns.

fibers [27]. This solution appears in field evolution dynamics with variety of initial conditions [56]. Moreover, the Peregrine solution can be considered as a universal structure emerging in any type of intensity localization of high power pulses [57]. We can consider it as an "elementary particle" of more complicated patterns that can appear on a plane wave background. As the NLSE is the envelope equation, the Peregrine solution may describe both the wave of elevation or a depression. The latter is known as the rogue wave hole [58]. Solutions similar to the Peregrine one can also be found in other physical systems [59–62]. Thus, the phenomenon of rogue wave is even more universal than we can imagine.

4 HIGHER-ORDER ROGUE WAVE SOLUTIONS

There are higher-order rational solutions of the NLSE that we can call rogue waves. The second order solution has been first presented in [14]. Several methods are known for constructing higher-order solutions of integrable equations [18, 32–34]. A hierarchy of rogue wave solutions with progressively increasing central amplitudes are presently known as Akhmediev-Peregrine (AP) breathers [63–66]. Their general form can be written as:

$$\psi_N(x,t) = \left[(-1)^N + \frac{G_N(x,t) + ixH_N(x,t)}{D_N(x,t)}\right]e^{ix}, \tag{6}$$

where G, H and D are polynomials, and N is the order of the solution. In the case of the Peregrine solution **(5)**, $N = 1$ and we have: $G_1(x,t) = 4$, $H_1(x,t) = 8$, $D_1(x,t) = 1 + 4x^2 + 4t^2$.

For the second order solution, the polynomials are given by [14, 67]:

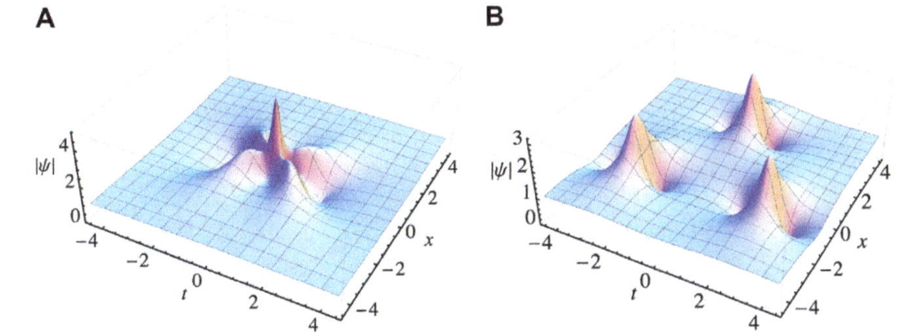

FIGURE 2 | (A) AP solution of the second order [15]. **(B)** Rogue wave triplet with parameters $\gamma = 200$ and $\beta = 0$ [73]. Here, the second order solution consists of three Peregrine waves.

$$G_2 = \frac{3}{8} - 3t^2 - 2t^4 - 9x^2 - 10x^4 - 12t^2x^2,$$

$$H_2 = \frac{15}{4} + 6t^2 - 4t^4 - 2t^4 - 2x^2 - 4x^4 - 8t^2x^2,$$

$$D_2 = \frac{1}{8}\left[\frac{3}{4} + 9t^2 + 4t^4 + \frac{16}{3}t^6 + 33x^2 + 36x^4 + \frac{16}{3}x^6 - 24t^2x^2 + 16t^4x^2 + 16t^2x^4\right].$$

This solutions is shown in **Figure 2A**. While for the first order solution shown in **Figure 1** the amplitude is 3 times the background, the maximum amplitude of the second order solution shown in **Figure 2A** is 5 times the background.

The second-order solution with free parameters has been given earlier in [32, 73]. The solution intensity $|\psi|^2$ is completely defined by the denominator $D_2(x,t)$. Namely, $|\psi_2(x,t)|^2 = 1 + [\log(D_2(x,t))]_{tt}$ [71]. The latter is now given by:

$$D_2(x,t) = \beta^2 + \gamma^2 + 64t^6 + 48t^4(4x^2+1) - 16\beta t^3 + 12t^2(16x^4 - 24x^2 + 4\gamma x + 9) + 12t\beta(1+4x^2) + 64x^6 + 432x^4 - 16\gamma x^3 + 396x^2 - 36\gamma x + 9$$

where β and γ are free real parameters of the solution. This solution is shown in **Figure 2B**. When $\beta = 0$ and $\gamma = 0$, the solution reduces to the second-order AP solution shown in 2(a). An interesting fact about the solution shown in **Figure 2B** is that despite being of the second order it contains three elementary rogue waves rather than two. This is a distinctive feature of multi-rogue wave solutions. They are different from multi-soliton solutions that have the number of solitons equal to the order of the solution [30].

The general form of the third-order rogue wave solution is cumbersome. It has several free parameters. Various forms of the third-order solution have been presented in [71, 73–75] and [76]. Instead of giving the general form, we restrict ourselves with a simpler one-parameter case. As the solution intensity $|\psi_3|^2$ is completely defined by the denominator $D_3(x,t)$, $|\psi_3(x,t)|^2 = 1 + [\log(D_3(x,t))]_{tt}$ [71], we only present the expression for $D_3(x,t)$ [37, 76]. It has a single free real parameter b:

$$D_3(x,t) = \sum_{j=0}^{12} d_j(T)(2x)^j, \tag{7}$$

where the polynomials $d_j(T = 2t)$ are:

$$d_0 = b^2(T^2+1) + T^{12} + 6T^{10} + 135T^8 + 2340T^6 + 3375T^4 + 12150T^2 + 2025,$$
$$d_1 = -10b(T^6 - 9T^4 - 45T^2 + 45),$$
$$d_2 = b^2 + 6(T^{10} - 15T^8 + 90T^6 + 2250T^4 - 6075T^2 + 15525),$$
$$d_3 = 10b(T^4 + 18T^2 + 9),$$
$$d_4 = 15(T^8 - 12T^6 - 90T^4 + 5220T^2 + 9585),$$
$$d_5 = 6b(3T^2 - 17),$$
$$d_6 = 20(T^6 + 3T^4 + 675T^2 + 765), d_7 = -2b,$$
$$d_8 = 15(T^4 + 18T^2 + 249),$$
$$d_9 = 0,$$
$$d_{10} = 6(T^2 + 21),$$
$$d_{11} = 0, d_{12} = 1.$$

The third-order solutions are shown in **Figure 3** for the cases $b = 0$ and $b = 2 \times 10^7$. The third-order solution consists of six Peregrine waves. When $b = 0$, all six are located at the origin leading to the central amplitude 7. For nonzero b, the solution splits into six components. Each of them is a Peregrine wave as can be seen from **Figure 3B**. One of them is located at the origin. Five others are at the corners of an equilateral pentagram. The total number of Peregrine waves is again higher than the order of the solution. Remarkably, the number of Peregrine waves in exact solutions cannot be equal to 2, 4, 5. We can consider the Peregrine wave as the elementary rogue wave solution. Equivalently, we can consider it as a rogue wave quantum or elementary particle of rogue wave patterns.

5 COMPLEX ROGUE WAVE PATTERNS

The Nth order rogue wave solution always contains $N(N+1)/2$ Peregrine waves. Namely, solution of the order $N = 1, 2, 3, 4, 5\cdots$ contains $1, 3, 6, 10, 15, \cdots$ Peregrine waves, respectively. These are known as 'triangular numbers' illustrated in **Figure 4**. They are defined as the total number of points in a regular pattern within a triangle with N points along its edges.

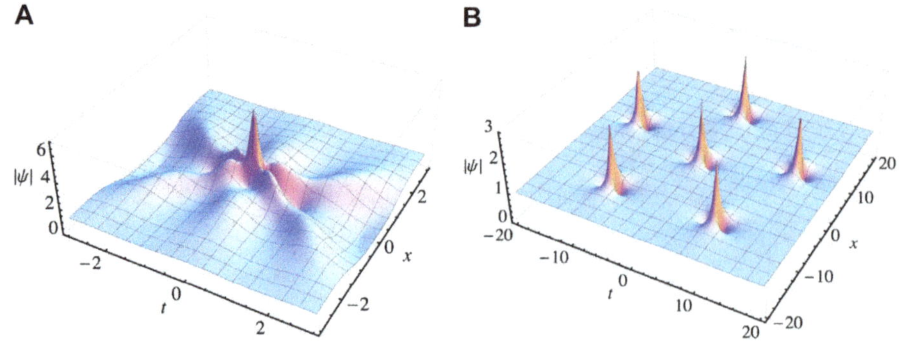

FIGURE 3 | (A) Third order rogue wave solution given by **Eq. (7)** with $b = 0$. **(B)** Third-order rogue wave solution with a circular pattern. It is given by **Eq. (7)** with $b = 2 \times 10^7$. Each 'elementary particle' in this structure is a single Peregrine wave.

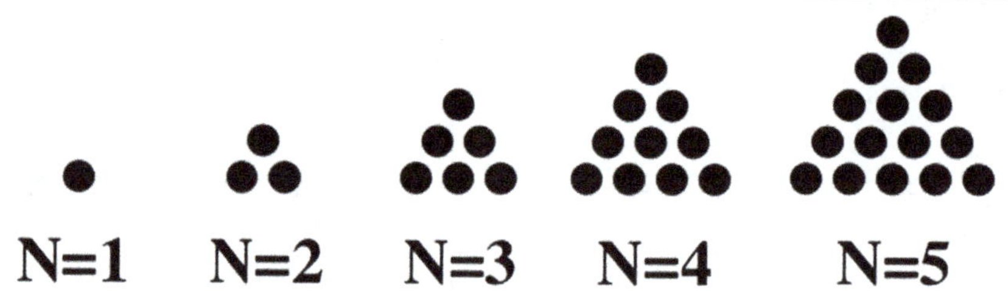

FIGURE 4 | The first five triangular numbers. They are represented by the number of solid dots within each triangle.

Rogue wave patterns are defined by the order of the solution, N, and may depend on additional free parameters. Two examples are given above. These free parameters split the solution into individual Peregrine waves. These parameters specify separations and relative positions of the individual components on the (x, t)-plane. Any solution of Nth order contains $N(N + 1)/2$ individual components that, when well-separated, can be identified as Peregrine waves. This is a fundamental result [35]: the number of Peregrine waves in a multi-rogue wave solution of the NLSE is given by a triangular number. No other number of Peregrine waves can be present in a general multi-rogue wave solution. This number does not depend on whether the "elementary particles" in the higher-order rogue wave are well-separated or partially separated. In the latter case, when the free parameters are small, the number of Peregrine waves is not visually obvious.

The maximal amplitude of a rogue wave is one of its main characteristics. The maximal amplitude of the higher-order AP solutions when all Peregrine waves are located at the origin is $(2N + 1)$ times the background, where N is the order of the solution [63, 68–70]. This is the highest possible amplitude for all imaginable rogue wave patterns.

A detailed classification of the rogue wave patterns has been given in [72]. The main results of this classification are shown in **Figure 5**. It shows the calculated rogue wave patterns that increase in order from top to bottom of table. A single Peregrine wave that is the solution of the first order is located in the top left cell. Higher-order AP solutions are located in the left column. The figure shows only solutions of up to 6-th order, although the list can be continued indefinitely. The second column represents regular triangular structures. The third column represents pentagram patterns. The fourth column provides examples of heptagram structures, i.e., patterns with seven Peregrine waves on one or several circles of different radii. Further columns have 9, 11, ... Peregrine waves in the outer and inner shells of the structure.

"Atomic-like" circular structures are located along the diagonal line. Starting from order 3, they consist of a central nucleus containing $N - 2$ elementary rogue waves in the form of an AP solution and $2N - 1$ Peregrine waves playing the role of the shell of "electrons". Patterns in inner cells of the table can be considered as "atomic structures" with several shells of "electrons". The larger variety of patterns for higher-order solutions is caused by the larger number of free parameters controlling the solution. The rogue wave patterns can also have a lower symmetry than in **Figure 5**. Some examples can be found in [36, 71].

There is no doubt that these patterns can be observed experimentally. Indeed, a single Peregrine wave has been observed in optics [27], in water waves [55] and in a multicomponent plasma [77]. It may soon be observed in Bose-Einstein condensate [78, 79]. The rogue wave triplet has

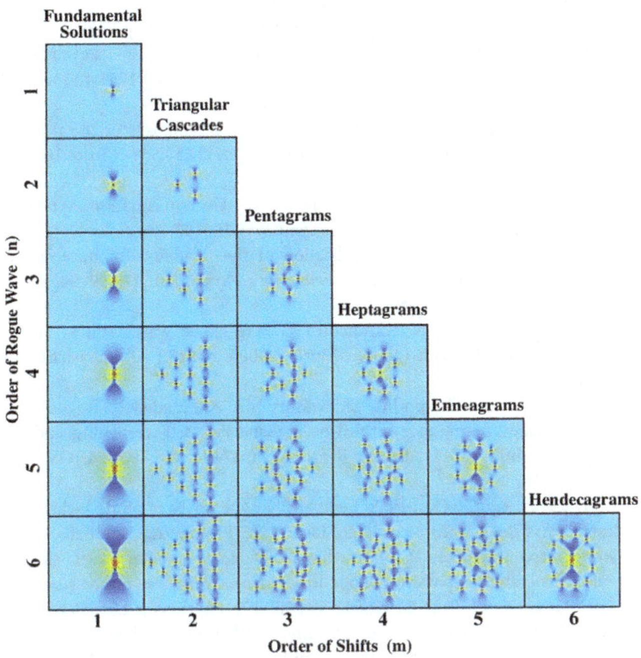

FIGURE 5 | Patterns of higher-order rogue wave solutions [72]. Only solutions of up to order six are shown. The Peregrine soliton is on the first row. Second-order solutions are of two types: the AP solution and the triplet. They are located in the second row. Variety of patterns increases with the order. For example, the third order solutions do exist in the AP, triangular and pentagram forms. This happens because of the larger number of free parameters in the solution. Patterns located along the diagonal line are circular. They have a ring of $(2N - 1)$ Peregrine solitons around a central rogue wave of order $N - 2$ (for $N > 2$). These are basically "atomic structures" with a nucleus and a shell of "electrons".

been observed in a water wave tank [80]. AP solutions up to fifth-order have also been observed in water waves [81]. As the NLSE has a wide range of applications, these solutions may appear in situations that we cannot even predict right now.

Can this approach be used for a description of elementary particles and atomic structures? Obviously, this will need more sophisticated equation than the NLSE. As mentioned, the phenomena described in this review are not unique to the NLSE although, perhaps, the NLSE case is the most studied of all. Rogue waves have been found in systems described by the Hirota and Maxwell-Bloch equations [42, 82], Sasa-Satsuma equation [83], Fokas-Lenells equation [84], in systems with self-steepening effect [85, 86] and even in the case of the complex KdV equation [87, 88]. Patterns of multi-rogue wave solutions similar to those in **Figure 5** have been found for the complex modified KdV equation [89], Kundu-Eckhaus equation [90], coupled nonlinear Schrödinger–Boussinesq equations [91], generalized derivative nonlinear Schrödinger equations [92, 93], generalized Landau-Lifshitz equation [48], Manakov equations [94], the three-wave resonant interaction equations [95, 96], discrete Ablowitz-Ladik equations [97] and even in the case of breather collisions [98, 99]. Most recent work [100] shows that the ideas of rogue waves now enter the field of elementary particles. Namely, patterns similar to those in **Figure 5** may appear in the case of coupled Higgs field equations describing nucleons interacting with neutral scalar mesons [100]. Further developments of the rogue wave theory along these lines may boost our vision of the complex world and how it is build out of simple fundamental particles.

AUTHOR CONTRIBUTIONS

The author confirms being the sole contributor of this work and has approved it for publication.

REFERENCES

1. Hasegawa A, Tappert F. Transmission of stationary nonlinear optical pulses in dispersive dielectric fibers. I. Anomalous dispersion. *Appl Phys Lett* (1973) 23: 142–4. doi:10.1063/1.1654836
2. Agrawal GP. *Nonlinear fiber optics* 4th Edn. NY: Academic Press (2007)
3. Zakharov VE. Stability of periodic waves of finite amplitude on a surface of deep fluid. *J Appl Mech Tech Phys* (1968) 9:190–4. doi:10.1007/BF00913182
4. Osborne A. *Nonlinear ocean waves and the inverse scattering transform* Academic Press (2010).
5. Moslem WM, Shukla PK, Eliasson B. Surface plasma rogue waves. *Europhys Lett* (2011) 96:25002. doi:10.1209/0295-5075/96/25002
6. El-Tantawy SA. Rogue waves in electronegative space plasmas: the link between the family of the KdV equations and the nonlinear Schrödinger equation. *Astrophys Space Sci* (2016) 361:164. doi:10.1007/s10509-016-2754-8
7. Stenflo L, Marklund M. Rogue waves in the atmosphere. *J Plasma Phys* (2010) 76:293. doi:10.1017/S0022377809990481
8. Gross EP. Structure of a quantized vortex in boson systems. *Nuovo Cimento* (1961) 20:454. doi:10.1007/BF02731494
9. Pitaevskii LP. Vortex lines in an imperfect Bose gas. *Zh Eksp Teor Fiz* (1961) 40:646.
10. Lidsey JE. Scalar field cosmologies hidden within the nonlinear Schrödinger equation (2013). Available at: https://arxiv.org/abs/1309.7181 (Accessed September 27 2013).
11. Bespalov VI, Talanov VI. Filamentary structure of light beams in nonlinear liquids. *Sov Phys JETP Lett* (1966) 3:307.
12. Benjamin TB, Feir J. The disintegration of wave trains on deep water: Part 1. *Theory J Fluid Mech* (1967) 27:417. doi:10.1017/S002211206700045X
13. Zakharov VE, Shabat AB. Exact theory of two-dimensional self-focusing and one-dimensional self-modulation of waves in nonlinear media. *J Exp Theor Phys* (1972) 34:62–9.
14. Akhmediev N, Eleonskii VM, Kulagin NE. Generation of periodic trains of picosecond pulses in an optical fiber: exact solutions. *Sov Phys JETP* (1985) 62(5):894–9.
15. Akhmediev N, Ankiewicz A, Taki M. Waves that appear from nowhere and disappear without a trace. *Phys Lett* (2009) 373:675–8. doi:10.1016/j.physleta.2008.12.036
16. Shrira VI, Geogjaev VV. What makes the Peregrine soliton so special as a prototype of freak waves? *J Eng Math* (2009) 67:11–22. doi:10.1007/s10665-009-9347-2
17. Conforti M, Mussot A, Kudlinski A, Trillo S, Akhmediev N. Doubly periodic solutions of the focusing nonlinear Schrödinger equation: recurrence, period doubling, and amplification outside the conventional modulation-instability band. *Phys Rev* (2020) 101:023843. doi:10.1103/PhysRevA.101.023843

18. Matveev VB, Salle MA. *Darboux transformations and solitons* Berlin-Heidelberg: Springer (1991).
19. Its OR, Kotlyarov V. Explicit formulas for solutions of the Schrödinger nonlinear equation. *Doklady Akad. Nauk Ukrainian SSR* (1976) 10:965–8.
20. Its AR, Rybin AV, Sall MA. Exact integration of nonlinear Schrodinger equation[original in Russian: teor. *Theor Math Phys* (1988) 69:29–32.
21. Akhmediev N, Korneev VI. Modulation instability and periodic solutions of nonlinear Schrödinger equation. *Theor Math Phys* (1986) 69:189–94.
22. Akhmediev N, Eleonskii VM, Kulagin NE. Exact first-order solutions of the nonlinear Schrödinger equation. *Theor Math Phys* (1987) 72:809.
23. Yang G, Wu FO, Lopez Aviles HE, Christodoulides D. Optical amplification and transmission of attenuated multi-soliton based on spectral characteristics of Akhmediev breather. *Optic Commun* (2020) 473:125899.
24. Andral U, Kibler B, Dudley JM, Finot C. Akhmediev breather signatures from dispersive propagation of a periodically phase-modulated continuous. *wave Motion* (2020) 95:102545.
25. Akhmediev N, Korneev VI, Mitskevich NV. N-modulation signals in a single-mode optical waveguide under nonlinear conditions. *Zh Exp Teor Fiz* (1988) 94:159–70.
26. Peregrine DH. Water waves, nonlinear Schrödinger equations and their solutions. *J Australian Math Soc Ser B* (1983) 25:16.
27. Kibler B, Fatome J, Finot C, Millot G, Dias F, Genty G, et al. The Peregrine soliton in nonlinear fibre optics. *Nat Phys* (2010) 6:790–5.
28. Gaillard P, Gastineau M. The Peregrine breather of order nine and its deformations with sixteen parameters solutions to the NLS equation. *Phys Lett* (2015) 379:1309–13.
29. Ablowitz MJ, Segur H. *Solitons and the inverse scattering transform* Philadelphia, Pa, USA: SIAM (1981).
30. Akhmediev N, Mitskevich NV. Extremely high degree of N-soliton pulse compression in an optical fiber. *IEEE J Quantum Electron QE* (1991) 27: 849–57.
31. Akhmediev N, Soto-Crespo JM, Ankiewicz A. Extreme waves that appear from nowhere: on the nature of rogue waves. *Phys Lett* (2009) 373:2137–45. doi:10.1016/j.physleta.2009.04.023
32. Dubard P, Gaillard P, Klein C, Matveev VB. On multi-rogue wave solutions of the NLS equation and positon solutions of the KdV equation. *Eur Phys J Spec Top* (2010) 185:247–58. doi:10.1140/epjst/e2010-01252-9
33. Ohta Y, Yang J. General high-order rogue waves and their dynamics in the nonlinear Schrödinger equation. *Proc R Soc A* (2012) 468:1716–40.
34. Gaillard P. Degenerate determinant representation of solutions of the nonlinear Schrödinger equation, higher order Peregrine breathers and multi-rogue waves. *J Math Phys* (2013) 54:013504. doi:10.1063/1.4773096
35. Ankiewicz A, Akhmediev N. Multi-rogue waves and triangular numbers. *Rom Rep Phys* (2017) 69:104.
36. Kedziora DJ, Ankiewicz A, Akhmediev N. Triangular rogue wave cascades. *Phys Rev E Stat Nonlinear Soft Matter Phys* (2012) 86:056602. doi:10.1103/PhysRevE.86.056602
37. Kedziora DJ, Ankiewicz A, Akhmediev N. Circular rogue wave clusters. *Phys Rev E Stat Nonlinear Soft Matter Phys* (2011) 84:056611. doi:10.1103/PhysRevE.84.056611
38. Satsuma J, Yajima N. Initial value problems of one-dimensional self-modulation of nonlinear waves in dispersive media. *Prog Theor Phys Suppl* (1974) 55:284.
39. Kedziora DJ, Ankiewicz A, Akhmediev N. Rogue waves and solitons on a cnoidal background. *Eur Phys J Spec Top* (2014) 223:43–62. 10.1140/epjst/e2014-02083-4.
40. Chen J, Pelinovsky DE, White RE. Rogue waves on the double-periodic background in the focusing nonlinear Schrödinger equation. *Phys Rev E* (2019) 100:052219. doi:10.1103/PhysRevE.100.052219
41. Zabusky NJ, Kruskal MD. Interaction of "solitons" in a collisionless plasma and the recurrence of initial states. *Phys Rev Lett* (1965) 15:240–3.
42. Ankiewicz A, Soto-Crespo JM, Akhmediev N. Rogue waves and rational solutions of the Hirota equation. *Phys Rev E Stat Nonlinear Soft Matter Phys* (2010) 81:046602. doi:10.1103/PhysRevE.81.046602
43. Bandelow U, Akhmediev N. Persistence of rogue waves in extended nonlinear Schrödinger equations: integrable Sasa-Satsuma case. *Phys Lett* (2012) 376: 1558–61. doi:10.1016/J.PHYSLETA.2012.03.032
44. Ohta Y, Yang J. Rogue waves in the Davey-Stewartson I equation. *Phys Rev E Stat Nonlinear Soft Matter Phys* (2012) 86:036604. doi:10.1103/PhysRevE.86.036604
45. Ohta Y, Yang JK. Dynamics of rogue waves in the Davey – stewartson II equation. *J Phys Math Theor* (2013) 46:105202. doi:10.1088/1751-8113/46/10/105202
46. Rao J, Cheng Y, He J. Rational and semirational solutions of the nonlocal Davey–Stewartson equations. *Stud Appl Math* (2017) 139:568–98. doi:10.1111/sapm.12178
47. Li R, Geng X. Rogue periodic waves of the sine—Gordon equation. *Appl Math Lett* (2020) 102:106147. doi:10.1016/j.aml.2019.106147
48. Li R, Geng X, Xue B. A generalization of the Landau-Lifschitz equation: breathers and rogue waves. *J Nonlinear Math Phys* (2020) 27:279–94. doi:10.1080/14029251.2020.1700636
49. Ankiewicz A, Devine N, Akhmediev N. Are rogue waves robust against perturbations? *Phys Lett* (2009) 373:3997–4000. doi:10.1016/j.physleta.2009.08.053
50. Ankiewicz A, Soto-Crespo JM, Chowdhury MA, Akhmediev N. Rogue waves in optical fibers in presence of third-order dispersion, self-steepening, and self-frequency shift. *J Opt Soc Am B* (2013) 30:87–94. doi:10.1364/JOSAB.30.000087
51. Zaviyalov A, Egorov O, Iliew R, Lederer F. Rogue waves in mode-locked fibre lasers. *Phys Rev* (2012) 85:013828. doi:10.1103/PhysRevA.85.013828
52. Soto-Crespo JM, Grelu P, Akhmediev N. Dissipative rogue waves: extreme pulses generated by passively mode-locked lasers. *Phys Rev E Stat Nonlinear Soft Matter Phys* (2011) 84:016604. doi:10.1103/PhysRevE.84.016604
53. Lynch P. The origins of computer weather prediction and climate modelling. *J Comput Phys* (2008) 227:3431–44. doi:10.1016/j.jcp.2007.02.034
54. Krishnamurti TN. Numerical weather prediction. *Annu Rev Fluid Mech* (1995) 27:195–225. doi:10.1146/annurev.fl.27.010195.001211
55. Chabchoub A, Hoffmann NP, Akhmediev N. Rogue wave observation in a water wave tank. *Phys Rev Lett* (2011) 106:204502. doi:10.1103/PhysRevLett.106.204502
56. Bertola M, Tovbis A. Universality for the focusing nonlinear Schrödinger equation at the gradient catastrophe point: rational breathers and poles of the tritronqué solution to Painlevé I, Commun. *Pure Appl Math* (2012) 66:678. doi:10.1002/cpa.21445
57. Tikan A, Billet C, El G, Tovbis A, Bertola M, Sylvestre T, et al. Universal Peregrine soliton structure in nonlinear pulse compression in optical fiber. *Phys Rev Lett* (2017) 119:033901. doi:10.1103/PhysRevLett.119.033901
58. Chabchoub A, Hoffmann NP, Akhmediev N. Observation of rogue wave holes in a water wave tank. *J Geophys Res* (2012) 117:C00J02. doi:10.1029/2011JC007636
59. Chen S, Grelu P, Soto-Crespo JM. Dark- and bright-rogue-wave solutions for media with long-wave-short-wave resonance. *Phys Rev E Stat Nonlinear Soft Matter Phys* (2014) 89:011201. doi:10.1103/PhysRevE.89.011201
60. Baronio F, Degasperis A, Conforti M, Wabnitz S. Solutions of the vector nonlinear Schrödinger equations: evidence for deterministic rogue waves. *Phys Rev Lett* (2012) 109:044102. doi:10.1103/PhysRevLett.109.044102
61. Baronio F. Akhmediev breathers and Peregrine solitary waves in a quadratic medium. *Opt Lett* (2017) 42:1756. doi:10.1364/OL.42.001756
62. Baronio F, Conforti M, Degasperis A, Lombardo S. Rogue waves emerging from the resonant interaction of three waves. *Phys Rev Lett* (2013) 111:114101. doi:10.1103/PhysRevLett.111.114101
63. Gaillard P. Other 2N-2 parameters solutions of the NLS equation and 2N+1 highest amplitude of the modulus of the N-th order Akhmediev-Peregrine breather. *J Phys Math Theor* (2015) 48:145203. doi:10.1088/1751-8113/48/14/145203
64. Dontsop PYG, Essama BGO, Doing JM, Dedzo MM, Atangana J, Yemele D, et al. Akhmediev-Peregrine rogue waves generation in a composite right/left-handed transmission line. *Opt Quant Electron* (2016) 48:59. doi:10.1007/s11082-015-0333-9
65. Chabchoub A, Fink M. Time-reversal generation of rogue waves. *Phys Rev Lett* (2014) 112:124101. doi:10.1103/PhysRevLett.112.124101
66. Vitanov NK, Chabchoub A, Hoffmann N. Deep-water waves: on the nonlinear Schrödinger equation and its solutions. *J Theor Appl Mech Sofia* (2013) 43: 43–54. doi:10.2478/jtam-2013-0013
67. Akhmediev N, Ankiewicz A. *Solitons, nonlinear pulses and beams* London: Chapman & Hall (1997).

68. Ankiewicz A, Clarkson P, Akhmediev N. Rogue waves, rational solutions, the patterns of their zeros and integral relations. *J Phys Math Theor* (2010) 43:122002.
69. Wang L, Yang C, Wang J, He J. The height of an n-th order fundamental rogue wave for the nonlinear Schrödinger equation. *Phys Lett* (2017) 381:1714–8. doi:10.1016/j.physleta.2017.03.023
70. Chin SA, Ashour OA, Nikolić SN, Belić MR. Maximal intensity higher-order Akhmediev breathers of the nonlinear Schrödinger equation and their systematic generation. *Phys Lett* (2016) 380:3625–9. doi:10.1016/j.physleta.2016.08.038
71. Ling L, Zhao LC. Simple determinant representation for rogue waves of the nonlinear Schrödinger equation. *Phys Rev E Stat Nonlinear Soft Matter Phys* (2013) 88:043201. doi:10.1103/PhysRevE.88.043201
72. Kedziora DJ, Ankiewicz A, Akhmediev N. Classifying the hierarchy of nonlinear-Schrödinger-equation rogue-wave solutions. *Phys Rev E - Stat Nonlinear Soft Matter Phys* (2013) 88:013207. doi:10.1103/PhysRevE.88.013207
73. Ankiewicz A, Kedziora DJ, Akhmediev N. Rogue wave triplets. *Phys Lett* (2011) 375:2782–5. doi:10.1016/j.physleta.2011.05.047
74. Gaillard P. Deformations of third-order Peregrine breather solutions of the nonlinear Schrödinger equation with four parameters. *Phys Rev E Stat Nonlinear Soft Matter Phys* (2013) 88:042903. doi:10.1103/PhysRevE.88.042903
75. Gaillard P. Degenerate determinant representation of solutions of the nonlinear Schrödinger equation, higher order Peregrine breathers and multi-rogue waves. *J Math Phys* (2013) 54:013504. doi:10.1063/1.4773096
76. Akhmediev N, Ankiewicz A, Soto-Crespo JM. Rogue waves and rational solutions of the nonlinear Schrödinger equation. *Phys Rev E - Stat Nonlinear Soft Matter Phys* (2009) 80:026601. doi:10.1103/PhysRevE.80.026601
77. Bailung H, Sharma SK, Nakamura Y. Observation of Peregrine solitons in a multicomponent plasma with negative ions. *Phys Rev Lett* (2011) 107:255005. doi:10.1103/PhysRevLett.107.255005
78. Bludov YV, Konotop VV, Akhmediev N. Matter rogue waves. *Phys Rev* (2009) 80:033610. doi:10.1103/PhysRevA.80.033610
79. Nguyen JHV, Luo D, Hulet RG. Formation of matter-wave soliton trains by modulational instability. *Science* (2017) 356:422–6. doi:10.1126/science.aal3220
80. Chabchoub A, Akhmediev N. Observation of rogue wave triplets in water waves. *Phys Lett* (2013) 377:2590–3. doi:10.1016/j.physleta.2013.07.027
81. Chabchoub A, Hoffmann N, Onorato M, Slunyaev A, Sergeeva A, Pelinovsky E, et al. Observation of a hierarchy of up to fifth-order rogue waves in a water tank. *Phys Rev E Stat Nonlinear Soft Matter Phys* (2012) 86:056601. doi:10.1103/PhysRevE.86.056601
82. Li C, He J, Porsezian K, Porseizan K. Rogue waves of the Hirota and the Maxwell-Bloch equations. *Phys Rev E Stat Nonlinear Soft Matter Phys* (2013) 87:012913. doi:10.1103/PhysRevE.87.012913
83. Gui M, Qin Z, Grimshaw R, Akhmediev N. Intricate dynamics of rogue waves governed by the Sasa-Satsuma equation. *Physica D* (2020) 402:132252. doi:10.1016/j.physd.2019.132252
84. He J, Xu S, Porsezian K. Rogue waves of the Fokas-Lenells equation. *J. Phys. Soc. Japan* (2012) 81:124007. doi:10.1143/JPSJ.81.124007
85. Liu C, Akhmediev N. Super-regular breathers in nonlinear systems with self-steepening effect. *Phys Rev E* (2019) 100:062201. doi:10.1103/PhysRevE.100.062201
86. Chen S, Pan C, Grelu P, Baronio F, Akhmediev N. Fundamental Peregrine solitons of ultrastrong amplitude enhancement through self-steepening in vector nonlinear systems. *Phys Rev Lett* (2020) 124:113901. doi:10.1103/PhysRevLett.124.113901
87. Ankiewicz A, Bokaeeyan M, Akhmediev N. Shallow-water rogue waves: an approach based on complex solutions of the Korteweg–de Vries equation. *Phys Rev E* (2019) 99:050201. doi:10.1103/PhysRevE.99.050201
88. Crabb M, Akhmediev N. Rogue wave multiplets in the complex KdV equation. *Rom Rep Phys* (2020) 72:118.
89. He J, Wang L, Li L, Porsezian K, Erdélyi R. Few-cycle optical rogue waves: complex modified Korteweg-de Vries equation. *Phys Rev E Stat Nonlinear Soft Matter Phys* (2014) 89:062917. doi:10.1103/PhysRevE.89.062917
90. Qiu D, He J, Zhang Y, Porsezian K. The Darboux transformation of the Kundu-Eckhaus equation. *Proc Royal Soc A Math Phys Eng Sci* (2015) 471:20150236. doi:10.1098/rspa.2015.0236
91. ZhangChen Y. General high-order rogue waves to nonlinear Schrödinger–Boussinesq equation with the dynamical analysis. *Nonlinear Dynam* (2018) 93:2169–84. doi:10.1007/s11071-018-4317-8
92. Yang B, Chen J, Yang J. Rogue waves in the generalised derivative nonlinear Schrödinger equations. *J. Nonlin. Science* (2020) 30:3027–56.
93. Yang B, Yang J. *Universal patterns of rogue waves* Available at: https://arxiv.org/abs/2009.06060 (Accessed September 13 2020).
94. Chen S, Mihalache D. Vector rogue waves in the Manakov system: diversity and compossibility. *J Phys Math Theor* (2015) 48:215202. doi:10.1088/1751-8113/48/21/215202
95. Chen S, Soto-Crespo JM, Grelu P. Watch-hand-like optical rogue waves in three-wave interactions. *Optic Express* (2015) 23:349–59. doi:10.1364/OE.23.000349
96. Wang X, Cao J, Chen Y. Higher-order rogue wave solutions of the three-wave resonant interaction equation via the generalized Darboux transformation. *Phys Scripta* (2015) 90:105201. doi:10.1088/0031-8949/90/10/105201
97. Ohta Y, Yang J. General rogue waves in the focusing and defocusing Ablowitz—Ladik equations. *J Phys Math Theor* (2014) 47:255201. doi:10.1088/1751-8113/47/25/255201
98. Kedziora DJ, Ankiewicz A, Akhmediev N. Second-order nonlinear Schrödinger equation breather solutions in the degenerate and rogue wave limits. *Phys Rev E Stat Nonlinear Soft Matter Phys* (2012) 85:066601. doi:10.1103/PhysRevE.85.066601
99. Wang L, He J, Xu H, Wang J, Porsezian K. Generation of higher-order rogue waves from multibreathers by double degeneracy in an optical fiber. *Phys Rev E* (2017) 95:042217. doi:10.1103/PhysRevE.95.042217
100. Zhaqilao. Dynamics of localized wave solutions for the coupled Higgs field equation. *Nonlinear Dynam* (2020) 101:1181–98.

8

Numerical Analysis of Turbulent Flow and Heat Transfer in Internally Finned Tubes

*Zhanwei Liu [1], Yanwei Yue [2], Luchao She [1] and Guangming Fan [1]**

[1] *Fundamental Science on Nuclear and Simulation Technology Laboratory, Harbin Engineering University, Harbin, China,*
[2] *Research and Development Center, China Nuclear Power Engineering Co., Ltd, Beijing, China*

Correspondence:
Guangming Fan
fanguangming007@hotmail.com

In this study, single-phase heat transfer enhancement in internally finned tubes is investigated numerically. The influence of fin number, helix angle, fin height, fin width, and shape on the flow and heat transfer characteristics is studied. The research results indicate that the resistance coefficient and Nusselt number both increase with the increment of these parameters, among which the helix angle has the largest impact on the heat transfer enhancement. In addition, the shape of fins also has a small effect on the flow and heat transfer, and the heat transfer effect of triangular fins is the best.

Keywords: internally finned tube, heat transfer enhancement, numerical simulation, single -phase flow, heat transfer coefficient

INTRODUCTION

Single-phase convection heat transfer enhancement techniques are widely applied in industries such as petroleum, chemical engineering, etc. In recent years, its applications in the field of nuclear technology has been an engaging research area of heat transfer augmentation. In the advanced nuclear power plant like AP1000, HPR1000 and CAP1400, passive residual heat removal system (PRHR) (Schulz, 2006; Xing et al., 2016; Zheng et al., 2016) can discharge residual heat through single-phase convection heat transfer under normal condition, therefore enhancement of its heat transfer efficiency is of great significance to the safe operation of nuclear reactors.

In order to strengthen the single-phase convection heat transfer technology, many heat exchanger elements have been developed, the internally finned tube is one of them. Ji et al. (2015) found that their thermal-hydraulic performance is the best among all enhanced tubes through a comprehensive literature review. Much work has been done to investigate the heat transfer characteristics of internally finned tubes in condensation (Seo and Kim, 2000; Kim et al., 2002; Yu et al., 2002; Yun et al., 2002) and evaporation (Chamra and Webb, 1996; Chamra et al., 2004, 2005; Chamra and Mago, 2006), but the heat transfer characteristics of single-phase in these tubes are not thoroughly understood for lack of research. The internally finned tube has high requirements for water quality, and the working fluid is required to be clean enough with almost no impurities, otherwise they may affect the heat transfer performance. Compared with other heat exchanger elements, the tremendous advantage of internally finned tubes is that its heat transfer effect is greater than the increase of pressure drop. Generally, this effect can be seen in turbulent state, but neither the enhancement of heat transfer nor that of pressure drop is obvious in laminar state, thus they are recommended to be used in turbulent flow (Al-Fahed et al., 1999; Copetti et al., 2004).

Previous experimental studies have been carried out on single-phase flow of different working media to investigate the influence of different geometrical sizes (fin height, fin width, helix angle, diameter, fin number) on the heat transfer characteristics. Wang et al. (1996) experimentally tested seven internally finned tubes with different geometric sizes such as diameter, fin height, helix angle and fin numbers, and established heat transfer and pressure drop correlations in a wide range of Reynold number. Han and Lee (2005) studied the single-phase heat transfer and flow characteristics of water in four internally finned tubes with different diameters, fin heights, helix angles and fin pitchs experimentally, and developed correlations for friction factor and heat transfer coefficients. Jensen and Vlakancic (1999) carried out experimental studies on turbulent flow in eight internally finned tubes with different geometrical parameters of helix angles, fin heights, fin numbers and fin widths, they obtained the correlations between friction coefficient and Nusselt number. Later Zdaniuk et al. (2008) determined and compared friction factors as well as heat transfer coefficients of eight internally finned tubes experimentally. Empirical correlations for one particular internally finned tubes

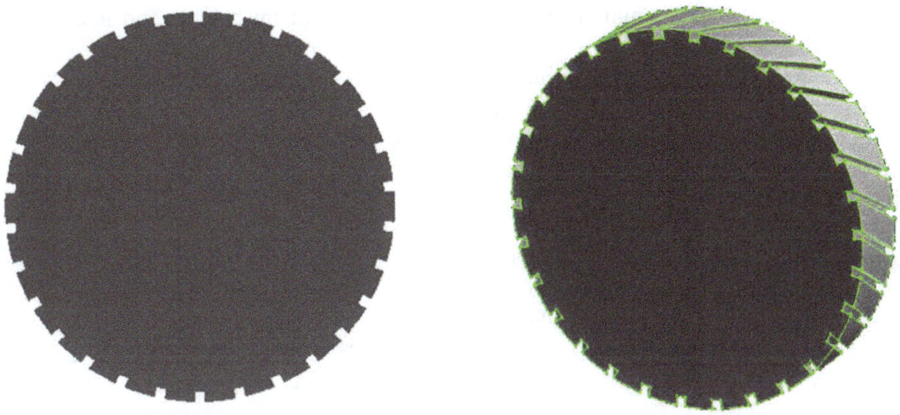

FIGURE 1 | Schematic diagram of cross section of helix internal finned tube.

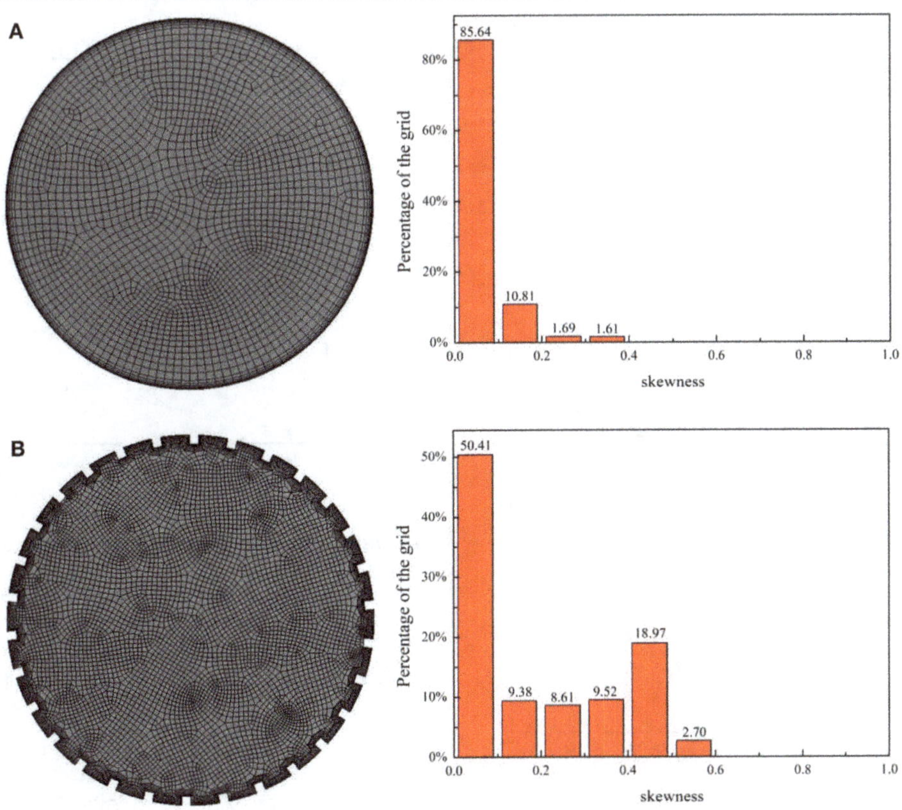

FIGURE 2 | Grid distribution and volume grid quality of heat transfer tube. **(A)** Smooth tube. **(B)** Internally finned tube.

was developed by Siddique and Alhazmy (2008) through experimental research. And a critical Reynolds number Re_{cr} was discovered by Li et al. (2007) through experimental study. Ji et al. (2011) measured the turbulent heat transfer and friction factors of sixteen internally finned tubes with various geometrical parameters including fin number, fin height, diameter, helix angle, and developed an equation for the heat transfer in the fully developed flow as an extension of the Gnielinski equation. Friction formula was also developed by Celen et al. (2013). Besides, Experimental studies of R134, R12 (Eiamsa-ard and Wongcharee, 2012) and the water-CuO nanofluid (Eiamsa-ard and Wongcharee, 2013) were also conducted.

Due to the limitations of manufacturing and experimental conditions, only a few tubes with specific geometrical parameters have been studied in previous experiment research, so it is difficult to obtain the detailed and comprehensive heat transfer characteristics in different tubes.

There are only a few numerical studies on the heat transfer characteristics of internally finned tubes. Agra et al. (2011) studied the heat transfer characteristics and pressure drop of five internally finned tubes in the turbulent state numerically, they compared their simulation results with the experimental data of Zdaniuk et al. (2008), and discovered that CFD predicted the experimental data more accurately than the Blasius equation. Their results also indicated that the internally finned tube has a higher heat transfer coefficient than the corrugated tube. Celen et al. (2014) studied the pressure drop in one internally finned tube with TiO_2-water nanofluids numerically, and described the average and local values of temperature, pressure, and velocity distributions of the tube. Dastmalchi et al. (2017) investigated the characteristics of laminar oil flow in internally finned tubes with different fin heights and helical angles numerically. They

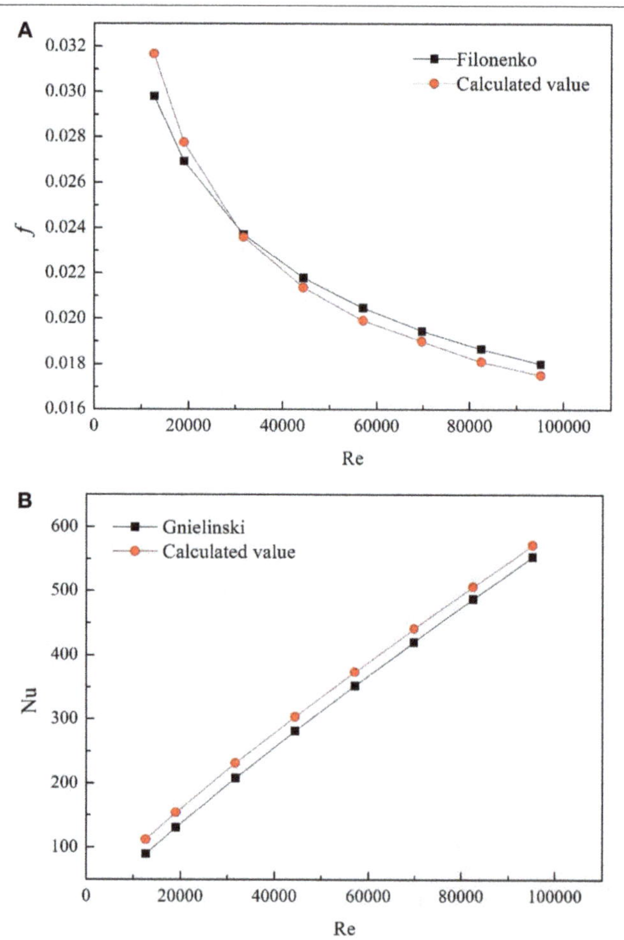

FIGURE 3 | Comparison between the calculated value and the theoretical value in smooth tubes. **(A)** f. **(B)** Nu.

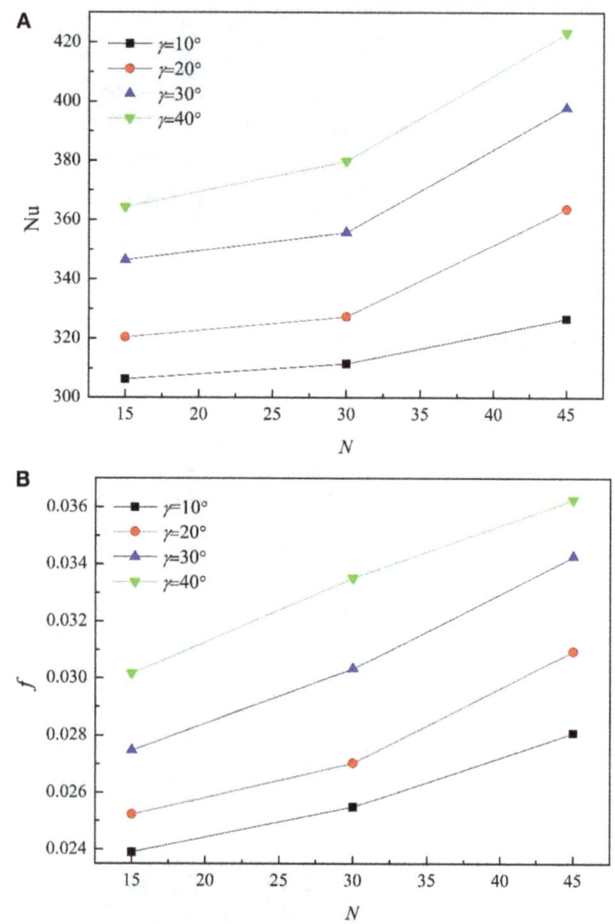

FIGURE 4 | The effect of the fin number on the Nu and f at different helix angles. **(A)** Nu. **(B)** f.

TABLE 1 | Increment of heat exchange area under different fin number and helix angle.

δ	$\gamma = 10°$	$\gamma = 20°$	$\gamma = 30°$	$\gamma = 40°$
$N = 15$	0.2407	0.2517	0.2722	0.3065
$N = 30$	0.4814	0.5034	0.5444	0.6130
$N = 45$	0.7221	0.7551	0.8166	0.9195

concluded that there is an optimal height and an optimal helix angle which leads to the maximum performance of heat transfer.

The foregoing literatures indicate that a comprehensive study on the flow and heat transfer characteristics considering a wide range of parameters including fin number, helix angle, fin height, fin width and shape of fins remains to be conducted. Influence of geometric parameters on characteristics of heat transfer could be implemented with modern visualization techniques and computational fluid dynamics (CFD) tools. Therefore, the purpose of the study is to conduct a comprehensive numerical study on the factors affecting the heat transfer efficiency of the internally finned tubes.

NUMERICAL METHOD AND PROCEDURE

Geometric Model

The fins are uniformly distributed in circumferential direction and along the axis with the same helix angle, so they are periodic in geometric structure. The rectangular fins studied have the diameter $D = 20$ mm, length $L = 20$ mm, fin number $N = 30$, the fin height $e = 0.5$ mm, the width $s = 0.5$ mm, the helix angle of 30°. The cross section is shown in **Figure 1**.

Using the periodicity of the internally finned tube and taking one pitch of the fully developed segment as the computational domain, the difficulty of grid generation is greatly reduced. The established model is shown in **Figure 1**.

Grid Generation and Boundary Conditions

In order to determine a proper grid for the numerical simulations, a grid independence study is carried out for the smooth and internally finned tubes. A section of circular tube with a diameter of 20 mm and a length of 20 mm is taken as the computing domain, and meshing is used to generate grids.

The mesh number chosen in this paper is the minimum of the mesh number that does not affect the simulation results. When the number of grids is lower than a certain value, the change of the number of grids will have an impact on the simulation results. The larger the number of grids, the closer to the stable result. When the number of grids is higher than a certain value, the encrypted grid will have no impact on the simulation results.

As shown in **Figures 2A,B**, the grid of smooth tube is made by independence verification. Grid quality measurement method is skewness, the value <0.7 is acceptable, so the grid generation is credible as most values are below 0.6.

Realizable k-epsilon model satisfies the constraint conditions of Reynolds stress, so it can be consistent with the real turbulence at Reynolds stress. This is something neither the standard k-epsilon model nor the RNG k-epsilon model can do. The advantage of this feature in the calculation is that it can more accurately simulate the diffusion velocity of plane and circular jets. Meanwhile, in the calculation of rotating flow, boundary layer with directional pressure gradient and separation flow, the calculation results are more in line with the real situation.

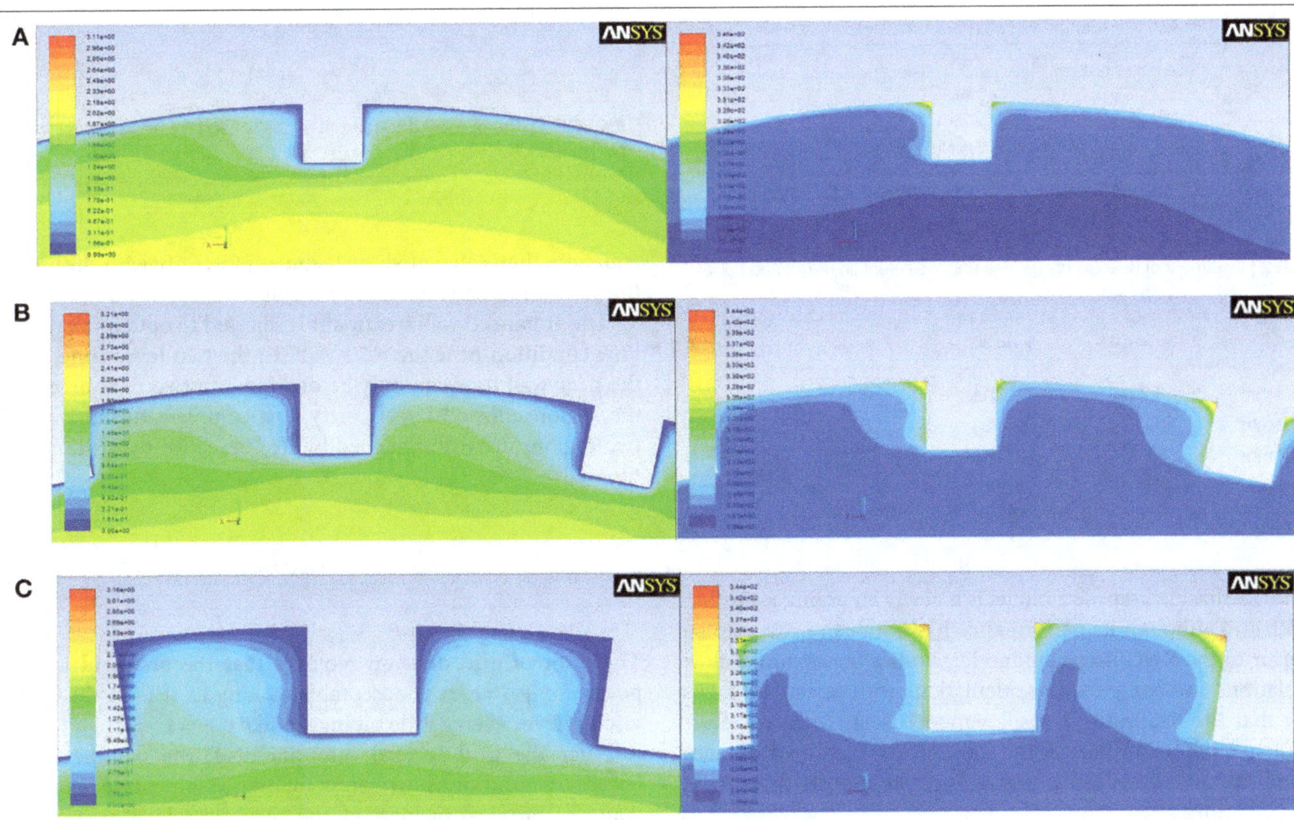

FIGURE 5 | The velocity cloud diagram and temperature cloud diagram with different fin numbers. **(A)** $N = 15$. **(B)** $N = 30$. **(C)** $N = 45$.

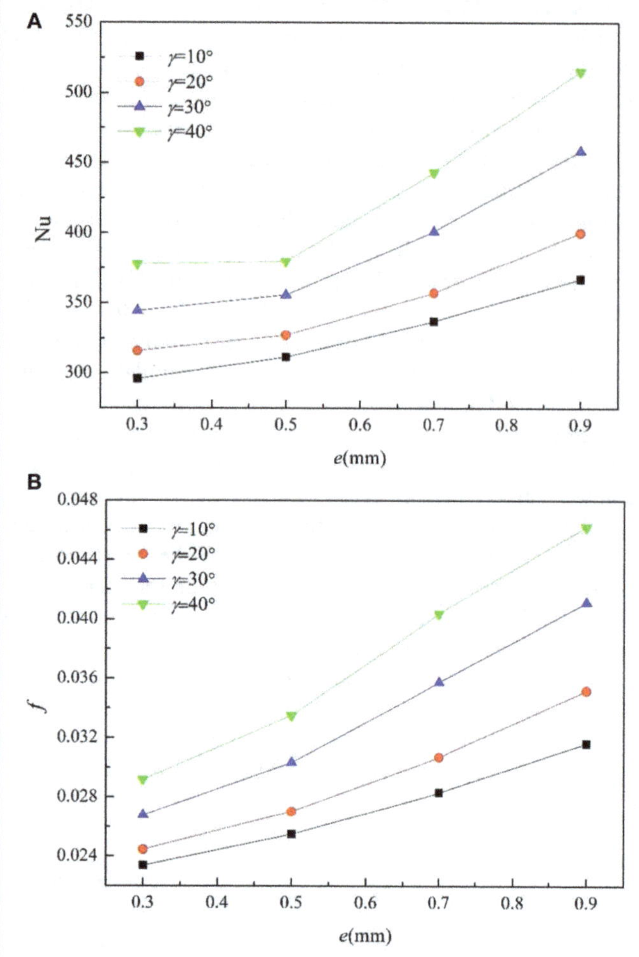

FIGURE 6 | The effect of the fin height on the Nu and f at different helix angles. **(A)** Nu. **(B)** f.

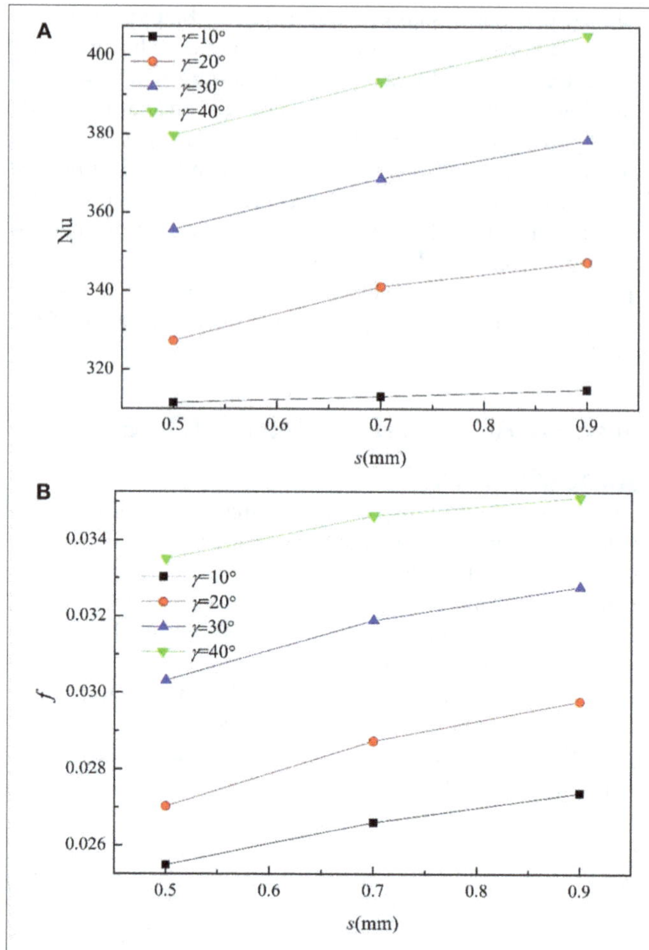

FIGURE 7 | The effect of the fin width on the Nu and f at different helix angles. **(A)** Nu. **(B)** f.

TABLE 2 | Increment of heat exchange area under different fin height and helix angle.

δ	γ = 10°	γ = 20°	γ = 30°	γ = 40°
e = 0.3 mm	0.2877	0.3011	0.3260	0.3677
e = 0.5 mm	0.4814	0.5034	0.5444	0.6130
e = 0.7 mm	0.6750	0.7053	0.7617	0.8561
e = 0.9 mm	0.8684	0.9067	0.9779	1.0973

The Realizable k-epsilon model is a newly emerging k-epsilon model, and although its performance has not been proven to be superior to the RNG k-epsilon model, studies on separated flow calculations and complex flow calculations with secondary flows show that the Realizable k-epsilon model is the most excellent turbulence model among all the models. Given the advantages of the Realizable k-epsilon model, this paper chooses this model for calculation.

We used the standard $k - \varepsilon$ model in fluent to solve the turbulent kinetic energy equation and turbulent dissipation rate equation, then calculated the turbulent viscosity with the value of the sum, and finally obtained the solution.

The enhanced wall treatment is selected to obtain the detailed flow condition near the tube wall. In the two-layer zone model, the near-wall flow can only be divided into two regions, namely the region affected by viscosity and complete turbulence. The two regions are distinguished by Re_y based on the distance y to the wall.

$$Re_y = \frac{\rho\sqrt{k}y}{\mu} \tag{1}$$

The optimal grid division requires that the first grid is at the position y+ =1. It is acceptable as long as it is in the bottom viscous layer even a little bit bigger, like y+ =4~5.

After the grid is read into FLUENT, the parameters and boundary conditions are set as follows: single phase water with constant physical properties, the constant heat flow boundary condition and the given heat flux is $q = 1{,}00{,}000\ \text{W/m}^2$, the countercurrent volume temperature is set as 300 K, the coupling solution of pressure and velocity is SIMPLEC algorithm, and the momentum and energy equations are solved by the second-order upwind scheme.

Verification of the Numerical Procedure

Based on the pressure drop value per unit length obtained, the average f can be calculated using Darcy formula:

$$f = \frac{2\Delta PD}{\rho u^2} \tag{2}$$

Based on the difference between the average wall temperature and the average fluid temperature, the average Nusselt number can be calculated:

$$Nu = \frac{qD}{\Delta T\lambda} \tag{3}$$

For forced convection heat transfer in smooth circular tubes, the resistance coefficient can be calculated using the Filonenko formula (Filonenko, 1954):

$$f_F = (1.82\lg Re - 1.64)^{-2} \tag{4}$$

The nussel number can be calculated according to the Gnielinski (Gnielinski, 1975) formula:

$$Nu = 0.012Pr^{0.4}(Re^{0.87} - 280) \tag{5}$$

The selected grid scheme is used for verification. The comparison results with the classical experimental formula are shown in **Figures 3A,B**. The calculated values of f and Nu are compared with the values of Filonenko formula (Filonenko, 1954) and Gnielinski formula (Gnielinski, 1975), respectively. It can be seen that the calculated value of the resistance coefficient and the Nusselt number are in good agreement with the theoretical value. The maximum error between the calculated value and the theoretical value of the resistance coefficient is 6.25%, most of which are within 3%.The maximum error between the calculated value and the theoretical value of nusserl number is 11.2%, most of which is about 5%. The maximum error occurs in the lower Reynolds number region. The results show that the application of periodic boundary conditions and the established calculation model are correct and can accurately simulate the turbulent flow and heat transfer in the tube.

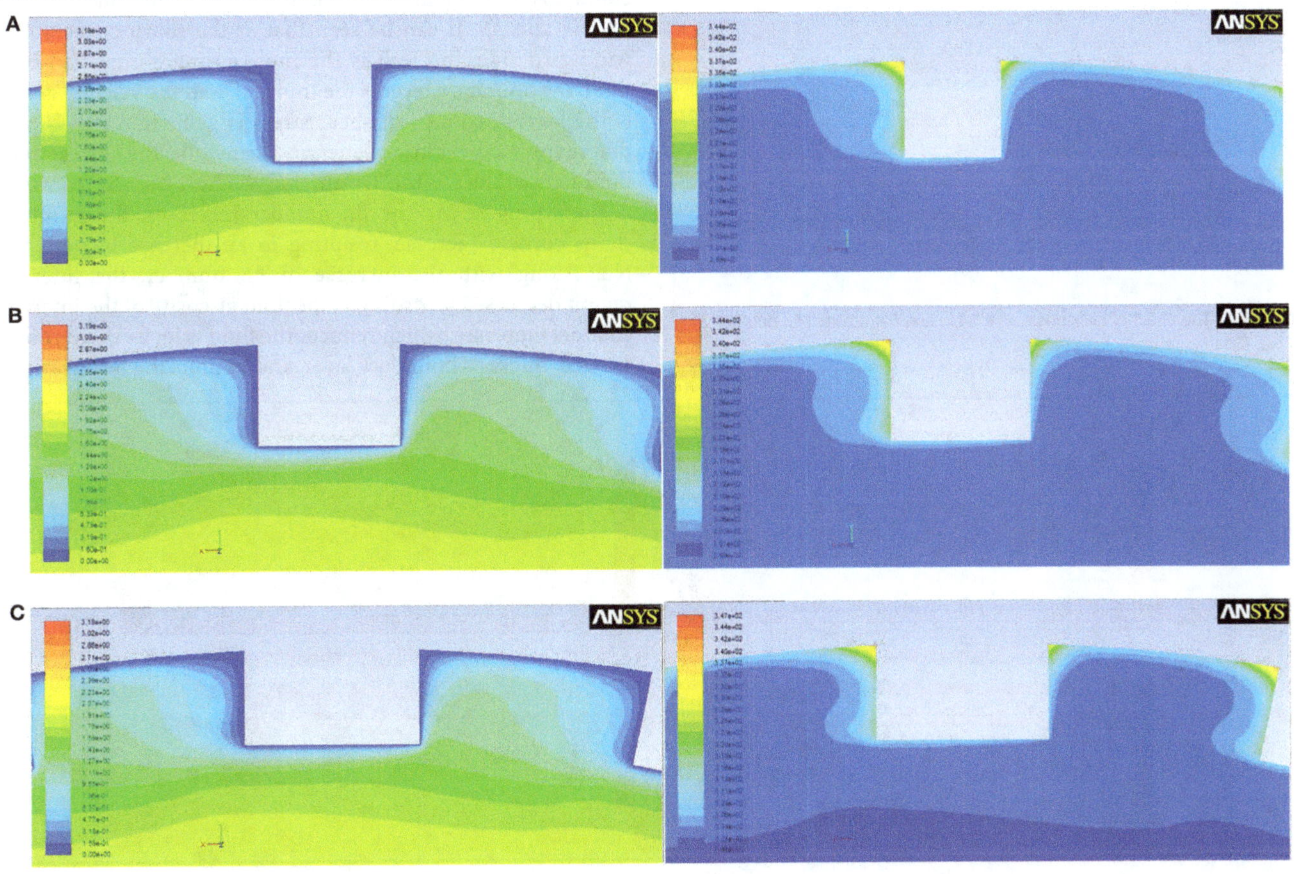

FIGURE 8 | The cloud picture of velocity and temperature with different fin width. **(A)** s = 0.5 mm. **(B)** s = 0.7 mm. **(C)** s = 0.9 mm.

RESULTS AND DISCUSSION

In this section, the effect of fin number, fin height, fin width, helix angle and the shape of fins on the heat transfer characteristics of internally finned tube are studied.

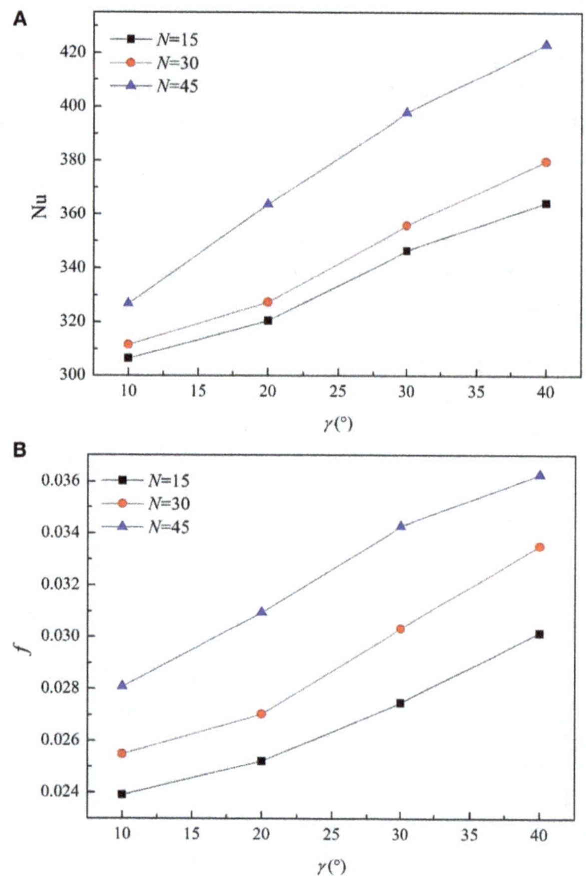

FIGURE 9 | The effect of helix angle on the Nu and f at different fin numbers. **(A)** Nu. **(B)** f.

Effect of Fin Number

Figure 4 shows the effect of the fin number on the Nu and f at different helix angles. The fin height and fin width are fixed at 0.5 mm, and the fin number are 15, 30, and 45, respectively. It can be seen that the Nu and f both increase with the augment of the fin number and helix angle.

Table 1 shows the variation of the increment of heat exchange area with fin number and the helix angle. Here the increment of heat exchange area is defined as:

$$\delta = \frac{A_{fin} - A_{smooth}}{A_{smooth}} \tag{6}$$

$$A_{smooth} = 2\pi DL \tag{7}$$

$$A_{fin} = \frac{(2\pi D + 2Ne)\,L}{\cos\gamma} \tag{8}$$

It can be seen that the rate at which Nu increases is bigger than the increase speed of heat exchange area and f, which indicates that the enhancement effect of heat transfer is greater than the increase of resistance and heat transfer area.

Figure 5 shows the velocity cloud diagram and temperature cloud diagram of internally finned tubes with fin number of 15, 30, and 45. It can be seen that at the sharp corners of the windward side, the wall is the most strongly impacted by the fluid, the boundary layer is the thinnest, and the velocity gradient is the largest, so a great shear stress is generated which means that there is a great loss of energy. Meanwhile, the corresponding fluid temperature is the lowest, and the heat transfer convection is the strongest. As the fin number increases, the number of sharp corners increase, resulting in an increase in both f and Nu. Along with the increase of fin number, the intercostal region decreases accordingly, the fluid viscosity of the intercostal channel increases, which reduces the fluid velocity of fin root and surface, thickens boundary layer, and inhibits the heat transfer.

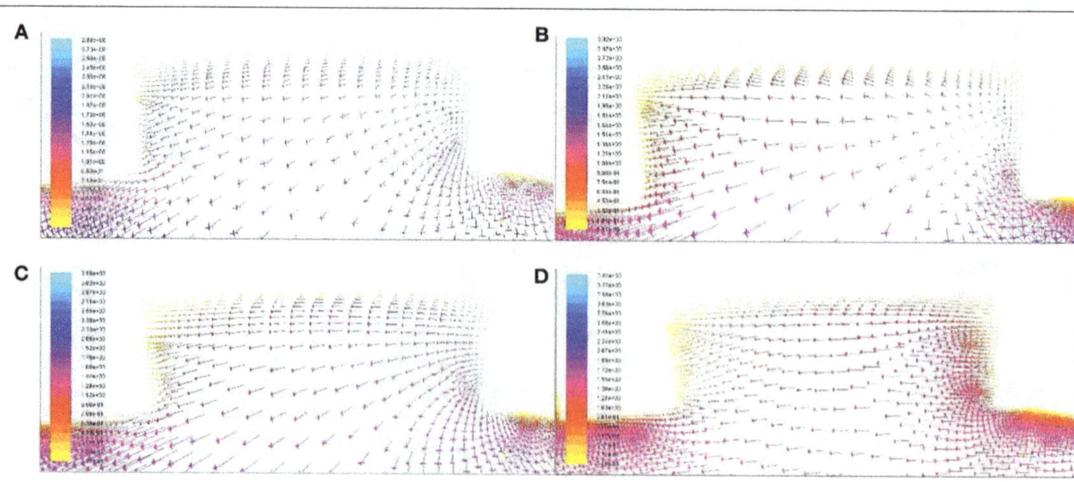

FIGURE 10 | The velocity vector of intercostal area with different helix angle. **(A)** 10°. **(B)** 20°. **(C)** 30°. **(D)** 40°.

Effect of Fin Height

Figure 6 shows the changes of Nu and f with the fin height under different helix angles. The fin number is 30, the fixed fin width is 0.5 mm, and the fin height is 0.3, 0.5, 0.7, and 0.9 mm, respectively (the dimensionless fin height H changes from 0.03 mm to 0.09 mm). It can be seen that the Nu and f both increase with the augment of the fin height, as it not only increases the heat transfer area on both fin sides, but also promotes the interactions between the fin top. When the helix angle is >20°, Nu and f grow significantly compared with small helix angle such as 10 and 20° with the increase of fin height, which indicates that the helix angle plays an important role in the heat transfer.

Table 2 shows the changes of heat exchange area with the fin height and helix angle. Nu and f both increase with fin height, because the higher the fin height, the greater the disturbance to the fluid near the wall surface. Therefore, the heat transfer coefficient is enhanced and the resistance coefficient is increased. Increasing the spiral angle will further enhance the heat transfer and the resistance.

The increment of the fin height increases the interaction between fins and fluid, resulting in rapid increase in f and Nu.

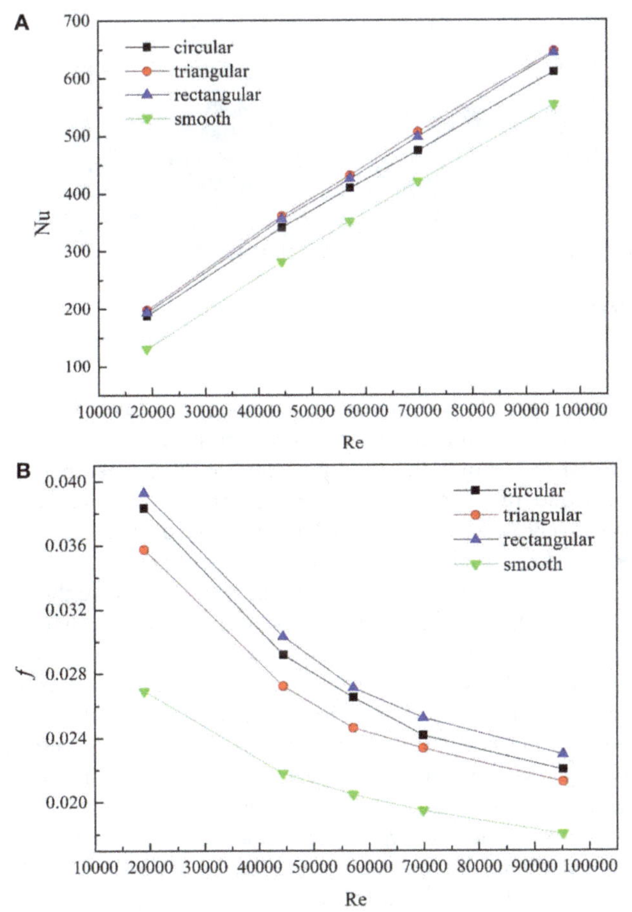

FIGURE 11 | Comparison between three fin shapes. **(A)** Nu. **(B)** f.

Effect of Fin Width

Generally, the fin width has a complex effect on the flow and heat transfer in internally finned tubes. The increment of fin width does not increase the heat transfer area, but it will cause the change of the size of the intercostal passage. Due to the limitations of time and conditions, only the effect of fin width within a small range between 0.5 and 0.9 mm is studied under the condition that the number of fins is 30.

Figure 7 shows the changes of Nu and f with the fin width. As we can see, with the increment of the fin width, Nu and f both increase slowly.

Figure 8 shows the velocity cloud and temperature cloud when the fin width is 0.5, 0.7, and 0.9 mm, respectively. It can be seen that the increase of fin width means that the area of the fin top increases, equivalently the area in contact with the fluid is larger, and fins are more impacted by the fluid, so heat transfer and friction resistance are both enhanced.

Similar to the increase of fin number, as fin width increases, the intercostal region decreases accordingly, the viscous effect of the fin side and the inner wall against the intercostal fluid also increases, which reduces the flow rate in the intercostal region, thickens the boundary layer and suppresses the heat transfer between the fin root and the wall, thus causes the decrease of Nu and f. The opposite effect of the two factors makes Nu and the f changed slowly.

Effect of Helix Angle

Figure 9 shows the changes of f and Nu with the helix angle under different fin numbers to study the effect of helix angle. The fixed fin width is 0.5 mm, and the fixed fin height is 0.5 mm. The reason that we chose the spiral angle of 10 to 40° is that the spiral Angle is usually 10 to 45° in industrial applications, so we chose the most representative parameters for numerical simulation.

As we can see, Nu and f both increase with helix angle dramatically. The bigger the helix angle is, not only the area of the fin exposed to the incoming flow windward increases, but also the change of fluid flow direction increases, that is, the greater the obstruction to the incoming flow. Therefore, the incoming flow has a stronger impact on the windward side of the fin, and f and Nu both increase.

Figure 10 shows the velocity vector of intercostal area. The arrows in the bottom parts of **Figures 10A–C** point to left bottom, while the arrows in the bottom part of **Figure 10D** point to left, this is because the larger the spiral angle is, the more the fluid near the wall is affected by the spiral angle, and the greater the included Angle with the mainstream is. So it tends to the horizontal.

TABLE 3 | Comparison of parameters of three kinds of fin tubes.

	Circular	Triangular	Rectangular
δ	0.4357	0.3565	0.5444
Nu/Nu_S	1.1267	1.2037	1.1834
f/f_S	1.2413	1.2002	1.2972

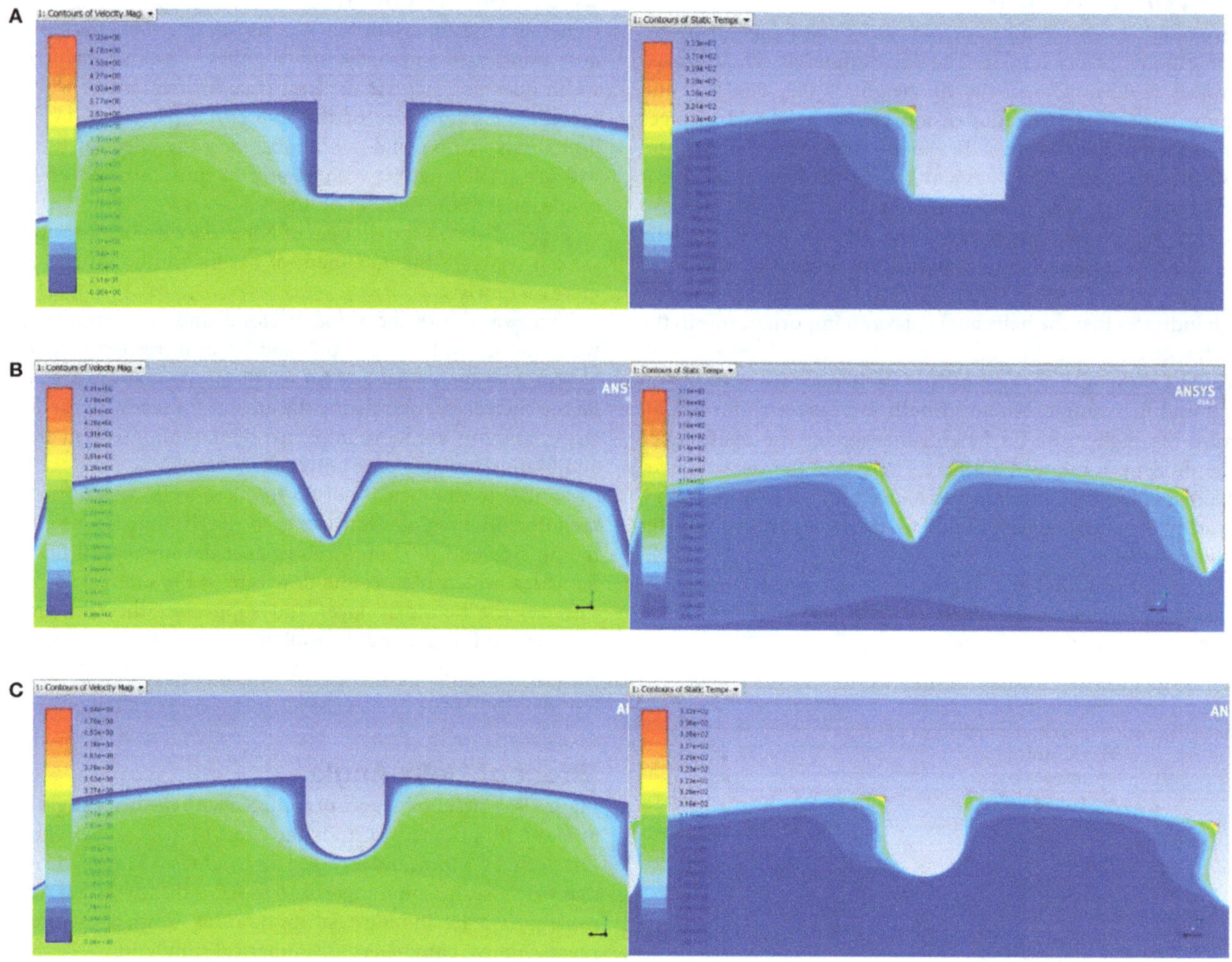

FIGURE 12 | The velocity field cloud and temperature field cloud of different fin shapes. **(A)** Rectangular fins. **(B)** Triangular fins. **(C)** Circular fins.

Effect of Fin Shape

Figure 11 shows the changes of f and Nu with three fin shapes, as shown in the picture, the average f in rectangular fins is the highest and that of triangular fins is the lowest, and that of circular fins is between them. The maximum difference between rectangular and triangular fins is 9.85%, and with the increase of Re, the difference between f gradually decreases. As can be seen from **Figure 11**, the Nu of rectangular and triangular fins is basically the same, and the triangular fin is slightly larger. With the increase of Re, the difference of Nu between the circular fin and the other two kinds of fins gradually increases, and the maximum difference is 5.9%.

Table 3 compares the increment in heat transfer area, Nu and f caused by the three types of fins in turbulent state. It can be seen from the table that the heat transfer area of rectangular fins increases the most, while that of triangular fins increases the least, and the difference of the three types of fins lies in the shape of fin top. For rectangular fins and triangular fins, the top has a sharp angle, while the top of circular fins is a smooth surface, which lead to different flow and heat transfer characteristics.

Figure 12 shows the velocity field cloud and temperature field cloud of rectangular fin, triangular fin and circular fin respectively when the Re is 70000 and ρ is 100000. As we can see, the sharp angles of rectangular fins and triangular fins are impacted by the incoming flow, which greatly enhance the heat transfer in the region. Therefore, the average Nu is higher than that of circular fins, and as the Re increases, the interaction between the fluid and the sharp angle of rectangular fins and triangular fins becomes more intense. It is noted that the Nu of the triangular fins is slightly higher than that of the rectangular fins, because compared with rectangular fins, the intercostal passage between adjacent fins of the tube in the triangular fins is larger, and the larger intercostal area could pump more turbulence into the root area of the fin, thinning the boundary layer, thus enhancing heat transfer. The heat transfer area of the triangular fins increases the least, f increases the least and Nu increases the most, so the heat transfer performance is the best.

CONCLUSION

In this study, FLUENT software and periodic boundary conditions are used to simulate the flow and heat transfer in internally finned tubes comprehensively, and to study the influence of various geometric parameters. The parameters are fin number (15–45), fin height (0.3–0.9 mm), fin width (0.5–0.9 mm), helix angle (10–40°) and the operating condition is Re = 44430. Meanwhile, the influence of the fin shapes on the heat transfer and the influencing mechanism are also explored. The conclusions are as follows:

(1) The geometrical parameters of internally finned tubes affect the characteristics of flow and heat transfer, and Nu and the f both increase with them.
(2) Compared with fin number, fin height has a greater influence on the flow and heat transfer of internally finned tubes.
(3) Fin width has no significant effect on the flow and heat transfer. The Nu and f both change slowly with fin width.
(4) Among all the geometrical parameters, the helix angle has the largest influence on the flow and heat transfer in the internally finned tube.
(5) The shape of the fin has no significant influence on the flow and heat transfer characteristics. The maximum difference of drag coefficient between the three finned shapes was within 10%, and the difference of nussel number was within 6%. If other parameters are the same, the heat transfer performance of the triangular fin is the best.

AUTHOR CONTRIBUTIONS

In the course of the completion of this thesis, all authors have substantial contributions to design of the work, and the contribution of YY is equivalent to that of the first author. GF provided the research direction of the subject, all authors worked together to define the methodology and procedures. LS built the geometric model. ZL carried out numerical simulation and obtained, collated, and analyzed the data. GF and LS also provided guidance and assistance in the process. ZL wrote the first draft of the paper and corrected it by GF and LS. YY provided important guidance in the process, and his contribution is equivalent to that of the first author. All authors approved the final version to be published. All authors agreed to be responsible for all aspects of the work.

ACKNOWLEDGMENTS

ZL would like to extend his sincere gratitude to his supervisor GF for his instructive advice and guidance on his thesis. ZL would also like to thank YY for his important guidance in this process, whose contribution is equivalent to that of the first author.

REFERENCES

1. Agra, O., Demir, H., Atayilmaz, S., Kantas, F., and Dalkiliç, A. S. (2011). Numerical investigation of heat transfer and pressure drop in enhanced tubes. *Int. Commun. Heat Mass Transf.* 38, 1384–1391. doi: 10.1016/j.icheatmasstransfer.2011.07.013
2. Al-Fahed, S., Chamra, L., and Chakroun, W. (1999). Pressure drop and heat transfer comparison for both microfin tube and twisted- tape inserts in laminar flow. *Exp. Thermal. Fluid Sci.* 18, 323–333. doi: 10.1016/S0894-1777(98)10037-7
3. Celen, A., Dalkilic, A. S., and Wongwises, S. (2013). Experimental analysis of the single phase pressure drop characteristics of smooth and microfin tubes. *Int. Commun. Heat Mass Transf.* 46, 58–66. doi: 10.1016/j.icheatmasstransfer.2013.05.010
4. Celen, A., Kayaci, N., Cebi, A., Demir, H., Dalkiliç, A. S., and Wongwises, S. (2014). Numerical investigation for the calculation of TiO2ewater nanofluids' pressure drop in plain and enhanced tubes. *Int. Commun. Heat Mass Transf.* 53, 98–108. doi: 10.1016/j.icheatmasstransfer.2014.02.022
5. Chamra, L. M., and Mago, P. J. (2006). Modeling of condensation heat transfer of refrigerant mixture in micro-fin tubes. *Int. J. Heat Mass Transf.* 49, 1915–1921. doi: 10.1016/j.ijheatmasstransfer.2005.11.006
6. Chamra, L. M., Mago, P. J., Tan, M. O., and Kung, C. C. (2005). Modeling of condensation heat transfer of pure refrigerants in micro-fin tubes. *Int. J. Heat Mass Transf.* 48, 1293–1302. doi: 10.1016/j.ijheatmasstransfer.2004.10.005
7. Chamra, L. M., Tan, M. O., and Kung, C. C. (2004). Evaluation of existing condensation heat transfer models in horizontal micro-fin tubes. *Exp. Thermal Fluid Sci.* 28, 617–628. doi: 10.1016/j.expthermflusci.2003.10.006
8. Chamra, L. M., and Webb, R. L. (1996). Advanced micro-fin tubes for condensation. *Int. J. Heat Mass Transf.* 39, 1839–1846. doi: 10.1016/0017-9310(95)00275-8
9. Copetti, J. B., Macagnan, M. H., de Souza, D., and Oliveski, R. D. C. (2004). Experiments with micro-fin tube in single phase. *Int. J. Refrig.* 27, 876–883. doi: 10.1016/j.ijrefrig.2004.04.015
10. Dastmalchi, M., Arefmanesh, A., and Sheikhzadeh, G. A. (2017). Sheikhzadeh.Numerical investigation of heat transfer and pressure drop of heat transfer oil in smooth and micro-finned tubes. *Int. J. Therm. Sci.* 121, 294–304. doi: 10.1016/j.ijthermalsci.2017.07.027
11. Eiamsa-ard, S., and Wongcharee, K. (2012). Single-phase heat transfer of CuO/water nanofluids in micro-fin tube equipped with dual twisted-tapes. *Int. Commun. Heat Mass Transf.* 39, 1453–1459. doi: 10.1016/j.icheatmasstransfer.2012.08.007
12. Eiamsa-ard, S., and Wongcharee, K. (2013). Heat transfer characteristics in micro-fin tube equipped with double twisted tapes: effect of twisted tape and micro-fin tube arrangements. *J. Hydrodyn. Ser. B* 25, 205–214. doi: 10.1016/S1001-6058(13)60355-8
13. Filonenko, G. K. (1954). Hydraulic resistance in pipes. *Teploenergetika* 1, 40–44. Gnielinski, V. (1975). *New Equations for Heat and Mass Transfer in the Turbulent Flow in Pipes and Channels.* NASA STI/Recon Technical Report A, 75.
14. Han, D. H., and Lee, K. J. (2005). Single-phase heat transfer and flow characteristics of microfin tubes. *Appl. Therm. Eng.* 25, 1657–1669. doi: 10.1016/j.applthermaleng.2004.10.015
15. Jensen, M.K., and Vlakancic, A. (1999). Experimental investigation of turbulent heat transfer and fluid flow in internally finned tubes. *Int. J. Heat Mass Transf.* 42, 1343–1351. doi: 10.1016/S0017-9310(98)00243-9

16. Ji, W. T., Jacobi, A. M., He, Y. L., and Tao, W. Q. (2015). Summary and evaluation on single-phase heat transfer enhancement techniques of liquid laminar and turbulent tube flow. *Int. J. Heat Mass Transf.* 88, 735–754. doi: 10.1016/j.ijheatmasstransfer.2015.04.008
17. Ji, W. T., Zhang, D. C., He, Y. L., and Tao, W. Q. (2011). Prediction of fully developed turbulent heat transfer of internal helically ribbed tubes-an extension of Gnielinski equation. *Int. J. Heat Mass Transf.* 55, 1375–1384. doi: 10.1016/j.ijheatmasstransfer.2011.08.028
18. Kim, Y., Seo, K., and Chung, J. T. (2002). Evaporation heat transfer characteristics of R-410A in 7 and 9.52 mm smooth/micro-fin tubes. *Int. J. Refriger.* 25, 716–730. doi: 10.1016/S0140-7007(01)00070-6
19. Li, X. W., Meng, J. A., and Li, Z. X. (2007). Experimental study of single-phase pressure drop and heat transfer in a micro-fin tube. *Exp. Therm. Fluid Sci.* 32, 641–648. doi: 10.1016/j.expthermflusci.2007.08.005
20. Schulz, T. L. (2006). Westinghouse AP1000 advanced passive plant. *Nuclear Eng. Design.* 236, 1547–1557. doi: 10.1016/j.nucengdes.2006. 03.049
21. Seo, K., and Kim, Y. (2000). Evaporation heat transfer and pressure drop of R-22 in 7 and 9.52 mm smooth/micro-fin tubes. *Int. J. Heat Mass Transf.* 43, 2869–2882. doi: 10.1016/S0017-9310(99)00338-5
22. Siddique, M., and Alhazmy, M. (2008). Experimental study of turbulent single-phase flow and heat transfer inside a micro-finned tube. *Int. J. Refrig.* 31, 234–241. doi: 10.1016/j.ijrefrig.2007.06.005
23. Wang, C. C., Chiou, C., and Lu, D. C. (1996). Single-phase heat transfer and flow friction correlations for microfin tubes. *Int. J. Heat Fluid Flow* 17, 500–508. doi: 10.1016/0142-727X(96)00048-3
24. Xing, J., Song, D., and Wu, Y. (2016). HPR 1000:advanced pressurized water reactor with active and passive safety. *Engineering* 2, 83–91. doi: 10.1016/J.ENG.2016.01.017
25. Yu, M., Lin, T., and Tseng, C. (2002). Heat transfer and flow pattern during two- phase flow boiling of R-134a in horizontal smooth and microfin tubes. *Int. J. Refriger.* 25, 789–798. doi: 10.1016/S0140-7007(01)00075-5
26. Yun, R., Kim, Y., Seo, K., and Kim, H. Y. (2002). A generalized correlation for evaporation heat transfer of refrigerants in micro-fin tubes. *Int. J. Heat Mass Transf.* 45, 2003–2010. doi: 10.1016/S0017-9310(01)00321-0
27. Zdaniuk, G. J., Chamra, L. M., and Mago, P. J. (2008). Experimental determination of heat transfer and friction in helically-finned tubes. *Exp. Thermal Fluid Sci.* 32, 761–775. doi: 10.1016/j.expthermflusci.2007.09.006
28. Zheng, M., Yan, J., Shentu, J., Tian, L., Wang, X., Qiu, Z., et al. (2016). The general design and technology innovations of CAP1400. Engineering 2, 101–106. doi: 10.1016/J.ENG.2016.01.018

9

A Primer on R for Numerical Analysis in Educational Research

Tricia R. Prokop[1] *and Michael Wininger*[1,2,3*]

[1] *Department of Rehabilitation Sciences, University of Hartford, West Hartford, CT, United States,* [2] *Department of Biostatistics, Yale School of Public Health, New Haven, CT, United States,* [3] *Cooperative Studies Program, Department of Veterans Affairs, West Haven, CT, United States*

***Correspondence:**
Michael Wininger
wininger@hartford.edu

Researchers engaged in the scholarship of teaching and learning seek tools for rigorous, quantitative analysis. Here we present a brief introduction to computational techniques for the researcher with interest in analyzing data pertaining to pedagogical study. Sample dataset and fully executable code in the open-source R programming language are provided, along with illustrative vignettes relevant to common forms of inquiry in the educational setting.

Keywords: education, code, numerical, analysis, statistics, programming, measurement, SoTL

INTRODUCTION

The scholarship of teaching and learning (SoTL), i.e., research as applied in a pedagogical setting (Witman and Richlin, 2007; Kanuka, 2011), has become increasingly prominent over the past 20 years (Boyer, 1997; Gilpin and Liston, 2009; Hutchings et al., 2011; Bishop-Clark and Dietz-Uhler, 2012). As the interest in SoTL grows, so, too, grows the need for robust tools in support of inquiry and analysis (Bishop-Clark and Dietz-Uhler, 2012; Mertens, 2014). The purpose of this article is to provide introduction to simple but effective computational techniques for application to SoTL. The objective of this article is to empower investigators to adopt heretofore non-traditional approaches for the advancement of SoTL.

In this article, we introduce R, a numerical computing environment that is now 25 years' matured beyond its first release (Thieme, 2018). R is derived from the S programming language, developed at Bell Labs in the 1970s, designed to provide a variety of statistical and graphical techniques, with an intentionally *extensionable* design, meaning that users can devise and publish code add-ons to enhance base-R's functionality. R is regularly ranked in the top-20 of the TIOBE index of programming language popularity and as of March, 2018 is the highest-ranked software with focal application to statistical analysis (cf. SAS, Stata, and SPSS, which are not typically ranked in TIOBE; TIOBE, 2018), and is the analytical tool of choice for major print media, social media, Federal agencies, and search engines (Willems, 2014). R is freeware and completely Open Source; it is employed in the writing of innumerable scientific manuscripts, and is being added to ever more undergraduate and graduate curricula.

Recognizing that numerical analysis and scripted computation are not typically the province of educational researchers, this article is intentionally basic, with an interest in introduction. Accordingly, this article mixes novice-level guidance with tutorial-style walkthroughs of a few realistic datasets, obtained through the authors' own scholarly activities in educational research. This article is completely self-contained: all datasets and code are printed within the manuscript, although condensed code and data files are available for convenient download as article **Supplements**.

FAMILIARIZATION

Acquisition and Operation

The R software package can be downloaded from its online repository (http://r-project.org), and typically installs in a minute or two. When opening R, the Command Window will open. The Command Window is "live" meaning that commands can be typed in, and will deliver a response. Typing `8 * 2 <ENTER>` will yield the answer `16`.

In general practice, analyses are executed using a "script," i.e., a file with .R extension that executes commands in order, displaying the results in the Command Window. When opening R, open the Script Window, either by File > New Document, or File > Open Document. Commands are written identically, whether in the Command Window, or Script Window, but the Script Window is not live: pressing <ENTER> will simply bring the cursor to the next line, as in any text editor. In order to run a single command from within the Script Window, place cursor on the relevant line, and type `<CTRL> + R` in Windows, or `<COMMAND> + <ENTER>` in Mac. It is possible to highlight multiple lines (or a portion of a line) and send this selection to the Command Window for execution.

Basic Syntax

Each coding language has its own particular way of creating and manipulating objects. These objects are called variables and the language-specific way the variable is created or manipulated is called syntax. Given the introductory nature of this article, only cursory overview can be given to the syntax here, but it is sufficient to give a few pointers to accelerate the reader along the learning curve.

Creating Variables

R is unique in its flexibility of assigning variables. R allows assignment "from the left" and from the right, and assignment can be accomplished with either the equals sign =, or a left- <- or right-arrow ->. For consistency, we shall employ only the equals sign. Variables can be named arbitrarily, except that the variable name must start with an alphabetic character (upper- or lower-case), and cannot start with a numerical character or special character (underscore, etcetera).

```
alpha = 1
beta_1 = 0.20
Score = 100
```

Here, we have created three variables, using three different styles of naming: all lower-caps, delimited enumeration, and capitalized first letter. R can handle all three; it is up to those writing the code to pick the style that suits.

Array Types

Generally, an array is a vehicle for storing data. There are three main array types in R: scalars, vectors, and matrices. A scalar is a single datum, e.g., alpha, beta_1, and Score, as above. A vector is an ordered set of variables

```
vect = c(1, 2, 3, 6, 7, 64, 81, 100, pi)
```

Note the use of the `c(...)` syntax: so named because it *combines* the data. Note further that we invoked `pi` in our vector. R contains a number of useful variables that can be called upon as needed.

Lastly, a matrix is a collection of vectors

```
matx = matrix(c(1, 2, 3, 4, 5, 6, 7, 8),
        nrow = 4)
```

This creates a 4 × 2 ("four by two"; rows then columns) matrix. By typing these commands into the R Command Window, these variables are created. By typing their names again at the command window, they will be displayed, e.g.,

```
> matx
          [ ,1] [,2]
    [1,]     1    5
    [2,]     2    6
    [3,]     3    7
    [4,]     4    8
```

The rows and columns of this matrix can be named:

```
colnames(matx) = c("col_1", "col_2")
rownames(matx) = c("a", "b", "c", "d")
```

yields

```
> matx
      col_1 col_2
    a     1     5
    b     2     6
    c     3     7
    d     4     8
```

Data Frames

One of the most convenient aspects of R is its data frame class. Data frames look like matrices, but allow for convenient utilization of many sophisticated functions. In many cases, it is desirable to convert a matrix into a data frame; in most (nearly all) cases, it is not undesirable to create a data frame. When in doubt, we recommend to convert a variable to a matrix-based data object to data frame class.

```
my_frame = data.frame(matx)
```

The two variables `my_frame` and `matx` will appear the same if called at the Command Window, but they are of different classes (verify via `class(matx)` and `class(my_frame)`). Data frames have the provision of intuitive variable selection. We can retrieve a desired column from a data frame using the dollar sign

```
> my_frame$col_1
    1 2 3 4
```

If at first this appears unnecessary or clunky, be assured that the many R functions that operate on data frames will provide ample evidence of their true convenience.

Arithmetic Operations

The most elementary activity that can be performed in a numerical coding environment is basic arithmetic. Within R, these operations are performed using the common special characters used in other computing applications, including the Google search engine:

```
> beta_1/2
    0.1
```

and

```
> Score^2
   10000
```

Naturally, we may want to create a new variable based on the manipulation of an old variable, or simply re-assign a new variable to the same variable. Both are easily done:

```
> gamma = 2*alpha
> alpha = 10
> gamma
    2
> alpha
    10
```

Arithmetic operations can also be performed over vectors and matrices

```
> matx^2
   col_1 col_2
 a     1    25
 b     4    36
 c     9    49
 d    16    64
```

We leave element-wise operations, e.g., matrix multiplication to the reader to explore, as these more sophisticated manipulations are beyond the scope of this primer.

Functions

The vast utility of a numerical computing approach is in the utilization of pre-packaged functions for executing intensive operations on data. Each coding language has thousands of functions, and most languages allow for the creation of user-defined functions. Writing one's own functions is typically the interest of the experienced coder, but are not out of reach of the beginner. However, we shall only discuss a few relevant pre-existing functions within R here. Consider, for example, taking the absolute value

```
> all_pos = abs(c(-1,0,2,3,-2))
> all_pos
    1 0 2 3 2
```

or computing the average and standard deviation

```
> mean(vect)
     29.6824
> sd(vect)
     40.0569
```

Within the parentheses, the argument specifies the variable to be operated on. These three functions all carry a single argument by default. But often, multiple arguments can be specified. For example, `round(mean(vect))` yields `30`, but `round(mean(vect),2)` provides two digits of rounding: `29.68`. Functions are useful not only for manipulating variables, but for writing/reading data to/from file, data visualization, and housekeeping within the workspace (i.e., clearing vestigial variables, or checking variable type). For instance `my_data=read.csv("my_file.csv")` will read a local file and store its contents in a data frame (`my_data`) for analysis within R, and `class(my_data)` will verify `data.frame`.

Notes

A few additional items bear mentioning in order to fully introduce R as appertaining to the examples that follow.

Sensitivities

R code is generally space-insensitive. In the case of horizontal space (as would be produced by pressing the space bar), R is insensitive in nearly all assignments, e.g., `a =1` and `a = 1` yield the same result; `round(vect,1)` and `round(vect, 1)` yield the same result, etcetera, with a notable exception of non-equals assignment operators (see section Assignment Operator, below). In the case of vertical space, R is also usually insensitive: an arbitrary number of carriage returns can be placed between lines of code. Vertical space does become complicated when interrupting a function, i.e., splitting the function across multiple lines, but this is generally acceptable, as long as each carriage return is preceded by the comma separating arguments, and the last line ends with the parenthesis:

```
read_data = read.csv("my_file.csv",
            stringsAsFactors=TRUE,
            header=TRUE)
```

The R language, however, is case-sensitive.

Strings

We have already seen here several functions where an argument was passed using the double-quote character. These are *string* variables; the quotes mark these variables as containing character information and to be treated as text, not as numerical. Setting `my_string = ""` tells R that this variable contains a character that does not yield to arithmetic manipulation: `my_string / 2` produces an error. Try this and see: `all_pos / 2 = 0.5 0 1 1.5 1`, but `"all_pos"/2` yields an error.

Assignment operator

In this Primer, all variable assignments use the equal sign (=). However, R has multiple operators for making assignments; perhaps the most common amongst R users is the left-ward (or right-ward) arrow: a dash next to a pointy-bracket. The three lines of code that follow all yield the same result

```
x = 100
y <- 100
100 -> z
```

This plurality of assignment approaches has some subtle, but illuminating consequences. For instance, we observe that R allows chained assignments:

```
a = b = 2
a <- b <- 2
```

both have the effect of creating two variables in one line of code, both equal to 2; other languages (Matlab, for example), do not allow this. However, note that this does not always work with a mix of assignments (i.e., `a <- b = 2` throws an error). Similarly, it can be confusing for the human coder or code reader to interpret whether `f<-2` is meant to convey "f is a variable with value negative two" vs. the assessment of the variables value, i.e., "is f < −2?" Placing those four characters in strict order, i.e., `f<-2` will yield the assignment; placing a space before the two will yield an evaluative statement (with output either `TRUE` or `FALSE`). These examples provide some insight into the way R's coding language is deconstructed by its parser. More can be learned about the hierarchy of syntactic operators in R's widely available documentations (refer: Operator Syntax and Precedence). Naturally, whether to force assignments by way of = or <- will often be determined by user preference.

Best practice

A code file should be written in such a way that everything is self-explanatory. This is critically important when code is put away for several months and then revived, or when passing code between collaborators. The three most essential practices here are: (1) intuitive naming of variables, (2) code commenting, and (3) code structure. In learning or re-learning a code segment, it is much more interpretable for a variable to be named in a way that conveys its meaning, e.g., `avg_gpa = mean(data$gpa)`. If this seems unnecessarily explicit at this juncture (since the right-hand clearly contains a grade point variable in a dataset), consider a downstream code line that converts scholastic performance from grade-point on a 4.0 scale to raw percentage: `avg_gpa = avg_gpa * 10 + 55`. If this variable were named more opaquely, e.g., `a = a * 10 + 55`, would be much less readily understood.

Likewise, every code line should be commented. All coding languages allow for commenting. In R, the hash character # tells the compiler that all characters to the right are to be ignored. Best practice is to have a comment section at the top of the file to give overview to the file's main actions and revision history, and comments throughout the body to explain specific actions.

```
# compute the mean grade point average
mean_gpa = mean(data$gpa)
# create a histogram of grade points range:
1.0 to 4.0 by 0.1
hist(data$gpa, breaks = seq(1,4,by=0.1))
```

It is atypical to use comments when working within the Command Window; it is vital to utilize comments when working within the Script Window.

Lastly, wherever the task is complex (see, for example, *nesting* discussed in sections Descriptive Statistics and Control Structures), artful code structure can make the difference between interpretable code and uninterpretable code. Comments require interpretation by the code author. But when the code is to be reviewed and de-bugged, the comments are only a guide, and can conceivably be misleading: the comment may be more aspirational, rather than actual, especially if the code is in draft form and the output hasn't been extensively checked. Ultimately, the code must speak for itself. The best way to facilitate this is to make the code easy to understand even without the comments: re-factor the code, limit the length of any single line of code, and break large code blocks into smaller sub-blocks as necessary. The coder should consider three aspects of their code document, whenever a modification is to be made: (1) the operation, (2) the assignment, and (3) the comment. The operation (the right-hand side of the equation, e.g., "`mean(data$gpa)`," above) should be reflected in the variable name (the left-hand side of the equation, i.e., "`mean_gpa`"), and also the comment ("`# compute the mean grade point average`"). If one of these three items is to change, all three items should change in synchrony, e.g.,

```
# compute the median grade point average
median_gpa = median(data$gpa)
```
or
```
# compute the mean skill score
mean_skill = mean(data$skill)
```

Outside of the code file, another best practice is to save code in a new file and/or a new folder at each session. Code files take up very little memory, and overwriting working code can result in painful losses when the revision is less functional than its predecessor. Whenever a functional code segment is achieved, store the file in a safe location, and Save As a new version. For more resources on best practices, which may differ by discipline, application, and setting, we point to a small number of resources (Martin, 2009; Prlić and Procter, 2012; Osborne et al., 2014; Wilson et al., 2014), advising that there are many more available.

Packages

The R package as downloaded (section Acquisition and Operation) is "Base R." It contains the vast majority of functions that are typically needed in most routine scenarios. However, the R community is diverse and highly engaged; users are constantly designing and sharing new functions that suit a particular application. These functions are contained in *packages* available for download from an R repository site. This can be from within R using the Package Installer. Packages do not load automatically in R, so if a certain function is needed, the package will have to be summoned via the library function. For instance, consider the `summaryBy` function which can be used to provide summary statistics within categorical groups using syntax similar to that used in R's ANOVA or linear modeling function families. At the command prompt, type `install.packages("doBy")` to acquire the package, and then `library(doBy)` within any R session where `summaryBy` is to be used. It is necessary to install the package only once per R install: the package will only ever need to be re-acquired if either a) a new version of R is installed on the investigator's computer, or b) a new version of the package is posted on the online repository and the user seeks to upgrade. Likewise, once an instance of R is created (i.e., once the software is opened for a session), the library needs only be called once. The library will not load again automatically when R is closed and re-opened, but re-loading the library is a simple, single command.

Directories

By default, R will point to its own location on the hard drive. This is rarely where the user wants to work. When opening an

TABLE 1 | Demographic data pertaining to students non-compliant with study protocol.

	ID	Cohort	Group	GPA_UG	GPA_G	Skill
1	20	II	T	3.50	3.20	83.000
2	23	II	C	3.40	3.40	90.000
3	24	I	T	3.71	3.03	18.875
4	27	I	C	3.38	3.14	18.210
5	32	II	T	3.73	3.68	94.000
6	34	I	C	3.68	3.25	18.880
7	58	I	C	3.93	3.92	19.500
8	63	II	T	3.25	3.09	85.000
9	78	I	T	3.76	3.70	19.580
10	83	II	T	3.40	3.10	88.000
11	87	I	T	3.90	3.85	19.460
12	89	I	C	3.55	3.00	18.460

Cohort, curriculum cohort; GPA, grade point average (Undergrad or Graduate). Table is captured in R dataframe "`comp_data`*."*

R session, set the working directory using the `setwd` command. Pass a string argument containing the folder where the work is to be performed:

```
setwd("/Users/Smith/Desktop/Research/
      Code_Files")
```

The forward slash character is the delimiter used by R in directories.

EXAMPLES

Descriptive Statistics

We implemented an internal study on student performance using two different pedagogical approaches. Students were randomized to either "treatment" group (T) or "control" (C). Twelve students signed the study's informed consent, but did not complete the full protocol, and therefore were removed from the mainline analysis. These students are described in **Table 1**.

Computing descriptive statistics for this sub-group is a straight-forward three-step process: (1) set the working directory, (2) read in the data from file, and (3) analyze.

```
# set the working directory
setwd("/Users/Docs/Frontiers_Manuscript/
       Code")
# read-in the data from file
comp_data=read.csv("Frontiers_1
       _SubSample.csv",
        head=TRUE, stringsAsFactors=TRUE)
# analyze and report
print(paste(nrow(comp_data),
       "non-compliant students"))
```

Note the two optional arguments in `read.csv`. By stating `head=TRUE`, R recognizes that the dataset contains a header, i.e., "ID," "Group," and so-on, are not mistaken for data points; and `stringsAsFactors=TRUE` forces the columns containing only character (and not numerical) data to be taken as a factor, as might be relevant to statistical analysis. This is helpful even in assessing descriptive statistics, given the `summaryBy` command (among others).

We note that .csv (comma-separated values) is a common file format for storing numerical data. While we refer to .csv files in this manuscript, R has the ability to open data in a wide variety of formats, included tab-delimited (.txt) and documents prepared in common spreadsheet softwares (i.e., .xls and .xlsx).

The last line of code contains one command *nested* in the other. This is an entirely legitimate maneuver, and may be quite common amongst experienced coders. The R compiler will read from the inside-out. The `nrow(data)` function counts the number of rows in the dataset. The `paste` command attaches a number variable (the number of rows) onto a string variable (the characters in quotes). The `print` command displays this information on the command line for easy viewing.

The result of this code is a single line report

```
"12 non-compliant students"
```

We report the average GPA in this group similarly:

```
# compute average GPA across sub-group
mean_gpa = round(mean(comp_data$GPA_UG), 1)
stdv_gpa = round(sd(comp_data$GPA_UG), 1)
print(paste("Average GPA:", mean_gpa,
        "+/-", stdv_gpa))
```

which yields

```
"Average GPA: 3.6 +/- 0.2"
```

In just a few commands, we can visualize these data. For example, suppose we wish to show grade-point distribution as a function of cohort

```
# show GPA as box plots (break-out by
   cohort)
boxplot(GPA_G~Cohort, data=comp_data)
title(main = "GPA by Cohort")
title(ylab = "Grade-Point Average")
```

The plot can be seen in **Figure 1A**. And while accurate, the visualization may not quite be suitable for publication. We can add a few additional arguments to increase the visual appeal

```
# modify box-plot for visualization
boxplot(GPA_G~Cohort, data = comp_data,
         names = c("Cohort I", "Cohort II"),
         ylim = c(1.9, 4.1),   cex.axis = 1.5,
         lwd = 3, col = c("grey50",
         "grey80"))
title(main = "GPA by Cohort", line = 1,
         cex.main = 2, font.main = 1)
title(ylab = "Grade-Point Average",
         line = 2.5, cex.lab = 2)
```

We leave it to the reader to thoroughly inspect the code, but summarize as follows: tick labels are customized for interpretability, the limits of the y-axis is re-defined, the tick fonts (`cex.axis`) are increased by 50%, the lines comprising the box

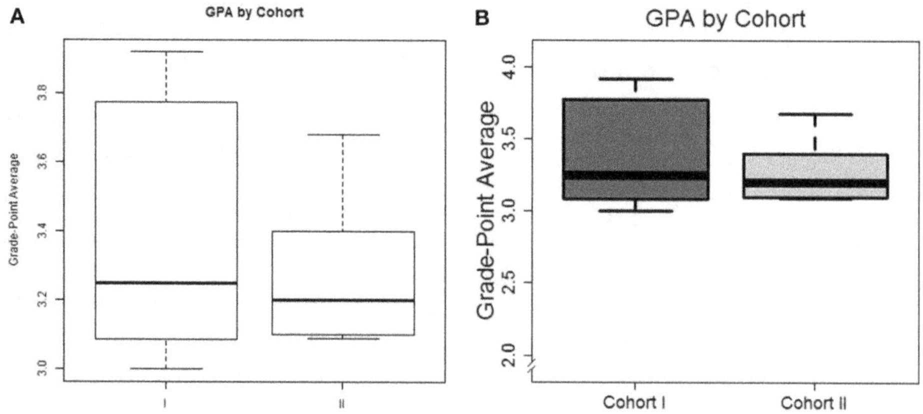

FIGURE 1 | Default plot **(A)** and modified plot **(B)** depicting the distribution of graduate GPA amongst non-compliant students in box plot format.

plot are darkened, and shades of gray are added to the boxes; the title and axis label are re-positioned and re-configured.

Lastly, we note that both the default axis and our own re-definition, places the axis away from zero. In many cases, it may be desirable to "force" the visualization that includes the zero-point. While Base-R does not provide this functionality, the `plotrix` library does.

```
library(plotrix)
axis.break(axis = 2, 1.925, style =
            "slash")
```

The final graph can be seen in **Figure 1B**. Note that this figure can be saved to file from within the R workspace by placing `png("Figure_1b.png")` ahead of the `boxplot` command, and `dev.off()` below the last plotting code.

Conditions

It is often desirable to subset the data somehow. Consider the set of data pertaining to non-compliant students (**Table 2**). Suppose we wanted to isolate just the students with Graduate GPA over 3.50. We first identify those students meeting the GPA criterion with a single line of interrogative code

```
# test for students exceeding undergrad GPA
threshold
comp_data$GPA_UG > 3.5
```

This will return a series of Boolean responses (binary True or False):

```
FALSE FALSE TRUE FALSE TRUE TRUE TRUE FALSE
TRUE FALSE TRUE TRUE
```

In order to extract the rows where this condition is met, we use the `which` function

```
# return rows reflecting to hi undergrad
GPAs
which(comp_data$GPA_UG > 3.5)
    3   5   6   7   9   11   12
```

To find items exactly meeting a criterion, use the double-equals (voices as "is equals to") ==

```
# find which students are in Cohort I
which(comp_data$Cohort == "I")
    3   4   6   7   9   11   12
```

To find items within a range, it is useful to leverage some of R's set operators (e.g., `intersect`) and apply multiple conditions

```
# find students with both GPAs above 3.4
intersect(which(comp_data$GPA_UG > 3.4),
which(comp_data$GPA_G > 3.4))
    5   7   9   11
```

Dataset Manipulation

There are several useful tools for creating and manipulating categorical variables in R. Consider the same example of students in **Table 1**. Suppose we wish to flag students with lower-tier scores in skills check. Firstly, we observe that the two Cohorts have scores reported in different scales. We re-scale using the Conditions methods, as in section Conditions:

```
# convert skill from 20-points to 100%
scale in Cohort I
inds_coh1 = which(comp_data$Cohort == "I")
comp_data$Skill[inds_coh1] =
comp_data$Skill[inds_coh1] * 5
```

We can then convert to a categorical variable via the `cut` function. Suppose we have three tiers: <85 points (Failure), between 85 and 90 points (Retest) and 90 or more points (Pass):

```
# convert skill to leveled variable
comp_data$SkillLevel = cut(comp_data$Skill,
      c(0,85,90,100), labels = c("Fail",
      "Retest", "Pass"))
```

Suppose further that we want to convert this into a 2-level category: those who require follow-up (a Red Flag) and those who do not. We can initialize an column of empty values (default: no follow-up required), and then

```
# re-level factor variable
comp_data$Category = rep("",
      nrow(comp_data))
```

TABLE 2 | Demographic data with adjusted skill scores, factored skill level, and re-sorted based on Red Flag status.

	ID	Cohort	Group	GPA_UG	GPA_G	Skill	SkillLevel	Category
1	23	II	C	3.40	3.40	90.000	Retest	REDFLAG
2	83	II	T	3.40	3.10	88.000	Retest	REDFLAG
3	20	II	T	3.50	3.20	83.000	Fail	REDFLAG
4	63	II	T	3.25	3.09	85.000	Fail	REDFLAG
5	24	I	T	3.71	3.03	94.375	Pass	
6	27	I	C	3.38	3.14	91.050	Pass	
7	32	II	T	3.73	3.68	94.000	Pass	
8	34	I	C	3.68	3.25	94.400	Pass	
9	58	I	C	3.93	3.92	97.500	Pass	
10	78	I	T	3.76	3.70	97.900	Pass	
11	87	I	T	3.90	3.85	97.300	Pass	
12	89	I	C	3.55	3.00	92.300	Pass	

```
comp_data$Category[comp_data$SkillLevel
      %in% c("Fail", "Retest")]
      = "REDFLAG"
```

The `%in%` operator, posed as `A %in% B` returns Boolean values according to whether each item in A appears in B:

```
TRUE TRUE FALSE FALSE FALSE FALSE FALSE
TRUE FALSE TRUE FALSE FALSE
```

Thus, four students require follow-up. To sort these to the top, we re-order the data-frame using

`order(comp_data$Category)` as a re-sorting vector:

```
# sort data-frame based on Red Flag status
comp_data = comp_data
             [order(comp_data$Category,
              decreasing = TRUE),]
```

The phrasing `data[inds,]` sorts all columns (the empty space between the command and the closing square bracket) according to the indices specified in `inds`. We can execute a multi-layer sort by adding additional items serially within order:

```
# sort on Red Flag; sub-sort on Skill Level
comp_data = comp_data
             [order(comp_data$Category,
              comp_data$SkillLevel,
              decreasing=TRUE),]
```

This modified dataset is shown in **Table 2**.

Diagnostics

Outliers

Take as a new example the data collected by our group in our studying the impact of video-guided self-reflection as an adjuvant strategy in supplement to the conventional pedagogical strategy in the graduate curriculum. Students from two graduate cohorts were recruited, yielding 25 students with complete results. Here, we show a subset of outcomes, viz. the Academic Locus of Control-Revised questionnaire [ALC-R (Curtis and Trice, 2013)], measured early- (ALC_1) and late (ALC_2) in the semester. Cohort, undergraduate GPA and graduate GPA were noted (**Table 3**).

Load in this dataset as follows:

```
# read-in alc dataset
alc_data = read.csv("Frontiers_2_
              ALCData.csv", head = TRUE,
              stringsAsFactors = TRUE)
```

It is typical to identify outliers in a dataset. R can do this efficiently via sub-functionality on the boxplot command:

```
# identify outliers
boxplot.stats(alc_data$GPA_G)$out
boxplot.stats(alc_data$GPA_UG)$out
```

where the dollar sign retrieves the outliers ("`out`") from the data frame that resulted from the `boxplot.stats` command. As it turns out, there are no outliers in our dataset with regard to GPA; R returns `numeric(0)` in both cases. If this seems unlikely, we can confirm by inspection of the data itself. Consider the "z-normalization" transform, i.e., standardization by subtraction of the mean and division by the standard deviation. R performs this via the `scale` command. Taking the `range` (min and max) of both datasets, we see:

```
# double-check outliers via standardization
range(scale(alc_data$GPA_G))
          -1.642978    1.880937
range(scale(alc_data$GPA_UG))
          -1.942531    1.458947
```

Thus, there are no GPA values that are more than two standard deviations departed from the mean. There are no outliers to remove.

Linearity

Single variables

While we might want to jump right into hypothesis testing, it is prudent to perform some diagnostics first. For instance: while we might wish to test for differential effects, but accounting

TABLE 3 | Responses to academic locus of control (ALC) questionnaire 15 weeks apart.

	ID	Cohort	GPA_UG	GPA_G	Group	ALC_1	ALC_2
1	72	I	3.37	3.44	T	3	7
2	55	I	3.8	3.7	T	7	5
3	42	I	3.65	3.26	C	2	2
4	90	I	3.8	3.5	C	4	5
5	70	I	3.52	2.92	T	6	5
6	29	I	3.9	3.59	T	6	4
7	67	I	3.9	3.85	T	6	7
8	8	I	3.5	2.96	C	5	3
9	25	I	3.2	3.5	T	2	4
10	81	I	3.5	2.92	T	4	5
11	28	I	3.93	3.63	T	3	4
12	92	I	3.2	3.11	T	4	5
13	17	I	3.43	3.44	C	4	4
14	7	I	3.5	3.6	C	7	5
15	84	II	3.1	2.99	C	6	8
16	14	II	3.54	3.31	T	9	9
17	26	II	3.44	3.11	C	6	8
18	97	II	3.87	3.64	T	4	3
19	94	II	3.31	3.45	T	10	9
20	85	II	3.8	3.5	T	4	6
21	86	II	3.8	3.5	T	4	3
22	79	II	3.34	3.4	T	4	7
23	5	II	3.5	3.1	T	6	8
24	52	II	3.8	3.31	C	3	3
25	11	II	3.65	3.11	C	5	4

Cohort, Curriculum cohort, GPA, grade point average (Undergrad or Graduate). Table is captured in R dataframe "`alc_data`*."*

for academic preparation as co-variate information. Our study collected both undergraduate- and graduate GPA; if they are highly collinear, it would be imprudent to account for them both. We can assess collinearity visually (assessment for a straight-line trend, by inspection), or numerically, via Pearson's correlation. Generating a plot is simple:

```
# generate a scatter plot of GPA datasets
plot(alc_data$GPA_UG, alc_data$GPA_G,
main = "",xlab = "",ylab = "")
title(main="Scatterplot of GPA")
title(xlab="Graduate GPA")
title(ylab="Undergraduate GPA")
```

This plot is shown in **Figure 2A**. Note that R will add default labels to the data unless they are specifically overwritten as empty values (`main="",xlab="",ylab=""`). It is possible to integrate these commands (specifying our custom labels within the `plot()` command; but we break these commands apart here for clarity. While the figure itself is very helpful, it is not quite publication-ready. We can plot with customized formats, and overlay the correlation value with a few embellishments:

```
# modify scatterplot for visualization
plot(alc_data$GPA_UG, alc_data$GPA_G,
     main = "", xlab = "", ylab = "",
     pch = 21, lwd = 2, cex = 2,
     col = "darkblue", bg = "lightgreen",
     ylim = c(1.9,4.1))
title(main = "Scatterplot of GPA", line=1,
     cex.main=2, font.main=1)
title(xlab="Graduate GPA", line=3,
     cex.lab=2, font.main=1)
title(ylab="Undergraduate GPA",
     line=2.5, cex.lab=2, font.main=1)
linmod=lm(GPA_G~GPA_UG,data= alc_data)
abline(linmod)
axis.break(axis = 2, 1.925, style =
    "slash")
rho_val=round(cor(alc_data$GPA_UG,
     alc_data$GPA_G), 2)
text(3.1, 4, paste("Correlation = ",
     rho_val), adj = c(0,0), cex=1.4)
```

This plot is shown as **Figure 2B**. Again, we leave it to the reader to work through the code item-by-item, but we summarize as follows: data are plotted as a scatter plot, with filled circles (#21 in the R plot character set, `pch`), with double the linewidth and double the size vs. default. A linear model (`lm`) is created to reflect Graduate GPA as a function of ("~") Undergraduate GPA, and then plotted via `abline`. Next, the Pearson correlation is

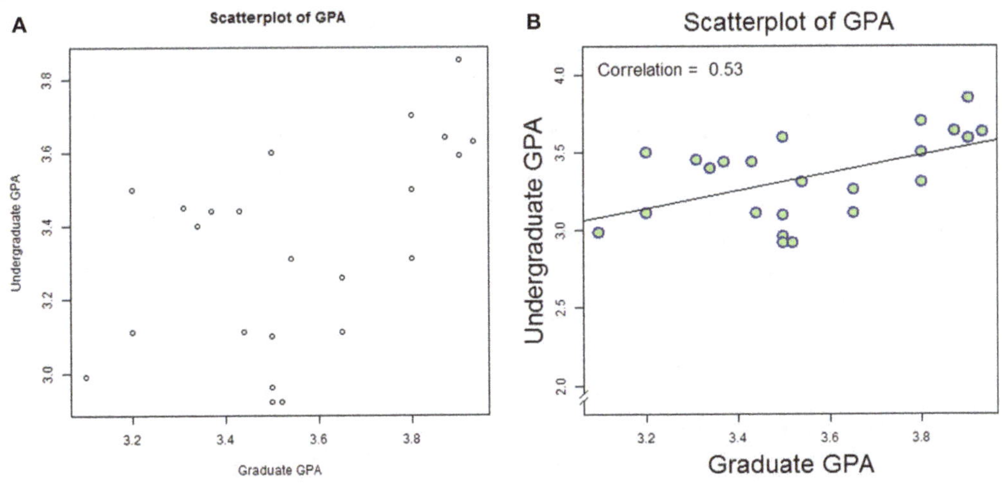

FIGURE 2 | Default plot **(A)** and modified plot **(B)** depicting scatter of relationship between undergraduate GPA and graduate GPA.

computed, rounded to 2 digits, and annotated on the plot at position $x = 3.1$, $y = 4.0$, with a 40% larger font than default; the text is adjusted so that the bottom-left of the first character coincides with the specified position.

Multi variable

There are many convenient tools for diagnostics on multi-variate data, as well. We extend the example of single-variate linearity assessment by fabricating a few additional variables. Let `GPA_R1`, `_R2`, `_R3`, and `_R4` be randomly-generated number sets meant to look like plausible GPA datasets.

```
# create two more GPA variables (for
exploration)
set.seed(311)
alc_data$GPA_R1=3.0 + runif(nrow(alc_data))
set.seed(311)
alc_data$GPA_R2=2.5 + runif(nrow(alc_data))
set.seed(331)
alc_data$GPA_R3=3.0 + runif(nrow(alc_data))
set.seed(156)
alc_data$GPA_R4=1.0 + runif(nrow(alc_data))
```

The random numbers are generated via `runif`; a random sample of uniform distribution and range zero to one is generated with compatible length (same as number of rows in `alc_data`). The additive constant (3.0, 2.5, or 1.0) is there merely to re-scale the data to a GPA-like dataset. Note the use of `set.seed`. Most programming languages allow for random numbers sets to be generated in a way that is reproducible. Soliciting `runif(100)` will generate a new set of 100 random numbers every time the command is entered. However, by setting a specific seed before invoking the random numbers, the numbers will be called from the same distribution each time. In this way, `GPA_R1` and `GPA_R2`, being of the same seed, will be identical; they will differ from `_R3` and `_R4`, which have different seeds.

We shall now subset the data for convenience:

```
# take as only those columns containing
"GPA" in their column names
gpa_data = alc_data[,grep("GPA",
           names(alc_data))]
```

We used the grep function (`grep` = "globally search a regular expression and print"). The `grep` function takes as its first argument the text to be searched *for*, and as its second argument the text to be searched *in*. Thus, within the names of `alc_data`, find where "GPA" appears. Given that the names of `alc_data` are `ID`, `Cohort`, `GPA_UG`, `GPA_G`, `Group`, `ALC_1`, `ALC_2`, `GPA_R1`, `GPA_R2`, `GPA_R3` and `GPA_R4`, the `grep` command returns {3, 4, 8, 9, 10, 11}. Thus, our subsetting command is equivalent to `alc_data[,c(3, 4, 8, 9, 10, 11)]`, and yields a 25×6 matrix of GPA values (some obtained empirically, some fabricated).

We can perform our collinearity assessment through visual means merely by plotting this data frame:

```
# show scatter
plot(gpa_data, main = "Collinearity Review
      Scatterplot")
```

This plot is shown in **Figure 3A**. Notice that the plot ticks are scattered: some plot panels show ticks along the top row, some along the bottom; some along the left side, some along the right. For this kind of descriptive assessment, we may prefer to turn off the axis ticks (`xaxt` and `yaxt`) altogether. We can add optional arguments to stylize the plot (**Figure 3B**):

```
# modify scatterplot for visualization
plot(gpa_data, pch = 21, lwd = 2,
     cex = 1.5, col = "darkred",
     bg = "pink", xaxt = "n", yaxt = "n")
title(main = "Collinearity Review
      Scatterplot", line = 2,
      cex.main = 2, font.main = 1)
```

Base-R provides the `cor` function for calculating the correlation matrix. However, the result is in matrix form, and may

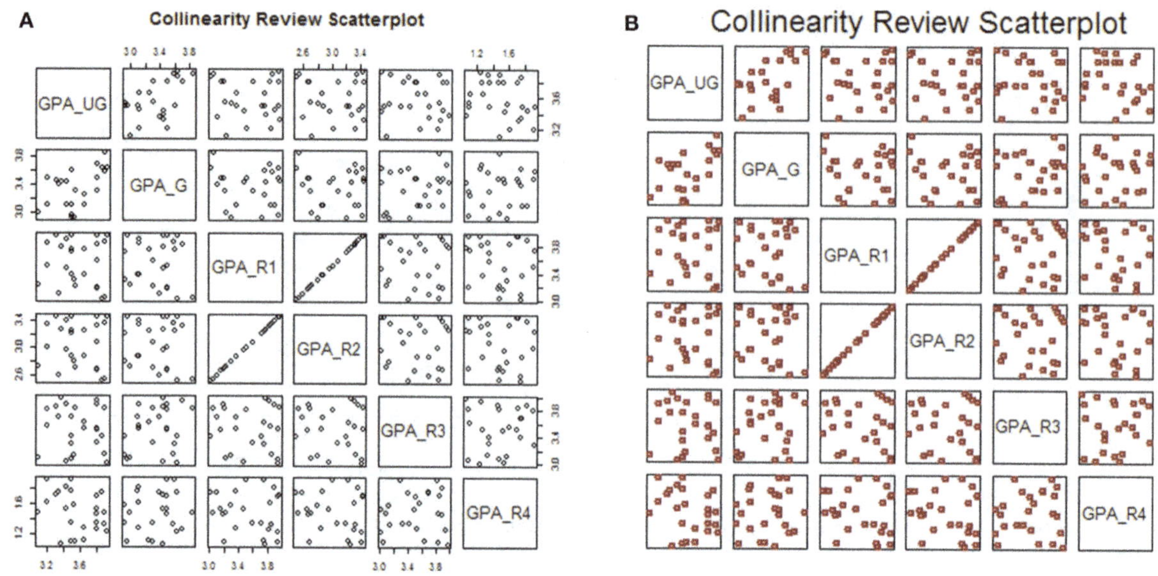

FIGURE 3 | Default plot **(A)** and modified plot **(B)** of multi-variate scatterplots for assessing collinearity.

not yield easy visualization, especially if there are many variables being compared simultaneously. We can leverage the `corrplot` package to facilitate visualization of high-dimensional correlation data.

```
# show correlation matrices
library(corrplot)
corrplot(cor(gpa_data))
title(main = "Correlation Heat Map")
```

This is shown in **Figure 4A**. We can add additional features to make the plot more print-worthy. Notice that we add a new library for the support of additional graphical devices, i.e., the color ramp palette, which ramps between any number of designated colors (here: red to white to blue)

```
# modify correlation plot for visualization
library(grDevices)
clrs=colorRampPalette(c("red", "white",
     "blue"))
corrplot(cor(gpa_data), col=clrs(10),
     addCoef.col = "white", number.digits
     = 2, number.cex = 0.75, tl.pos = "d",
     tl.col = "black", type="upper")
title(main="Correlation Heat Map", line=1,
     cex.main=2, font.main=1)
```

Again we leave it to the reader to fully explore the code as presented, but summarize as follows: a three-color palette was defined (`clrs`), and correlation coefficients were added in white text with two-digit precision at 75% default font size. The text label (`tl`) position was defined to appear along the diagonal ("`d`") in black text, and only the upper-triangle of the (symmetric) matrix is shown. The modified correlation matrix is shown in **Figure 4B**.

Hypothesis Testing

Univariate

A common approach to hypothesis testing for assessment of a single factor is the *t*-test. Consider a pre- and post-test, for example the Achievement Goal Questionnaire-Revised [AGQ-R (Elliot and Murayama, 2008)], as shown in **Table 4**.

A first line of inquiry might be: do students –irrespective of cohort– show an improvement in the AGQ-R after a 15 weeks period. We read-in the data, and can assess statistical significance straight away:

```
# read-in agq dataset
agq_data = read.csv("Frontiers_3_
           AGQData.csv", head = TRUE,
           stringsAsFactors = TRUE)
# paired t-test
pre=agq_data$AGQ_1
post=agq_data$AGQ_2
t.test(pre, post, paired=TRUE)
```

Note that the optional argument "`paired=TRUE`" enforces that this is not merely a two-sample *t*-test with identical sample sizes, but that the measurements are paired. The output reports the t-value (2.24), the degrees of freedom (here: 24, i.e., 25 minus 1), and *p*-value (0.034). Thus, by frequentist conventions, there is a statistically significant difference post- vs. pre. For more informative view, we compute the summary statistics:

```
# report differences
print(paste("AGQ Pre = ",round(mean(pre),
      1), "+/-", round(sd(pre),1)))
print(paste("AGQ Post = ",round(mean(post),
      1), "+/-", round(sd(post),1)))
```

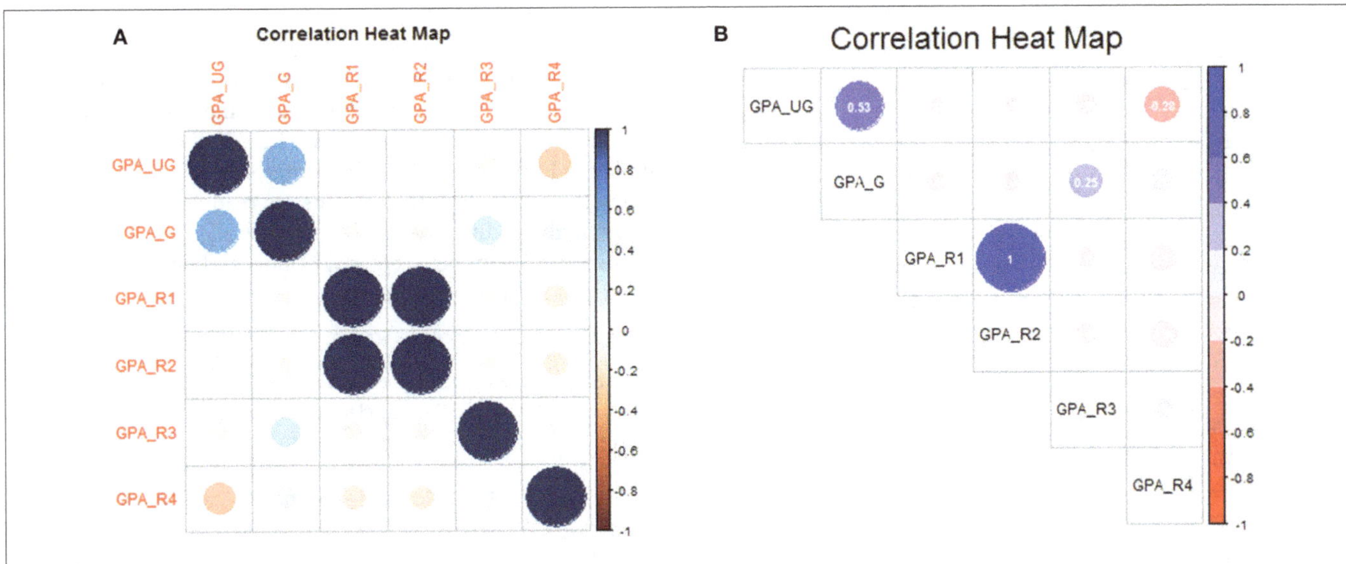

FIGURE 4 | Default plot **(A)** and modified plot **(B)** of correlation matrices of multi-variate GPA data.

Which yields

```
"AGQ Pre = 46.2 +/- 7.6"
"AGQ Post = 43.6 +/- 6.8"
```

This is a curious result, so we assess each treatment group independently:

```
pre=agq_data$AGQ_1[which(agq_data$Group ==
    "C")]
post=agq_data$AGQ_2[which(agq_data$Group
    == "C")]
t.test(pre, post, paired=TRUE)
```

and

```
pre=agq_data$AGQ_1[which(agq_data$Group ==
    "T")]
post=agq_data$AGQ_2[which(agq_data$Group
    == "T")]
t.test(pre, post, paired=TRUE)
```

yields, in both cases, a difference that is not statistically significant: for the control group: difference on means of 4.0, 95% Confidence Interval {−0.46, 8.46}, $P = 0.07$; for the treatment group: 1.88 (−1.31–5.06), $P = 0.23$. This suggests there may be some value to a multi-variate analysis.

Multivariate

Adding new dimensions to the data can provide greater insight in cases where univariate analysis proves limiting. As a first step, we modify our dataset slightly in order to make for efficient coding:

```
# convert dataset for multi-variate
analysis
agq_data$Change = agq_data$AGQ_2 -
agq_data$AGQ_1
```

The only change here is in defining a single variable to report change in score (post-minus-pre). For comfort, verify that the single-sample *t*-test yields the same result as shown for the same data in the paired-test configuration, i.e., difference on means of 2.64, $P = 0.034$.

```
# demonstrate that t-test is the same as
paired test
t.test(agq_data$Change)
```

Now, consider re-configuring as a one-way ANOVA. We use the same functional notation: `aov(Change~Group,data=agq_data)`, which produces its own output report. But often, the information available through the `summary` function is more direct, so we wrap `aov` in the `summary` command, and name it so that we may extract the *p*-value efficiently:

```
# report p-value
aov_summ=summary(aov(Change~Group,
          data=agq_data))
print(paste("P-value, for AGQ change=",
      round(aov_summ[[1]][1,5], 3)))
```

Which yields $P = 0.398$. We note that the summary of an ANOVA in R is of class type `list`. Lists may contain matrices, but the matrix contents cannot be extracted directly, the list item has to be extracted first. The list item is extracted by double braces (first list item by `[[1]]`; nth item by `[[n]]...`). Since the first item is a matrix (data frame, really), we can then reference it with the conventional matrix references, i.e., [row, column]. Since the first row, fifth column is the *P*-value associated with the group term, we extract it as `aov_summ[[1]][1,5]`.

In order to add new information, consider merging the co-variates provided through the ALC-R dataset. R contains several

TABLE 4 | Responses to achievement goal questionnaire (AGQ) 15 weeks apart.

	ID	Group	AGQ_1	AGQ_2
1	42	C	47	41
2	90	C	53	54
3	8	C	38	40
4	17	C	57	45
5	7	C	48	46
6	84	C	38	35
7	26	C	49	51
8	52	C	60	46
9	11	C	39	35
10	72	T	39	39
11	55	T	55	53
12	70	T	48	48
13	29	T	48	51
14	67	T	37	36
15	25	T	60	60
16	81	T	36	37
17	28	T	48	33
18	92	T	41	45
19	14	T	51	42
20	97	T	40	42
21	94	T	36	39
22	85	T	45	48
23	86	T	55	40
24	79	T	39	37
25	5	T	48	46

Cohort, curriculum cohort, GPA, grade point average (Undergrad or Graduate). Table is captured in R dataframe "`agq_data`*."*

helpful tools for merging datasets. We highlight one here, the `merge` function, available through base R.

```
# merge AGQ dataset onto ALC dataset for
co-variates
agq_data=merge(agq_data, alc_data)
```

We can now assess this more informative dataset with an ANOVA where each variable is accounted for:

```
# test with full anova
aov(Change~Cohort+Group+GPA_UG+GPA_G,
    data=agq_data)
```

We see from the summary:

```
# report summary of the anova
summary(aov(Change~Cohort+Group+
        GPA_UG+GPA_G,data=agq_data))
```

```
        Df  Sum Sq  Mean Sq  F value  Pr(>F)
Cohort  1   16.1    16.10    0.455    0.507
Group   1   25.7    25.74    0.728    0.404
GPA_UG  1   76.8    76.80    2.172    0.156
GPA_G   1   6.0     5.98     0.169    0.685
```

Thus it is evident that there are no main effects of significance (P-values for all terms >0.05). It is beyond the scope of this article to address the discrepancy between the two tests; rather, the reader is referred on to more foundational statistics texts for the assumptions of t-tests vs. ANOVA, and encouraged that R has tools built in to help decide which test is most appropriate on a given dataset. Rather, we emphasize here that ANOVA is easily applied in R, and the results easily interpreted.

Lastly, we note that interactions are readily assessed in R, by swapping the additive syntax (+) to multiplicative (*): including an interaction term between, say, Cohort and Graduate GPA is a simple re-arrangement of terms:

```
# test with anova with an interaction term

summary(aov(Change~Cohort+GPA_UG+
        Group*GPA_G,data=agq_data))
```

which yields

```
        Df  Sum Sq  Mean Sq  F value  Pr(>F)
Cohort  1   16.1    16.10    0.441    0.514
GPA_UG  1   67.2    67.15    1.840    0.191
Group   1   35.4    35.38    0.970    0.337
GPA_G   1   6.0     5.98     0.164    0.690
Group:
GPA_G   1   13.8    13.81    0.378    0.546
```

Where the last row in the summary reports the summary for the interaction term.

Categorical

There are a multitude of tools in R for facilitating hypothesis testing in categorical data. For brevity, we discuss a simple example, that of a 2 × 2 table. Consider the AGQ-R data, with a grouping threshold of those who decreased in AGQ-R vs. those who remained at the same score or improved over time. We start by identifying students who meet each of the strata

```
# identify the students in each stratum
cohort_1=which(agq_data$Cohort == "I")
cohort_2=which(agq_data$Cohort == "II")
agq_incr=which(agq_data$Change < 0)
agq_decr=which(agq_data$Change >= 0)
```

and prepare each cell in the contingency table:

```
# determine cell values for a contingency
table
agq_cell1=length(intersect(cohort_1,
          agq_incr))
agq_cell2=length(intersect(cohort_2,
          agq_incr))
agq_cell3=length(intersect(cohort_1,
          agq_decr))
agq_cell4=length(intersect(cohort_2,
          agq_decr))
```

Lastly, we create the contingency table, furnish row- and column-names for the sake of clarity, and perform our hypothesis test:

```
# create the contingency table, test two
ways
agq_table=matrix(c(agq_cell1,agq_cell2,
          agq_cell3,agq_cell4),nrow=2)
```

```
rownames(agq_table) = c("Cohort_I",
          "Cohort_II")
colnames(agq_table) = c("Increase",
          "Decrease")
chisq.test(agq_table)
fisher.test(agq_table)
```

Our contingency table had the structure of

```
            Increase      Decrease
Cohort_I    6             8
Cohort_II   7             4
```

And neither the Chi-Square test ($P = 0.5293$), nor its small-sample alternative, Fisher's Exact Test ($P = 0.4283$) yielded significance. This matches expectation based on the highly non-diagonalized nature of the contingency table.

Extraction of odds ratios and confidence intervals are straightforward. We note also that R can handle n-way contingency tables (3×3, 4×4, etcetera).

Modeling

Linear Regression

Perhaps the most commonly-used methodology for modeling the relationship between two- or more variables is through regression. Here, we demonstrate regression models on these datasets. Firstly, consider **Figure 2**, showing the relationship between Undergraduate and Graduate GPA. These variables appear to have an approximately linear relationship, so we model as a linear regression:

```
# read-in alc dataset
alc_data = read.csv("Frontiers_2_
          ALCData.csv", head = TRUE,
           stringsAsFactors = TRUE)
# perform linear regression
linmod = lm(GPA_UG~GPA_G,data=alc_data)
```

Calling this variable from the command prompt yields an output with two items: a summary of the call function, and a listing of the coefficients. Since we specified a simple regression equation, there are only two coefficients: the slope (0.494) and the intercept (1.917). Similarly to the ANOVA, a more informative review of the model can be obtained through the `summary` command:

```
# summarize the linear regression
summary(linmod)
```

which yields, among other things,

```
            Estimate Std.   t     Pr(>|t|)
                     Error  value
(Intercept) 1.9174   0.5482 3.498 0.00194
GPA_G       0.4940   0.1630 3.031 0.00594
```

At the bottom of the output from the `summary`, it should be noted that the p-value for the model is $p = 0.005939$. To connect this to our first look at this dataset, recall that we computed the correlation coefficient via `cor` (yielding 0.53). We can obtain the statistical significance for that p-value via `cor.test(alc_data$GPA_G, alc_data$GPA_UG)`: $p = 0.005939$, i.e., the same value obtained through linear regression.

Of course, correlation is a one-dimensional calculation; regression can include many parameters through use of the addition (+) or multiplication (*) notation, see again: section on multi-way ANOVA including interaction terms (above).

Logistic Regression

Logistic regression is used to test predictors of a binary outcome. Consider the very first example presented here, i.e., the sub-group of non-compliant students. Suppose we wanted to determine whether there were any factors that might predict participation vs. non-compliance. We start by reading in the data brand new:

```
# merge compliant versus non-compliant
data_nonc=read.csv("Frontiers_1_
          SubSample.csv", head = TRUE,
stringsAsFactors = TRUE)
data_comp=read.csv("Frontiers_2_
          ALCData.csv", head = TRUE,
          stringsAsFactors = TRUE)
```

and perform a brief quality-control assessment to ensure that there are no common students between the group that participated in the study vs. those that did not:

```
# verify that there is no overlap between
datasets
olap=intersect(data_nonc$ID, data_comp$ID)
if (length(olap)==0){
      print("Data Quality Check OK")
}else{
      print("Data Error: Non-distinct
            datasets")
}
```

We note that this quality check used a new technique: control structure. More about that in a subsequent section (see below).

Because we want to test compliance as a single parameter shared between these two datasets, we must merge them into a single dataset. Firstly, create a new parameter (call it "Status")

```
# bind an indicator for compliant versus
non-compliant
data_nonc$Status = "Noncompliant"
data_comp$Status = "Compliant"
```

and subset the two datasets to the same variables

```
# re-configure variable lineups
data_nonc=subset(data_nonc,
select=c("ID","Cohort","Group","GPA_UG",
        "GPA_G","Status"))
data_comp=subset(data_comp,
select=c("ID","Cohort","Group","GPA_UG",
        "GPA_G","Status"))
```

These datasets are now ready for merger by row-binding:

```
# merge datasets
data_stat=rbind(data_nonc, data_comp)
```

The `Status` variable is more likely than not going to be viewed by R as a character variable, given that it was constructed by entering strings. But a logistic regression requires the dependent variable to be factor. Force this class through `as.factor`:

```
# enforce classes
data_stat$Status=as.factor(data_stat$Status)
```

Now, a logistic regression is possible.

```
# set up a logistic regression
logmod=glm(Status~GPA_G+GPA_UG,
         data=data_stat,family=binomial)
```

which yields (through `summary(logmod)`):

	Estimate	Std. Error	z value	Pr(>|t|)
(Intercept)	−2.3298	5.6412	−0.413	0.680
GPA_G	−0.1772	1.5685	−0.113	0.910
GPA_UG	0.6109	1.9398	0.315	0.753

from which it can be concluded that neither undergraduate, nor graduate GPA is significant in its association with compliance in participation.

Mixed Effects

One additional approach worthy of mention is mixed effects modeling, where a process is believed to contain both fixed and random effects. Consider, for example, that our logistic regression model contains graduate students from a diverse set of backgrounds, i.e., various undergraduate institutions and degree programs, and perhaps the students had various levels of engagement during their undergraduate careers. In this setting, and following the constructs of LaMotte (Roy LaMotte, 2006) it is reasonable to believe that the undergraduate GPA might be considered a random effect in slope. We load the library containing mixed effects resources, and reconstruct the logistic regression model with a slight rephrasing of the `GPA_UG` term:

```
# load library for mixed effects modeling
library(lme4)
# set up a mixed effects logistic
regression
logmod2=glmer(Status~GPA_G+(1|GPA_UG),
         data=data_stat,
family=binomial)
```

where the parentheses in the `(1|GPA_UG)` term flag that variable for random effects, and the `1|` indicates that the variable is random in slope (`0|` would indicate random in intercept). Otherwise, the syntax for performing and obtaining results from mixed models is generally the same as for conventional models. Note that the syntax for a mixed effects models is almost identical to that of a conventional linear- or logistic regression: the first argument in the Response ~ Predictor(s), but with the addition of a term in parentheses indicating random effects.

NOTES

Control Structures

While beyond the intended scope of this article, we briefly follow-up on a concept introduced in section Logistic Regression, i.e., control structures. It is possible to provide R with a set of instructions, and for R to carry out only one action, according to pre-defined criteria. The generic structure is as follows: `if (condition){action #1}else{action #2}`, however more elaborate, multi-conditional structures are possible using `else if`. Consider a simplistic example

```
# simple conditional control structure
gpa_val=3.5
if (gpa_val < 3.6){
      print("Med")
}else if (gpa_val < 3.0){
      print("Low")
}else{
      print("Hi")
}
```

Surely this loop would have no conceivable utility in an actual analytical code, but as an example, it shows proper use of if-else, a basic command structure in every coding language. The criteria does not need to be numerical; any logical/Boolean cue will work, `"g" %in% c("a","b","c","d","e","f","g","h")` will yield a `TRUE`.

Perhaps the most common iterative loop structure is the for-loop. The basic structure is `for (sequence){action}`. A simple example would be

```
# simple for-loop
for (i in 1:5){
    print(paste("This is loop number ", i,
sep=""))
}
```

```
     This is loop number 1
     This is loop number 2
     This is loop number 3
     This is loop number 4
     This is loop number 5
```

The for-loop can be very powerful: it is possible to perform the same action many times through use of a for-loop, especially when a conditional logic is nested within. For instance:

```
# scroll through each record and report
red-flag students
for (i in 1:nrow(comp_data)){
     if (comp_data$Category[i] == "REDFLAG"){
          print(paste("Student ",
                 comp_data$ID[i],
                 ": Red Flag", sep=""))
     }
}
```

One last control structure bearing discussion is the while-loop: `while (condition){action}`. This loop is, in some respects, equivalent to a hybrid between the if-loop and the for-loop: it iterates (like a for-loop), until a condition has been met (like an if-loop). The major caveat here is that for a for-loop there is a finite terminus: once the control sequence is exhausted, the loop will cease; for a while-loop, it is possible for the criterion to never be realized, leading the loop to continue *ad infinitum* (a "runaway"). A runaway loop will seize R; the only remediation is to stop the calculation through the STOP button in the R console, or to brute-force stop the software, e.g., Task Manager.

RStudio

R as described here is intended to convey primarily command-line programming within the traditional R computing environment. The reader may take interest, however, in some of the tool suites available as augmentation or substitute for R, including RStudio. RStudio offers the same functionality of R in computation, but much more accessibility in terms of user interface. Personal license for single-use RStudio or RStudio Server remains free of charge, but there are Commercial packages that unlock additional features including Priority Support.

Other Softwares

It is expected that new users to R will be "immigrating" from other software platforms, or may have need to interface with team members who are utilizing other softwares. There are options for this kind of transfer or exchange. Firstly, there are many guides offering syntactic cross-walk between R and SPSS, SAS, Matlab, etcetera (Muenchen and Hilbe, 2010; Muenchen, 2011; Kleinman and Horton, 2014). But there are also some built-in codeworks that allow for ease of transition. For example, the `SASxport` library and sub-functions within the `Hmisc` library. While some functions shall remain -by design- the singular domain of their proprieter, a great many functionalities available in the commonly available statistical softwares can be found in R.

We note, also, that R is not the only open-source computing environment that is popular among data scientists. For instance, Python, is widely used, particularly in the engineering and physical sciences. Historically, Python is considered a more general-purpose language; R is almost always used for statistical analysis. However, both packages have flexibility into both analytical and technical application, and either present many options for those seeking to add a computational edge to their work. For proprietary environments, Matlab has grown its statistical packages incredibly over the past 15 years, and has an incredible plotting engine, and is a de rigueur skills set for nearly every graduating engineer. SAS is considered by many to be the mainline software in clinical trials design and primary analyses, although the US Food and Drug recently clarified that they do not explicitly require the use of any specific software for statistical analysis (Statistical Software Clarifying Statement, 2015). We find that among the commercially available softwares used for statistical analysis, STATA and SPSS are popular with colleagues in the fields of education and allied health. We urge that every software has its plusses-and minuses, but that the open access of R, and its thriving users community, make it highly attractive to the intrepid investigator seeking a useful and portable analytical tool.

CONCLUSION

In this article, we provide educational researchers with an introduction to the statistical program, R. The examples provided illustrate the applicability of R for use in educational research; these vignettes are designed to be extensible to a broader set of problems that interest those in educational research. While numerical programming is not typically the province of the educational researcher, we urge that it is easy to get started, and that there is excellent support for those wishing to integrate R into their analytical repertoire.

ETHICS STATEMENT

Data presented in this manuscript were collected in a protocol approved by the Human Subjects Committee at the University of Hartford. All participants gave written informed consent prior to study participation.

AUTHOR CONTRIBUTIONS

Code creation: Led by MW, reviewed and approved by TP. Data collection: Led by TP, consulting with MW. Manuscript creation: Tandem effort by MW and TP.

REFERENCES

1. Bishop-Clark, C., and Dietz-Uhler, B. (2012). *Engaging in the Scholarship of Teaching and Learning: A Guide to the Process, and How to Develop a Project From Start to Finish*. 1st edn. Sterling, VA: Stylus Publishing, LLC.
2. Boyer, E. L. (1997). *Scholarship Reconsidered: Priorities of the Professoriate*. 1st edn. Princeton, NJ: Carnegie Foundation for the Advancement of Teaching.
3. Curtis, N. A., and Trice, A. D. (2013). A Revision of the Academic Locus of Control Scale for College Students. *Percept. Mot. Skills* 116, 817–829. doi: 10.2466/08.03.PMS.116.3.817-829
4. Elliot, A. J., and Murayama, K. (2008). On the measurement of achievement goals: Critique, illustration, and application. *J. Educ. Psychol.* 100, 613–628. doi: 10.1037/0022-0663.100.3.613
5. Gilpin, L. S., and Liston, D. (2009). Transformative education in the scholarship of teaching and learning: an analysis of SoTL literature. *Int. J. Scholarsh. Teach. Learn.* 3:11. doi: 10.20429/ijsotl.2009.030211
6. Hutchings, P., Huber, M., and Ciccone, A. (2011). Feature essays: getting there: an integrative vision of the scholarship of teaching and learning. *Int. J. Scholarsh. Teach. Learn.* 5:31. doi: 10.20429/ijsotl.2011.050131
7. Kanuka, H. (2011). Keeping the scholarship in the scholarship of teaching and learning. *Int. J. Scholarsh. Teach. Learn.* 5:3. doi: 10.20429/ijsotl.2011. 050103
8. Kleinman, K., and Horton, N. J. (2014). *SAS and R: Data Management, Statistical Analysis, and Graphics*. 2nd edn. Boca Raton, FL: CRC Press, Taylor & Franciss Group.
9. Martin, R. C. (Ed.) (2009). *Clean Code: A Handbook of Agile Software Craftsmanship*. Upper Saddle River, NJ: Prentice Hall.
10. Mertens, D. M. (2014). *Research and Evaluation in Education and Psychology: Integrating Diversity With Quantitative, Qualitative, and Mixed Methods, 4th Edn.* Sage Publications, Inc.
11. Muenchen, R. A. (2011). *R for SAS and SPSS Users*. 2nd edn. New York, NY: Springer.

12. Muenchen, R. A., and Hilbe, J. M. (2010). *R for Stata Users*. New York, NY: Springer.
13. Osborne, J. M., Bernabeu, M. O., Bruna, M., Calderhead, B., Cooper, J., Dalchau, N., et al. (2014). Ten simple rules for effective computational research. *PLoS Comput. Biol.* 10:e1003506. doi: 10.1371/journal.pcbi. 1003506
14. Prlic´, A., and Procter, J. B. (2012). Ten simple rules for the open development of scientific software. *PLoS Comput. Biol.* 8:e1002802. doi: 10.1371/journal.pcbi.1002802
15. Roy LaMotte, L. (2006). "Fixed-, random-, and mixed-effects models," in *Encyclopedia of Statistical Sciences*, eds. S. Kotz, C. B. Read, N. Balakrishnan, B. Vidakovic, and N. L. Johnson (Hoboken, NJ: John Wiley & Sons, Inc.).
16. Statistical Software Clarifying Statement. (2015). *United States: Food & Drug Administration*. Available online at: https://www.fda.gov/downloads/ForIndustry/DataStandards/StudyDataStandards/UCM587506.pdf (Accessed August 13, 2018).
17. Thieme, N. (2018). R generation. *Significance* 15, 14–19. doi: 10.1111/j.1740-9713.2018.01169.x
18. TIOBE (2018). *TIOBE Index for March 2018 (Eindhoven, Netherlands)*. Available online at: https://www.tiobe.com/tiobe-index/ (Accessed March 19, 2018).
19. Willems, K. (2014). *What is the Best Statistical Programming Language? Infograph*. Available online at: https://www.datacamp.com/community/tutorials/statistical-language-wars-the-infograph (Accessed March 19, 2018).
20. Wilson, G., Aruliah, D. A., Brown, C. T., Chue Hong, N. P., Davis, M., Guy, R. T., et al. (2014). Best practices for scientific computing. *PLoS Biol.* 12:e1001745. doi: 10.1371/journal.pbio.1001745
21. Witman, P. D., and Richlin, L. (2007). The status of the scholarship of teaching and learning in the discipline. *Int. J. Scholarsh. Teach. Learn.* 1:14. doi: 10.20429/ijsotl.2007.010114

10

Numerical Analysis of Zipper Fracturing using a Non-Planar 3D Fracture Model

Zhen Wang[1,2]*, *Lifeng Yang*[1,2], *Rui Gao*[1,2], *Guanshui Xu*[3], *Zhe Liu*[1,2], *Shaoyuan Mo*[1,2], *Meng Fan*[1,2] *and Xin Wang*[1,2]

[1]*CNPC Key Laboratory of Oil and Gas Reservoir Stimulation, Langfang, China,* [2]*Research Institute of Petroleum Exploration and Development, PetroChina, Beijing, China,* [3]*University of California, Riverside, Riverside, CA, United States*

Correspondence:
Zhen Wang
wangzhen69@petrochina.com.cn

Zipper fracturing has become one of the routine reservoir stimulation methods for developing unconventional resources such as shale oil and gas. A non-planar 3D fracture propagation model is used to study the fracture propagation behavior of zipper fracturing. The effects of stage time lag, staggered perforation cluster spacing, and horizontal principal stress contrast on the fracture geometry are analyzed through numerical simulation. The simulation results show that increasing the stage time lag can increase the fracture length and improve inter-well stimulation. For the formation with higher horizontal principal stress contrast, the staggered perforation cluster spacing has a minimum effect on the fracture configuration of zipper fracturing. For the formation with lower horizontal principal stress contrast, the fractures follow the curved paths and intersect with each other, which potentially has adverse effects on the stimulation of the subsequent stages.

Keywords: zipper fracturing, stress interference, frac hit, 3D fracture propagation, proppant transport

1 INTRODUCTION

The development of hydraulic fracturing technology for unconventional petroleum resources has been a subject of considerable interest in recent years (Soliman et al., 2008; Li et al., 2017; Zhang et al., 2017; Zhang et al., 2019; Xie et al., 2020; Zhang et al., 2021a; Zhang et al., 2021b). Multistage, multicluster fracturing (MMF) of horizontal wells has become widely used to produce hydrocarbon from unconventional reservoirs, such as shales and tight sandstones. As shown in **Figure 1A**, the consecutive-fracturing sequence is the most commonly used fracturing sequence of MMF for a single horizontal well. In consecutive fracturing, fractures are created sequentially stage by stage along the same wellbore. Another fracturing sequence of MMF in a single horizontal well, referred to as "alternate fracturing," was proposed to decrease fracture spacing and generate additional fracture complexity (Roussel and Sharma 2011). This strategy places the second fracture at the location of what would traditionally be the third fracture, followed by the propagation of a "middle fracture" (**Figure 1B**). However, due to the limitations of fracturing tools and fracturing technique levels, the alternate fracturing method is rarely used for practical hydraulic-fracturing treatments. Zipper fracturing is also proposed as a method to improve production performance in wells in pad-drilling applications. Specifically, this method refers to fracturing two or more parallel horizontal wells. The fracturing treatment is performed stage by stage alternatively along the two wells. The method saves the pump-down and plug-and-perf standby time and, therefore improves operational efficiency compared with consecutive fracturing in one well. It is also expected that the treatment further

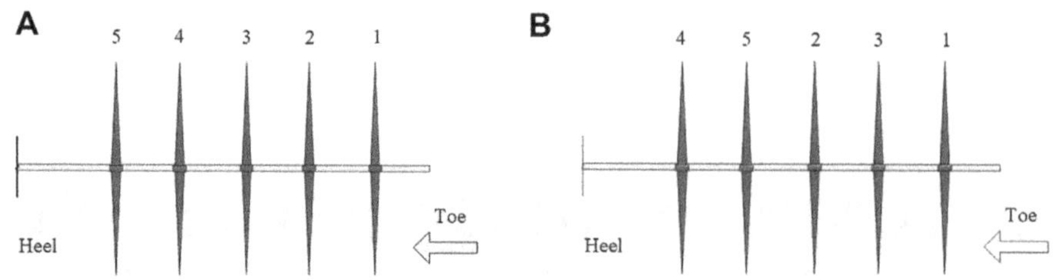

FIGURE 1 | Sketch of two fracturing sequence in the single horizontal well. **(A)** Consecutive sequence (1-2-3-4-5); **(B)** alternate sequence (1-3-2-4-5).

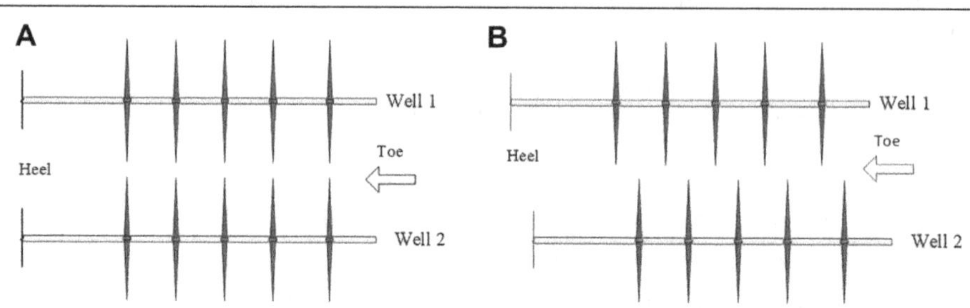

FIGURE 2 | Sketch of two completion scenarios of zipper fracturing in two horizontal wells. **(A)** Tip-to-tip zipper fracturing configuration; **(B)** staggered zipper fracturing configuration.

promotes fracture network complexity to improve production performance (Qiu et al., 2015). **Figure 2** illustrates the two completion scenarios of zipper fracturing. The so called "tip-to-tip" zipper fracturing is shown in **Figure 2A**, and staggered zipper fracturing is shown in **Figure 2B.** In a practical stimulation treatment, it is nearly impossible to guarantee the precise tip-to-tip fracture placement in two horizontal wells. Therefore, we consider the tip-to-tip configuration as the special case of staggered zipper fracturing. Only staggered zipper fracturing is studied in this paper.

Various mathematical models have been used to evaluate the effectiveness and performance of zipper fracturing. Rafiee et al. (2012) estimated both tip-to-tip and staggered zipper in terms of the fracture geometry, stress interference, and production performance through an analysis of different predefined planar fractures. The results demonstrated that the staggered zipper fracturing improves the performance of fracturing treatment compared with the tip-to-tip zipper fracturing due to increased contact area. Using a poroelastic model, Manchanda et al. (2014) studied the time-dependent stress interference behavior of induced unpropped fractures in both consecutive and zipper fracturing. The results showed that the time interval between adjacent fractures in a wellbore is a key parameter for the design of treatments. The longer time interval leads to stress shadow shrinking from the closure of the induced unpropped fractures, resulting in less fracture interference and improved performance. The model also assumed the predefined planar fracture in the numerical analysis. Shen (2014) investigated the influence of neighboring stimulation stages on generating the stimulated reservoir volume (SRV) with a continuum damage model based on a hydro-mechanical finite element method (FEM). The results indicated that the SRV generated by the sequential injection method is significantly less than that generated by the simultaneous injection method. Nagel et al. (2014) presented a study of multiwell completion using a hydromechanical coupled discrete element model. The results showed that the improvement in well stimulation using the zipper-fracturing completion technique is highly dependent on the *in situ* pore pressure, *in situ* stresses, and natural fracture characteristics. Sierra and Mayerhofer (2014) reviewed and discussed numerical reservoir-modeling results of zipper fracturing and demonstrated that the benefits of zipper fracturing could be maximized by generating proper fracture geometry and properly staggering the perforation/fracture system. Qiu et al. (2015) studied zipper fracturing using an unconventional fracture model capable of simulating branched fracture propagation. The quantitative results show that zipper fracturing may not deliver an obvious production benefit compared with sequential fracturing, depending on well spacing and perforation cluster spacing. Shi et al. (2018) applied the extended finite element method (XFEM), which is based on the cohesive zone method (CZM), to model fracture propagation at a reservoir with different mechanical homogeneities. The numerical studies confirmed that a staggered zipper fracturing scenario could assist hydraulic fracture propagation along a straight path.

The displacement discontinuity method (DDM) is widely used in dealing with fracture deformation problems for its high computational efficiency (Tang et al., 2018; Tang et al., 2019; Cong et al., 2021a; Cong et al., 2021b; Li et al., 2021). In this paper, we adopt a nonplanar 3D fracture model, based on 3D DDM, to investigate the fracture propagation behavior of zipper fracturing. The effects of the stage time lag, staggered perforation cluster spacing, and stress contrast will be studied in detail. Here the stage time lag refers to the time interval between the adjacent two fracturing stages. The nonplanar 3D fracture model simulates fracture propagation, fluid flow, and proppant transportation in a fully coupled hydromechanical manner (Xu and Wong 2013). The special feature of this model includes fracture closure in shut-in and curved fracture intersection, which makes quantitative analysis of time-dependent fracture interference and intersection possible.

2 NON-PLANAR 3D FRACTURE MODEL

Mathematical formulations

The adopted non-planar 3D fracture model assumes that all fractures are vertical and can turn in the horizontal direction. The fractures are discretized into structured rectangular elements, and all field variables, such as fracture opening, shear displacement, fluid pressure, and proppant volumetric concentration, are registered and associated with these elements. Fractures are assumed to be filled with slurry (the mixture of fracturing fluid and proppants), and fluid lag at the crack tip region is neglected.

Specifically, the fracture deformation is described by the 3D displacement discontinuity method (Shou 1994):

$$\begin{cases} -\sigma_{sl}^{i} = \sum_{k=1}^{N} K_{sl,sl}^{ik} D_{sl}^{k} + \sum_{k=1}^{N} K_{sl,sh}^{ik} D_{sh}^{k} + \sum_{k=1}^{N} K_{sl,nn}^{ik} D_{nn}^{k} \\ -\sigma_{sh}^{i} = \sum_{k=1}^{N} K_{sh,sl}^{ik} D_{sl}^{k} + \sum_{k=1}^{N} K_{sh,sh}^{ik} D_{sh}^{k} + \sum_{k=1}^{N} K_{sh,nn}^{ik} D_{nn}^{k} \\ p - \sigma_{nn}^{i} = \sum_{k=1}^{N} K_{nn,sl}^{ik} D_{sl}^{k} + \sum_{k=1}^{N} K_{nn,sh}^{ik} D_{sh}^{k} + \sum_{k=1}^{N} K_{nn,nn}^{ik} D_{nn}^{k} \end{cases} \tag{1}$$

where D_{nn} is the normal displacement discontinuity; D_{sl}, D_{sh} is the shear displacement discontinuity in the fracture length and height direction, respectively; p is the fluid pressure; and σ_{nn}, σ_{sl}, and σ_{sh} are the effective formation normal and shear stress, which can be calculated from the remote *in situ* stress through the principle of superposition. The nine matrices Ks are the boundary influence matrices related to the element size, location, and orientation, as well as the elastic constants. The superscripts i and k represent the element index; N is the total number of elements.

The fluid flow and proppant transport within a fracture are represented by the volume conservation equations (Adachi et al., 2007):

$$\frac{\partial w}{\partial t} = \nabla \cdot \left[w(1-c)\mathrm{v}^{f} + wc\mathrm{v}^{p} \right] + \delta(x,y)q_0 - q_l \tag{2}$$

$$\frac{\partial (cw)}{\partial t} = \nabla \cdot (cw\mathrm{v}^{p}) + \delta(x,y)c_0 q_0 \tag{3}$$

where t is the time; w is the fracture width, namely, normal displacement discontinuity D_{nn}; c is the volume concentration of the proppants; q_0 is the fluid injection rate; c_0 is the specified proppant volume concentration in fluid injection; $\delta(x, y)$ is the Dirac delta function; and q_l corresponds to the fluid leak-off rate following a one-dimensional Carter type leak-off model:

$$q_l = \frac{2C_L}{\sqrt{t - \tau(x,y)}} \tag{4}$$

where $\tau(x, y)$ is the time at which the surface position is first exposed to the fracturing fluid, and C_L is the Carter leak-off coefficient.

v^f and v^p are volume averaged velocities of the fracturing fluid and proppant, respectively, described as:

$$\mathrm{v}^{f} = \frac{w^2}{12\mu^f} Q^f\left(c, \frac{w}{a}\right)\left(\nabla p - \rho^f \mathbf{g}\right) \tag{5}$$

$$\mathrm{v}^{p} = Q^p\left(c, \frac{w}{a}\right)\left(\mathrm{v}^{f} + \mathrm{v}^{s}\right) \tag{6}$$

where p is the fluid pressure, $\mathbf{g}$ is the gravity vector, a is the proppant radius, μ^f is the apparent viscosity of fracturing fluid, ρ^f is the fracturing fluid density, and v^s is the proppant settling velocity, given by:

$$\mathrm{v}^{s} = \frac{a^2}{12\mu^f}\left(\rho^p - \rho^f\right)\mathbf{g} \tag{7}$$

where ρ^p is the proppant density.

Function $Q^f(c, \frac{w}{a})$ reflects the effects of proppant concentration on the fracturing fluid viscosity and captures the transition from Poiseuille's flow to Darcy's filtration flow when the concentration reaches the maximum value, which can be written as (Dontsov and Peirce 2015):

$$Q^f\left(c, \frac{w}{a}\right) = \left(1 - \frac{c}{c_{max}}\right)^{\beta} + \frac{a^2}{w^2}\frac{c}{c_{max}}\tilde{D} \tag{8}$$

where $\beta = 1.5$, $\tilde{D} = 8(1 - c_{max})^{\alpha}/3c_{max}$, $\alpha = 4.1$, and $c_{max} = 0.585$.

Blocking function $Q^p(c, \frac{w}{a})$ is equal to 0 when proppant bridging appears or proppant concentration arrives at the maximum value, and 1 otherwise, which can be evaluated from experimental data for specific fracturing materials.

To include the limited entry technique, the perforation pressure drop p_r is modeled by the Bernoulli equation (Long and Xu 2017):

$$p_r = \frac{8\rho}{\pi^2 C_d^2 d_p}\left(\frac{q_{ri}}{n}\right)^2 \tag{9}$$

where C_d is the dimensionless discharge coefficient; d_p is the perforation diameter; q_{ri} is the volumetric injection rate at a specific injection location; n the number of perforations at the specific injection location; ρ is the slurry density, and $\rho = (1-c)\rho^f + c\rho^p$. It is assumed that the pressure at the perforation in the wellbore is equal everywhere.

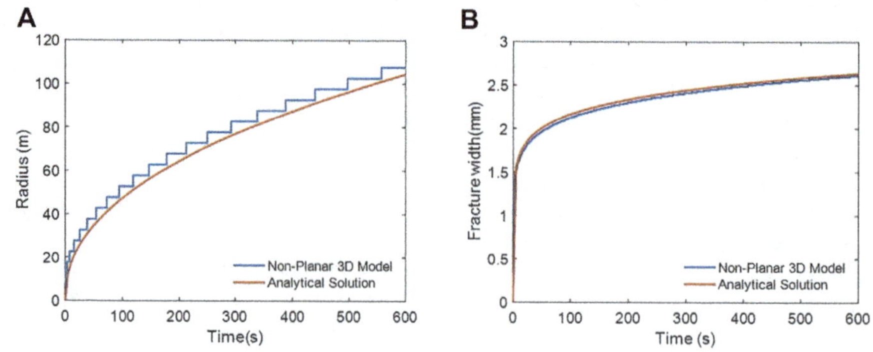

FIGURE 3 | Comparisons between nonplanar 3D model and analytical solutions. **(A)** Radius; **(B)** fracture width.

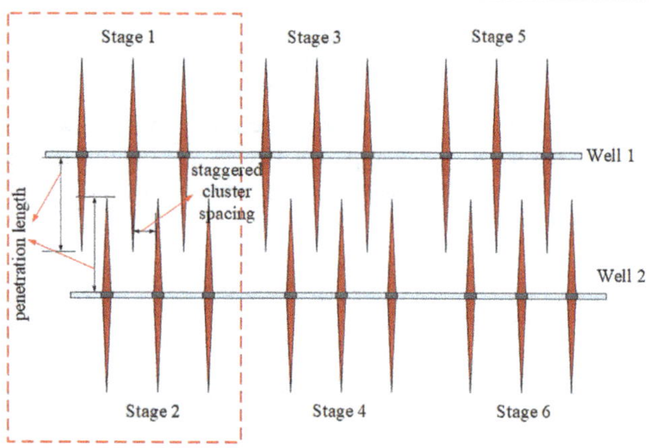

FIGURE 4 | Sketch of zipper fracturing.

Numerical procedure

The volume conservation **Eqs. 2** and **3** for fluid and proppant are discretized by the finite volume method as follows:

$$w^{t+\Delta t} = w^t + \left(\sum_M l_{nm} q_{nm}\right)\frac{\Delta t}{A_n} + q_0\frac{\Delta t}{A_n} - 4C_L\left(\sqrt{t+\Delta t - t_0} - \sqrt{t-t_0}\right) \quad (10)$$

$$c^{t+\Delta t} = c^t + \left(\sum_M l_{nm} q^p_{nm}\right)\frac{\Delta t}{w_n A_n} + c_0 q_0 \frac{\Delta t}{w_n A_n} \quad (11)$$

where Δt is the time step, A_n is the area of element n, t_0 is the time at which the element position is first exposed to the fracturing fluid, l_{nm} is the length of the intersection line between elements n and m, and M is the number of surrounding elements of element n. The subscript nm indicates the interface between element n and m. The interface slurry flux q_{nm} is expressed as:

$$q_{nm} = q^f_{nm} + q^p_{nm} \quad (12)$$

The interface fluid flux q^f_{nm} is given by:

$$q^f_{nm} = (1 - c_{nm})\frac{w^3_{nm}}{12\mu^f_{nm}} Q^f_{nm}\left(c, \frac{w}{a}\right)\left(\frac{p_n - p_m}{D_{nm}} - \rho^f_{nm} g_{nm}\right) \quad (13)$$

The interface proppant flux q^p_{nm} is:

$$q^p_{nm} = c_{nm} Q^p_{nm}\left(c, \frac{w}{a}\right)\left(\frac{w^3_{nm}}{12\mu^f_{nm}} Q^f_{nm}\left(c, \frac{w}{a}\right)\left(\frac{p_n - p_m}{D_{nm}} - \rho^f_{nm} g_{nm}\right) + w_{nm}\frac{a^2_{nm}}{12\mu^f_{nm}}\left(\rho^p_{nm} - \rho^f_{nm}\right) g_{nm}\right) \quad (14)$$

where D_{nm} is the distance between the two central nodes of element n and m. The interface variables a, c, w, ρ^f, ρ^p, μ^f are determined by upwind scheme.

The elasticity **Eq. 1** and volume conservation **Eqs 10–11** form a set of transient nonlinear coupled equations, with the fracture width w, fluid pressure p, and proppant

TABLE 1 | Basic input parameters.

Fluid properties		**Proppant properties**	
Viscosity (Pa.s)	**Density (kg/m³)**	**Diameter (mm)**	**Density (kg/m³)**
0.1	1,000	0.2	2,500
Formation properties			
Young's modulus (GPa)	Poission's ratio	Toughness (MPa·m$^{0.5}$)	Leak-off (m/s$^{0.5}$)
20	0.25	1	2×10^{-5}
Injection parameters			
Perforation diameter (mm)		Dimensionless discharge coefficient Cd	No. of perforations
10		0.6	12

TABLE 2 | Pumping schedule.

Injection rate (m^3/min)	Injection time (min)	Proppant volume fraction	Proppant weight fraction (kg/m^3)	Stage
12	10	0	0	1
12	10	0.05	131.579	1
12	10	0.1	277.778	1
12	10	0.15	441.176	1
12	10	0.2	625	1
12	10	0.25	833.333	1
0	60	0	0	1
12	10	0	0	2
12	10	0.05	131.579	2
12	10	0.1	277.778	2
12	10	0.15	441.176	2
12	10	0.2	625	2
12	10	0.25	833.333	2
0	120	0	0	2

concentration c as the primary variables. The coupled nonlinear equations are solved in an explicit time-stepping method with an adaptive time integration algorithm. For a given time step Δt, substituting the known w, p, and c of the last time step into **Eqs. 1, 10,** and **11**, the primary variables of the current time step can be easily obtained.

Fracture advances when the mode I (opening) stress intensity factor K_I of the crack tip element exceeds the fracture toughness K_{Ic} of the rock, namely:

$$K_I \geq K_{Ic} \tag{15}$$

where $K_I = 0.806\frac{\sqrt{\pi}E}{4(1-\upsilon^2)\sqrt{\Delta a}}D_{nn}$, and Δa is the crack tip element length; E is Young's modulus, and υ is Poisson's ratio.

The direction of fracture growth is oriented along with the maximum horizontal principal stress, which is expressed by:

$$\theta = \begin{cases} 2\tan^{-1}\left(\frac{1}{4}\frac{K_I}{K_{II}} + \frac{1}{4}\sqrt{\left(\frac{K_I}{K_{II}}\right)^2 + 8}\right), & when\ \frac{K_I}{K_{II}} < 0 \\ 2\tan^{-1}\left(\frac{1}{4}\frac{K_I}{K_{II}} - \frac{1}{4}\sqrt{\left(\frac{K_I}{K_{II}}\right)^2 + 8}\right), & when\ \frac{K_I}{K_{II}} > 0 \end{cases} \tag{16}$$

where θ is the propagation angle, which is in the polar coordinate system with the origin at the crack tip; K_{II} is the stress intensity factor of mode II (shearing) in the length direction and given by $K_{II} = 0.806\frac{\sqrt{\pi}E}{4(1-\upsilon^2)\sqrt{\Delta a}}D_{sl}$.

The model is validated by comparing with existing asymptotic analytical solutions of penny-shaped hydraulic fracture propagation (Dontsov 2016), where the viscosity-dominated fracture configuration is described by:

$$r(t) = 0.6944\left(\frac{q_0^3 E t^4}{12\mu(1-\upsilon^2)}\right)^{\frac{1}{9}}$$
$$w_{in}(t) = 0.1901\left(\frac{(12\mu)^2(1-\upsilon^2)^2 q_0^3 t}{E^2}\right)^{\frac{1}{9}} \tag{17}$$

where r is the fracture radius, and w_{in} is the fracture width at the inlet. The input parameters are listed as $q_0 = 5\ m^3/min$, $\mu = 5$ mPa·s, $E = 30$ GPa, $\nu = 0.20$, $K_{Ic} = 0.2\ MPa{\cdot}m^{0.5}$, $C_L = 0\ m/s^{0.5}$, $t = 10$ min. The results (**Figure 3**) show a good agreement between our non-planar 3D model and analytical solutions.

3 MODELING DETAILS AND RESULTS

Basic inputs

This section presents a detailed study of the effects of various parameters on zipper fracturing, including the stage time lag, perforation cluster spacing, and stress contrast. A typical two-horizontal-well zipper fracturing with staggered fractures is considered as shown in **Figure 4**. Each stage has three perforation clusters. Without loss of generality, only stage 1 and stage 2 are selected to be analyzed. Here we define the length of the part of a fracture between two horizontal wells as penetration length, which can reflect the interwell stimulation effectiveness to some extent. The staggered perforation cluster spacing is also shown in **Figure 4** as the spacing between two adjacent perforation clusters in the two horizontal wells. We assume that the perforation cluster has an even spacing in the single horizontal well.

The basic inputs of the simulations, such as treatment, reservoir and completion parameters, are listed in **Table 1**. The formation is assumed to have three layers. The thickness of the perforated zone is 60 m and sandwiched by two stress-barrier zones with 5 MPa *in situ* stress confinement. The stress confinement contains the vertical height of the fracture; as a result, we can focus on the fracture-width distribution and length

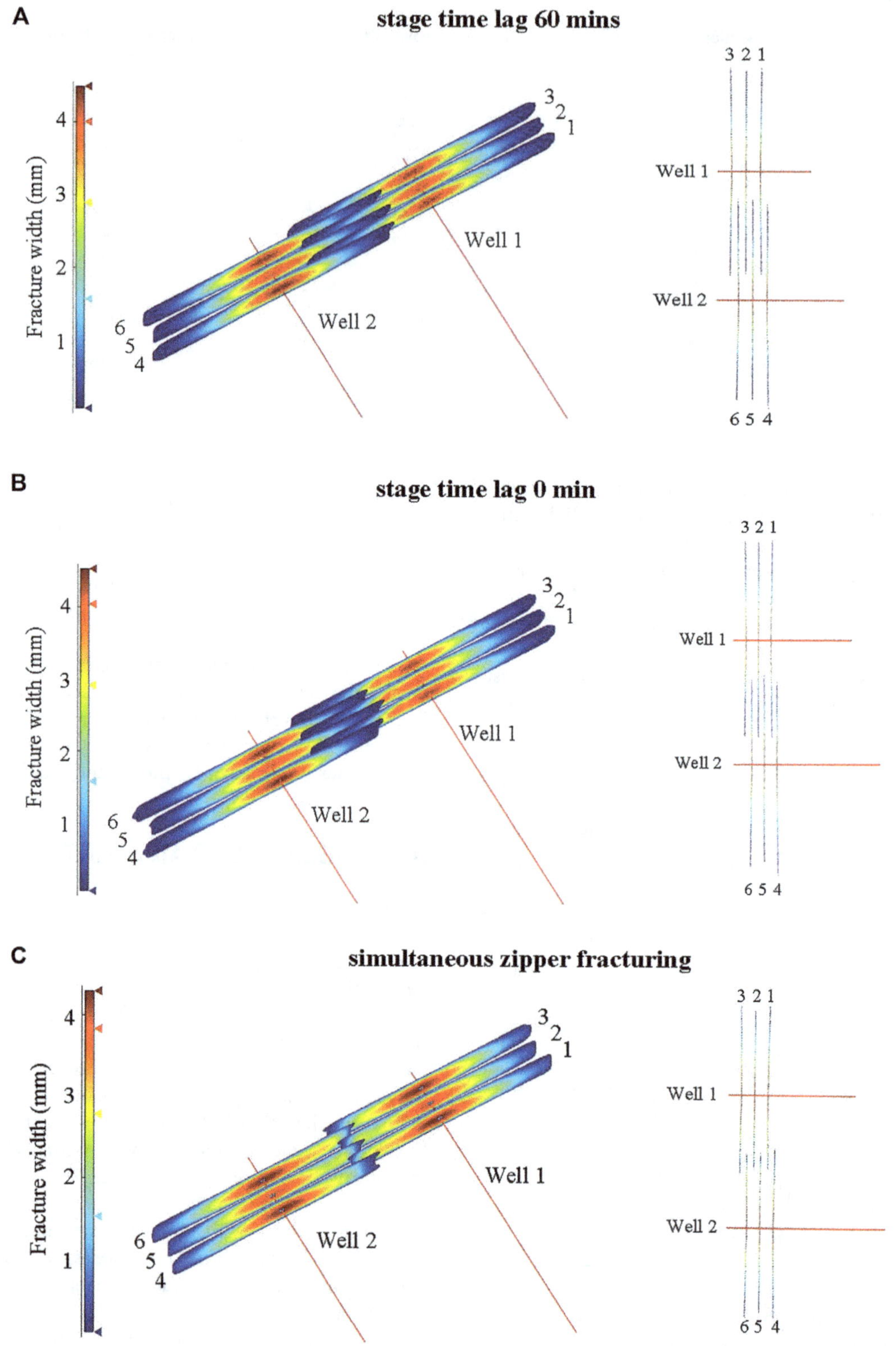

FIGURE 5 | Fracture configuration of different well completions. **(A)** Side view (left) and top view (right) of sequential zipper fracturing, stage time lag 60 min; **(B)** side view (left) and top view (right) of sequential zipper fracturing, stage time lag 0 min; **(C)** side view (left) and top view (right) of simultaneous zipper fracturing.

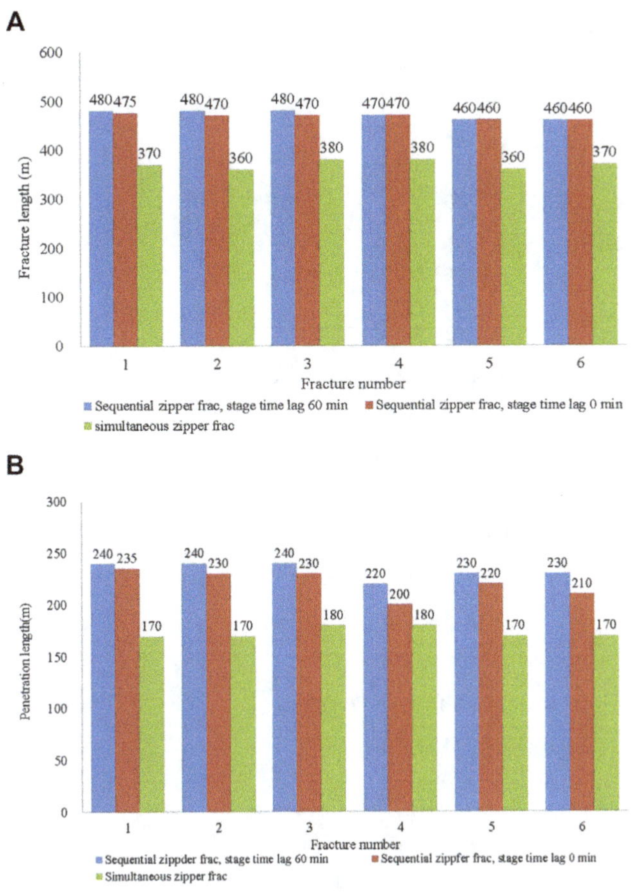

FIGURE 6 | Fracture length and penetration length of different well completions. **(A)** Fracture length; **(B)** penetration length.

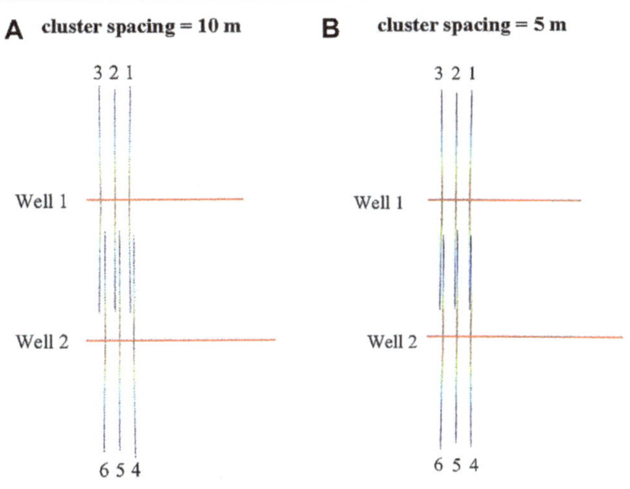

FIGURE 7 | Fracture configuration of irregular staggered zipper fracturing. **(A)** Top view of fracture configuration with staggered perforation cluster spacing 10 m; **(B)**: top view of fracture configuration with staggered perforation cluster spacing 5 m.

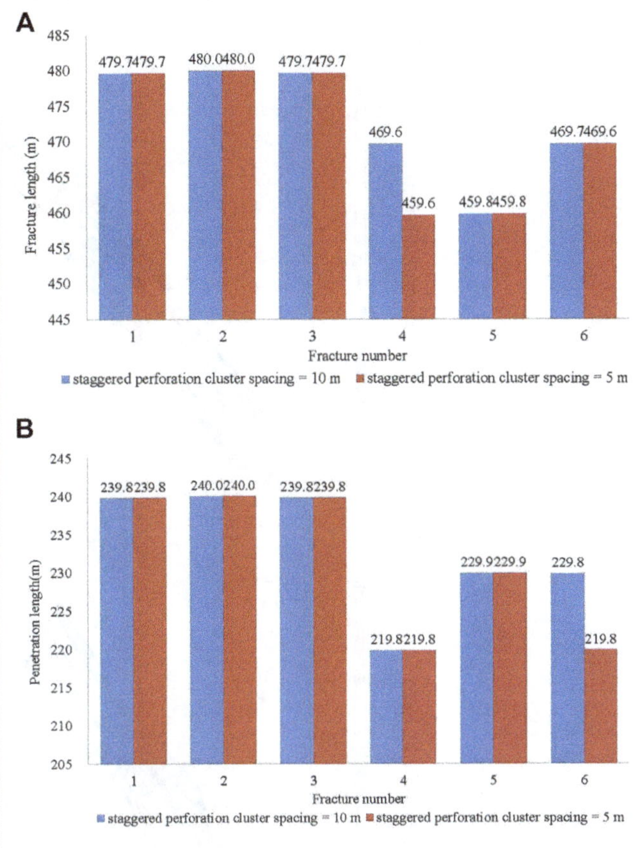

FIGURE 8 | Fracture size of irregular staggered zipper fracturing. **(A)** Fracture length; **(B)** penetration length.

when performing sensitivity analysis. The horizontal principal stress contrast is 10 MPa, and the well spacing is 300 m. The perforation cluster spacing of both wells is 30 m, and the staggered perforation cluster spacing is 15 m. **Table 2** presents the pumping schedule. The slurry injection rate is 12 m^3/min, and the total injection time is 60 min for one stage with a total proppant of 277 tons. The time lag between stage 1 in well 1 and stage 2 in well 2 is 60 min. The two stages use the same pumping schedule. After slurry injection, a 120-min shut-in time ensures that the fracturing fluid can leak off adequately and release the fluid pressure thoroughly. If not specified, the above-mentioned parameters are used for all the simulations.

Influence of stage time lag

Three zipper fracturing scenarios are presented and analyzed to investigate the influence of the stage time lag of a two-horizontal-well zipper fracturing on fracture growth. **Figure 5** shows the fracture configurations of different well completions. **Figure 5A** is side and top view of the basic model of sequential zipper fracturing with a 60-min stage time lag, as shown in the section 3.1. **Figure 5B** is side and top view of a sequential zipper fracturing without the stage time lag. **Figure 5C** is side and top view of a case of simultaneous zipper fracturing. Especially, for simultaneous zipper fracturing, the pumping schedule merges to one stage for the two wells, and the

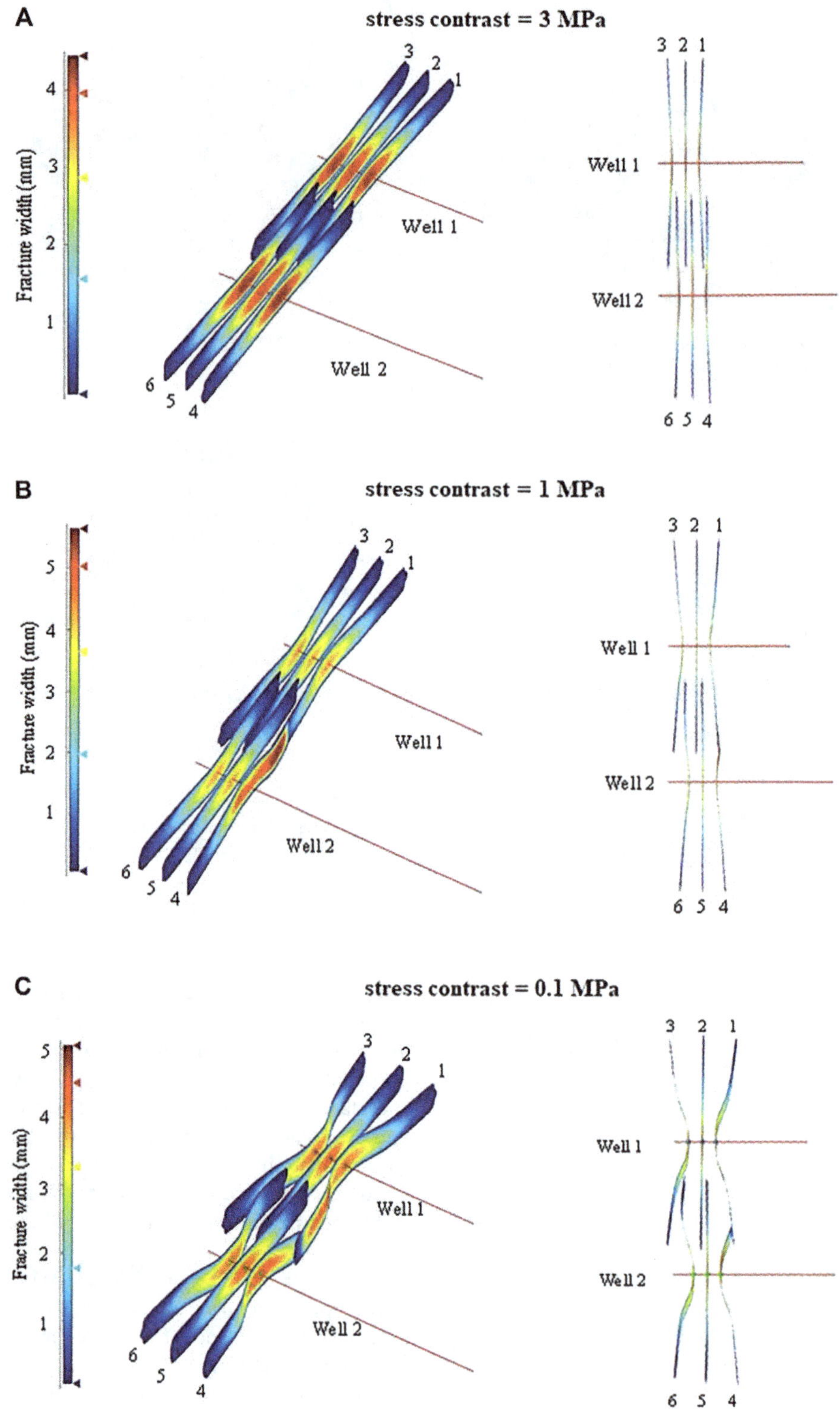

FIGURE 9 | Fracture configuration of different stress contrast scenarios. **(A)** Side view (left) and top view (right), stress contrast 3 MPa; **(B)** side view (left) and top view (right), stress contrast 1 MPa; **(C)** side view (left) and top view (right), stress contrast 0.1 MPa.

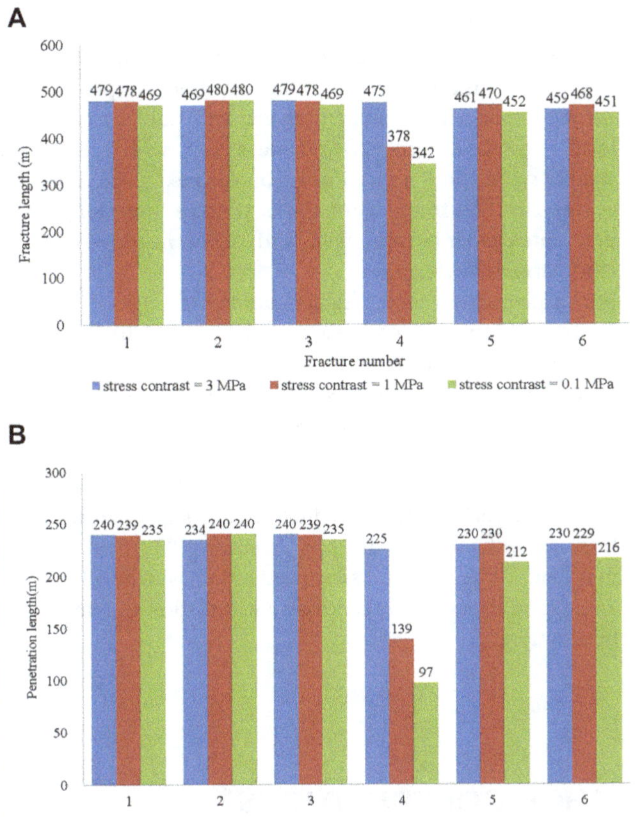

FIGURE 10 | Fracture size under different stress contrast scenarios. **(A)** Fracture length; **(B)**: penetration length.

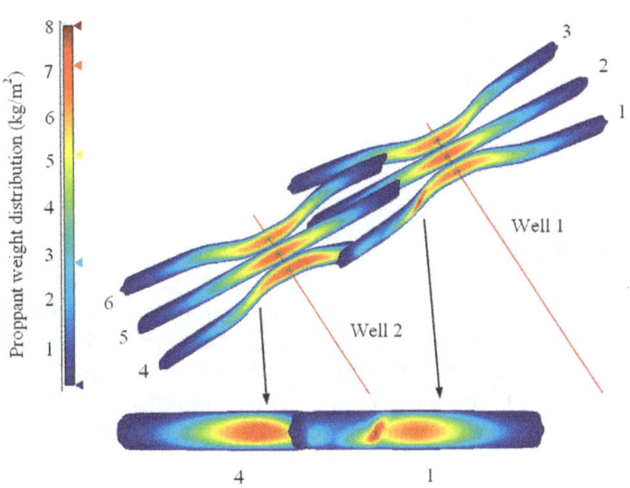

FIGURE 11 | Proppant weight distribution in fractures under 0.1-MPa stress contrast condition.

injection time doubles. **Figure 6** presents the fracture length and penetration length of hydraulically induced fractures of the corresponding well completion.

FIGURE 12 | Rod tubing filled with proppant sand during a frac-hit; cited from Jacobs (2017).

The results in **Figures 5**, **6** indicate that the stage time lag substantially influences fracture configuration. For the sequential zipper fracturing, as the stage time lag reduces, both the fracture length and the penetration length of well 1 decrease, as shown in **Figures 5A,B**, **6**. This implies that during the stage time lag, the fractures in well 1 continue propagating while leak-off, and this is caused by the rebalancing of fluid pressure in the fracture even though there is no fluid injection. Although the two stages employ the same pumping schedule, the induced fractures of stage 1 changed the local stress field around fractures, so the hydraulically induced fractures of stage 2 in well 2 are suppressed, to some extent, by the fractures of stage 1 in well 1. When the stage time lag reduces to zero, the fracture growth of stage 2 influences the rebalancing of fluid pressure of fractures in stage 1 and, thus, suppresses the further growth of fractures in stage 1. This is why the fracture length of well 1 with no stage time lag is less than the fracture length of well 1 with a 60-min stage time lag as shown in **Figure 6**. It is noteworthy that interwell/interstage interference is mutual, i.e., all of the fracture propagation in stage 1 and stage 2 are influenced by each other. The fracture length of well 2 does not seem to change for the two scenarios of the sequential zipper fracturing, but the penetration length decreases for the well with less stage time lag. The fracture tends to propagate away from the interwell region, which may result in inefficient interwell stimulation. The simultaneous zipper fracturing has the most severe interwell/interstage interference; the fracture length and penetration length decrease about 20% compared with sequential zipper fracturing, as shown in **Figures 5C**, **6**. The corresponding inter-well stimulation effectiveness is worst.

Influence of staggered perforation cluster spacing

To study the influence of staggered perforation cluster spacing on zipper fracturing, we decrease the staggered perforation cluster spacing to 10m and 5 m. Compared with **Figure 5A**, **Figure 7** shows that the fracture configuration still keeps nearly straight under the 10MPa horizontal principal stress contrast, even for the 5m perforation cluster spacing. Compared with **Figure 6**, **Figure 8** shows that the fracture length and penetration length remain almost the same in the three scenarios. Decreasing the staggered perforation cluster spacing has virtually no effect on the fracture configuration of zipper fracturing in this stress contrast condition. Consequently, we may conclude that one can design small perforation cluster spacing in one stage and small staggered perforation cluster spacing of two wells to further improve interwell stimulation effectiveness under high stress contrast conditions.

Influence of horizontal principal stress contrast

To study the influence of horizontal principal stress contrast on zipper fracturing, we decreased the horizontal principal stress contrast to 3MPa, 1MPa, and 0.1 MPa. Compared with **Figure 5A**, **Figures 9, 10** show that fracture-induced stress interference and inter-well interference significantly affect the fracture configuration. As shown in **Figure 9**, the two outer fractures of well 1 curve more as the stress contrast decreases. After some curved distance, the two outer fractures propagate again along the direction of in situ maximum horizontal principal stress. This is due to the less stress interference with the increase in fracture spacing. The inner fracture always keeps straight propagation because of the balance of two opposite stress interference caused by the outer fractures. Overall, the three fractures of well 1 for different scenarios present a uniform propagation pattern. The fracture length difference is less than 4% as shown in **Figure 10A**. The propagation of the three fractures of well 2 are affected by the interwell interference of well 1 and their own interfracture interference. Hence, the fractures grow shorter than the ones of well 1 and curve noticeably. It is worth noting that fracture 4 intersects with fracture 1 of well 1 under 1 and 0.1 MPa stress contrast, resulting in "frac hits" between the two wells. The fracture width and proppant weight distribution (**Figures 9C, 11**) show that after frac hits, the fracturing fluid and proppant of fracture 4 enter fracture 1, which causes fracture 1 to reopen and proppant transport toward the wellbore of well 1. This may result in a negative effect on the successive stimulation treatment and production. The simulation results in this section also confirm the proppant sand production phenomenon during a frac-hit in a practical treatment, as shown in **Figure 12** cited from Jacobs (2017).

4 CONCLUSION

In this paper, we investigate the effects of the stage time lag, staggered perforation cluster spacing, and horizontal principal stress contrast on the behavior of zipper fracturing. Based on our analyses, we have the following specific conclusions:

1) The stage time lag significantly affects the fracture configuration of zipper fracturing. The stress interference of simultaneous zipper fracturing is most severe, and hence, the fracture length and penetration length are affected the most, leading to the worst inter-well stimulation effectiveness. With the increase in the stage time lag, the pressurized fractures of the previous stage have more time to complete the final fracture propagation through rebalance of fluid pressure and fluid leak-off. The local stress field recovers to some extent, which leads to a weaker stress interference impact on the next stage. This may ensure that the fractures of the next stage grow longer and penetrate deeper between wells and, hence, improve interwell stimulation effectiveness.
2) For a formation with higher horizontal principal stress contrast, stress contrast dominates the fracture propagation. Decreasing the staggered perforation cluster spacing has almost no effect on the fracture configuration of zipper fracturing. One may design small perforation cluster spacing to improve interwell stimulation effectiveness.
3) For a formation with lower horizontal principal stress contrast, interfracture stress interference significantly affects fracture configurations of zipper fracturing. The treatment job may cause "frac hits" and sand production, which will result in an adverse effect on the successive stimulation treatment and production.

AUTHOR CONTRIBUTIONS

ZW and LY proposed the research. RG and ZL prepared figures and tables, and interpreted the structural data. ZW, LY, and XW developed the main ideas. GX, SM, and MF participated in the numerical simulation work. ZW, LY, and GX contributed to the writing the original manuscript. SM, MF, and XW contributed to revising the manuscript. All co-authors actively contributed to manuscript with comments, ideas, and suggestions.

FUNDING

This work was financially supported by Research and Application of Key Technical Equipment for Major Engineering (Phase II) of CNPC Scientific Research and Technology Development Project (2018E-2105), Software Development Project of CNPC on Volume Fracturing Optimization & Design (2020B-4118).

ACKNOWLEDGMENTS

The authors would like to thank the editor and the reviewers of this article for their valuable comments that helped improve the quality of the paper.

REFERENCES

1. Adachi, J., Siebrits, E., Peirce, A., and Desroches, J. (2007). Computer Simulation of Hydraulic Fractures. *Int. J. Rock Mech. Mining Sci.* 44 (5), 739–757. doi:10.1016/j.ijrmms.2006.11.006
2. Cong, Z., Li, Y., Pan, Y., Liu, B., Shi, Y., Wei, J., et al. (2021b). Study on CO2 Foam Fracturing Model and Fracture Propagation Simulation. *Energy* 238, 121778, doi:10.1016/j.energy.2021.121778
3. Cong, Z., Li, Y., Liu, Y., and Xiao, Y. (2021a). A New Method for Calculating the Direction of Fracture Propagation by Stress Numerical Search Based on the Displacement Discontinuity Method. *Comput. Geotechnics* 140, 104482. doi:10.1016/j.compgeo.2021.104482
4. Dontsov, E., and Peirce, A. (2015). Proppant Transport in Hydraulic Fracturing: Crack Tip Screen-Out in KGD and P3D Models. *Int. J. Sol. Structures.* 63, 206–218, doi:10.1016/j.ijsolstr.2015.02.051
5. Dontsov, E. V. (2016). An Approximate Solution for a Penny-Shaped Hydraulic Fracture that Accounts for Fracture Toughness, Fluid Viscosity and Leak-Off. *R. Soc. Open Sci.* 3, 160737. doi:10.1098/rsos.160737
6. Jacobs, T. (2017). Oil and Gas Producers Find Frac Hits in Shale wells a Major challenge. *J. Pet. Tech.* 69 (04), 29–34. doi:10.2118/0417-0029-jpt
7. Li, J., Liu, Y., and Wu, K. (2021). A New Higher Order Displacement Discontinuity Method Based on the Joint Element for Analysis of Close-Spacing Planar Fractures. *SPE J.* doi:10.2118/208614-PA
8. Li, Y., Jia, D., Rui, Z., Peng, J., Fu, C., and Zhang, J. (2017). Evaluation Method of Rock Brittleness Based on Statistical Constitutive Relations for Rock Damage. *J. Pet. Sci. Eng.* 153, 123–132. doi:10.1016/j.petrol.2017.03.041
9. Long, G B., and Xu, G S. (2017), The Effects of Perforation Erosion on Practical Hydraulic-Fracturing Applications, *SPE J.*, 22 (2). 645–659, SPE-185173-PA. doi:10.2118/185173-pa
10. Manchanda, R., Sharma, M. M. M., and Holzhauser, S. (2014). Time-Dependent Fracture-Interference Effects in Pad Wells. *SPE Prod. Operations* 29 (04), 274–287. doi:10.2118/164534-pa
11. Nagel, N., Sheibani, F., Lee, B., Agharazi, A., and Zhagn, F. (2014). "Fully-Coupled Numerical Evaluations of Multiwell Completion Schemes: The Critical Role of *In-Situ* Pressure Changes and Well Configuration," in SPE Hydraulic Fracturing Technology Conference (The Woodlands, Texas: Society of Petroleum Engineers) USASPE-168581-MS. doi:10.2118/168581-ms
12. Qiu, F., Porcu, M. M., Xu, J., Malpani, R., Pankaj, P., and Pope, T. L. (2015). "Simulation Study of Zipper Fracturing Using an Unconventional Fracture Model," in SPE/CSUR Unconventional Resources Conference (Calgary, Alberta, Canada: Society of Petroleum Engineers). SPE-175980-MS. doi:10.2118/175980-MS
13. Rafiee, M., Soliman, M. Y., and Pirayesh, E. (2012). "Hydraulic Fracturing Design and Optimization: A Modification to Zipper Frac," in SPE Annual Technical Conference and Exhibition (San Antonio, Texas, USA: Society of Petroleum Engineers). SPE-159786-MS. doi:10.2118/159786-ms
14. Roussel, N. P., and Sharma, M. M. (2011). Optimizing Fracture Spacing and Sequencing in Horizontal-Well Fracturing. *Spe Prod. Operations* 26 (02), 173–184. doi:10.2118/127986-pa
15. Shen, X. (2014). "Numerical Analysis on the Interaction between Two Zipper Frac Wells Using the Continuum Damage Method," in Offshore Technology Conference-Asia (Kuala Lumpur, Malaysia: Offshore Technology Conference). OTC-24975-MS. doi:10.2118/24975-ms
16. Shi, X., Li, D. J., Liu, Y., Xu, H. X., and Du, Y. K. (2018). "Hydraulic Fracture Propagation in Horizontal Wells With Modified Zipper Fracturing in Heterogeneous Formation," in 52nd U.S. Rock Mechanics/Geomechanics Symposium (Seattle, Washington: American Rock Mechanics Association). ARMA-2018-189.
17. Shou, K. (1994). Three-Dimensional Displacement Discontinuity Method with Application to Bonded Half-Space Problems. Ph.D. Dissertation. University of Minnesota.
18. Sierra, L., and Mayerhofer, M. (2014). "Evaluating the Benefits of Zipper Fracs in Unconventional Reservoirs," in SPE Unconventional Resources Conference, April 1–3, 2014 (The Woodlands, Texas, USA: Society of Petroleum Engineers). SPE-168977-MS.
19. Soliman, M. Y., East, L. E., and Adams, D. L. (2008), 23. Geomechanics Aspects of Multiple Fracturing of Horizontal and Vertical Wells, *SPE Drilling & Completion*, 217–228. SPE-86992-PA. doi:10.2118/86992-PA
20. Tang, J., Wu, K., Zeng, B., Huang, H., Hu, X., Guo, X., et al. (2018). Investigate Effects of Weak Bedding Interfaces on Fracture Geometry in Unconventional Reservoirs. *J. Pet. Sci. Eng.* 165, 992–1009. doi:10.1016/j.petrol.2017.11.037
21. Tang, J., Wu, K., Zuo, L., Xiao, L., Sun, S., and Ehlig–Economides, C. (2019). Investigation of Rupture and Slip Mechanisms of Hydraulic Fractures in Multiple-Layered Formations. *SPE J.* 24 (05), 2292–2307. doi:10.2118/197054-PA
22. Xie, J., Tang, J., Yong, R., Fan, Y., Zuo, L., Chen, X., et al. (2020). A 3-D Hydraulic Fracture Propagation Model Applied for Shale Gas Reservoirs with Multiple Bedding Planes. *Eng. Fracture Mech.* 228, 106872. doi:10.1016/j.engfracmech.2020.106872
23. Xu, G. S., and Wong, S. W. (201317043). "Interaction of Multiple Non-planar Hydraulic Fractures in Horizontal wells," in International Petroleum Technology Conference, 26–28, 2013 (Beijing, China: IPTC). doi:10.2523/17043-ms
24. Zhang, F., Damjanac, B., and Maxwell, S. (2019). Investigating Hydraulic Fracturing Complexity in Naturally Fractured Rock Masses Using Fully Coupled Multiscale Numerical Modeling. *Rock Mech. Rock Eng.* 52 (12), 5137–5160. doi:10.1007/s00603-019-01851-3
25. Zhang, F., Dontsov, E., and Mack, M. (2017). Fully Coupled Simulation of a Hydraulic Fracture Interacting with Natural Fractures with a Hybrid Discrete-Continuum Method. *Int. J. Numer. Anal. Meth. Geomech.* 41 (13), 1430–1452. doi:10.1002/nag.2682
26. Zhang, F., Wang, X., Tang, M., Du, X., Xu, C., Tang, J., et al. (2021b). Numerical Investigation on Hydraulic Fracturing of Extreme Limited Entry Perforating in Plug-And-Perforation Completion of Shale Oil Reservoir in Changqing Oilfield, China. *Rock Mech. Rock Eng.* 54, 2925–2941. doi:10.1007/s00603-021-02450-x
27. Zhang, J., Li, Y., Pan, Y., Wang, X., Yan, M., Shi, X., et al. (2021a). Experiments and Analysis on the Influence of Multiple Closed Cemented Natural Fractures on Hydraulic Fracture Propagation in a Tight Sandstone Reservoir. *Eng. Geology.* 281, 105981. doi:10.1016/j.enggeo.2020.105981

11

Probabilistic Assessment of Investment Options in Honey Value Chains in Lamu County, Kenya

Joshua Wafula [1], Yusuf Karimjee [1], Yvonne Tamba [1], Geoffrey Malava [1], Caroline Muchiri [1], Grace Koech [1], Jan De Leeuw [1,2], Josephat Nyongesa [1], Keith Shepherd [1] and Eike Luedeling [1,3]*

[1] World Agroforestry Centre (ICRAF), Nairobi, Kenya, [2] ISRIC World Soil Information, Wageningen University, Wageningen, Netherlands, [3] Department of Horticultural Sciences, University of Bonn, Bonn, Germany

***Correspondence:**
Joshua Wafula
jokumu56@yahoo.com

Designing and implementing biodiversity-based value chains can be a complex undertaking, especially in places where outcomes are uncertain and risks of project failure and cost overruns are high. We used the Stochastic Impact Evaluation (SIE) approach to guide the Intergovernmental Authority on Development (IGAD) on viable investment options in honey value chains, which the agency considered implementing as an economic incentive for communities along the Kenya-Somalia border to conserve biodiversity. The SIE approach allows for holistic analysis of project cost, benefit, and risk variables, including those with uncertain and missing information. It also identifies areas that pose critical uncertainties in the project. We started by conducting a baseline survey in Witu and Awer in Lamu County, Kenya. The aim of the survey was to establish the current farm income from beekeeping as a baseline, against which the prospective impacts of intervention options could be measured. We then developed an intervention decision model that was populated with all cost, benefit and risk variables relevant to beekeeping. After receiving training in making quantitative estimates, four subject-matter experts expressed their uncertainty about the proposed variables in the model by specifying probability distributions for them. We then used Monte Carlo simulation to project decision outcomes. We also identified variables that projected decision outcomes were most sensitive to, and we determined the value of information for each variable. The variable with the highest information value to the decision-maker in Witu was the honey price. In Awer, no additional information on any of the variables would change the recommendation to invest in honey value chains in the region. The analysis demonstrates a novel and comprehensive approach to decision-making for different stakeholders in a project where decision outcomes are uncertain.

Keywords: value chains, probabilistic projection, decision outcomes, uncertainity, value of information

INTRODUCTION

How to approve and prioritize among projects that aim at biodiversity conservation has been highlighted as one of the most critical decisions that conservation planners face [1]. This is not surprising, because conservation outcomes are often achieved through complex mechanisms, and the success of conservation actions is rarely guaranteed, with many uncertainties preventing precise impact prediction. Success is even harder to predict, when conservation agencies aim to strengthen biodiversity indirectly, e.g., by supporting livelihoods and economies of local people as an incentive for them to conserve biodiversity outcomes [2]. Investing in biodiversity based value chains does not necessarily result in positive biodiversity outcomes. Negative impacts can arise, when value chain development results in depletion of the biodiversity that forms the resource base, on which the value chain depends (e.g. fisheries or non-timber forest products).

The production of honey is an example of a biodiversity based value chain that strengthens rather than erodes the conservation of biodiversity [3]. This is because honey producers have an interest to conserve the vegetation and plant species that produce the nectar and pollen that supports the value chain. The development of honey value chains typically revolves around a combination of introducing improved bee keeping and honey production techniques and improved access to markets for honey [4]. Yet, while attractive at first sight, such improved techniques are not always easily adopted [5]. An important reason for this is uncertainty among farmers about the financial outcomes of their investment in improved honey production techniques.

A detailed cost-benefit analysis on beekeeping projects can be considered to reduce the perceived uncertainty. However, there are rarely sufficient data on all relevant aspects of an investment decision to allow precise, purely data-driven projections to support decision-making [6]. Given such a lack of perfect knowledge, decision-makers need appropriate tools for handling uncertainties, and for identifying and prioritizing knowledge gaps, whose narrowing would reduce their chance of selecting a suboptimal decision option [7, 8]. Furthermore, decision-makers need improved capabilities to quantify risks surrounding proposed interventions, because failure to adequately account for risk can lead to high chances of project failure [9].

The Stochastic Impact Evaluation (SIE) approach allows for a structured decision analysis that incorporates all relevant variables, even those with uncertain and missing information [10]. It considers risk factors that may compromise project success or affect project performance. The approach incorporates Value of Information analysis that prioritizes critical uncertainties in a project, where further research has the greatest potential of enhancing clarity on the decisions. The present study uses the SIE approach to assess investment decisions in honey value chains for the Intergovernmental Authority on Development (IGAD) in its program on Biodiversity Management (BMP).

Study Background

IGAD-BMP partnered with the World Agroforestry Centre (ICRAF) to implement the program's biodiversity-based interventions along the Kenya-Somalia border. During the project inception phase, stakeholders were consulted, and they proposed participatory honey value chain development as one of the economic incentives for biodiversity conservation. Communities from Witu and Awer were selected to pilot the beekeeping project in Lamu County. One hundred farmers from both communities were selected and trained in beekeeping techniques to boost their honey production knowledge and improve their access to formal markets.

The training was held in May 2015 in Malindi, Kenya. Malindi was selected for training due to its proximity to the Arabuko-Sokoke Forest, where surrounding communities intensively practice modern beekeeping. This provided an opportunity for the trainees to learn from established beekeepers. The training brought together different honey value chain actors, including representatives from the Ministry of Livestock and Fisheries in Kenya. During the workshop, stakeholders agreed that honey value chains had potential for being a viable investment option for Witu and Awer communities. A baseline survey (**Figure 1**) was commissioned to establish the current net income from beekeeping, the actual number of beekeepers among the trained farmers, beekeeping practices and risks associated with beekeeping.

In Awer, the average net income from beekeeping was 82 thousand Kenyan Shillings (Ksh) per year, approximately Ksh 225 per day (**Table 1**). Traditional beekeeping was widely practiced and characterized by high productivity and well-established markets. In Witu, the average net income from beekeeping per farmer was Ksh 39 thousand (approximately Ksh 107 per day). Very few farmers were currently practicing beekeeping. The region also had very low honey productivity and low farm gate prices, although a higher percentage of interviewed farmers were educated compared to Awer.

The baseline survey informed the project implementation approach for both communities. In Awer, it seemed sensible to support traditional beekeeping, since the region was characterized by high honey productivity and favorable farm gate prices. Establishment of a honey collection center and support of value addition activities, such as packaging, branding and processing of other bee products, were the most promising investment options. In Witu, the favored approach was to invest in modern beekeeping with emphasis on intensive farmer training and engagement of all honey value chain actors to boost productivity. Bridging existing gaps between beekeepers and formal markets was also a priority for the implementer to ensure favorable farm gate prices for the beekeepers.

While clear strategies for HVC development in the two regions thus emerged from the initial consultations and baseline survey, IGAD-BMP was still lacking certainty that the investment would raise farm incomes. To clarify the prospects for the intervention, the analysis aimed at (i) projecting impacts on beekeeping farms using a probabilistic impact evaluation approach, (ii) identifying and mitigating critical uncertainties in the project, and (iii) providing recommendations for project monitoring to reduce the chance of negative outcomes.

MATERIALS AND METHODS

We used ICRAF's Stochastic Impact Evaluation approach (**Figure 2**), which is based on the principles of Applied Information Economics (AIE) [9, 11], to project the impact of the decision on different stakeholders in the project. The method makes use of all available information, including expert

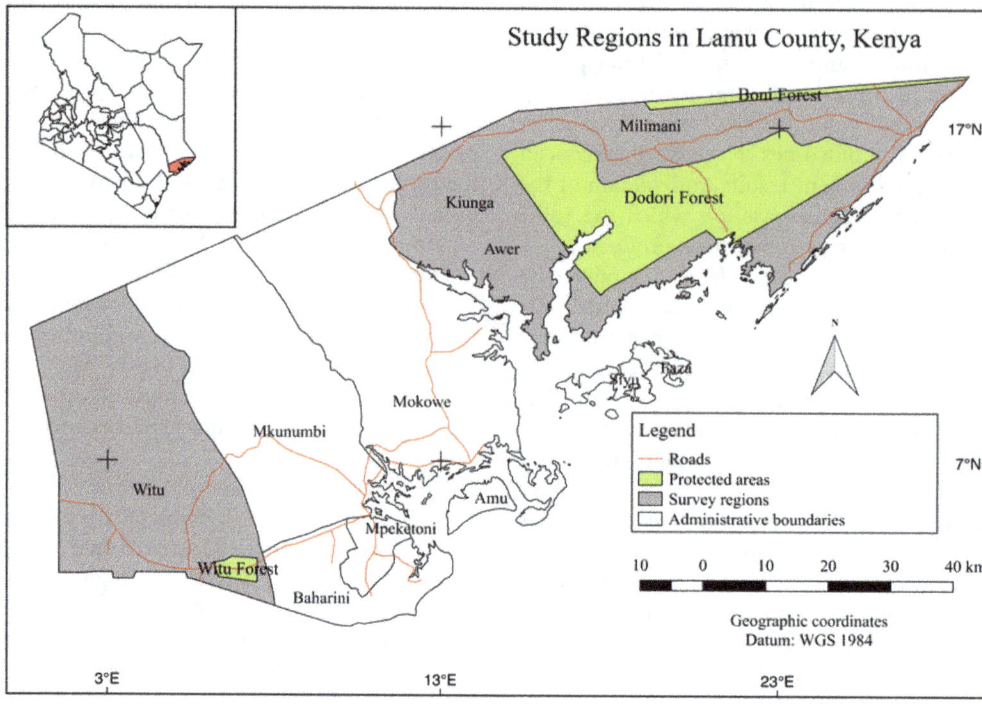

FIGURE 1 | Geographical context of IGAD's honey value chain intervention project.

TABLE 1 | Summary of the baseline survey results for net income from beekeeping in Witu and Awer.

Variable	Category	Awer Number (%)	Witu Number (%)
Gender	Men	37 (66)	37 (64)
	Women	19 (34)	21 (36)
Level of education	No education	32 (57)	13 (22)
	Primary	22 (39)	21 (36)
	Secondary	2 (4)	15 (26)
	Higher education	0 (0)	9 (16)
Bee keeping methods	None	2 (4)	43 (74)
	Traditional	29 (52)	8 (14)
	Modern	14 (25)	1 (2)
	Both	11 (19)	6 (10)
Hives per household		2 to 4	1 to 3
Honey per harvest (liters)		5 to 12	2 to 6
Income per beekeeper (thousand Ksh)		−19 to 299	−28 to 171

knowledge, in making impact projections. The SIE approach can be applied even in the absence of perfect data. We quantified all risks, costs and benefits based on the current state of knowledge about them, considering the uncertainties about intervention outcomes that result from this. We also embraced the concept of Value of Information, which is useful for determining decision-specific knowledge gaps that decision-supporting research should address [12].

Modeling Process

The SIE process begins by thoroughly defining the decision to be made. According to Hubbard [9], a decision may merit structured decision analysis if it has two or more realistic alternatives, invokes some form of uncertainty or dilemma, has potentially negative consequences if it turns out that the wrong position was taken, and involves a decision-maker. For this study, HVC development was identified as a viable investment option during the consultative workshop, but stakeholders were still unsure that the decision to invest in HVCs would benefit them. This led to the question: *"Should IGAD-BMP invest in the proposed honey value chain intervention as an economic incentive for Witu and Awer communities to conserve biodiversity?"*

To clearly define this question, a decision analysis team was constituted with the aim of clarifying whether IGAD-BMP was interested in return on investment, who exactly the agency was targeting to raise incomes from beekeeping, and how the project was going to be implemented. The team that was convened to develop the HVC decision model consisted of 8 members: IGAD-BMP's HVC intervention project manager, the biodiversity management program coordinator, two decision analysis experts, one participatory modeling scientist and four research assistants. The initial conceptual model (**Table 2**), which aimed to reflect the overall structure of the decision at hand, was then sent to a principal bee health scientist from the International Centre of Insect Physiology and Ecology (icipe) for review. The updated model (**Figure 3**) was translated into a set of equations that represented the team's understanding of the decision's impact pathway [13].

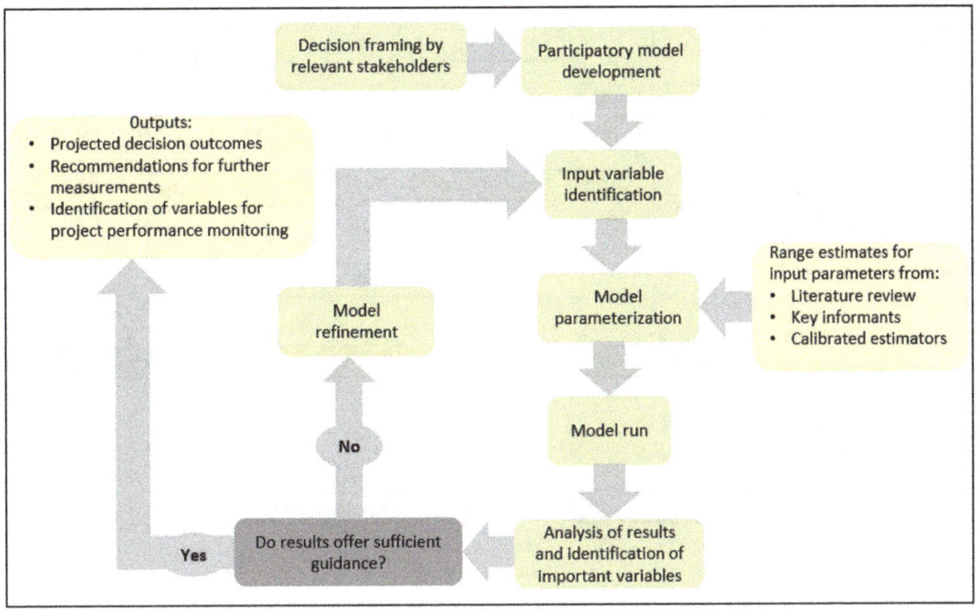

FIGURE 2 | Sequence of activities in the SIE decision modeling approach (modified from [8]).

Decision outcomes were projected using the decisionSupport package [14] for the R programming environment [15]. For this, the model equations were coded as a welfare function in R programming language (Data Sheets 3, 4). The model was parameterized using estimated probability distributions for variable values (Data Sheets 1, 2) (often specified as 90% confidence intervals), which were obtained from literature review and five established beekeepers from Mwingi in Kitui County, Kenya. The process required the team to undergo calibration training—a process that improves the capacity to make range estimates for which one is 90% confident that the actual value lies within the provided range [9]. This was done for all variables, including those for which no other information was available. Using the estimated variables as inputs, the decision model was run as a Monte Carlo simulation [16, 17]. This approach produced a distribution of possible decision outcomes by running the model a large number of times, each time fed with a different set of random draws from the defined distributions for the variables in the equations [18]). To identify variables that were most uncertain in the analysis, we used Partial Least Squares (PLS) regression [19], which was also implemented in the decisionSupport package [14].

The most uncertain variables in a project, which outcome projections are most sensitive to, are not necessarily most important to the decision-maker, because new information on them may not be able to change the recommendation emerging from the decision model. The value of additional information on a variable for decision-makers is determined by whether this information has the potential to change the sign of the expected value of the decision, which would change the preferred decision option [9].

To identify high-value variables, the Expected Value of Perfect Information (EVPI) was calculated by detecting correlations

TABLE 2 | List of variables used to populate the beekeeping model.

Benefits	Costs	Risks
Direct benefits • Revenue from sale of honey • Revenue from sale of wax Indirect benefits • Increased farm yields (from improved pollination)	Initial investment costs • Cost of labor to establish apiary • Transport of equipment cost • Cost of beehives • Cost of setting up collection centers • Cost of farmer packages • Other initial costs Recurring costs • Harvesting cost • Wax processing cost • Honey/wax transportation cost • Honey processing cost • Hive maintenance materials cost • Opportunity cost • Staff salaries • Farmer training cost • Other recurring costs	• Insecurity • Land conflicts • Fire outbreaks • Theft

between input and output variables and identifying those that were of importance to a decision-maker, i.e., the variables, whose measurement could help reduce the expected opportunity loss of the decision. Since there was no basis for deciding on the functional form of the relationship between test and outcome variable, a non-parametric test was used to detect these relationships. We used Spearman's rank correlation, using the usual statistical cutoff criterion of $\alpha < 0.05$ to exclude all variables that were not correlated with projected decision outcomes. For those that were not correlated, the value of information was zero, since the variable had no significant relationship with the outcome. This criterion excluded most

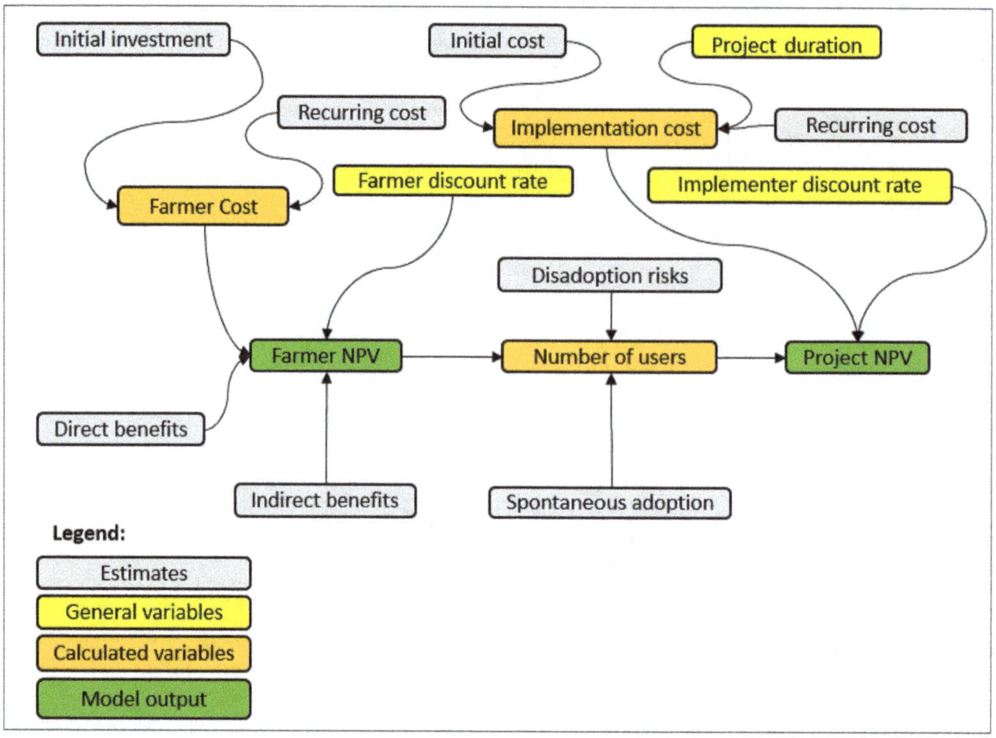

FIGURE 3 | Structure of the decision model.

variables, and discriminated well between informative and obviously uninformative variables.

We then sorted the dataset by values of each remaining input variable. In the sorted dataset, it was not possible to identify a clear threshold, where the expected decision outcomes transition from positive to negative. This was because the data was still too "noisy." To more clearly expose the effect of uncertain variables on the decision recommendation, we processed the dataset using a second-order low-pass Butterworth filter, with a critical frequency of one divided by one tenth of the number of values in the Monte Carlo output. This resulted in a smooth dataset, in which it was often possible to identify a threshold, where the sign of projected decision outcomes was reversed, e.g. for variable values above the threshold, positive outcomes would be expected, while values below the threshold would likely lead to negative outcomes (note that using a signal processing filter is a pragmatic way to solve the computational challenge of calculating the EVPI, but it introduces a small amount of inaccuracy into the threshold identification). With this threshold identified, the EVPI was computed as the sum of all outcomes with a sign that did not correspond to that of the expected value (e.g., all negative outcomes, when the analysis produced a positive expected value), multiplied by their respective chance of occurrence. This EVPI procedure was applied to all output variables.

RESULTS

Due to the economic and biophysical disparities between Awer and Witu, the two regions were modeled separately. The outcome of the analysis was expressed as the net present value (NPV) of the intervention for a farmer and for the overall project. Emphasis was placed on the farmer NPV and overall project NPV, since the implementer's direct objective was to raise farmer incomes from beekeeping. Cash flows were also illustrated, and variables that had information value for the decision maker were identified.

Witu Community

In Witu, the median of the modeled distribution of average annual monetary income per farmer practicing beekeeping was Ksh 65 thousand, with 90% confidence that the actual income lay within the range of Ksh −36 thousand to 140 thousand for a farmer who would continuously practice beekeeping (**Figure 4**). The chance of a negative average monetary NPV for a farmer in this region was 24%.

The honey price, the amount of honey produced per hive, the number of hives per farmer, the number of harvesting seasons per year and the honey processing cost were the most important variables according to the PLS analysis. The distribution of total farmer NPV, including indirect benefits from beekeeping, e.g. through improved pollination of crops, had a median of Ksh 72 thousand, with a 90% confidence interval of Ksh-30–251 thousand.

Since both positive and negative farmer outcomes appeared plausible according to the simulation results, the EVPI analysis indicated that additional information about four variables could potentially change the recommendation emerging from the SIE process. These were the honey price, with a value of information of about Ksh 2750, followed by honey production per hive (Ksh

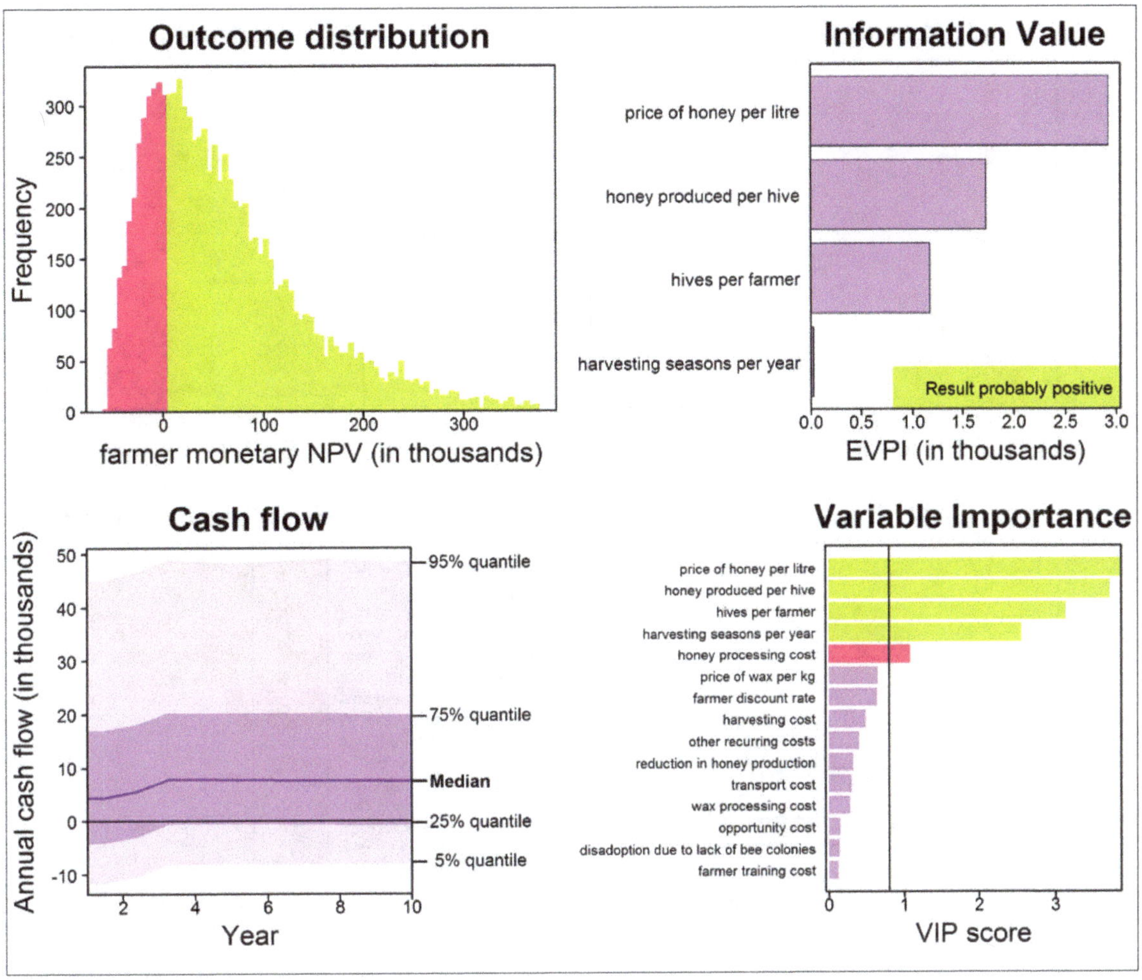

FIGURE 4 | Projected outcome distribution (top left), high-value variables (EVPI; top right), project cash flow (bottom left) and important variables (determined by PLS regression; bottom right) for a farmer practicing beekeeping in Witu, Kenya. Results were produced through Monte Carlo simulation (with 10,000 model runs) of project performance over 10 years. Red and green bars in the outcome distribution indicate positive and negative values, respectively. In the PLS plot, green bars indicate positive correlations of uncertain variables with the outcome variable, while red bars indicate negative correlations.

1750), the number of hives per farmer (Ksh 1250) and the number of harvesting seasons per year (Ksh 50).

The distribution of the projected NPV for the project in Witu had a median of Ksh 36 million, with 90% confidence that the actual NPV for the project lay within the range of Ksh-14–150 million. The model responded most sensitively to the amount of honey produced per hive. Seven other variables also had important impact on projected outcome values (**Figure 5**).

The distribution of projected decision outcomes included both positive and negative outcomes, and additional information on some individual uncertain variables had potential to change the decision recommendation. EVPI analysis indicated that information on the honey price in Witu was the most valuable to the project implementers, with a value of about Ksh 1.1 million. The amount of honey produced per hive (Ksh 0.6 million) and the number of hives per farmer (Ksh 0.4 million) were also of value to the project implementers.

Awer Community

In Awer, the distribution of projected average monetary income per farmer practicing beekeeping had a median of Ksh 130 thousand per year, with 90% confidence that the actual value lay within the range of Ksh 1.5–340 thousand per year. The chance of loss for a farmer in this region was 4.6%. Total farmer NPV including non-monetary benefits had a median of Ksh 140 thousand with a 90% confidence interval of Ksh 7.2–340 thousand. The price of honey was identified as the most uncertain variable by the PLS analysis, alongside five other variables (**Figure 6**).

EVPI analysis indicated that none of the variables had any information value for the farmer. This meant that, if the NPV was the main criterion for evaluating the attractiveness of the HVC intervention, then farmers would be well-advised to engage in beekeeping, and no additional information on any of the variables would change this recommendation.

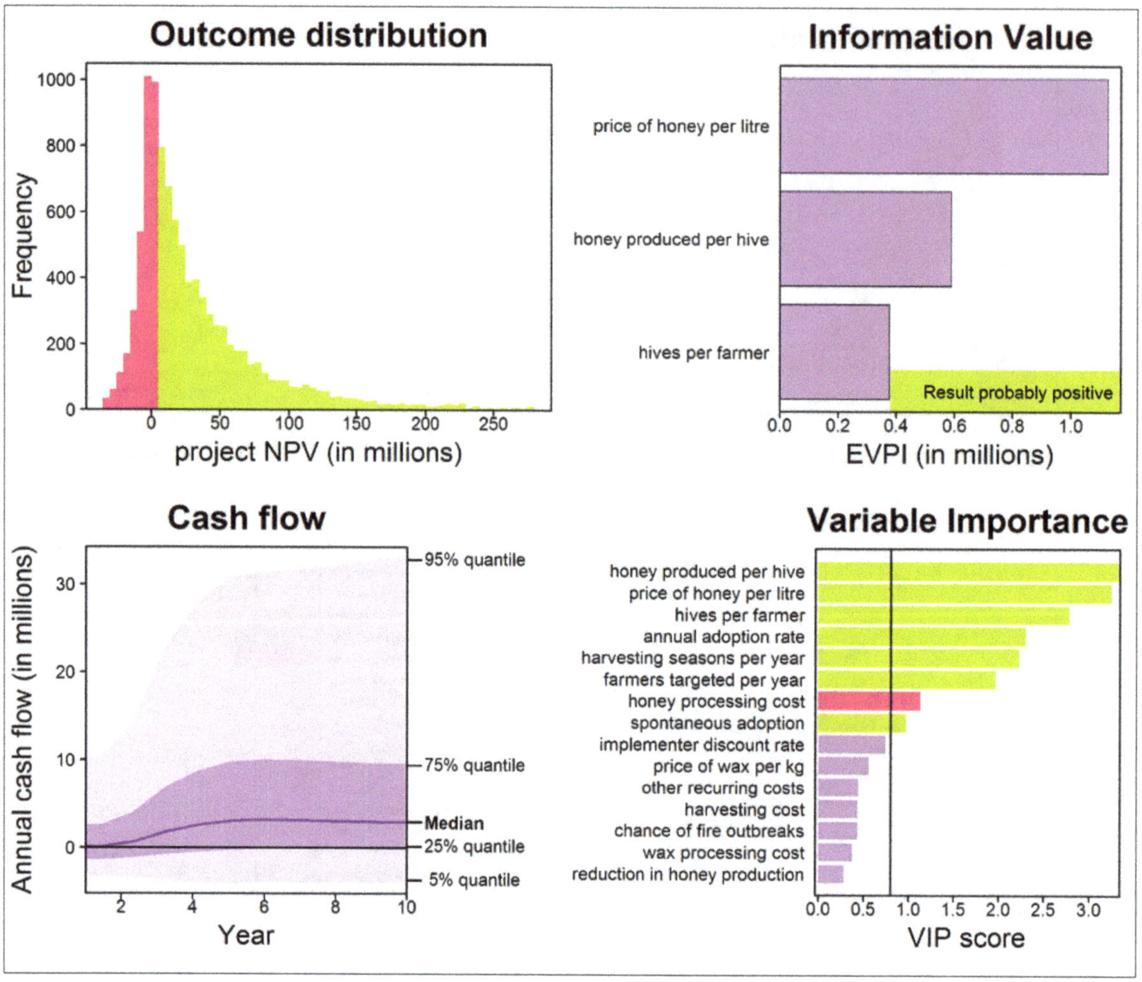

FIGURE 5 | Projected outcome distribution (top left), high-value variables (EVPI; top right), project cash flow (bottom left) and important variables (determined by PLS regression; bottom right) for the overall project in Witu, Kenya. Results were produced through Monte Carlo simulation (with 10,000 model runs) of project performance over 10 years. Red and green bars in the outcome distribution indicate positive and negative values, respectively. In the PLS plot, green bars indicate positive correlations of uncertain variables with the outcome variable, while red bars indicate negative correlations.

The overall project NPV had a median of Ksh 38 million per year, with 90% confidence that the actual value lay within the range of Ksh-0.9–120 million. PLS analysis indicated that the honey price was the most uncertain variable in the projection of project NPV in Awer (**Figure 7**). Although PLS indicated that projected project outcomes responded to variation in a number of variables, no additional information on any of these variables had the potential to change the recommendation that the project implementer should invest in honey value chains in the region.

DISCUSSION

Beekeeping is a high-risk, high-return venture that requires a well thought-out project design to maximize returns [20]. Introduction of the HVC intervention as an economic incentive for rural communities along the Kenya-Somalia border to conserve the environment may generate significant impact in terms of raising income (average of Ksh 65 thousand per farmer for Witu and Ksh 130 thousand per farmer for Awer). However, the success of this intervention, especially for regions such as Witu, requires close monitoring by the implementing agency to minimize the chance of loss (24%).

In Witu, the most critical variable was the price of honey, which individual beekeepers have relatively little influence on. Current honey price, based on the baseline survey, is within the range of 3–7 hundred Ksh per liter. At this price, returns may be too low for a farmer who pursues beekeeping as the only source of income. For a farmer who invests in beekeeping as a supplementary source of income, the current honey price looks promising. Analyzing the impact of different factors that may affect the honey price in the future, e.g., packaging, honey quality, access to formal markets and exploitation by middlemen, will provide useful information to support farmers' decisions on investments in honey value chains. The amount of honey produced per hive is also a critical variable to farmers in Witu (**Figure 4**). Since productivity is a function that depends on many

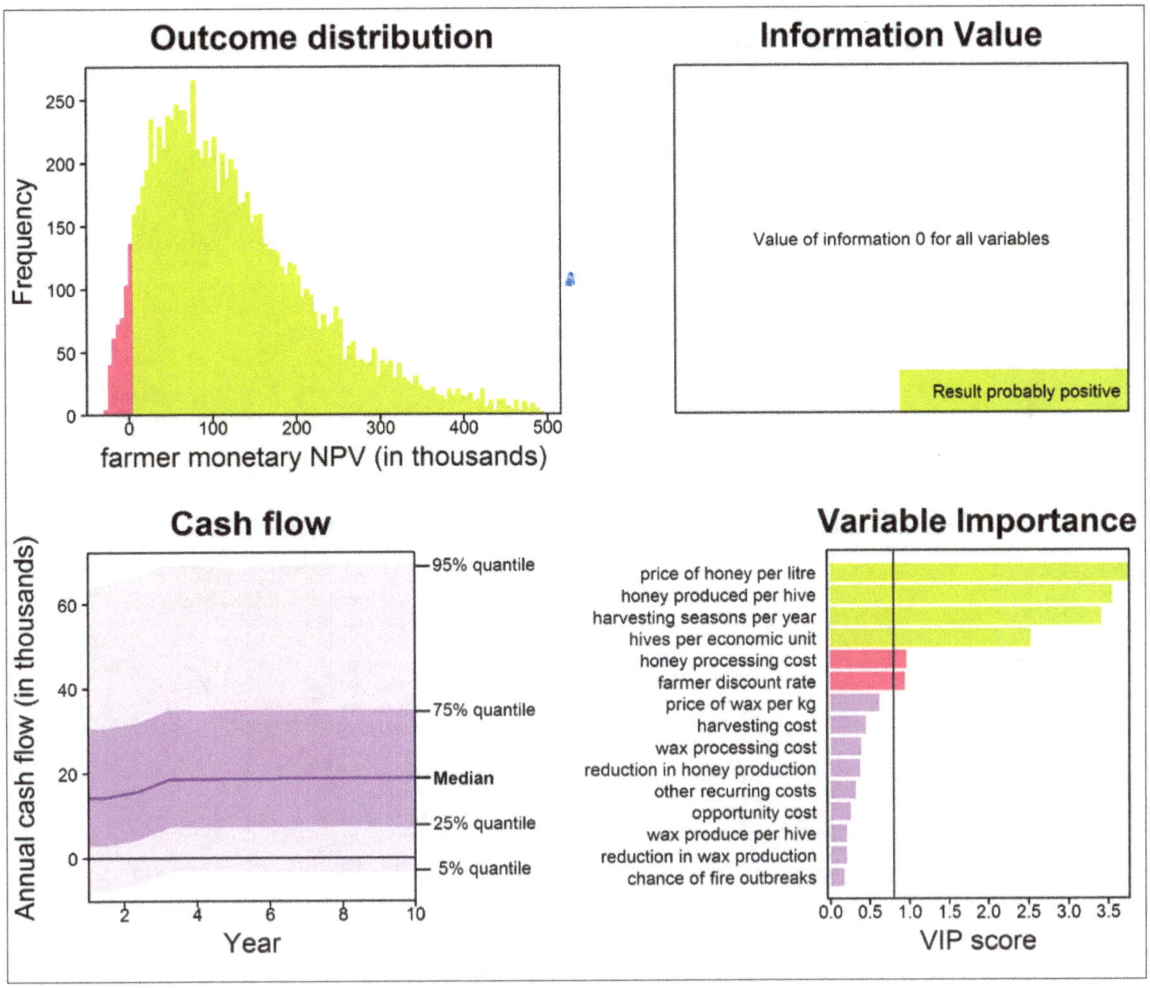

FIGURE 6 | Projected outcome distribution (top left), high-value variables (EVPI; top right), project cash flow (bottom left) and important variables (determined by PLS regression; bottom right) for a farmer practicing beekeeping in Awer, Kenya. Results were produced through Monte Carlo simulation (with 10,000 model runs) of project performance over 10 years. Red and green bars in the outcome distribution indicate positive and negative values, respectively. In the PLS plot, green bars indicate positive correlations of uncertain variables with the outcome variable, while red bars indicate negative correlations.

factors, such as good apiary management, access to extension services, technology deployed and bee forage availability, studies on these factors promise to reduce decision uncertainty for farmers. Modifications to the original project design that address the influence of these variables on the farmer NPV could reduce the chance of losses for beekeepers.

Most model runs for project NPV in Witu indicate a positive outcome, although there is still a chance (23%) of not achieving the goal of raising incomes for farmers in the region. This is because beekeeping is a relatively new venture for most farmers. The value chain is not well developed and a lot of investment has to go into farmer training to increase productivity. The honey price, the amount of honey produced per hive and the number of hives per farmer are the most critical variables for the project implementer. It is therefore important for the implementer to work with farmers to acquire more information on these variables and how they will influence their respective NPVs.

In Awer, the HVC intervention can increase monetary income for beekeepers by about Ksh 48 thousand based on the baseline survey. This is slightly higher than in Witu (Ksh 26 thousand), but this is expected, since most farmers in the region were already practicing beekeeping. The project plan to support existing traditional beekeeping methods, establish a honey collection center and support value addition activities such as packaging, branding and processing of other bee products, was likely to be more effective in raising incomes, since it would focus on bridging gaps in the chain. In Witu, the focus is on introducing the HVC to the region.

The EVPI analysis indicated that no additional information on any of the variables could change the recommendation that farmers should invest in beekeeping (**Figure 6**). This is not surprising, since the region is characterized by high honey productivity and well-defined markets. The overall project NPV in Awer suggests that introducing HVC is economically

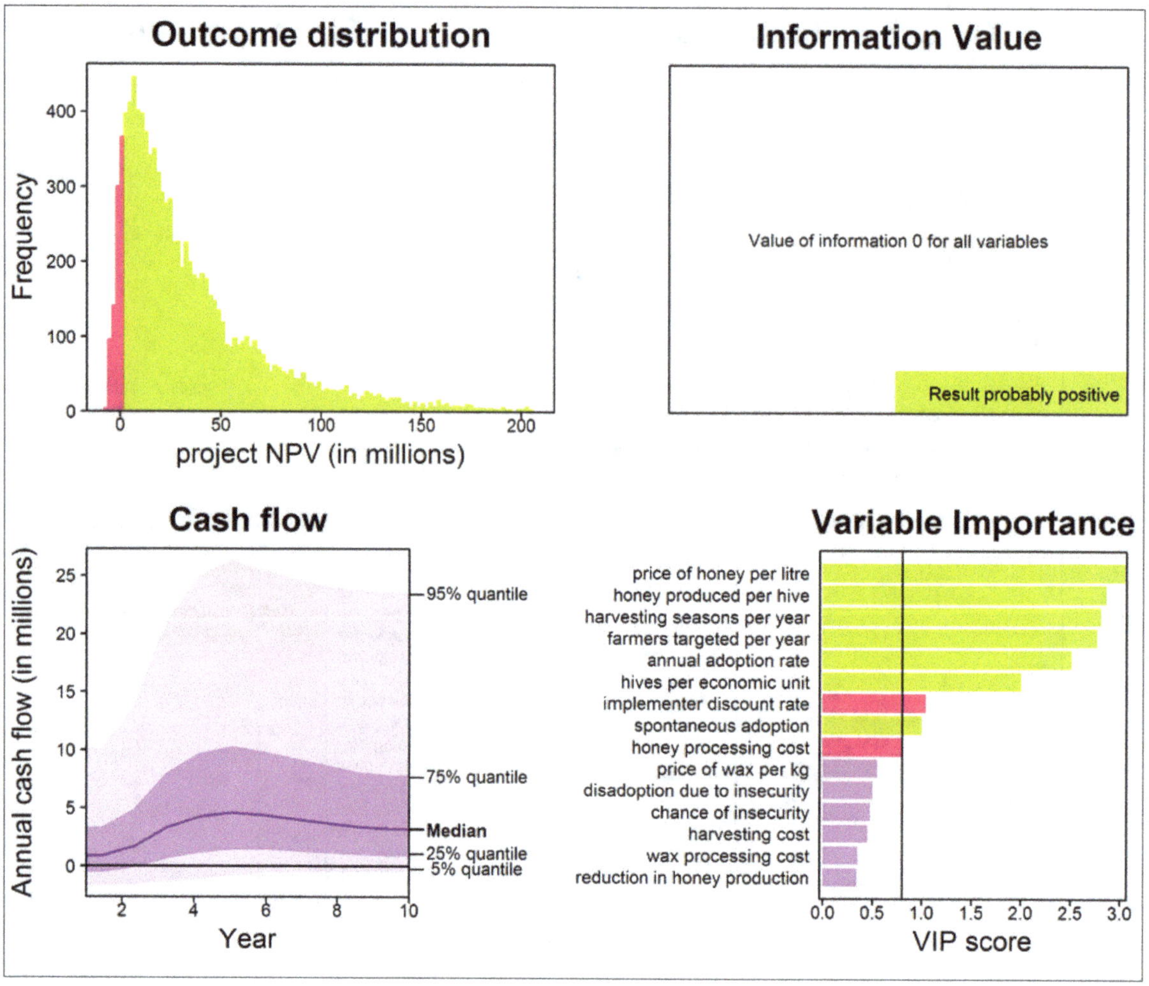

FIGURE 7 | Projected outcome distribution (top left), high-value variables (EVPI; top right), project cash flow (bottom left) and important variables (determined by PLS regression; bottom right) for the overall project in Awer, Kenya. Results were produced through Monte Carlo simulation (with 10,000 model runs) of project performance over 10 years. Red and green bars in the outcome distribution indicate positive and negative values, respectively. In the PLS plot, green bars indicate correlations of uncertain variables with the outcome variable, while red bars indicate negative correlations.

viable, even when fully considering the cost incurred by the implementing agency (**Figure 7**). The effort to improve certainty on the business case may not influence the decision to invest in HVC, since no additional information on any of the variables had value to the decision-maker.

Impact of the Intervention on Biodiversity Conservation Efforts

Biodiversity conservation is linked to income from beekeeping, because conserving the surrounding forests provides flowers for bees to forage on, thus ensuring high honey productivity throughout the year. However, raising incomes for communities within conservation areas, which is the goal of the HVC intervention, does not necessarily mean that these communities will conserve biodiversity [21]. Therefore, for IGAD-BMP to maximize its impact in terms of biodiversity conservation, the program must ensure that rural communities not only generate sustainable incomes from honey production, but also perceive relationships between these incomes and biodiversity conservation, i.e., that high honey production correlates with success in preserving biodiversity. A study on how increased incomes from honey value chain interventions would affect biodiversity conservation outcomes could produce valuable information on the prospective impacts of the intervention on biodiversity conservation.

Recommendations for Project Monitoring

Although the intervention looks promising, a few areas need to be monitored to reduce the chance of negative outcomes. In Witu, the price of honey has to be monitored to ensure farm gate prices can provide sufficient income to bee-keeping farmers. To reduce the chance of negative outcomes as a result of low farm gate prices, there is a need to link farmers to formal markets, promote collective marketing and invest in value addition. The amount of honey produced per hive should also be monitored, since it is critical in determining whether farmers investing in HVCs are likely to reap net benefits. Adequate farmer training and mitigation of risks, such as theft, land conflicts and fire outbreaks, will help reduce the chance of farmers abandoning honey production. Further, there is a need to identify the minimum number of hives per farmer that would provide substantial increase in income. The implementing agency would then focus on ensuring that most of the farmers have this minimum number to motivate them to continue contributing to HVCs.

In Awer, not much can be done in terms of project monitoring to reduce negative outcomes, since the chance of such negative outcomes was very low. However, the implementing agency can focus on monitoring how these positive outcomes affect biodiversity conservation efforts. There are no clear linkages between increase in income and biodiversity conservation. Clarifying and generating awareness on these linkages, e.g., through research and education on the need to conserve biodiversity to increase honey production, will help the implementing agency to better understand—as well as enhance the impact of—the project on biodiversity conservation.

Future Application of the SIE Approach

The costs, benefits and risks of biodiversity-based value chains, and agricultural interventions in general, typically have high levels of uncertainty, especially when considering the long-term and off-site effects of proposed interventions. In light of these uncertainties, the single most important task facing agricultural intervention planners is perhaps to determine the best way to make decisions [22]. Uncertainties and risks have to be quantified, policy preferences clarified and priority measurements for supporting their decisions identified.

The SIE approach can provide plausible solutions to this. The first critical step in the SIE process is to clearly define the decision to be taken. This step often does not receive sufficient attention [23], but it is instrumental in understanding what has to be measured. In this analysis, we spent about 30% of the entire effort on clarifying questions such as whether the implementing agency would cater for the full project implementation cost or whether these costs would be shared with farmers, who would be the actual beneficiaries of the project and whether the implementer was interested in return on investment. Clarifying these questions early in the implementation planning process enabled IGAD-BMP stakeholders to easily identify relevant benefits, costs, risks and external factors that would affect project performance.

The SIE approach allowed identification and prioritization of critical uncertainties in the project. Variables prioritized for further measurement in one of the regions (Witu) appeared easy to quantify with small efforts in data collection. In the other region (Awer), no measurements on any of the variables would change recommendations to farmers and the implementing agency to invest in HVCs. This demonstrates that the SIE approach has potential to substantially enhance the cost effectiveness of decision-supporting measurement campaigns for agricultural intervention planners. The approach presents a clear business case for or against investment in a project where outcomes are uncertain. This can provide critical information to investors aiming to support agricultural interventions with limited resources.

CONCLUSIONS

Investment in HVC as an economic incentive to conserve biodiversity requires thorough analysis of investment options to maximize returns to different stakeholders. This can be achieved through probabilistic stakeholder-disaggregated outcome projections for biodiversity-based value chain interventions. Clarifying the decision question turned out to be critical for the decision making process. This greatly facilitated the development of a decision's impact pathway, which could then be easily converted into a quantitative decision outcome projection model.

Tailoring interventions to meet economic, social and environmental requirements of rural communities is very important, so decision-makers need approaches that allow holistic ex-ante analysis of investment options. The cost-benefit and risk analyses for biodiversity-based value chains should consider all factors that are relevant for the implementation of the decision, even those that initially appear "intangible" or for which no data are available. Incorporation of local and expert knowledge into decision making using the SIE framework can significantly improve the quality of decisions.

Value of Information analysis can provide indications of what needs to be measured to support intervention decisions. While many uncertainties usually exist in all decisions that affect complex systems, only those uncertainties that are of value to the decision maker should be prioritized for further measurement. This can substantially reduce the cost of data collection aimed at informing decisions.

STUDY ETHICAL APPROVAL AND CONSENT

The analysis relied on information obtained from literature review, key informants and subject matter experts. This did not require ethical approval as per institutional and national guidelines, but a written and informed consent was obtained from key informants and subject matter experts. The baseline survey outcome in the study background was generated from an open access dataset, where the data was properly anonymized [24]. Written and informed consent was also obtained from all survey participants at the time of original data collection.

AUTHOR CONTRIBUTIONS

JW: The first author. YK: Contributed to the decision model development. YT: Contributed to the decision model development. GM: Contributed to the decision model development. CM: Contributed to the decision model development. GK: Contributed to the designing of the questionnaire for the baseline survey. Also provided insights on implementation of biodiversity based value chains. JD: Provided expertise on impact of biodiversity based value chains. Also contributed to the editing of the manuscript before submission. JN: Provided vital insights in the framing of the investment decision. Also contributed to the variable estimation process. KS: Provided expertise in decision modeling. EL: Provided expertise in decision modeling and reviewed the manuscript before submission. Also provided insights on the future application of the SIE approach.

ACKNOWLEDGMENTS

We acknowledge funding by IGAD's Biodiversity Management Program (IGAD-BMP) and the continuous support of the CGIAR Research Program on Water Land and Ecosystems (WLE). We also appreciate the support accorded to us by the International Centre of Insect Physiology and Ecology (icipe) through the African Reference Laboratory for Bee Health, led by Suresh Raina. We acknowledge the input by established beekeepers in Mwingi, who provided valuable insights to the modeling team. We also appreciate local communities in Witu and Awer for taking their time to answer questions during the baseline survey.

REFERENCES

1. Carwardine J, O'Connor T, Legge S, Mackey B, Possingham HP, Martin TG. Prioritizing threat management for biodiversity conservation. *Conserv Lett.* (2012) **5**:196–204. doi: 10.1111/j.1755-263X.2012.00228.x
2. De Leeuw J, Carsan S, Koech G, Yayé AD, Josephat N. *A Review of Best Practices for selected biodiversity value chains that promote pro-poor conservation in the Horn of Africa.* Nairobi: The World Agroforestry Centre (ICRAF) (2017).
3. Agera SIN. Role of beekeeping in the conservation of forests. *Global J Agric Sci.* (2011) **10**:27–32.
4. Figueiredo HS, Meuwissen MPM, Van der Lans IA, Oude Lansink AGJM. Identifying successful strategies for honey value chains in Brazil: a conjoint study. *Br Food J.* (2016) **118**:1800–20. doi: 10.1108/BFJ-01-2016-0026
5. Uaiene R, Arndt C, Masters W. Determinants of agricultural technology adoption in Mozambique. In: *National Directorate of Studies and Policy Analysis.* Maputo: Ministry of Planning and Development, Republic of Mozambique (2009). p. 1–31.
6. Arzberger P, Schroeder P, Beaulieu A, Bowker G, Casey K, Laaksonen L, et al. Promoting access to public research data for scientific, economic, and social development. *Data Sci J.* (2004) **3**:135–52. doi: 10.2481/dsj.3.135
7. Peterman RM, Anderson JL. Decision analysis: a method for taking uncertainties into account in risk-based decision making. *Hum Ecol Risk Assess Int J.* (1999) **5**:231–44. doi: 10.1080/10807039991289383
8. Luedeling E, Oord AL, Kiteme B, Ogalleh S. Malesu M, Shepherd KD, et al. Fresh groundwater for Wajir—ex-ante assessment of uncertain benefits for multiple stakeholders in a water supply project in Northern Kenya. *Front Environ Sci.* (2015), **3**:16. doi: 10.3389/fenvs.2015.00016
9. Hubbard DW. *How to Measure Anything: Finding the Value of Intangibles in Business.* Hoboken, NJ: Wiley (2014).
10. Luedeling E, Shepherd K. Decision-focused agricultural research. *Solutions* (2016) **7**:46–54.
11. Hubbard D. Applied information economics : a new method for quantifying IT value an executive overview. *Dec Anal.* (2004) **6**.
12. Koops MA. Reliability and the value of information. *Anim Behav.* (2004) **67**:103–11. doi: 10.1016/j.anbehav.2003.02.008
13. Thabrew L, Wiek A, Ries R. Environmental decision making in multi-stakeholder contexts: applicability of life cycle thinking in development planning and implementation. *J Clean Prod.* (2009) **17**:67–76. doi: 10.1016/j.jclepro.2008.03.008
14. Luedeling E, Göhring L. *decisionSupport: Quantitative Support of Decision Making under Uncertainty, CRAN archive.* Available online at: https://cran.r-project.org/web/packages/decisionSupport/ (2017).
15. Core Team R. *R: A Language and Environment for Statistical Computing.* Vienna: R Foundation for Statistical Computing (2017).
16. Velez-Pareja I. Including the risk in the analysis: montecarlo simulation (Análisis De Inversiones Bajo Riesgo: Simulación). *Soc Sci Res Netw.* (2007). doi: 10.2139/ssrn.986972
17. Rosenstock TS, Mpanda M, Rioux J, Aynekulu E, Kimaro AA, Neufeldt H, et al. Targeting conservation agriculture in the context of livelihoods and landscapes. *Agric Ecosyst Environ.* (2014) **187**:47–51. doi: 10.1016/j.agee.2013.11.011
18. Chen Y, Yang Y, Luo J. Uncertainty analysis of coalbed methane economic assessment with montecarlo method. *Pro Environ Sci.* (2012) **12**:640–5. doi: 10.1016/j.proenv.2012.01.329
19. Wold S, Sjöström M, Eriksson L. PLS-regression: a basic tool of chemometrics. In: *Chemometr Intell Lab Syst.* (2001) **58**:109–30. doi: 10.1016/S0169-7439(01)00155-1
20. Mujuni A, Natukunda K, Kugonza DR. Factors affecting the adoption of beekeeping and associated technologies in Bushenyi District, Western Uganda. *Livestock Res Rural Dev.* (2012) **24**.
21. Adams WM. Biodiversity conservation and the eradication of poverty. *Science* (2004) **306**:1146–9. doi: 10.1126/science.1097920
22. Lindenmayer DB, Gibbons P, Bourke M, Burgman M, Dickman CR, Ferrier S, et al. Improving biodiversity monitoring. *Aust Ecol.* (2012) **37**:285–94. doi: 10.1111/j.1442-9993.2011.02314.x
23. Rosenzweig P. *What Makes Strategic Decisions Different.* Boston, MA: Harvard Business Review (2013).
24. Wafula J, Koech G, Nyongesa J, Shepherd K, Luedeling E, De Leeuw J. Raw data for net income from bee-keeping in Lamu County, Kenya. *ICRAF Dataverse* (2017). doi: 10.7910/DVN/JNOEZM

12

Convergence Analysis and Approximate Optimal Temporal Step Sizes for Some Finite Difference Methods Discretising Fisher's Equation

Koffi Messan Agbavon [1]***, Appanah Rao Appadu*** [2*]***, Bilge Inan*** [3] ***and Herve Michel Tenkam*** [1]

[1] Department of Mathematics and Applied Mathematics, North-West University Potchefstroom Campus, Potchefstroom, South Africa, [2] Department of Mathematics and Applied Mathematics, Nelson Mandela University, Gqeberha, South Africa, [3] Department of Basic Sciences of Engineering, Faculty of Engineering and Natural Sciences, İskenderun Technical University, Hatay, Turkey

***Correspondence:**
Appanah Rao Appadu
rao.appadu@mandela.ac.za

In this study, we obtain a numerical solution for Fisher's equation using a numerical experiment with three different cases. The three cases correspond to different coefficients for the reaction term. We use three numerical methods namely; Forward-Time Central Space (FTCS) scheme, a Nonstandard Finite Difference (NSFD) scheme, and the Explicit Exponential Finite Difference (EEFD) scheme. We first study the properties of the schemes such as positivity, boundedness, and stability and obtain convergence estimates. We then obtain values of L_1 and L_∞ errors in order to obtain an estimate of the optimal time step size at a given value of spatial step size. We determine if the optimal time step size is influenced by the choice of the numerical methods or the coefficient of reaction term used. Finally, we compute the rate of convergence in time using L_1 and L_∞ errors for all three methods for the three cases.

Keywords: Fisher's equation, FTCS, NSFD, EEFD, optimal, convergence estimate, rate of convergence, coefficient of reaction

1. INTRODUCTION

The most enthralling recent progress of nonlinear science in particular mathematical science of partial differential equations, theoretical physics, chemistry, and engineering sciences has been a growth of strategy or procedure to try to find exact solutions for nonlinear differential equations. This is substantial due to the fact that countless mathematical models are described by nonlinear differential equations. To mention few among others the inverse scattering transform [1], the singular manifold method [2], the transformation method [3], the tanh-function method [4], and the Weierstrass function method [5] are subservient in many applications and known as stunning techniques to look for solutions of exactly solvable nonlinear partial differential equations.

In Kudryashov [6] developed a new numerical method for the solution of nonlinear partial differential equations. Amajor novelty of that technique is the utilization of finite Fourier series for the numerical approximation of the spatial derivative terms of the equations. It was proved that the precision of this method is that the order of accuracy is greater than that found by approximating the spatial derivative terms by finite-difference methods [6]. The numerical performance of this method, which was called the Accurate Space Derivatives (ASD) approach, can be run accurately by the utilization of the Fast Fourier Transform (FFT) algorithm [7]. In Kudryashov [6], the ASD approach was used to obtain the solution of nonlinear hyperbolic equations depicting convective fluid flows. Furthermore, the ASD approach was used in Gazdag and Canosa [8] to solve Fisher's equation, a nonlinear diffusion equation portraying the rate of advance of a new advantageous gene within a population of constant density inhabiting a one dimensional habitat [9]. An outstanding and compendious debate of Fisher's equation in the framework of the genetic problem can be found in Moran [10] and Kendall [11].

Kolmogorov [12] used the traveling waves with wave speed c to solve the problem. They showed that by assuming that when the initial condition belongs to the interval $[0,1]$, the speed of propagation c of the waves is superior to two and the solution is in the form $u(x,t) = w(\xi)$ where $\xi = x - ct$. They further demonstrated that there are no solutions for $c \in [0,1)$.

Fisher's equation has a limitless number of traveling wave solutions and each wave propagates at a characteristic speed, $c > 2$. This result appears to point out that the velocity of gene advance is undefined. Gazdag and Canosa [8] studied the problem of the indeterminacy of the diffusion speed of Fisher's equation which has not been plainly investigated in Kolmogorov [12]. Furthermore, a modification was made by Fisher [9] to his original model, he demonstrated that the rate of gene advance became the minimum one when $c = 2$. Kendall [11] investigated a linear model portraying a population that undergoes a Brownian motion and spreads geometrically at the same time. Canosa [13] demonstrated that all waves are stable against local perturbations but are linearly unstable against general perturbations of limitless magnitude. It is worthy to emphasize that the traveling wave profiles of Fisher's equation are similar to some of the steady-state solutions of the Korteweg-de Vries-Burgers equation which is a third-order nonlinear partial differential equation integrating diffusive and dispersive effects which have been found useful to represent blood flow through an artery, shallow water waves and plasma shocks disseminating perpendicularly to a magnetic field [13, 14].

Kudryashov [6] showed that a simple stability analysis enables us to see the estimation which is unstable against the roundoff errors growing up at the right tail of the waves. This is due to the physical nature of the problem depicted by the equation, not to the numerical method utilized and moreover entailed an exponential growth of the solution when roundoff errors are exponentially small. This simple issue makes it hard to do a strict simulation of the solutions of Fisher's equation. Kudryashov [6] went on with the removal of the forward tail of the wave of advance. This removal is necessary for the numerical stability of the Accurate Space Derivatives (ASD) approach and is physically conclusive because it is approximately equivalent to assuming that the role of long-distance dissipation in the spread of the gene is insignificant and probably effective for some species but not for others. Other numerical computations present how fast the asymptotic minimum speed wave is reached from an initial step function and confirm the local stability analysis of Kudryashov [6] which unveils that local perturbations are flattened very rapidly, even from superspeed waves. Another amazing result of the estimation is obtained for an initial dispensation localized in space which further gives rise to two identical waves of minimum speed evolution cases, one disseminating to the right and the other to the left.

1.1. Some Generalized and Conserved Fisher's Equation

Fisher's equation can be represented as generalized or conserved forms. Fisher's equation is the elementary model of spatial dynamics, in which competitive interactions between individuals happen locally. In Kudryashov and Zakharchenko [15], the generalized form is written as

$$u_t = u_x(u^l\, u_x) + u^a\,(1 - u^b) \tag{1}$$

where t stands for time, x stands for spatial coordinate, u is a population density, and a, b, and l are all positive parameters.

Fisher's equation can also forecast circumstances where population regulation happens globally due to the existence of a secondary agent (the controller agent is itself dispersed over a scale significantly greater than the dispersal distance of the individuals themselves). It is, thus, of notice to envisage a simple model of spatial population dynamics in which the total population size is controlled *via* a nonlocal mechanism. In that case in Newman et al. [16], the conserved equation is written as

$$u_t = D\,\nabla^2 u + r(t)\,u(t) - K(t)\,u(t)^2 \tag{2}$$

where D stands for the mobility of the individuals, r stands for the reproduction rate in the absence of competition. K is a parameter representing the carrying capacity of the system and regulating the population density through competition. The auxiliary equation related to Equation (2) is

$$r(t) = K(t) \int d^d\, x\, u(x,t) \tag{3}$$

It is worth mentioning that Fisher's equation belongs to the class of partial differential equations called Reaction-Diffusion equations. This class of equations has broad applications in science and engineering for instance transport of air, adsorption of pollutants in soil, food processing, and modeling of biological and ecological systems [17, 18]. Several reaction-diffusion equations involve traveling waves fronts yielding a fundamental role in the understanding of physical, chemical, and biological phenomena [19]. Reaction-diffusion systems clarify how the condensation of one or more substances diffused in space varies by the impact of two operations: first, it is local chemical reactions in which the substances are modified into each other, and second, it is the diffusion that sustains the substances to smear over a surface in space [20]. Reaction-diffusion systems are regularly used in chemistry. Nonetheless, the system can also portray the dynamical processes of non-chemical nature. Reaction-diffusion systems have mathematically the form of semi-linear parabolic partial differential equations. They are often written in the form of

$$u_t = D\,\nabla^2 u + R(u), \tag{4}$$

where each component of vector $u(x,t)$ stands for the concentration of one substance, x is the space variable, and t is the time. D is the diffusion coefficient and R represents the reaction term. The solution of reaction-diffusion equations shows an ample scale of behaviors, enclosing the formation of traveling waves. These waves are like phenomena and self-organized patterns which are e.g., stripes, hexagons, or more complicated fabrics such as dissipative solutions [20]. Reaction-diffusion equations are also grouped as one component, two components, or more component diffusion equations counting upon the component involved in the reaction. The basic reaction-diffusion equation regarding the concentration u of a mere substance in one spatial dimension is

$$u_t = D\,u_{xx} + R(u). \tag{5}$$

If the reaction $R(u)$ term goes away, then the equation gives a pure diffusion process and if the thermal diffusivity term appears instead of diffusion term D then the equation will turn into a parabolic partial differential equation in one dimensional space [20]. The choice $R(u) = u\,(1-u)$ yields Fisher's equation. Those reaction-diffusion equations arise also in flame propagation, the branching Brownian motion process, and nuclear reactor theory [20]. Many methods such as Adomian Decomposition [21], Variational Iteration [22], Factorization [23], Nonstandard Finite Difference, and Exponential Finite Difference methods [24, 25] are used to solve Fisher's equation.

Anguelov et al. [26] investigated the same Fisher's equation by the means of a periodic initial data with θ-non standard approach and found that the Nonstandard Finite Difference approach is elementary stable in the limit case of space independent variable, stable in regard to the boundedness and positivity property. Finally also stable in regard to the conservation of energy in the stationary case.

Let us consider simple Fisher's equation given by

$$u_t = u_{xx} + R(u), \tag{6}$$

where $R(u) = u(1-u)$ and $x \in \mathbb{R}$, t positive. The boundary and initial conditions are as follows

$$\lim_{x\to+|\epsilon|\infty} u(x,t) = \begin{cases} 1 & \text{if } \epsilon = 1 \\ 0 & \text{if } \epsilon = -1, \end{cases} \tag{7}$$

$$u(x,0) = u_0(x). \tag{8}$$

Hagstrom and Keller [27] revealed that when a positive function is taken as an initial condition satisfying

$$u_0(x) \sim \exp(-\alpha) \quad \text{when} \quad x \to \infty, \tag{9}$$

then the solution u develops a traveling wave speed in function of α which is

$$c(\alpha) = \begin{cases} \alpha + \frac{1}{\alpha}, & \alpha \le 1, \\ 2, & \alpha \ge 1. \end{cases} \tag{10}$$

2. ORGANIZATION OF THE ARTICLE

The organization of this article is as follows. In Section 3, we present the general form of the exact solution of Fisher's equation and in Section 4, we describe the numerical experiment [28]. In Section 5, we make use of Forward in Time Central Space (FTCS) in order to discretize Fisher's equation, study the stability and consistency and we also obtain error estimates. Sections 6, 7 discuss stability, consistency, and error estimates for NSFD and EEFD schemes. In Section 8, we conclude by presenting the important highlights of this article. The computations are performed by making use of MATLAB R2014a software on an intel core2 as CPU.

3. EXACT SOLUTION

In this section, we present the exact solution of generalized Fisher's equations as described in Kudryashov and Zakharchenko [15]. The nonlinear evolutional equation of that generalized Fisher's equation gives one dimensional diffusion models (for insect, animal dispersal, and invasion) as

$$u_t = u_x(u^l\,u_x) + u^a\,(1-u^b), \tag{11}$$

where t stands for time, x stands for spatial coordinate, u is population density, and a, b, and l are positive parameters. The first term, $u_x(u^l\,u_x)$ on the right-hand side of Equation (11) stands for the growth of population. The term u^l represents the diffusion process depending on the population density.

Let us consider $l \neq 0$ and $u(x,t) = v(\xi)$ and $\xi = s\,x - ct$, $s \neq 0$. Equation (11) gives the following nonlinear ordinary differential equation

$$s^2 \frac{d}{d\xi}\left(v^l \frac{dv}{d\xi}\right) + v^a - v^{a+b} + c\,\frac{dv}{d\xi} = 0. \tag{12}$$

For $l \neq 0$, $v^l = w$. Replacing v by $w^{\frac{1}{l}}$ in Equation (12) gives

$$\frac{s^2}{l} w_\xi^2 + s^2\,w\,w_{\xi\xi} - l\,w^{\frac{a+b+l-1}{l}} + w^{\frac{a+l-1}{l}} + \xi\,w_\xi = 0. \tag{13}$$

Using the Q function method as it is in Kudryashov [29], one has

$$w(\xi) = \sum_{j=0}^{P} P_j\,Q^j(\xi), \quad Q(\xi) = \frac{1}{1+e^{\xi-\xi_0}} \tag{14}$$

where P stands for the pole order and ξ_0 stands as an arbitrary constant. $Q(\xi)$ is the solution of

$$Q_\xi = Q - Q^2. \tag{15}$$

Using Equation (15), we obtain w_ξ and $w_{\xi\xi}$ by using polynomials of Q. Replacing $w \simeq Q^P$ into Equation (15), we have for $b \geq 0$, the following equality

$$\frac{a+b+l-1}{l} = 2 + \frac{2}{P}. \tag{16}$$

To have an integer value $\frac{a+b+l-1}{l}$, P should be 1 or 2. In that case

$$w(\xi) = \begin{cases} P_0 + P_1\, Q(\xi) \\ P_0 + P_1\, Q(\xi) + P_2\, Q^2(\xi) \end{cases} \tag{17}$$

For $P = 1$, $a = 3\,l + 1 - b$, Equation (13) becomes

$$\frac{s^2}{l}\, w_\xi^2 + s^2\, w\, w_{\xi\xi} - l\, w^4 + w^{4-\frac{b}{l}} + \xi\, w_\xi = 0. \tag{18}$$

We have the following solutions

$$w(x,t) = \begin{cases} \text{No exact solutions,} \quad b = l \text{ and } b = 3\,l, \\ \pm 1 \pm 2\, Q(\pm \frac{2\,l}{\sqrt{2l+1}}\, x \pm \frac{2\,l}{2l+1}\, t), \quad b = 2\,l, \\ \pm 1 \pm 2\, Q(\pm \frac{2\,l}{\sqrt{2l+1}}\, x \pm \frac{4l\,(l+1)}{2l+1}\, t), \quad b = 4\,l, \\ \text{or} \\ \pm i \pm 2\, i\, Q(\pm \frac{2\,l}{\sqrt{2l+1}}\, x \pm \frac{4l\,(l+1)}{2l+1} i\; t), \; b = 4\,l. \end{cases} \tag{19}$$

We finally get as exact solution

$$u(x,t) = \begin{cases} \sqrt[l]{\pm 1 \pm 2\, Q\left(\pm \frac{2\,l}{\sqrt{2l+1}}\, x \pm \frac{2\,l}{2l+1}\, t\right)}, \quad V_1 = \pm \frac{1}{\sqrt{2l+1}}, \\ \sqrt[l]{\pm 1 \pm 2\, Q\left(\pm \frac{2\,l}{\sqrt{2l+1}}\, x \pm \frac{4l\,(l+1)}{2l+1}\, t\right)}, \quad V_2 = \pm \frac{2\,(l+1)}{\sqrt{2\,l+1}}, \\ \text{or} \\ \sqrt[l]{\pm i \pm 2\, i\, Q\left(\pm \frac{2\,l}{\sqrt{2l+1}}\, x \pm \frac{4l\,(l+1)}{2l+1} i\; t\right)}, \quad V_2 = \pm \frac{2\,(l+1)}{\sqrt{2l+1}}, \end{cases} \tag{20}$$

where V_1 and V_2 stand for velocity.

For $P = 2$, $a = 2\,l + 1 - b$. With the same reasoning as above, we have

$$u(x,t) = \begin{cases} \text{No exact solutions,} \quad b = 2\,l \text{ and } b = 3\,l, \\ \sqrt[l]{\frac{2(3\,l+2)}{l+1}\left(Q\left(\pm \frac{l}{\sqrt{l+1}} i\, x\right) - Q^2\left(\pm \frac{l}{\sqrt{l+1}} i\, x\right)\right)}, \; b = l. \end{cases} \tag{21}$$

The case of $l = 0$, $a = 1$ and $b = 2$, one obtains the Burgers-Huxley equation and the case of $l = 0$, $a = 1$, $b = 1$, we have Fisher's equation. The exact solution of Equation (11) is described in Li et al. [28] as a scaled Fisher's equation in the form

$$u_t = u_{xx} + \rho u(1-u), \tag{22}$$

with $x \in \mathbb{R}$, t positive, and ρ is a positive constant. The Equations (7) and (8) stand for boundary and initial conditions, respectively. The traveling exact solution to this problem as presented in Polyanim and Zaitsev [30] is

$$u(x,t) = \left[1 + c\, \exp\left(\sqrt{\frac{\rho}{6}}x - \frac{5\rho}{6}t\right)\right]^{-2}, \tag{23}$$

where $c = 5\sqrt{\rho/6}$ stands for the wave speed with the minimum value, $2\sqrt{\rho}$.

4. NUMERICAL EXPERIMENTS

We consider the following problem from Qiu and Sloan [31].

Solve

$$u_t = u_{xx} + \rho\, u(1-u),$$

for $x \in [-0.2, 0.8]$ and $t \in [0,\ T_{max}]$ where $T_{max} = 2.5 \times 10^{-3}$. The initial data is given by

$$u(x,0) = \left[1 + \exp\left(\sqrt{\frac{\rho}{6}}x\right)\right]^{-2}. \tag{24}$$

The exact solution is given by

$$u(x,t) = \left[1 + \exp\left(\sqrt{\frac{\rho}{6}}x - \frac{5\rho}{6}t\right)\right]^{-2}. \tag{25}$$

The boundary conditions are as follows

$$u(-0.2,t) = \left[1 + \exp\left(-0.2\sqrt{\frac{\rho}{6}} - \frac{5\rho}{6}t\right)\right]^{-2} \quad \text{and}$$
$$u(0.8,t) = \left[1 + \exp\left(0.8\sqrt{\frac{\rho}{6}} - \frac{5\rho}{6}t\right)\right]^{-2}.$$

We consider three cases ρ namely; 1, 10^2, 10^4, and obtain a solution at time, $t = T_{max}$.

DEFINITION 1. *Miyata and Sakai [32]. For a vector* $\overline{x} \in \mathbb{R}^N$, $\| \overline{x} \|_1 = \sum_{i=1}^N |\overline{x}^i|$ *and* $\| \overline{x} \|_\infty = \max\{|\overline{x}^i|,\ i = 1, \cdots, N\}$.

DEFINITION 2. *Sutton [33]. Suppose* $\{t^n\}_0^N$ *forms a partition of* $[0, T]$, *with* $t_n = n\Delta t$ *for* $n = 0, \cdots, N$, *where* $\Delta t = T/N$. *Suppose a vector* $\overline{x} \in \mathbb{R}^N$, *defined by*

$$\| \overline{x} \|_{L^P(0,t^n)} = \begin{cases} \left(\| \overline{x} \|_{L^P(0,t^{n-1})} + \tau(\overline{x}^n)^p\right)^{\frac{1}{p}} & \text{for } p \in [0,\infty), \\ \max\{\| \overline{x} \|_{L^P(0,t^{n-1})}, \overline{x}^n\} & \text{for } p = \infty. \end{cases} \tag{26}$$

The rate of convergence with respect to time is defined by

$$rate_i(t) = \frac{\log(\overline{x}^i(t)) - \log(\overline{x}^{i-1}(t))}{\log(\Delta t^i) - \log(\Delta t^{i-1})}.$$

5. FORWARD IN TIME CENTRAL SPACE

The discretization of Equation (22) using the FTCS method gives [34].

$$\frac{u_m^{n+1}-u_m^n}{k}=\frac{u_{m+1}^n-2u_m^n+u_{m-1}^n}{h^2}+\rho u_m^n(1-u_m^n), \quad (27)$$

which leads to

$$u_m^{n+1}=(1-2R)u_m^n+k\rho u_m^n(1-u_m^n)+R(u_{m+1}^n+u_{m-1}^n), \quad (28)$$

or

$$u_m^{n+1}=(1-2R+k\,\rho)\,u_m^n-k\,\rho\,(u_m^n)^2+R\,u_{m+1}^n+R\,u_{m-1}^n, \quad (29)$$

where $R=\frac{k}{h^2}$.

5.1. Stability

The investigation regarding the stability of the scheme given by Equation (28) was done in Agbavon et al. [35]. Nevertheless, we highlight briefly some points. The stability of finite difference methods discretizing nonlinear partial differential equations is not straightforward. Subsequently, freezing the coefficients is needed before using Von Neumann stability analysis [36].

THEOREM 1. *Agbavon et al. [35]. The FTCS scheme given by Equation (28) is conditionally stable, and the stability region is*

$$k\le\frac{h^2}{2} \quad (30)$$

for given spatial step size $h>0$ and the time step size $k>0$. FTCS is first order and second order accurate in time and space, respectively.

Proof. The stability region of the Zabusky and Kruskal scheme using the Korteweg de Vries (KdV) equation was found by Taha and Ablowitz [37] by using the freezing coefficients method and Von Neumann Stability Analysis. The derived scheme by Zabusky and Kruskal [38] for the KdV equation, $u_t+6uu_x+u_{xxx}=0$ is

$$\frac{u_m^{n+1}-u_m^{n-1}}{2k}+6\left(\frac{u_{m+1}^n+u_m^n+u_{m-1}^n}{3}\right)\left(\frac{u_{m+1}^n-u_{m-1}^n}{2h}\right) \quad (31)$$
$$+\frac{1}{2h^3}\left(u_{m+2}^n-2u_{m+1}^n+2u_{m-1}^n-u_{m-2}^n\right)=0.$$

Taha and Ablowitz [37] express $u\,u_x$ as $u_{max}\,u_x$ and utilize the ansatz
$u_m^n=\xi^n e^{Imw}$ where w stands for the phase angle. They obtain

$$(\xi-\xi^{-1})\,(2k)^{-1}+(h)^{-1}\,(6u_{max})\mathrm{I}\sin(w)+(2h^3)^{-1}(e^{2Iw}$$
$$-2e^{Iw}+2e^{-Iw}-e^{-2Iw})=0,$$

which can be rewritten as

$$\xi=\xi^{-1}-(h)^{-1}\,(12ku_{max})\mathrm{I}\sin(w)-(h^3)^{-1}k(e^{2Iw}$$
$$-2e^{Iw}+2e^{-Iw}-e^{-2Iw}) \quad (32)$$

where $u_{max}=\max|u(x,t)|$. The requirement for the linear stability is

$$(h)^{-1}k\left|2u_{max}-(h^2)^{-1}\right|\le 2(3\sqrt{3})^{-1}. \quad (33)$$

For obtaining the stability region of the FTCS scheme discretizing Equation (28), we rewrite Equation (28) using the same idea as

$$u_m^{n+1}=\left(1-(h^2)^{-1}2k\right)u_m^n+(h^2)^{-1}(k)(u_{m+1}^n+u_{m-1}^n)$$
$$+k\,\rho\,u_m^n-k\,\rho\,(u_m^n)^2. \quad (34)$$

Utilization of Fourier series analysis on Equation (34), gives the amplification factor

$$\xi=1-(h^2)^{-1}(2k)(1-\cos(w))+k\rho(1-u_{max}), \quad (35)$$

where u_{max} is frozen coefficient. For the numerical experiment considered, we have $u_{max}=1$, and therefore,

$$\xi=1-(h^2)^{-1}(4k)\sin^2\left(\frac{w}{2}\right). \quad (36)$$

The stability is obtained by solving $|\xi|\le 1$ for $w\in[-\pi,\pi]$, and we obtain $k\le\frac{h^2}{2}$.
Using Taylor series expansion about the point (n,m) of Equation (28), we get

$$u+ku_t+\frac{k^2}{2}u_{tt}+\frac{k^3}{6}u_{ttt}+O(k^4)$$
$$=\left(1-(h^2)^{-1}(2k)+k\rho\right)u-k\rho u^2$$
$$+(h^2)^{-1}(k)\left(u+hu_x+\frac{h^2}{2}u_{xx}+\frac{h^3}{6}u_{xxx}+\frac{h^4}{24}u_{xxxx}+O(h^5)\right)$$
$$+(h^2)^{-1}(k)\left(u-hu_x+\frac{h^2}{2}u_{xx}-\frac{h^3}{6}u_{xxx}+\frac{h^4}{24}u_{xxxx}+O(h^5)\right), \quad (37)$$

which can be written as

$$u_t-u_{xx}-\rho u(1-u)=-\frac{k}{2}u_{tt}-\frac{k^2}{6}u_{ttt}$$
$$+\frac{h^2}{12}u_{xxxx}+O(k^4)+O(h^5). \quad (38)$$

Hence, FTCS is first order and second order accurate in time and space, respectively.

5.2. Error Estimates

THEOREM 2. *Let $u\in C^{4,2}(Q)$, Q defined by $Q=\{(x,t)/\ a\le x\le b,\ 0<t\le T,\ a,b\in\mathbb{R}\}$. If spatial step size, h and time*

step size, k are such that the stability condition (30) holds, then the error estimate, E_m^n for Equation (28) is given by

$$E_m^n \leq (1+3\,k\,\rho)^n\,E_m^0 + \frac{1}{9}\frac{M\,h^2}{\rho\,k}\left[(1+3\,k\,\rho)^n - 1\right] \qquad (39)$$

where $M = \max_{\{(x,t)\}\in Q}\left\{|u_{xxxx}(x,t)|,\ |u_{tt}(x,t)|\right\}$ *and* Θ *such that* $\Theta(k,h)u_{ttt} = O(k,h) \to 0$, *for* $k,\ h \to 0$.

Proof.
Forward in Time Central Space scheme is given by

$$u_m^{n+1} = (1-2R+k\ \rho)\,u_m^n - k\,\rho\,(u_m^n)^2 + R\,u_{m+1}^n + R\,u_{m-1}^n.(40)$$

Taylor series expansion about $(n,\ m)$ gives

$$\begin{aligned}&7v + kv_t + \frac{k^2}{2}v_{tt} + \frac{k^3}{6}v_{ttt} + O(k^4)\\&= \left(1-(h^2)^{-1}(2k)+k\rho\right)v - k\rho v^2\\&+(h^2)^{-1}(k)\left(v + hv_x + \frac{h^2}{2}v_{xx} + \frac{h^3}{6}v_{xxx} + \frac{h^4}{24}v_{xxxx} + O(h^5)\right)\\&+(h^2)^{-1}(k)\left(v - h\,v_x + \frac{h^2}{2}\,v_{xx} - \frac{h^3}{6}v_{xxx} + \frac{h^4}{24}\,v_{xxxx} + O(h^5)\right),\end{aligned} \qquad (41)$$

which can be rewritten as

$$v_t - v_{xx} - \rho\,v\,(1-v) = -\frac{k}{2}\,v_{tt} - \frac{k^2}{6}\,v_{ttt} + \frac{h^2}{12}\,v_{xxxx} + ..., (42)$$

and let $\Theta(k,h)v_{ttt} = O(k,h) = -\frac{k^2}{6}\,v_{ttt} \to 0$ for $k,\ h \to 0$. The exact discrete equation is

$$\begin{aligned}u_m^{n+1} &= (1-2R)\,u_m^n + k\rho\,u_m^n(1-u_m^n) + R(u_{m+1}^n + u_{m-1}^n)\\&+\frac{k}{2}\,u_{tt}(x,\tau_n) - \frac{h^2}{12}\,u_{xxxx}(X_m,t)\end{aligned} \qquad (43)$$

where $x_m < X_m < x_{m+1}$ and $t_n < \tau_n < t_{n+1}$. We define

$$e_m^n = u_m^n - v_m^n \implies e_m^{n+1} = u_m^{n+1} - v_m^{n+1}.$$

It follows that

$$\begin{aligned}e_m^{n+1} &= (1-2R)\,(u_m^n - v_m^n) + \ k\,\rho\,u_m^n(1-u_m^n)\\&-k\,\rho\,v_m^n(1-v_m^n) + R\,(u_{m+1}^n + u_{m-1}^n)\\&-R\,(v_{m+1}^n + v_{m-1}^n) + \frac{k}{2}\,u_{tt}(x,\tau_n) - \frac{h^2}{12}\,u_{xxxx}(X_m,t).\end{aligned} \quad (44)$$

We have

$$\begin{aligned}e_m^{n+1} &= (1-2R)\,(u_m^n - v_m^n) + \ k\,\rho\,u_m^n - k\,\rho\,(u_m^n)^2 - k\,\rho\,v_m^n\\&+k\,\rho\,(v_m^n)^2 + R\,(u_{m+1}^n - v_{m+1}^n)\\&-R\,(u_{m-1}^n - v_{m-1}^n) + \frac{k}{2}\,u_{tt}(x,\tau_n) - \frac{h^2}{12}\,u_{xxxx}(X_m,t),\end{aligned} \qquad (45)$$

which can be rewritten as

$$\begin{aligned}e_m^{n+1} &= (1-2R)\,(u_m^n - v_m^n) + \ k\,\rho\,(u_m^n - v_m^n)\\&-k\,\rho\,(u_m^n - v_m^n)(u_m^n + v_m^n) + R\,(u_{m+1}^n - v_{m+1}^n)\\&-R\,(u_{m-1}^n - v_{m-1}^n) + \frac{k}{2}\,u_{tt}(x,\tau_n) - \frac{h^2}{12}\,u_{xxxx}(X_m,t).\end{aligned} \quad (46)$$

Using the properties of absolute values $|a+b| \leq |a| + |b|$ for $a,\ b \in \mathbb{R}$, we have

$$\begin{aligned}|e_m^{n+1}| &\leq |1-2R|\,|e_m^n| + \ |k\,\rho|\,|e_m^n| + |k\,\rho|\,|e_m^n||u_m^n + v_m^n|\\&+R\,|e_{m+1}^n|\\&+R\,|e_{m-1}^n| + M\left(\frac{k}{2} + \frac{h^2}{12}\right),\end{aligned} \qquad (47)$$

where $M = \max_{\{(x,t)\in Q\}}\left\{|u_{xxxx}(x,t)|,\ |u_{tt}(x,t)|\right\}$. Since $0 \leq u_m^n \leq 1$ and $0 \leq v_m^n \leq 1$, based on numerical experiment chosen, we have

$$\begin{aligned}|e_m^{n+1}| &\leq |1-2R|\,|e_m^n|\\&+k\,\rho\,|e_m^n| + 2\,k\,\rho\,|e_m^n| + R\,|e_{m+1}^n|\\&+R\,|e_{m-1}^n| + M\left(\frac{k}{2} + \frac{h^2}{12}\right).\end{aligned} \qquad (48)$$

Let $E_m^n = \max_{0<m<N}\left\{|e_m^n|)|\right\}$. We have

$$\begin{aligned}|e_m^{n+1}| &\leq (|1-2R| + k\,\rho + 2\,k\,\rho + 2R)\,|E_m^n|\\&+M\left(\frac{k}{2} + \frac{h^2}{12}\right),\end{aligned} \qquad (49)$$

and for stability $R \leq 1/2$, therefore, $|1-2\,R| = 1-2\,R \geq 0$. We finally obtain

$$\begin{aligned}|e_m^{n+1}| &\leq (1+3\,k\,\rho)\,E_m^n\\&\quad + M\left(\frac{k}{2} + \frac{h^2}{12}\right).\end{aligned} \qquad (50)$$

Let $E_m^{n+1} = (1+3\,k\,\rho)\,E_m^n + \left(\frac{k}{2} + \frac{h^2}{12}\right)\,M$. We have

For $n = 0$, $E_m^1 = (1+3\,k\,\rho)\,E_m^0 + \left(\frac{k}{2} + \frac{h^2}{12}\right)\,M$.
For $n = 1$, we have

$$E_m^2 = (1+3\,k\,\rho)\,E_m^1 + \left(\frac{k}{2} + \frac{h^2}{12}\right)M \qquad (51)$$

$$= (1+3\,k\,\rho)^2\,E_m^0 + (1+3\,k\,\rho)^1\left(\frac{k}{2} + \frac{h^2}{12}\right)M + (1+3\,k\,\rho)^0\left(\frac{k}{2} + \frac{h^2}{12}\right)M \qquad (52)$$

$$= (1+3\,k\,\rho)^2\,E_m^0 + \left[(1+3\,k\,\rho)^1 + (1+3\,k\,\rho)^0\,\right]\left(\frac{k}{2} + \frac{h^2}{12}\right)M \qquad (53)$$

For $n = 2$, we have

$$E_m^3 = (1 + 3\,k\,\rho)^3\, E_m^2 + \left(\frac{k}{2} + \frac{h^2}{12}\right) M \quad (54)$$

$$= (1 + 3\,k\,\rho)^3\, E_m^0 + (1 + 3\,k\,\rho)^2\, (\frac{k}{2} + \frac{h^2}{12}\Big)\, M + (1 + 3\,k\,\rho)^1 \left(\frac{k}{2} + \frac{h^2}{12}\right) M + (1 + 3\,k\,\rho)^0 \left(\frac{k}{2} + \frac{h^2}{12}\right) M \quad (55)$$

$$= (1 + 3\,k\,\rho)^3\, E_m^0 + [\,(1 + 3\,k\,\rho)^2 + (1 + 3\,k\,\rho)^1 + (1 + 3\,k\,\rho)^0\,]\left(\frac{k}{2} + \frac{h^2}{12}\right) M \quad (56)$$

For n, we have

$$E_m^n = (1 + 3\,k\,\rho)\, E_m^{n-1} + \left(\frac{k}{2} + \frac{h^2}{12}\right) M$$
$$= (1 + 3\,k\,\rho)^n\, E_m^0 + \big[(1 + 3\,k\,\rho)^{n-1} + (1 + 3\,k\,\rho)^{n-2} + \cdots + (1 + 3\,k\,\rho)^1 + (1 + 3\,k\,\rho)^0\,\big]\left(\frac{k}{2} + \frac{h^2}{12}\right) M$$
$$= (1 + 3\,k\,\rho)^n\, E_m^0 + \left[\sum_{i=0}^{n-1} (1 + 3\,k\,\rho)^i\right]\left(\frac{k}{2} + \frac{h^2}{12}\right) M$$
$$= (1 + 3\,k\,\rho)^n\, E_m^0 + \left[\frac{1 - (1 + 3\,k\,\rho)^n}{1 - (1 + 3\,k\,\rho)}\right]\left(\frac{k}{2} + \frac{h^2}{12}\right) M$$
$$= (1 + 3\,k\,\rho)^n\, E_m^0 - \frac{1}{3\,k\,\rho}\left[1 - (1 + 3\,k\,\rho)^n\right]\left(\frac{k}{2} + \frac{h^2}{12}\right) M \quad (57)$$

Hence, for $k \leq \frac{h^2}{2}$, we can also write

$$E_m^n \leq (1 + 3\,k\,\rho)^n\, E_m^0 + \frac{1}{9}\frac{M\,h^2}{\rho\,k}\left[(1 + 3\,k\,\rho)^n - 1\right] \quad (58)$$

6. NONSTANDARD FINITE DIFFERENCE SCHEME (NSFD)

Over the past decade, the NSFD has been used extensively and often abbreviated as NSFD. The method was introduced by Mickens for the approximation of solutions of partial differential equations and is largely based on the concept of dynamical consistency [39] which plays a significant role in the construction of discrete models whose numerical solution can be complicated to compute. The dynamical consistency is bound to a precise property of a physical system (and varies according to the systems). To mention few among others these properties include the preservation of positivity, boundedness, monotonicity of the solutions, and stability of fixed-points [39]. The main advantage of this method was the dismissal of the primary numerical instabilities [40] caused by the use of standard methods. In order to reduce numerical sensitivities appearing using the classical finite difference methods, these NSFD were developed.

For practical use, the construction of NSFD methods is based on the following basic rules [39]:

(1) The order of discrete derivatives should be equal to the order of corresponding derivatives appearing in the differential equation.
(2) Discrete representation for derivatives, in general, have non trivial denominator functions, e.g.,

$$u_t \approx \frac{u_m^{n+1} - u_m^n}{\phi(\Delta t, \lambda)} \quad (59)$$

where

$$\phi(\Delta t, \lambda) = \Delta t + O(\Delta t^2). \quad (60)$$

6.1. Example of the Definition of the Function ϕ

Consider the following decay equation and logistic growth equation, respectively as in Anguelov et al. [26]

$$\begin{cases} u' = \lambda\, u, & u(0) = u_0,\ \lambda \neq 0, \\ u' = \lambda\, u(1 - u), & u(0) = u_0,\ \lambda > 0, \end{cases} \quad (61)$$

and the respective solutions at the time $t = t_{n+1}$ are

$$\begin{cases} u(t_{n+1}) = u_0\, e^{\lambda\, t_{n+1}}, \\ u(t_{n+1}) = \frac{u_0}{e^{-\lambda\, t_{n+1}} + (1 - e^{-\lambda\, t_{n+1}})\, u_0}. \end{cases} \quad (62)$$

Let $u(t_n) = u^n$. We have

$$\begin{cases} \frac{u^{n+1} - u^n}{(e^{\lambda\,\Delta t} - 1)\lambda^{-1}} = \lambda\, u^n, \\ \frac{u^{n+1} - u^n}{(e^{\lambda\,\Delta t} - 1)\lambda^{-1}} = \lambda\, u^n(1 - u^n). \end{cases} \quad (63)$$

The Equation (63) is called the exact scheme. The function ϕ can be then defined as

$$\phi(\Delta t, \lambda) = \frac{e^{\lambda\,\Delta t} - 1}{\lambda} \text{ or } \phi(\Delta t, \lambda) = \frac{1 - e^{-\lambda\,\Delta t}}{\lambda}$$

(3) Nonlocal discrete representations of nonlinear terms. For instance

$$u_m^2 \approx u_m\, u_{m+1},\ ,u_m^2 \approx \left(\frac{u_{m-1} + u_m + u_{m-1}}{3}\right) u_m, \quad (64)$$

and

$$u^3 \approx 2u_m^3 - u_m^2 u_{m+1},\ u_m^3 \approx u_{m-1} u_m u_{m+1}. \quad (65)$$

In Agbavon et al. [35], followed by the rule of Mickens [39] a **NSFD** for Equation (22) is

$$\frac{u_m^{n+1}-u_m^n}{\phi(\Delta t)} = \frac{u_{m+1}^n - 2u_m^n + u_{m-1}^n}{(\Delta x)^2} + \rho u_m^n - \rho\left(\frac{u_{m+1}^n + u_m^n + u_{m-1}^n}{3}\right) u_m^{n+1}, \quad (66)$$

where

$$\phi(\Delta t) = \phi(k) = \frac{1-e^{-\lambda k}}{\lambda}; \ (\Delta x)^2 = h^2. \quad (67)$$

The Equation (66) gives the following single expression

$$u_m^{n+1} = \frac{\left(1-\frac{2\phi(k)}{h^2}+\rho\,\phi(k)\right)u_m^n + \frac{\phi(k)}{h^2}\left(u_{m+1}^n + u_{m-1}^n\right)}{1+\frac{\rho\phi(k)}{3}\left(u_{m+1}^n + u_m^n + u_{m-1}^n\right)}. \quad (68)$$

6.2. Positivity and Boundedness

In this section, the dynamical consistency and some useful relationship between time and space step-sizes of NSFD are presented.

THEOREM 3. *The dynamical consistency (positivity and boundedness) of NSFD constructed in Equation (68) holds for Equation (22) and for relevant time step k, spatial step h if the following conditions hold*

(a) $\phi(k) \le \frac{h^2}{2-\rho h^2}[1-\Gamma]$ *with* $\Gamma = 1-2(h^2)^{-1}\phi(k)+\rho\,\phi(k)$,

(b) For $u_m^i \in [0,1], \forall\, i.\ \Gamma = (h^2)^{-1}(\phi(k)) = \frac{1}{2}\left[\frac{1}{1-\frac{\rho h^2}{2}}\right]$ *and* $\Gamma' = \sigma\,\Gamma$.

Proof.
We assume $u(x,0)=h(x)\in[0,1]$. We have, therefore, $u(x,t)\in[0,1]$ [24]. We assume also $u_m^n \ge 0$. The quantity u_m^{n+1} from Equation (68) is positive ($u_m^{n+1}\ge 0$) if only

$$\Gamma = 1-2(h^2)^{-1}\phi(k)+\rho\,\phi(k) \ge 0, \quad (69)$$

It follows that

$$0 \le 1-\Gamma = (2(h^2)^{-1}-\rho)\phi(k) \le 1. \quad (70)$$

Hence, in Mickens [24], the condition required for positivity is

$$\phi(k) \le \frac{h^2}{2-\rho h^2}[1-\Gamma] \ \text{ and } \ 0\le\Gamma<1,\ \rho h^2 \ne 2. \quad (71)$$

We investigate next the boundedness by assuming $u_m^n \in [0,1]$. Equation (68) is rewritten as follows

$$u_m^{n+1} = \frac{\Gamma u_m^n + R\left(u_{m+1}^n + u_{m-1}^n\right)}{1+\left(\frac{\rho\phi(k)}{3}\right)\left(u_{m+1}^n + u_m^n + u_{m-1}^n\right)}. \quad (72)$$

where $\Gamma = 1-2(h^2)^{-1}\phi(k)+\rho\,\phi(k),\ R=\frac{\phi(k)}{h^2}$.
Following the idea of Mickens [24], Equation (72) takes the symmetric form if $\Gamma = R$. Therefore, it follows that

$$\Gamma = \frac{\phi(k)}{h^2} = \frac{1}{3} + \frac{\rho\,\phi(k)}{3}. \quad (73)$$

We also have from Equation (71)

$$\phi(k) \le \frac{h^2}{2-\rho h^2}[1-\Gamma] \implies \frac{\phi(k)}{h^2} \le \frac{1}{2}\left[\frac{1}{1-\frac{\rho h^2}{2}}\right] \quad (74)$$

Based on the symmetric condition, we can take

$$\Gamma = \frac{\phi(k)}{h^2} = \frac{1}{2}\left[\frac{1}{1-\frac{\rho h^2}{2}}\right] \quad (75)$$

With regard to the symmetric condition (Equations 73), Equation (72) can be rewritten as

$$u_m^{n+1} = \frac{\Gamma\left(u_m^n + u_{m+1}^n + u_{m-1}^n\right)}{1+\left(\frac{\rho\,\phi(k)}{3}\right)\left(u_{m+1}^n + u_m^n + u_{m-1}^n\right)}. \quad (76)$$

We know by the assumption that $u_j^n \in [0,1], \forall j$. We have

$$0 \le \frac{u_m^n + u_{m+1}^n + u_{m-1}^n}{3} \le 1. \quad (77)$$

By multiplying Equation (77) by $1-\frac{\rho h^2}{2}$ and dividing by $1-\frac{\rho h^2}{2}$ and expanding, we have

$$\frac{u_m^n + u_{m+1}^n + u_{m-1}^n}{3\left[1-\frac{\rho h^2}{2}\right]} - \left[\frac{\rho h^2}{2}\right]\frac{u_m^n + u_{m+1}^n + u_{m-1}^n}{3\left[1-\frac{\rho h^2}{2}\right]} \le 1. \quad (78)$$

From Equation (78), we have

$$\frac{u_m^n + u_{m+1}^n + u_{m-1}^n}{3\left[1-\frac{\rho h^2}{2}\right]} \le 1+\rho\frac{h^2}{2}\frac{1}{3}\left[\frac{1}{1-\frac{\rho h^2}{2}}\right]\left(u_m^n + u_{m+1}^n + u_{m-1}^n\right). \quad (79)$$

Let $\Gamma' = \sigma\,\Gamma, \sigma \ne 0$. Then

$$\Gamma' = \frac{\sigma}{2}\left[\frac{1}{1-\frac{\rho h^2}{2}}\right] = \sigma\frac{\phi(k)}{h^2} \quad (80)$$

If $\frac{\sigma}{2}=\frac{1}{3}$, then $\sigma=\frac{2}{3}$ and using Equation (79), we have

$$\Gamma'\left(u_m^n + u_{m+1}^n + u_{m-1}^n\right) \le 1+\frac{\rho\,\phi(k)}{3}\left(u_m^n + u_{m+1}^n + u_{m-1}^n\right). \quad (81)$$

Hence,

$$0 \le u_m^{n+1} = \frac{\Gamma'\left(u_m^n + u_{m+1}^n + u_{m-1}^n\right)}{1 + \left(\frac{\rho\,\phi(k)}{3}\right)\left(u_{m+1}^n + u_m^n + u_{m-1}^n\right)} \le 1. \quad (82)$$

Thus, the boundedness of u_m^{n+1}.

6.3. Error Estimate

THEOREM 4. *Let $u \in C^{4,2}(Q)$ where Q is defined by*

$$Q = \{(x,t)/\ a \le x \le b,\ 0 < t \le T,\ T > 0,\ a,\ b \in \mathbb{R}\}.$$

Assume h and k are such that the Theorem 3 is satisfied and $e_m^n = u_m^n - v_m^n$, is the defined error of the scheme constructed. NSFD is consistent, and the estimate error e_m^n is defined by

$$|e_m^n| \le E_m^n = (3\,\Gamma)^n\,E_m^0 + \left(\frac{1-(3\,\Gamma)^n}{1-3\,\Gamma}\right)$$
$$\left[\left(\frac{\phi(k)}{2} + \rho\frac{\phi(k)^2}{2} + \frac{h^2}{12} + \rho\frac{h^4}{36}\right) M + \rho\, h^4 \frac{\phi(k)^2}{72} M^2\right] \quad (83)$$

where Γ defined in 3 M is defined by $M = \max_{\{(x,t)\}\in Q}\left\{|u_{xxxx}(x,t)|,\ |u_{tt}(x,t)|\right\}$, *and* Θ_i, $i = 1, 2, 3$ *such that*

$$\Theta_1(\phi(k),h) = \rho\,\frac{h^2}{3}\,u - \rho\, h^2\,\frac{\phi(k)^2}{6}\,u_{tt}$$
$$\Theta_2(\phi(k),h) = -\frac{\phi(k)^2}{6} + \rho\frac{\phi(k)}{3}$$
$$\left[\frac{\phi^2(k)}{2}\,u + h^2\,\frac{\phi^2}{6}\,u_{xx} + h^4\,\frac{\phi(k)^2}{72} u_{xxxx}\right]$$
$$\Theta_3(\phi(k),h) = -\phi(k)\,v - \rho\, h^2\,\frac{\phi(k)}{3} - \rho\, h^4\frac{\phi(k)}{36}\,u_{xxxx} \quad (84)$$

and $\Theta_1(\phi(k),h)\,u_{xx} + \Theta_2(\phi(k),h)\,u_{ttt} + \Theta_3(\phi(k),h)\,u_t = O\left(\phi(k),h\right) \to 0$ *when* $\phi(k) \to 0$ *and* $h \to 0$.

Proof.

$$v_m^{n+1} = \frac{\Gamma v_m^n + R\left(v_{m+1}^n + v_{m-1}^n\right)}{1 + \left(\frac{\rho\phi(k)}{3}\right)\left(v_{m+1}^n + v_m^n + v_{m-1}^n\right)}, \quad (85)$$

where $\Gamma = 1 - 2\frac{\phi(k)}{h^2} + \rho\phi(k)$, $R = \frac{\phi(k)}{h^2}$. Taylor series expansion of Equation (85) about (t_n, x_m) gives

$$\left(v + \phi(k)\,v_t + \frac{(\phi(k))^2}{2} v_{tt} + \frac{(\phi(k))^3}{6} v_{ttt} + O\left((\phi(k))^4\right)\right)$$
$$\left(1 + \rho\frac{\phi(k)}{3}\left\{\begin{array}{l} v + v + h\,v_x + \frac{h^2}{2}\,v_{xx} + \frac{h^3}{6}\,v_{xxx} + \frac{h^4}{24}\,v_{xxxx} \\ +v - h\,v_x + \frac{h^2}{2}\,v_{xx} - \frac{h^3}{6}\,v_{xxx} + \frac{h^4}{24}\,v_{xxxx} \end{array}\right\}\right)$$
$$= \left(1 - \frac{2\phi(k)}{h^2} + \rho\,\phi(k)\right) v$$
$$+ \frac{\phi(k)}{h^2}\left\{\begin{array}{l} v + h\,v_x + \frac{h^2}{2}\,v_{xx} + \frac{h^3}{6}\,v_{xxx} + \frac{h^4}{24}\,v_{xxxx} \\ +v - h\,v_x + \frac{h^2}{2}\,v_{xx} - \frac{h^3}{6}\,v_{xxx} + \frac{h^4}{24}\,v_{xxxx} \end{array}\right\}$$

This gives

$$v + \phi(k)\,v_t + \frac{(\phi(k))^2}{2} v_{tt} + \frac{(\phi(k))^3}{6} v_{ttt}$$
$$+\rho\frac{\phi(k)}{3}\left(v + \phi(k)\,v_t + \frac{(\phi(k))^2}{2} v_{tt} + \frac{(\phi(k))^3}{6} v_{ttt}\right)$$
$$\left(3\,v + h^2\,v_{xx} + \frac{h^4}{12}\,v_{xxxx}\right)$$
$$= v + \left(-\frac{2\phi(k)}{h^2} + \rho\,\phi(k)\right) v + \frac{\phi(k)}{h^2}\left(2\,v + h^2\,v_{xx} + + \frac{h^4}{12}\,v_{xxxx}\right). \quad (86)$$

It follows after division by $\phi(k)$ and simplification, we have

$$v_t + \frac{(\phi(k))}{2} v_{tt} + \frac{(\phi(k))^2}{6} v_{ttt}$$
$$+\frac{\rho}{3}\left(\phi(k)\,v_t + \frac{(\phi(k))^2}{2} v_{tt} + \frac{(\phi(k))^3}{6} v_{ttt}\right)$$
$$\left(3\,v + h^2\,v_{xx} + \frac{h^4}{12}\,v_{xxxx}\right)$$
$$+\frac{\rho}{3}\,v(3\,v) + \frac{\rho}{3}\,v\left(h^2\,v_{xx} + \frac{h^4}{12}\,v_{xxxx}\right)$$
$$= v_{xx} + \rho\,v + \frac{h^2}{12}\,v_{xxxx}. \quad (87)$$

It follows that

$$v_t - v_{xx} + \rho\,v^2 - \rho\,v$$
$$= \left\{\begin{array}{l} \left(\frac{h^2}{12} - \rho\frac{h^4}{36}\,v\right) v_{xxxx} - \left(\frac{\phi(k)}{2} + \rho\, h^4\frac{\phi(k)^2}{72}\,v_{xxxx} + \rho\,\frac{\phi^2(k)}{2}\,v\right) v_{tt} \\ +\left(\rho\,\frac{h^2}{3}\,v - \rho\, h^2\,\frac{\phi(k)^2}{6} v_{tt}\right) v_{xx} \\ +\left(-\frac{\phi(k)^2}{6} + \rho\frac{\phi(k)}{3}\left[\frac{\phi(k)^2}{2}\,v + h^2\,\frac{\phi^2(k)}{6}\,v_{xx} + h^4\,\frac{\phi(k)^2}{72}\,v_{xxxx}\right]\right) v_{ttt} \\ +\left(-\phi(k)\,v - \rho\, h^2\,\frac{\phi(k)}{3} - \rho\, h^4\frac{\phi(k)}{36}\,v_{xxxx}\right) v_t \end{array}\right\} \quad (88)$$

If $\phi(k) \to 0$ and $h \to 0$, Equation (88) reduces to $v_t - v_{xx} + \rho\,v^2 - \rho\,v \to 0$. Hence, the consistency.

For the simplicity of the proof, we consider the function Θ_i, $i = 1, 2, 3$ such that

$$\Theta_1(\phi(k),h) = \rho\,\frac{h^2}{3}\,v - \rho\, h^2\,\frac{\phi(k)^2}{6} v_{tt}$$
$$\Theta_2(\phi(k),h) = -\frac{\phi(k)^2}{6} + \rho\frac{\phi(k)}{3}$$
$$\left[\frac{\phi^2(k)}{2}\,v + h^2\,\frac{\phi^2}{6}\,v_{xx} + h^4\,\frac{\phi(k)^2}{72} v_{xxxx}\right]$$
$$\Theta_3(\phi(k),h) = -\phi(k)\,v - \rho\, h^2\,\frac{\phi(k)}{3} - \rho\, h^4\frac{\phi(k)}{36}\,v_{xxxx} \quad (89)$$

and $\Theta_1(\phi(k),h)\,v_{xx} + \Theta_2(\phi(k),h)\,v_{ttt} + \Theta_3(\phi(k),h)\,v_t = O\left(\phi(k),h\right) \to 0$ when $\phi(k) \to 0$ and $h \to 0$.

The exact discrete equation is

$$u_m^{n+1}=\frac{\Gamma u_m^n+R\left(u_{m+1}^n+u_{m-1}^n\right)}{1+\frac{\rho\phi(k)}{3}(u_{m+1}^n+u_m^n+u_{m-1}^n)}+\left(\frac{\phi(k)}{2}+\rho\,h^4\frac{\phi(k)^2}{72}\,u_{xxxx}(\varepsilon_m,t)+\rho\,\frac{\phi^2(k)}{2}\,u\right)u_{tt}(x,\tau_n)-\left(\frac{h^2}{12}-\rho\frac{h^4}{36}\,u\right)u_{xxxx}(\varepsilon_m,t) \tag{90}$$

where $\Gamma=1-2\frac{\phi(k)}{h^2}+\rho\phi(k)$, $R=\frac{\phi(k)}{h^2}$, and $x_m<\varepsilon_m<x_{m+1}$ and $t_n<\tau_n<t_{n+1}$.

We define $e_m^n=u_m^n-v_m^n\equiv e_m^{n+1}=u_m^{n+1}-v_m^{n+1}$. It follows by considering symmetry condition $\Gamma=R$

$$u_m^{n+1}-v_m^{n+1}=\left\{\begin{array}{l}\frac{\Gamma\left(u_{m+1}^n+u_m^n+u_{m-1}^n\right)}{1+\frac{\rho\phi(k)}{3}(u_{m+1}^n+u_m^n+u_{m-1}^n)}-\frac{\Gamma\left(v_{m+1}^n+v_m^n+v_{m-1}^n\right)}{1+\frac{\rho\phi(k)}{3}(v_{m+1}^n+v_m^n+v_{m-1}^n)}\\+\left(\frac{\phi(k)}{2}+\rho\,h^4\frac{\phi(k)^2}{72}\,u_{xxxx}(\varepsilon_m,t)+\rho\,\frac{\phi^2(k)}{2}\,u\right)u_{tt}(x,\tau_n)\\-\left(\frac{h^2}{12}-\rho\frac{h^4}{36}\,u\right)u_{xxxx}(\varepsilon_m,t)\end{array}\right\}. \tag{91}$$

It follows

$$e_m^{n+1}=\frac{\Gamma\left(e_{m+1}^n+e_m^n+e_{m-1}^n\right)}{\left(1+\frac{\rho\phi(k)}{3}(u_{m+1}^n+u_m^n+u_{m-1}^n)\right)\left(1+\frac{\rho\phi(k)}{3}(v_{m+1}^n+v_m^n+v_{m-1}^n)\right)}+\left(\frac{\phi(k)}{2}+\rho\,h^4\frac{\phi(k)^2}{72}\,u_{xxxx}(\varepsilon_m,t)+\rho\,\frac{\phi(k)^2}{2}\,u\right)u_{tt}(x,\tau_n)-\left(\frac{h^2}{12}-\rho\frac{h^4}{36}\,u\right)u_{xxxx}(\varepsilon_m,t) \tag{92}$$

Let $M=\max_{\{(x,t)\}\in Q}\left\{|u_{xxxx}(x,t)|,\ |u_{tt}(x,t)|\right\}$ and $E_m^n=\max_{0<m<N}\{|e_m^n|\}$. We have

$$\left(1+\frac{\rho\phi(k)}{3}(u_{m+1}^n+u_m^n+u_{m-1}^n)\right)\left(1+\frac{\rho\phi(k)}{3}(v_{m+1}^n+v_m^n+v_{m-1}^n)\right)>1,$$

$\forall\, u_i^n,\ v_i^n\in[0,1],\ i=m-1,\ m,\ m+1$
and by using Theorem 3, we have

$$|e_m^{n+1}|\le|\Gamma|(|e_{m+1}^n|+|e_m^n|+|e_{m-1}^n|)+\left|\left(\frac{\phi(k)}{2}+\rho\,h^4\frac{\phi(k)^2}{72}\,u_{xxxx}(\varepsilon_m,t)+\rho\,\frac{\phi(k)^2}{2}\,u\right)\right||u_{tt}(x,\tau_n)|+\left|-\left(\frac{h^2}{12}-\rho\frac{h^4}{36}\,u\right)\right||u_{xxxx}(\varepsilon_m,t)|$$
$$\le 3\,\Gamma\,E_m^n+\left(\frac{\phi(k)}{2}+\rho\,\frac{\phi(k)^2}{2}+\frac{h^2}{12}+\rho\frac{h^4}{36}\right)M+\rho\,h^4\frac{\phi(k)^2}{72}\,M^2 \tag{93}$$

Let

$$E_m^{n+1}=3\,\Gamma\,E_m^n+\left(\frac{\phi(k)}{2}+\rho\,\frac{\phi(k)^2}{2}+\frac{h^2}{12}+\rho\frac{h^4}{36}\right)M+\rho\,h^4\frac{\phi(k)^2}{72}\,M^2.$$

We have

$$|e_m^{n+1}|\le E_m^{n+1}=3\,\Gamma\,E_m^n+\left(\frac{\phi(k)}{2}+\rho\,\frac{\phi(k)^2}{2}+\frac{h^2}{12}+\rho\frac{h^4}{36}\right)M+\rho\,h^4\frac{\phi(k)^2}{72}\,M^2 \tag{94}$$

For $n=0$, $E_m^1=3\,\Gamma\,E_m^0+\left(\frac{\phi(k)}{2}+\rho\,\frac{\phi(k)^2}{2}+\frac{h^2}{12}+\rho\frac{h^4}{36}\right)M+\rho\,h^4\frac{\phi(k)^2}{72}\,M^2$
For $n=1$, we have

$$E_m^2=3\,\Gamma\,E_m^1+\left(\frac{\phi(k)}{2}+\rho\,\frac{\phi(k)^2}{2}+\frac{h^2}{12}+\rho\frac{h^4}{36}\right)M+\rho\,h^4\frac{\phi(k)^2}{72}\,M^2$$
$$=3\Gamma\left(3\,\Gamma\,E_m^0+\left(\frac{\phi(k)}{2}+\rho\,\frac{\phi(k)^2}{2}+\frac{h^2}{12}+\rho\frac{h^4}{36}\right)M+\rho\,h^4\frac{\phi(k)^2}{72}\,M^2\right)+\left(\frac{\phi(k)}{2}+\rho\,\frac{\phi(k)^2}{2}+\frac{h^2}{12}+\rho\frac{h^4}{36}\right)M+\rho\,h^4\frac{\phi(k)^2}{72}\,M^2$$
$$=(3\,\Gamma)^2\,E_m^0+(1+3\,\Gamma)\left[\left(\frac{\phi(k)}{2}+\rho\,\frac{\phi(k)^2}{2}+\frac{h^2}{12}+\rho\frac{h^4}{36}\right)M+\rho\,h^4\frac{\phi(k)^2}{72}\,M^2\right] \tag{95}$$

For $n=2$, we have

$$E_m^3=3\,\Gamma\,E_m^2+\left(\frac{\phi(k)}{2}+\rho\,\frac{\phi(k)^2}{2}+\frac{h^2}{12}+\rho\frac{h^4}{36}\right)M+\rho\,h^4\frac{\phi(k)^2}{72}\,M^2$$
$$=3\Gamma\left(3^2\,\Gamma^2+(1+3\,\Gamma)\left[\left(\frac{\phi(k)}{2}+\rho\,\frac{\phi(k)^2}{2}+\frac{h^2}{12}+\rho\frac{h^4}{36}\right)M+\rho\,h^4\frac{\phi(k)^2}{72}\,M^2\right]\right)+\left(\frac{\phi(k)}{2}+\rho\,\frac{\phi(k)^2}{2}+\frac{h^2}{12}+\rho\frac{h^4}{36}\right)M+\rho\,h^4\frac{\phi(k)^2}{72}\,M^2$$
$$=(3\,\Gamma)^3\,E_m^0+(1+3\,\Gamma+(3\,\Gamma)^2)\left[\left(\frac{\phi(k)}{2}+\rho\,\frac{\phi(k)^2}{2}+\frac{h^2}{12}+\rho\frac{h^4}{36}\right)M+\rho\,h^4\frac{\phi(k)^2}{72}\,M^2\right] \tag{96}$$

By recurrence for n, we have

$$E_m^{n+1} = (3\,\Gamma)^n E_m^0 + \left(\sum_{i=0}^{i-1}(3\,\Gamma)^i\right)$$
$$\left[\left(\frac{\phi(k)}{2} + \rho\,\frac{\phi(k)^2}{2} + \frac{h^2}{12} + \rho\frac{h^4}{36}\right) M + \rho\, h^4 \frac{\phi(k)^2}{72} M^2\right]$$
$$= (3\,\Gamma)^n E_m^0 + \left(\frac{1-(3\,\Gamma)^n}{1-3\,\Gamma}\right)$$
$$\left[\left(\frac{\phi(k)}{2} + \rho\,\frac{\phi(k)^2}{2} + \frac{h^2}{12} + \rho\frac{h^4}{36}\right) M + \rho\, h^4 \frac{\phi(k)^2}{72} M^2\right] \tag{97}$$

7. EXPLICIT EXPONENTIAL FINITE DIFFERENCE SCHEME

The EFFD method was developed by Bhattacharya [41] (primarily called the Bhattacharya method) for the numerical solution of the heat equation. The Exponential Finite Difference method was utilized to solve Burgers' equation and generalized Huxley and Burgers-Huxley equations [42–44]. Later, Macías-Díaz and İnan [45] used modified exponential methods to obtain the solution of the Burgers' equation. Furthermore, Inan et al. [46] utilized the EEFD method for numerical solutions for the Newell-Whitehead-Segel type equations which are very useful in biomathematics. They showed convergence, consistency, and stability of the method.

In this section, we obtain numerical solutions of the equation by **EEFD** method. The solution domain are discretized into cells as (x_m, t_n) in which $x_m = mh$, $(m = 0, 1, 2, ..., N)$ and $t_n = nk$, $(n = 0, 1, 2, ...)$, $h = \Delta x = \frac{b-a}{N}$ is the spatial mesh size and $k = \Delta t$ is the time step, u_m^n denotes the **EEFD** approximation and $u(x, t)$ denotes the exact solution.

Dividing Equation (6) by u gives

$$\frac{\partial \ln u}{\partial t} = \frac{1}{u}\left(u\,(1-u) + \frac{\partial^2 u}{\partial x^2}\right). \tag{98}$$

Using the finite difference approximations for derivatives, Equation (98) gives

$$u_m^{n+1} = u_m^n \exp\left\{k\left(1-u_m^n\right) + R\left(\frac{u_{m+1}^n - 2u_m^n + u_{m-1}^n}{u_m^n}\right)\right\} \tag{99}$$

where $R = \frac{k}{h^2}$. Equation (99) gives the expression for the **EEFD** method for Fisher's equation.

7.1. Convergence and Estimate Error

For stability analysis, we require non-iterative exponential time-linearization and iterative exponential quasilinearization techniques for Equation (6) which are found in the discretization of the time derivative, the freezing of the coefficients of the resulting linear ordinary differential equations, and the piecewise analytical solution of these ordinary differential equations. These techniques give three-point finite difference expressions that depend in an exponential manner on either the diffusion, reaction, and transient terms or the diffusion and reaction terms. Following the idea of Ramos [25], we transform (Equations 6, 99) into a linear ordinary differential equation by discretizing the time derivative by means of θ-method [25] and linearizing the nonlinear source term, $u(1-u)$, with respect to either the previous time level or the previous iteration with Jacobian, $J = \frac{d(u(1-u))}{du} = 1 - 2\,u$:

a) If the linearization is performed with respect to the previous time level, one obtains

$$\frac{u_m^{n+1} - u_m^n}{k} = \theta\,\frac{u_{m-1}^{n+1} - 2\,u_m^{n+1} + u_{m+1}^{n+1}}{h^2}$$
$$+(1-\theta)\,\frac{u_{m-1}^n - 2\,u_m^n + u_{m+1}^n}{h^2}$$
$$+u_m^n(1-u_m^n)$$
$$+\theta\, J_m^n\,(u_m^{n+1} - u_m^n) \tag{100}$$

which yields a non-iterative time linearization method.

b) If the linearization is performed with respect to the previous iteration, one obtains

$$\frac{u_m^{i+1} - u_m^n}{k} = \theta\,\frac{u_{m-1}^{i+1} - 2\,u_m^{i+1} + u_{m+1}^{i+1}}{h^2}$$
$$+(1-\theta)\,\frac{u_{m-1}^n - 2\,u_m^n + u_{m+1}^n}{h^2}$$
$$+(1-\theta)\,u_m^n(1-u_m^n) + \theta\,u_m^{n+1}(1-u_m^{i+1})$$
$$+\theta\, J_m^{i+1}\,(u_m^{i+1} - u_m^n) \tag{101}$$

which corresponds to an iterative quasilinear technique and $i = 1,\ 2, \cdots, n$.

Equations (100) and (101) can be solved in closed form in (x_{m-1}, x_{m+1}) subject to the following conditions:

$$u(x_{m-1}) = u_{m-1},\ \ u(x_m) = u_m,\ \ u(x_{m+1}) = u_{m+1} \tag{102}$$

and yield exponential solutions in (x_{m-1}, x_{m+1}) which are analytical in that interval and continuous everywhere. Since Equations (100) and (101) are very similar, we will only present in detail exponential methods for Equation (100) in the following subsections.

7.1.1. Time-Linearized Full Exponential Techniques

The piecewise analytical solution of Equation (100) can be rewritten as

$$\left(\theta\,J_m^n - \frac{1}{k}\right)\,u_m^{n+1} + \theta\,\frac{u_{m-1}^{n+1} - 2\,u_m^{n+1} + u_{m+1}^{n+1}}{h^2}$$
$$= \left(\theta\,J_m^n - \frac{1}{k}\right)\,u_m^n - (1-\theta)\,\frac{u_{m-1}^n - 2\,u_m^n + u_{m+1}^n}{h^2}$$
$$-u_m^n(1-u_m^n). \tag{103}$$

TABLE 1 | L_1 and L_∞ errors at some different values of time-step size, k with $\rho = 1$ at time, $T_{max} = 2.5 \times 10^{-3}$ with spatial mesh size, $h = 0.01$ using three methods.

Values of k	FTCS		NSFD		EEFD	
	L_1 error	L_∞ error	L_1 error	L_∞ error	L_1 error	L_∞ error
$T_{max}/52$	5.4192×10^{-9}	6.3334×10^{-9}	3.2264×10^{-8}	3.7216×10^{-8}	4.1091×10^{-9}	5.1492×10^{-9}
$T_{max}/100$	2.8854×10^{-9}	3.3694×10^{-9}	1.7026×10^{-8}	1.9624×10^{-8}	2.0683×10^{-9}	2.6091×10^{-9}
$T_{max}/200$	1.5135×10^{-9}	1.7645×10^{-9}	8.7757×10^{-9}	1.0098×10^{-8}	9.6251×10^{-10}	1.2354×10^{-9}
$T_{max}/300$	1.0563×10^{-9}	1.7645×10^{-9}	6.0260×10^{-9}	6.0260×10^{-9}	5.9431×10^{-10}	7.7761×10^{-10}
$T_{max}/400$	8.2780×10^{-10}	9.6234×10^{-10}	4.6512×10^{-9}	5.3361×10^{-9}	4.1024×10^{-10}	5.4763×10^{-10}
$T_{max}/500$	6.9066×10^{-10}	8.0194×10^{-10}	3.8264×10^{-9}	4.3837×10^{-9}	2.9973×10^{-10}	4.1032×10^{-10}
$T_{max}/600$	5.9924×10^{-10}	6.9502×10^{-10}	3.2766×10^{-9}	3.7488×10^{-9}	2.2612×10^{-10}	3.1904×10^{-10}
$T_{max}/700$	5.3392×10^{-10}	6.1863×10^{-10}	2.8838×10^{-9}	3.2954×10^{-9}	1.7352×10^{-10}	2.5366×10^{-10}
$T_{max}/800$	4.8495×10^{-10}	5.6136×10^{-10}	2.5894×10^{-9}	2.9554×10^{-9}	1.3402×10^{-10}	2.0476×10^{-10}
$T_{max}/900$	4.4686×10^{-10}	5.1681×10^{-10}	2.3602×10^{-9}	2.6910×10^{-9}	1.0334×10^{-10}	1.6694×10^{-10}
$T_{max}/1,000$	4.1639×10^{-10}	4.8117×10^{-10}	2.1769×10^{-9}	2.4795×10^{-9}	7.8784×10^{-11}	1.3668×10^{-10}
$T_{max}/1,100$	3.9144×10^{-10}	4.5199×10^{-10}	2.0269×10^{-9}	2.3064×10^{-9}	5.8705×10^{-11}	1.1192×10^{-10}
$T_{max}/1,200$	3.7066×10^{-10}	4.2769×10^{-10}	1.9019×10^{-9}	2.1622×10^{-9}	4.1956×10^{-11}	9.1419×10^{-11}
$T_{max}/1,300$	3.5308×10^{-10}	4.0714×10^{-10}	1.7962×10^{-9}	2.0402×10^{-9}	3.0241×10^{-11}	7.4118×10^{-11}
$T_{max}/1,400$	3.3801×10^{-10}	3.8953×10^{-10}	$\mathbf{1.7056 \times 10^{-9}}$	$\mathbf{1.9356 \times 10^{-9}}$	2.3908×10^{-11}	5.9401×10^{-11}
$T_{max}/1,500$	3.2496×10^{-10}	3.7427×10^{-10}	1.6270×10^{-9}	1.8449×10^{-9}	2.1174×10^{-11}	4.6738×10^{-11}
$T_{max}/1,600$	3.1353×10^{-10}	3.6091×10^{-10}	1.5583×10^{-9}	1.7656×10^{-9}	$\mathbf{2.0904 \times 10^{-11}}$	$\mathbf{3.5802 \times 10^{-11}}$
$T_{max}/1,700$	3.0345×10^{-10}	3.4913×10^{-10}	1.4976×10^{-9}	1.6956×10^{-9}	2.2344×10^{-11}	4.2194×10^{-11}
$T_{max}/1,800$	2.9449×10^{-10}	3.3866×10^{-10}	1.4438×10^{-9}	1.6335×10^{-10}	2.4984×10^{-11}	4.8536×10^{-11}
$T_{max}/2,000$	$\mathbf{2.9449 \times 10^{-10}}$	$\mathbf{3.3866 \times 10^{-10}}$	$\mathbf{1.3521 \times 10^{-9}}$	$\mathbf{1.5279 \times 10^{-10}}$	3.2504×10^{-11}	5.9412×10^{-11}

Bold values indicate the lowest errors.

TABLE 2 | L_1 and L_∞ errors at some different values of time-step size, k with $\rho = 10^2$ at time, $T_{max} = 2.5 \times 10^{-3}$ with spatial mesh size, $h = 0.01$ using three methods.

Values of k	FTCS		NSFD		EEFD	
	L_1 error	L_∞ error	L_1 error	L_∞ error	L_1 error	L_∞ error
$T_{max}/52$	3.2613×10^{-5}	6.9142×10^{-5}	6.3178×10^{-5}	1.3471×10^{-4}	2.2654×10^{-5}	7.2263×10^{-5}
$T_{max}/100$	1.7120×10^{-5}	3.6652×10^{-5}	3.3705×10^{-5}	7.2133×10^{-5}	1.1628×10^{-5}	3.7401×10^{-5}
$T_{max}/200$	8.7183×10^{-6}	1.9060×10^{-5}	1.7709×10^{-5}	3.8414×10^{-5}	5.6602×10^{-6}	1.8584×10^{-5}
$T_{max}/300$	5.9159×10^{-6}	1.3196×10^{-5}	1.2371×10^{-5}	2.7303×10^{-5}	3.6732×10^{-6}	1.2324×10^{-5}
$T_{max}/400$	4.5154×10^{-6}	1.0268×10^{-5}	9.7020×10^{-6}	2.1809×10^{-5}	2.6794×10^{-6}	9.1874×10^{-6}
$T_{max}/500$	3.6749×10^{-6}	8.5169×10^{-6}	8.1001×10^{-6}	1.8545×10^{-5}	2.0833×10^{-6}	7.3251×10^{-6}
$T_{max}/600$	3.1146×10^{-6}	7.3489×10^{-6}	7.0319×10^{-6}	1.6379×10^{-5}	1.6864×10^{-6}	6.0864×10^{-6}
$T_{max}/700$	2.7143×10^{-6}	6.5146×10^{-6}	6.2689×10^{-6}	1.4842×10^{-5}	1.4024×10^{-6}	5.2014×10^{-6}
$T_{max}/800$	2.4141×10^{-6}	5.8925×10^{-6}	5.6966×10^{-6}	1.3689×10^{-5}	1.1903×10^{-6}	4.5376×10^{-6}
$T_{max}/900$	2.1806×10^{-6}	5.4093×10^{-6}	5.2515×10^{-6}	1.2802×10^{-5}	1.0244×10^{-6}	4.0219×10^{-6}
$T_{max}/1,000$	1.9941×10^{-6}	5.0227×10^{-6}	4.8954×10^{-6}	1.2093×10^{-5}	8.9174×10^{-7}	3.6143×10^{-6}
$T_{max}/1,100$	1.8416×10^{-6}	4.7064×10^{-6}	4.6040×10^{-6}	1.1513×10^{-5}	7.8356×10^{-7}	3.2846×10^{-6}
$T_{max}/1,200$	1.7144×10^{-6}	4.4428×10^{-6}	4.3612×10^{-6}	1.1030×10^{-5}	7.0876×10^{-7}	3.0099×10^{-6}
$T_{max}/1,300$	1.6068×10^{-6}	4.2197×10^{-6}	4.1557×10^{-6}	1.0621×10^{-5}	6.5558×10^{-7}	2.7774×10^{-6}
$T_{max}/1,400$	1.5153×10^{-6}	4.0298×10^{-6}	3.9797×10^{-6}	1.0274×10^{-5}	6.1557×10^{-7}	2.5767×10^{-6}
$T_{max}/1,500$	1.4371×10^{-6}	3.8658×10^{-6}	3.8279×10^{-6}	9.9737×10^{-6}	5.8466×10^{-7}	2.4043×10^{-6}
$T_{max}/1,600$	1.3698×10^{-6}	3.7222×10^{-6}	3.6960×10^{-6}	9.7113×10^{-6}	5.6048×10^{-7}	2.2528×10^{-6}
$T_{max}/1,700$	1.3113×10^{-6}	3.5956×10^{-6}	3.5803×10^{-6}	9.4798×10^{-6}	5.4168×10^{-7}	2.1186×10^{-6}
$T_{max}/1,800$	1.2601×10^{-6}	3.4830×10^{-6}	3.4781×10^{-6}	9.2740×10^{-6}	5.2654×10^{-7}	2.0034×10^{-6}
$T_{max}/2,000$	$\mathbf{1.1748 \times 10^{-6}}$	$\mathbf{3.2916 \times 10^{-6}}$	$\mathbf{3.3055 \times 10^{-6}}$	$\mathbf{8.9242 \times 10^{-6}}$	$\mathbf{5.0458 \times 10^{-7}}$	$\mathbf{1.807 \times 10^{-6}}$

Bold values indicate the lowest errors.

TABLE 3 | L_1 and L_∞ errors at some different values of time-step size, k with $\rho = 10^4$ at time, $T_{max} = 2.5 \times 10^{-3}$ with spatial mesh size, $h = 0.01$ using three methods.

Values of k	FTCS		NSFD		EEFD	
	L_1 error	**L_∞ error**	**L_1 error**	**L_∞ error**	**L_1 error**	**L_∞ error**
$T_{max}/52$	3.6440×10^{-1}	1.6012	1.1960×10^{-1}	9.1526×10^{-1}	over flow	over flow
$T_{max}/100$	6.5143×10^{-2}	6.9877×10^{-1}	7.0119×10^{-2}	7.1253×10^{-1}	1.3594×10^{-2}	1.5978×10^{-1}
$T_{max}/200$	3.2980×10^{-2}	3.9927×10^{-1}	3.6170×10^{-2}	4.1776×10^{-1}	1.0084×10^{-2}	1.1786×10^{-1}
$T_{max}/300$	2.0726×10^{-2}	2.5613×10^{-1}	2.3034×10^{-2}	2.7854×10^{-1}	8.9264×10^{-3}	1.0476×10^{-1}
$T_{max}/400$	1.4262×10^{-2}	1.7830×10^{-1}	1.6098×10^{-2}	1.9602×10^{-1}	8.3519×10^{-3}	9.8394×10^{-2}
$T_{max}/500$	1.0268×10^{-2}	1.2943×10^{-1}	1.1810×10^{-2}	1.4691×10^{-1}	8.0087×10^{-3}	9.4591×10^{-2}
$T_{max}/600$	7.5556×10^{-3}	9.5604×10^{-2}	8.8965×10^{-3}	1.1270×10^{-1}	7.7810×10^{-3}	9.2071×10^{-2}
$T_{max}/700$	5.5928×10^{-3}	7.1037×10^{-2}	6.7882×10^{-3}	8.7558×10^{-2}	7.6182×10^{-3}	9.0294×10^{-2}
$T_{max}/800$	4.1067×10^{-3}	5.2473×10^{-2}	5.1924×10^{-3}	6.8387×10^{-2}	7.4974×10^{-3}	8.8949×10^{-2}
$T_{max}/900$	2.9425×10^{-3}	3.7988×10^{-2}	3.9442×10^{-3}	5.3326×10^{-2}	7.4020×10^{-3}	8.7921×10^{-2}
$T_{max}/1,000$	2.0059×10^{-3}	2.6389×10^{-2}	2.9442×10^{-3}	4.1202×10^{-2}	7.3264×10^{-3}	8.7094×10^{-2}
$T_{max}/1,100$	1.2403×10^{-3}	1.6901×10^{-2}	2.1334×10^{-3}	3.1242×10^{-2}	7.2643×10^{-3}	8.6421×10^{-2}
$T_{max}/1,200$	6.2582×10^{-4}	9.0012×10^{-3}	1.4824×10^{-3}	2.2920×10^{-2}	7.2120×10^{-3}	8.5864×10^{-2}
$T_{max}/1,300$	**1.9347×10^{-4}**	**2.3239×10^{-3}**	9.7801×10^{-4}	1.5866×10^{-2}	7.1694×10^{-3}	8.5384×10^{-2}
$T_{max}/1,400$	4.2509×10^{-4}	4.3457×10^{-3}	6.3252×10^{-4}	9.8126×10^{-3}	7.1310×10^{-3}	8.4982×10^{-2}
$T_{max}/1,500$	8.3336×10^{-4}	8.7691×10^{-3}	**4.5235×10^{-4}**	**4.5623×10^{-3}**	7.0992×10^{-3}	8.4631×10^{-2}
$T_{max}/1,600$	1.1913×10^{-3}	1.3138×10^{-2}	5.0611×10^{-4}	5.3708×10^{-3}	7.0703×10^{-3}	8.4321×10^{-2}
$T_{max}/1,700$	1.5077×10^{-3}	1.7003×10^{-2}	8.4385×10^{-4}	7.6874×10^{-3}	7.0451×10^{-3}	8.4052×10^{-2}
$T_{max}/1,800$	1.7895×10^{-3}	2.0446×10^{-2}	1.1467×10^{-3}	1.0154×10^{-2}	7.0231×10^{-3}	8.3812×10^{-2}
$T_{max}/2,000$	1.5538×10^{-2}	1.8402×10^{-1}	1.6628×10^{-3}	1.5383×10^{-2}	**6.9852×10^{-3}**	**8.3399×10^{-2}**

Bold values indicate the lowest errors.

Let $\left(\theta J_m^n - \frac{1}{k}\right) u_m^{n+1} + \theta \frac{u_{m-1}^{n+1} - 2u_m^{n+1} + u_{m+1}^{n+1}}{h^2} = z_m^{n+1}$. Equation (103) becomes

$$z_m^{n+1} = \left(\theta J_m^n - \frac{1}{k}\right) u_m^n - (1-\theta)\frac{u_{m-1}^n - 2u_m^n + u_{m+1}^n}{h^2} - u_m^n(1-u_m^n) \tag{104}$$

which accounts for diffusion, reaction, and transient effects in the differential operator [25]. The solution of Equation (104) depends on the sign of $D_m^n \equiv \theta J_m^n - \frac{1}{k}$. We have the following three cases:

a) If $D_m^n = 0$, the solution of Equation (104) subject to the condition (102) gives the finite difference expression

$$z_m^{n+1} = \frac{u_{m-1}^n - 2u_m^n + u_{m+1}^n}{h^2}. \tag{105}$$

b) If $D_m^n = -\theta(\lambda_m)^2 < 0$,the solution of Equation (104) subject to the condition (102) gives the three-point finite difference expression

$$z_m^{n+1} = \frac{D_m^n}{2}\left(\frac{u_{m-1}^n - 2\cosh(\lambda_m h)\, u_m^n + u_{m+1}^n}{1-\cosh(\lambda_m h)}\right). \tag{106}$$

c) If $D_m^n = \theta(\lambda_m)^2 > 0$, the solution of Equation (104) subject to the condition (102) gives the three-point finite difference expression

$$z_m^{n+1} = \frac{D_m^n}{2}\left(\frac{u_{m-1}^n - 2\cos(\lambda_m h)\, u_m^n + u_{m+1}^n}{1-\cos(\lambda_m h)}\right). \tag{107}$$

REMARK 1. *We can make the following remarks*

a) The values $\theta = \frac{1}{2}$, 1, corresponding to the time linearization methods.

b) The quasilinear full exponential corresponds to the iterative solution of

$$z_m^{i+1} = -(1-\theta)\frac{u_{m-1}^n - 2u_m^n + u_{m+1}^n}{h^2} - (1-\theta)\, u_m^n(1-u_m^n) - \theta\, u_m^i(1-u_m^i) + \theta J_m^i u_m^i - \frac{u_m^n}{k} \tag{108}$$

for $i = 1, 2, \cdot\cdot, n-1$, have analogous solutions to those reported in time-linearized full exponential techniques section and the cases $\theta = \frac{1}{2}$ and 1, corresponding to the quasilinear methods.

7.1.2. Time-Linearized Exponential Techniques

The time-linearized exponential techniques presented in this section consider the following differential operator.

$$\theta J_m^n u_m^{n+1} + \theta\frac{u_{m-1}^{n+1} - 2u_m^{n+1} + u_{m+1}^{n+1}}{h^2} = \theta J_m^n u_m^n - (1-\theta)\frac{u_{m-1}^n - 2u_m^n + u_{m+1}^n}{h^2} + \frac{u_m - u_m^n}{k} - u_m^n(1-u_m^n) \tag{109}$$

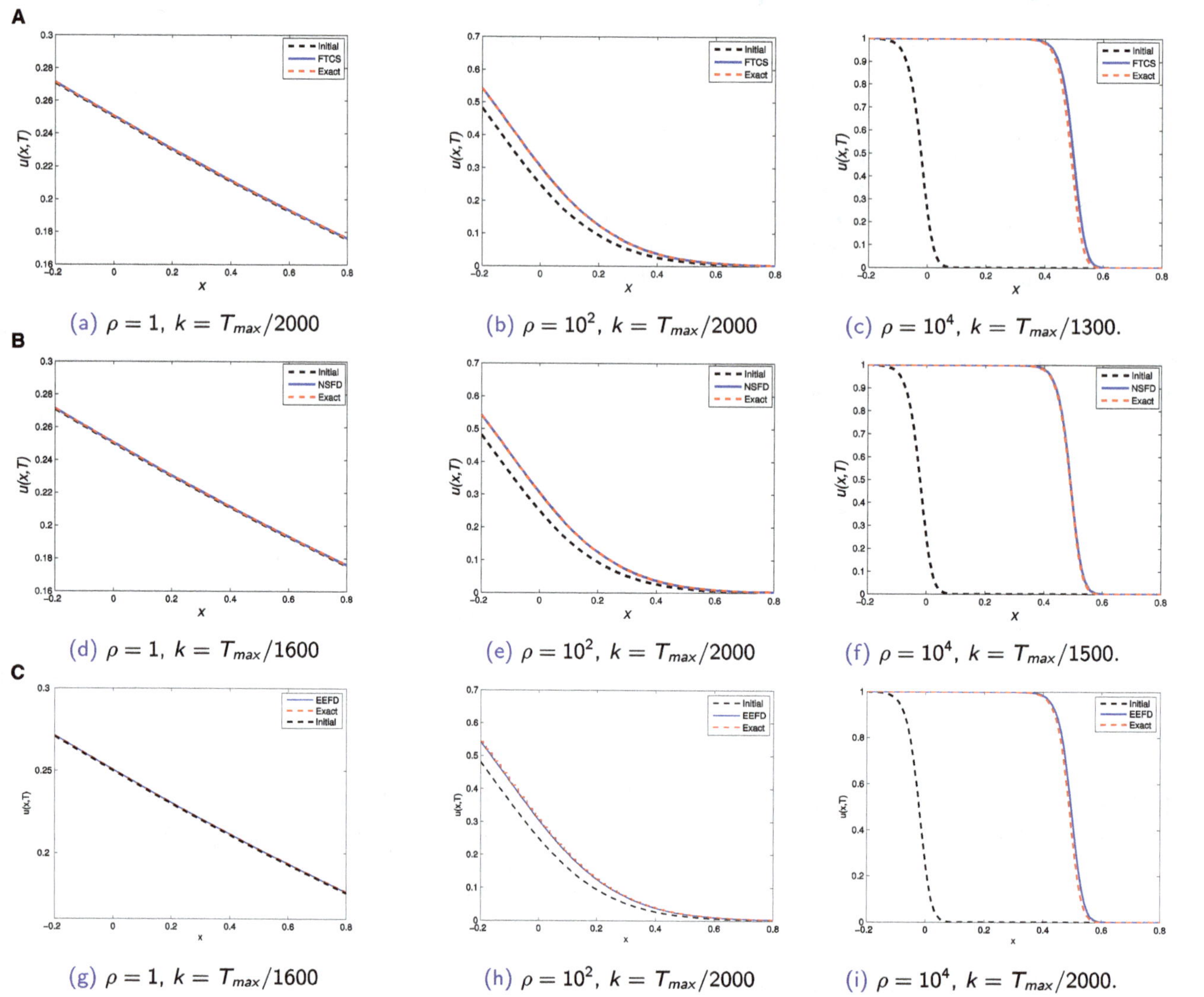

FIGURE 1 | Plot of initial solution, numerical solution of (FTCS), NSFD, and EEFD respectively in **(A–C)** and exact solution against x at time, $T_{max} = 2.5 \times 10^{-3}$ using $h = 0.01$ and optimal value of time step size of k for three different values of ρ namely 1, 10^2, and 10^4.

Let $\theta\, J_m^n\, u_m^{n+1} + \theta\, \frac{u_{m-1}^{n+1} - 2\, u_m^{n+1} + u_{m+1}^{n+1}}{h^2} = z_m^{n+1}$. The Equation (109) becomes

$$z_m^{n+1} = \theta\, J_m^n\, u_m^n - (1-\theta)\, \frac{u_{m-1}^n - 2\, u_m^n + u_{m+1}^n}{h^2} + \frac{u_m - u_m^n}{k} - u_m^n(1 - u_m^n) \tag{110}$$

which only accounts for reaction and diffusion processes and whose solutions depend on the sign of J_m^n. We have therefore the solution of Equation (110) subject to the condition (Equation 102) gives the solution in form of:

a) If $J_m^n = 0$

$$u_{m-1} - \left(2 + \frac{h^2}{k}\right) u_m + u_{m+1} = -\frac{h^2}{k}\left(u_m^n + k\, u_m^n\,(1 - u_m^n)\right), \quad t^n < t < t^{n+1}. \tag{111}$$

b) If $J_m^n = -(\lambda_m)^2 < 0$

$$u_{m-1} - \frac{2 + (k\, J_m^n - 1)(e^{-\lambda_m h} + e^{\lambda_m h})}{k\, J_m^n}\, u_m + u_{m+1} = \frac{2 - (e^{-\lambda_m h} + e^{\lambda_m h})}{k\, J_m^n}\left[-u_m + k\,(J_m^n\, u_m - u_m^n\,(1 - u_m^n))\right], \quad t^n < t < t^{n+1}. \tag{112}$$

c) If $J_m^n = (\lambda_m)^2 > 0$

$$u_{m-1} - 2\frac{1 + (kJ_m^n - 1)cos(\lambda_m h)}{kJ_m^n} u_m + u_{m+1} = 2\frac{1 - (cos(\lambda_m h)}{kJ_m^n} [-u_m + k(J_m^n u_m - u_m^n(1 - u_m^n))],$$

$$t^n < t < t^{n+1}. \tag{113}$$

REMARK 2. *The quasilinear full exponential corresponds to the iterative solution of*

$$\theta J_m^i u_m^{n+1} + \theta \frac{u_{m-1}^{n+1} - 2u_m^{n+1} + u_{m+1}^{n+1}}{h^2} = \theta J_m^n u_m^n - (1-\theta)\frac{u_{m-1}^n - 2u_m^n + u_{m+1}^n}{h^2} + \frac{u_m - u_m^n}{k} - (1-\theta)u_m^n(1-u_m^n) - \theta u_m^i(1-u_m^i) \tag{114}$$

for $i = 1, 2 \cdots n-1$, have analogous solutions to those reported in time-linearized exponential techniques Section 7.1.2.

REMARK 3. *The principal primacy of the full exponential techniques is that they reckon for the reaction, diffusion, and transient terms in finding the homogeneous solution of Equation (103). Notwithstanding, this technique also has the disadvantage that the time step cannot be chosen carelessly because, if $k \ll \frac{1}{\theta J_i^n}$ then the reaction terms do not influence the values of λ_m despite the fact that they do influence the particular solution of Equation (103) through z_m^{n+1}. On top of that, the portion of the transient term influences the ordinary differential operator and, therefore, the homogeneous solution, though the other part influences the particular solution. These obstacles are fully suppressed with the exponential techniques displayed in the time-linearized exponential techniques and the quasilinear full exponential in Section 7.1.2.*

THEOREM 5. *Ramos [25]. The schemes displayed in Equations (100) and (101) are convergent and convergence is reached when*

$$(e_m^n)^2 = \frac{1}{N}\sum_{j=1}^{N}(u_j^{i+1} - u_j^i)^2 \le 10^{-|\alpha|} \tag{115}$$

where $i = 1, 2 \cdots n$ and $|\alpha|$ is an integer obtained from numerical computation, N denotes the number of grid points, and e_m^n is the error which is defined by $e_m^n = u_m^n - v_m^n$.

Proof. The full proof of this theorem is detailed in Ramos [25].

8. NUMERICAL RESULTS

The stability region of FTCS is $k \le \frac{h^2}{2}$. For $h = 0.01$, we obtain $k \le 5 \times 10^{-5}$. In the case of NSFD, the condition for positivity gives $\phi(k) \le \frac{h^2}{2}$ where $\phi(k) = \frac{1-e^{\lambda k}}{\lambda}$. We tabulate L_1 and L_∞ errors at certain values of k using $\rho = 1$, $h = 0.01$, at time, $T_{max} = 2.5 \times 10^{-3}$ using FTCS, NSFD, and EEFD schemes at certain various values of time-step size as chosen from $T_{max}/52$, $T_{max}/100$, $T_{max}/200$, $\cdots$, $T_{max}/1{,}800$, $T_{max}/1{,}900$ and $T_{max}/2{,}000$. The errors are shown in **Table 1**.

TABLE 4 | Rate of convergence in time for FTCS using different values of ρ at time, $T_{max} = 2.5 \times 10^{-3}$ with $h = 0.01$.

Values of ρ	Values of k	Rate of convergence in time using L_1	Rate of convergence in time using L_∞ error
1	$T_{max}/100$		
	$T_{max}/200$	0.9309	0.9332
	$T_{max}/400$	0.8705	0.8746
	$T_{max}/800$	0.7714	0.7776
10^2	$T_{max}/100$		
	$T_{max}/200$	0.9736	0.9433
	$T_{max}/400$	0.9492	0.8924
	$T_{max}/800$	0.9034	0.8012
10^4	$T_{max}/100$		
	$T_{max}/200$	0.9820	0.8075
	$T_{max}/400$	1.2094	1.1631

TABLE 5 | Rate of convergence in time for NSFD using different values of ρ at time, $T_{max} = 2.5 \times 10^{-3}$ with $h = 0.01$.

Values of ρ	Values of k	Rate of convergence in time using L_1	Rate of convergence in time using L_∞ error
1	$T_{max}/100$		
	$T_{max}/200$	0.9561	0.9585
	$T_{max}/400$	0.9159	0.9202
10^2	$T_{max}/100$		
	$T_{max}/200$	0.9284	0.9090
	$T_{max}/400$	0.8681	0.8167
10^4	$T_{max}/100$		
	$T_{max}/200$	0.9769	0.7702

TABLE 6 | Rate of convergence in time for EEFD using different values of ρ at time, $T_{max} = 2.5 \times 10^{-3}$ with $h = 0.01$.

Values of ρ	Values of k	Rate of convergence in time using L_1	Rate of convergence in time using L_∞ error
1	$T_{max}/100$		
	$T_{max}/200$	1.1030	1.0795
	$T_{max}/400$	1.2306	1.1727
10^2	$T_{max}/100$		
	$T_{max}/200$	1.0384	1.0091
	$T_{max}/400$	1.0789	1.0162
	$T_{max}/800$	1.1714	1.0176
10^4	$T_{max}/100$		
	$T_{max}/200$	0.4317	0.4388

In the case of FTCS, NSFD, minimum L_1, and L_∞ errors occur at $k = T_{max}/2{,}000$ while in the case of EEFD, the errors are least $k = T_{max}/1{,}600$. The least error is of order 10^{-9} and 10^{-10} in the case of NSFD and FTCS respectively while the least error is of order 10^{-11} in the case of EEFD. EEFD is a better scheme than FTCS at all values of k used.

We obtain values for L_1 and L_∞ errors at certain values of k using $\rho = 10^2$, $h = 0.01$, at time, $T_{max} = 2.5 \times 10^{-3}$ using the three methods in **Table 2**. The schemes behave differently. In all the three methods FTCS, NSFD, and EEFD, the L_1 and L_∞ errors keep on decreasing as the values of k are decreased gradually from $k = T_{max}/52$ to $k = T_{max}/2{,}000$.

L_1 and L_∞ errors for the third case using $\rho = 10^4$, $h = 0.01$ are displayed in **Table 3**. Again, the schemes behave differently from each other. Optimal k using FTCS occurs when $k \approxeq T_{max}/1{,}300$. The optimal k using NSFD and EEFD are $k \approxeq T_{max}/1{,}500$ and $k \approxeq T_{max}/2{,}000$, respectively. Once that optimal is reached the error starts increasing again. The least L_1 and L_∞ errors using NSFD are 4.5235×10^{-4} and 4.5623×10^{-3}. The corresponding errors are 1.9347×10^{-4} and 2.3239×10^{-3} when FTCS is used. In the case of EEFD, least L_1 is 6.9852×10^{-3} and least L_∞ is 8.3399×10^{-2}.

We obtain plots of numerical solution vs. x at time, $T_{max} = 2.5 \times 10^{-3}$ using three methods FTCS, NSFD, and EEFD in **Figure 1**.

9. CONCLUSION

We have investigated in this paper the spectral analysis and optimal step sizes for some finite difference methods discretising Fisher's equation. We used three methods namely; FTCS, NSFD, and EEFD in order to solve Fisher's equation with a coefficient of reaction being 10, 10^2, and 10^4. We studied the properties of the methods such stability, positivity, and boundedness. This is the one of rare article which includes the estimate errors for the methods studied. Numerical results are displayed at optimal time step size with $h = 0.01$ for the three cases for the three methods used. We also obtained numerically the rate of convergence as shown in **Tables 4–6**. We have shown from **Tables 1–3** that all the three methods (FTCS, NSFD, and EEFD) perform well for the small coefficient of reaction. It is worthy mentioning that freezing coefficient technique with Von Neumann Stability Analysis only present an approximate stability region for standard methods discretising non linear partial differential equations which might give reason to the standard method in **Table 1** to perform better than the NSFD. Furthermore the NSFD in regard to the discrete representation derivative in Mickens [39] rule has nontrivial denominator function and make use of positivity and boundedness conditions. Finally the results are dependent on initial conditions used. For $\rho = 1$, the difference in L_1 and L_∞ errors from the three methods is very small which lead to conclude that the best methods are FTCS and NSFD. Also for $\rho = 10^2$ and $\rho = 10^4$, the best method are EEFD and NSFD, respectively. Our results matched with the one found in Lubuma and Roux [47] for numerical experiment for small reaction term. Moreover, Lubuma and Roux [47] proved that NSFD is elementary stable. As NSFD methods, the EEFD displayed in this article do not require any knowledge of the exact solution of the differential equation. Contrast to that, the best finite difference scheme is stable for large grid sizes but costly in inaccuracies at the propagation front.

AUTHOR CONTRIBUTIONS

The plan of the paper was prepared by AA. The coding was done by KA and BI. All authors helped in writing the manuscript, contributed to the article, and approved the submitted version.

ACKNOWLEDGMENTS

KA is grateful to the North-West University for the Postdoctoral Research Fellowship (PDRF). AA thank Nelson Mandela University for the financial support. Finally, all the authors are grateful to the two reviewers who provided feedback which enabled to improve the paper considerably.

REFERENCES

1. Gardner CS, Greene JM, Kruskal MD, Miura RM. Method for solving the Korteweg-de Vries equation. *Front Neurosci Phys Rev Lett.* (1967) 19:1095. doi: 10.1103/PhysRevLett.19.1095
2. Weiss J, Tabor M, Carnevale G. The Painlevé property for partial differential equations. *J Math Phys.* (1983) 155:522–6. doi: 10.1063/1.525721
3. Kudryashov NA. On types of nonlinear nonintegrable equations with exact solutions. *Phys Lett A.* (1991) 24:269–75. doi: 10.1016/0375-9601(91)90481-M
4. Lou SY, Huang GX, Ruan HY. Exact solitary waves in a convecting fluid. *J Phys A Math Gen.* (1991) 24:L587. doi: 10.1088/0305-4470/24/11/003
5. Kudryashov NA. Exact solutions of a family of Fisher equations. *Theor Math Phys.* (1993) 94:211–218. doi: 10.1007/BF01019332
6. Kudryashov NA. Numerical convective schemes based on accurate computation of space derivatives. *J Comput Phys.* (1973) 13:100–13. doi: 10.1016/0021-9991(73)90128-9
7. Cooley JW, Lewis P, Welch P. The finite Fourier transform. *IEEE Trans Audio Electroacoust.* (1969) 17:77–85. doi: 10.1109/TAU.1969.1162036

8. Gazdag J, Canosa J. Numerical solution of Fisher's equation. *J Appl Probab.* (1974) 11:445–57. doi: 10.2307/3212689
9. Fisher RA. The wave of advance of advantageous genes. *Ann Hum Genet.* (1937) 17:355–69. doi: 10.1111/j.1469-1809.1937.tb02153.x
10. Moran PAP. *The Statistical Processes of Evolutionary Theory.* Oxford: Clarendon Press; Oxford University Press (1962).
11. Kendall DG. *A Form of Wave Propagation Associated With the Equation of Heat Conduction.* Cambridge: Cambridge University Press (1948).
12. Kolmogorov AN. Étude de l'équation de la diffusion avec croissance de laquantité de matière et son application à un problème biologique. *Bull Univ Moskow Ser Internat A.* (1937) 1:1–25.
13. Canosa J. On a nonlinear diffusion equation describing population growth. *IBM J Res Dev.* (1937) 17:307–13. doi: 10.1147/rd.174.0307
14. Jeffrey A, Kakutani T. Weak nonlinear dispersive waves: a discussion centered around the Korteweg-de Vries equation. *Siam Rev SIAM.* (1972) 14:582–643. doi: 10.1137/1014101
15. Kudryashov NA, Zakharchenko AS. A note on solutions of the generalized Fisher equation. *Appl Math Lett.* (2014) 32:53–6. doi: 10.1016/j.aml.2014.02.009
16. Newman TJ, Kolomeisky EB, Antonovics J. Population dynamics with global regulation: the conserved Fisher equation. *Phys Rev Lett.* (2004) 92:228103. doi: 10.1103/PhysRevLett.92.228103
17. Chen BM, Kojouharov HV. Non-standard numerical methods applied to subsurface biobarrier formation models in porous media. *Bull Math Biol.* (1999) 61:779–98. doi: 10.1006/bulm.1999.0113
18. Yatat V, Couteron P, Dumont Y. Spatially explicit modelling of tree-grass interactions in fire-prone savannas: a partial differential equations framework. *Ecol Complex.* (2018) 36:290–313. doi: 10.1016/j.ecocom.2017.06.004
19. Wang XY. Exact and explicit solitary wave solutions for the generalised Fisher equation. *Phys Lett A.* (1988) 131:277–9. doi: 10.1016/0375-9601(88)90027-8
20. Chandraker V, Awasthi A, Jayaraj S. A numerical treatment of Fisher equation. *Procedia Eng.* (2015) 131:1256–62. doi: 10.1016/j.proeng.2015.11.481
21. Ismail HN, Raslan K, Rabboh AAA. Adomian decomposition method for Burger's-Huxley and Burger's-Fisher equations. *Appl Math Comput.* (2004) 159:291–301. doi: 10.1016/j.amc.2003.10.050
22. Moghimi M, Hejazi FSA. Variational iteration method for solving generalized Burger-Fisher and Burger equations. *Chaos Solitons Fractals.* (2007) 33:1756–61. doi: 10.1016/j.chaos.2006.03.031
23. Fahmy ES. Travelling wave solutions for some time-delayed equations through factorizations. *Chaos Solitons Fractals.* (2008) 38:1209–16. doi: 10.1016/j.chaos.2007.02.007
24. Mickens RE. Relation between the time and space step-sizes in nonstandard finite-difference schemes for the Fisher equation. *Numer Methods Partial Differ Equ.* (1997) 13:51–5. doi: 10.1002/(SICI)1098-2426(199701)13:1<51::AID-NUM4>3.0.CO;2-L
25. Ramos JI. Exponential methods for one-dimensional reaction-diffusion equations. *Math Methods Appl Sci.* (2005) 170:380–98. doi: 10.1016/j.amc.2004.12.003
26. Anguelov R, Kama P, Lubuma JMS. On non-standard finite difference models of reaction-diffusion equations. *J Comput Appl Math.* (2005) 175:11–29. doi: 10.1016/j.cam.2004.06.002
27. Hagstrom T, Keller HB. The numerical calculation of traveling wave solutions of nonlinear parabolic equations. *SIAM J Scientific Stat Comput.* (1986) 7:978–88. doi: 10.1137/0907065
28. Li S, Petzold L, Ren Y. Stability of moving mesh systems of partial differential equations. *SIAM J Scientific Comput.* (1998) 20:719–38. doi: 10.1137/S1064827596302011
29. Kudryashov NA. One method for finding exact solutions of nonlinear differential equations. *Commun Nonlinear Sci Numer Simulat.* (2012) 17:2248–53. doi: 10.1016/j.cnsns.2011.10.016
30. Polyanim AD, Zaitsev ZF. *Handbook of Nonlinear Partial Differential Equations. 2nd ed.* Chapman and Hall/CRC Press. (2011). p. 1912.
31. Qiu Y, Sloan DM. Numerical solution of Fisher's equation using a moving mesh method. *J Comput Phys.* (1998) 146:726–746. doi: 10.1006/jcph.1998.6081
32. Miyata T, Sakai Y. Vectorized total variation defined by weighted L infinity norm for utilizing inter channel dependency. In: *2012 19th IEEE International Conference on Image Processing.* Orlando, FL: IEEE (2012).
33. Sutton OJ. Long time $L^{\infty}(L^2)$ a posteriori error estimates for fully discrete parabolic problems. *arXiv[Preprint].arXiv:180303207.* (2018). doi: 10.1093/imanum/dry078
34. Chen-Charpentier BM, Kojouharov HV. An unconditionally positivity preserving scheme for advection-diffusion reaction equations. *Math Comput Model.* (2013) 57:2177–85. doi: 10.1016/j.mcm.2011.05.005
35. Agbavon K, Appadu A, Khumalo M. On the numerical solution of Fisher's equation with coefficient of diffusion term much smaller than coefficient of reaction term. *Adv Diff Equat.* (2019) 2019:1–33. doi: 10.1186/s13662-019-2080-x
36. Durran DR. *Numerical Methods for Fluid Dynamics (With Applications to Geophysics).* Springer Science & Business Media (2010).
37. Taha TR, Ablowitz MI. Analytical and numerical aspects of certain nonlinear evolution equations, III. Numerical, Korteweg-de Vries equation. *J Comput Phys.* (1984) 55:231–53. doi: 10.1016/0021-9991(84)90004-4
38. Zabusky NJ, Kruskal MD. Interaction of solitons in a collisionless plasma and the recurrence of initial states. *Phys Rev Lett.* (1965) 15:240. doi: 10.1103/PhysRevLett.15.240
39. Mickens RE. Dynamic consistency: a fundamental principle for constructing nonstandard finite difference schemes for differential equations. *J Diff Equat Appl.* (2005) 11:645–53. doi: 10.1080/10236190412331334527
40. Kumar Verma A, Kayenat S. On the stability of Micken's type NSFD schemes for generalized Burgers Fisher equation. *Chaos Solitons Fractals.* (2019) 25:1706–37. doi: 10.1080/10236198.2019.1689236
41. Bhattacharya MC. An explicit conditionally stable finite difference equation for heat conduction problems. *Int J Numer Methods Eng.* (1985) 21:239–65. doi: 10.1002/nme.1620210205
42. Handschuh RF, Keith TG. Applications of an exponential finite-difference technique. *Numer Heat Transfer A Appl.* (1992) 22:363–78. doi: 10.1080/10407789208944773
43. İnan B, Bahadı r AR. An explicit exponential finite difference method for the Burger's equation. *Eur Int J Sci Technol.* (2013) 2:61–72.
44. İnan B, Bahadı r AR. Finite difference methods for the generalized Huxley and Burgers-Huxley equations. *Kuwait J Sci.* (2017) 44:20–7.
45. Macías-Díaz JE, İnan B. Numerical efficiency of some exponential methods for an advection-diffusion equation. *Int J Comput Math.* (2019) 96:1005–29. doi: 10.1080/00207160.2018.1478416
46. Inan B, Osman M, Ak T, Baleanu D. Analytical and numerical solutions of mathematical biology models: the newell-whitehead-segel and allen-cahn equations. *Math Methods Appl Sci.* (2020) 43:2588–600. doi: 10.1002/mma.6067
47. Lubuma J, Roux A. An improved theta-method for systems of ordinary differential equations. *J Diff Equat Appl.* (2003) 9:1023–35. doi: 10.1080/1023619031000146904

13

Statistical Analysis of fNIRS Data: Consideration of Spatial Varying Coefficient Model of Prefrontal Cortex Activity Changes During Speech Motor Learning in Apraxia of Speech

Rachel Johnson [1], Jennifer Matthews [2], Norou Diawara [3]* and Rachel Carroll [4]

[1] Department of Communication Disorders and Special Education, Old Dominion University, Norfolk, VA, United States, [2] Department of Mathematics, U.S. Naval Academy, Annapolis, MD, United States, [3] Department of Mathematics and Statistics, Old Dominion University, Norfolk, VA, United States, [4] Department of Mathematics and Statistics, University of North Carolina Wilmington, Wilmington, NC, United States

***Correspondence:**
Norou Diawara
ndiawara@odu.edu

Apraxia of speech is an impairment in the planning and programming of speech typically accompanied by aphasia (language impairment) secondary to a left hemisphere stroke. It is unknown if the structural and functional connections to the damaged area implicate the integrity of the cognitive functions of the prefrontal cortex (PFC). The present study examines the feasibility of measuring hemodynamic activity in the PFC in response to the structure of practice and during treatment. This multiple-baseline single case-design study involving two individuals with chronic acquired apraxia of speech measured the hemodynamic changes in PFC activity during treatment across the intervention period using functional near-infrared spectroscopy (fNIRS). Two models—a generalized linear model and a spatial varying coefficient model—are used to distinguish the repeated measures of PFC activity differences corresponding to the stage of practice and time of intervention. There were significant differences in the pattern of PFC activity associated with the structure of practice and the time of intervention. The outcomes from this pilot study demonstrate the utility of fNIRS to identify cognitive effort during speech motor learning. The implications include consideration for statistical methods used for fNIRS analysis and its potential use as a clinical tool to complement behavior changes to guide patient-directed intervention to optimize patient outcomes.

Keywords: fNIRS, prefrontal cortex activity, generalized linear model, spatial varying, Moran statistics, apraxia of speech

INTRODUCTION

The complex neurobiology of speech involves distributed networks in cortical and subcortical brain regions [1]. A compromise to networks of interconnected regions following a stroke can result in apraxia of speech (AOS; is a sensorimotor impairment in the planning and programming for speech), aphasia (language disorder), and/or dysarthria (a disorder of speech execution and control) [2, 3]. The anatomical location and size of the lesion can affect the processing of regions functionally and structurally connected to the damaged area, [3–5] including the prefrontal cortex (PFC) [6]. Consequently, post-stroke communication deficits often involve concomitant speech, language and cognitive impairments. In general, behavioral interventions for AOS are efficacious; however, there is limited evidence to support a particular treatment [7–9].

One of the contributing factors to the limited evidence for AOS, treatment methods is the incomplete understanding of the underlying neural mechanisms responsible for motor planning and programming for typical and disordered speech. Advancements in functional neuroimaging and computational modeling have resulted in two widely recognized models for speech motor control, Directions into Velocities of Articulators (DIVA), and speech sequencing, Gradient Order DIVA (GODIVA) [10, 11]. The core characteristics of AOS, as explained in the DIVA model, are the result of impaired feedforward control and difficulty integrating feedback to correct or refine motor programs [12]. The GODIVA model explains suprasegmental characteristics of AOS, result from errors within the higher-level mechanisms involving temporary phonological buffer (working memory) and sequential activation in longer utterances for fluent speech [12]. Although the combined models have a limited explanation of motor planning, they do provide a framework to guide the interpretation of behavior and neuroimaging studies for AOS cases involving multiple sites.

Functional near-infrared spectroscopy (fNIRS) is an established neuroimaging technique to non-invasively monitor concentration changes of oxygenated hemoglobin (HbO) and deoxygenated hemoglobin (HbR) evoked neural activity in the cerebral cortex using light in the infrared spectrum [13]. The wavelength absorption signal correlates spatially and temporally with the functional magnetic resonance imaging (fMRI) blood oxygen-level dependent signal during cognitive tasks [14–16]. The resistance to motion artifact and the ability to use in a natural therapeutic setting makes fNIRS ideal for repeated measures of the brain's underlying cognitive processes during learning and performance across the course of treatment [17]. Investigations using fNIRS during language and cognitive tasks across the age span report changes in PFC activity between persons with aphasia, persons without aphasia, and healthy adults during confrontation naming tasks, [18] verbal fluency, [19, 20], and memory learning [21] are related to age and cortical integrity. To our knowledge there are no investigations involving persons with AOS during a speech, cognitive, or learning task over time.

The cognitive functions and its implications in the rehabilitation in persons with post-stroke acquired AOS is an understudied area, despite its potential influence on patient outcomes. The dual-process learning theory explains that a high level of bilateral PFC activity indicates a reliance on cognitive control processes during the initial stages of practice [22]. The areas of the prefrontal cortex associated with the cognitive control processes for cognitive flexibility are the dorsal lateral prefrontal cortex (DLPFC) and medial prefrontal cortex (MPFC). Specifically, Brodmann's Area (BA) 9, 10, and 46 are associated with tasks involving salience detection and attention, working memory, inhibition, and task switching for efficient adaption and response to changing environments [23, 24]. In the case of AOS, a heavy reliance on the PFC cognitive control processes are likely for speech motor learning during rehabilitation [5]. This information has clinical benefits to structure patient-centered interventions to maximize treatment outcomes.

One treatment protocol for AOS, Motor Learning Guided (MLG), structures the practice and schedules the feedback based on on the principles of motor learning framework found to foster acquisition and learning of a limb motor skill [25]. The three-stage treatment protocol uses a distributed, random order, practice schedule of variable whole task stimuli, and provides delayed terminal summary knowledge of results feedback at a 20% frequency rate. The measure of speech motor changes is the retention of learning from one session to the next. At the beginning of each treatment session prior to initiating the treatment protocol, productions are elicited using a written prompt of the stimuli. The rating of these productions using a Likert scale is the behavior indicator for speech motor learning. Multiple single-case treatment studies report positive outcomes [26–29]. A recent single-case treatment study involving two individuals with chronic AOS and aphasia, report positive treatment effects on treated items while comparing two methods to measure speech motor learning [27]. This study highlights the challenges clinicians face to measure changes in speech production in the presence of aphasia.

This same study investigated the utility of using fNIRS to detect PFC activity associated with the structure of practice during the treatment across the intervention period. In this pilot study, we use a computational modeler framework to identify brain activity measured using fNIRS in these two individuals with chronic acquired AOS and aphasia during each treatment day across the intervention period. First, a generalized linear model (GLM) was used to detect differences in participants' PFC neural activity across the intervention period [30–33]. Upon detection, a novel statistical method involving spatial correlation presents the structural connectivity in the brain [34].

METHODS

Participants

The participants in this study are a 68-year-old male (P1), 91 months post-onset of a left hemisphere stroke, and a 61-year-old male (P2), 86 months post-onset of a left hemisphere stroke. Both participants are right-handed. They met the inclusion criteria of normal or corrected-to-normal vision, sufficient unaided hearing, and functional reading competency to participate in the study. Both participant's speech characteristics include sound distortions and distorted sound substitutions increasing with articulatory complexity or increased rate of speech. They both had an overall slow rate of speech with longer segmental and intersegmental durations and equal stress across syllables. Please see Johnson et al. [27] for a complete description of participants, stimuli, and behavioral outcome measures. Prior to participation in this study, written informed consent according to IRB approval was obtained.

Experimental Procedure

Stimuli

The stimuli practiced during each treatment session consisted of 5 phrases (Appendix). A structured interview identified the specific content, and the severity of AOS determined the length and complexity of the stimuli.

Experimental Design

This was a multiple baseline single-case design study across participants and behaviors performed according to recommended procedures [35]. Treatment sessions occurred for 30 min, two times a week, across 9 weeks for a total of 18 treatment sessions.

Participants were seated comfortably at ~5 m distance from a computer monitor. Each session began with an oral reading task used as a measure of speech motor learning, proceeded by the intervention following the Motor Learning Guided (MLG) treatment protocol, as previously described [36]. There are three stages in the MLG protocol, each stage differing by practice structure. Each practice session presented stimuli in random order and imposed a two-minute break between each stage of practice. The fNIRS recorded the prefrontal cortex activity during each treatment session. Each stage began with a blank screen and the 30 s fNIRS baseline measure. After the baseline measure, the written prompt for the stimuli was provided one at a time using a programmed PowerPoint presentation with a set-pause time between each production (**Figure 1**).

The differences between the stages of practice in the MLG protocol are the method to elicit productions and the delay time between productions. Stage 1 elicits productions using a written prompt and clinician verbal model, while Stage 2 and Stage 3 elicit productions using a written prompt only. The delay imposed between productions in Stage 1 and Stage 2 is 4 s, while in Stage 3, it is 10 s. The 18 days of intervention trained stimuli according to behavior outcomes; P1 practiced set 1 for days 1–12 and set 2 for days 13–18, whereas P2 practiced the same set for the 18 days [27].

Data Acquisition

A 16-channel continuous-wave fNIRS system (fNIR Imager Model 1100; fNIR Devices, LLC) recorded changes in hemodynamic activity of oxygenated (HbO) and deoxygenated (HbR) hemoglobin levels of the anterior PFC [37, 38], as previously described [39] and shown in **Figure 2**. The data were recorded and processed using COBI Studio software version 1.5.0.51 and fNIRSoft version 4.9.6716.18813 (BioPac TM systems).

The positioning of the headpiece followed the standard sensor placement procedure [37, 40] and secured using an adjustable headband and a dark cap to reduce ambient light. The sensor has a source-detector distance of 2.5 cm and a sampling rate of 2 Hz. Manual markers indicate the beginning and end of each stimulus presentation and participant production. The markers ware labeled by phrase presentation to account for the random presentation of stimuli.

fNIRS Data Processing

The modified Beer-Lambert law converted the fNIRS raw light intensity signals to changes in oxygenated hemoglobin (HbO) and deoxygenated hemoglobin (HbR) concentrations. The sum of HbO and HbR yields the change in total hemoglobin (HbT) described as an indicator of variations in the regional cerebral blood volume. The automated sliding-window artifact rejection algorithm uses a coefficient-of-variation approach to assess the signal quality and reject problematic channels with bad contact or saturated raw light intensity [13]. The algorithm resulted in the rejection of 6.3% of the data secondary to artifacts. Following the application of the sliding-window artifact rejection algorithm, the changes in HbO and HbR time series for each channel were bandpass filtered at 0.1 Hz to reduce the slow signal drift and remove the physiological artifacts. Lastly, Correlation Based Signal Improvement was used to improve signal quality [41]. For statistical analysis, the mean value [HbO, HbR, HbT] was calculated by averaging the recorded [HbO, HbR, HbT] values during each stimulus block, and subtracting the mean change in HbO, HbR, and HbT of the recorded global baseline (first 30 s) for each stage of the treatment protocol. Imputation from the previous and next values' average replaced the missing values. With more than 600 observations per subject, we had sufficient statistical power to study the PFC activity.

Analysis

The goal of this analysis is to identify any significant differences between the participant's intensity and response amplitude by patterns of activity during practice and the number of days performing the task. The locations of interest in the PFC included the left and right hemisphere and isolated Brodmann area (BA) 9, 10, and 46. The channels assigned to each group were determined based on the Montreal Neurological Institute (MNI) cortical virtual spatial registration for the 16-channels for older adults [42]. The first grouping associated with the DLPFC (BA 9 and 46), contained channels 1, 2, and 15, and the second grouping of the MPFC (BA 10) contained channels 3–14 and 16 (**Figure 3**). The two models considered, are a GLM and a Spatial Varying Coefficients Model. These analyses were performed in SAS using proc GLM and in R using the spatstat package.

Generalized Linear Model

In a conventional study, effects of participants, days, and location, a GLM is based on the fact that intensity, is a function of many sources, including nuisance or errors. Many authors have proposed the GLM model for similar PFC data [30–33]. A mixed-effect model can also be used. Such a model can be described as follows:

$$Y_{itkl} = \mu + \alpha_i + \gamma_t + \tau_k + \delta_l + \epsilon_{itkl},$$

- where:
- μ is the overall intensity
- α_i represents the effect due to location, $i = 1, \dots, I$
- γ_t represents the effect due to day, $t = 1, \dots, T$
- τ_k represents the effect due to participant, $k = 1, \dots, K$

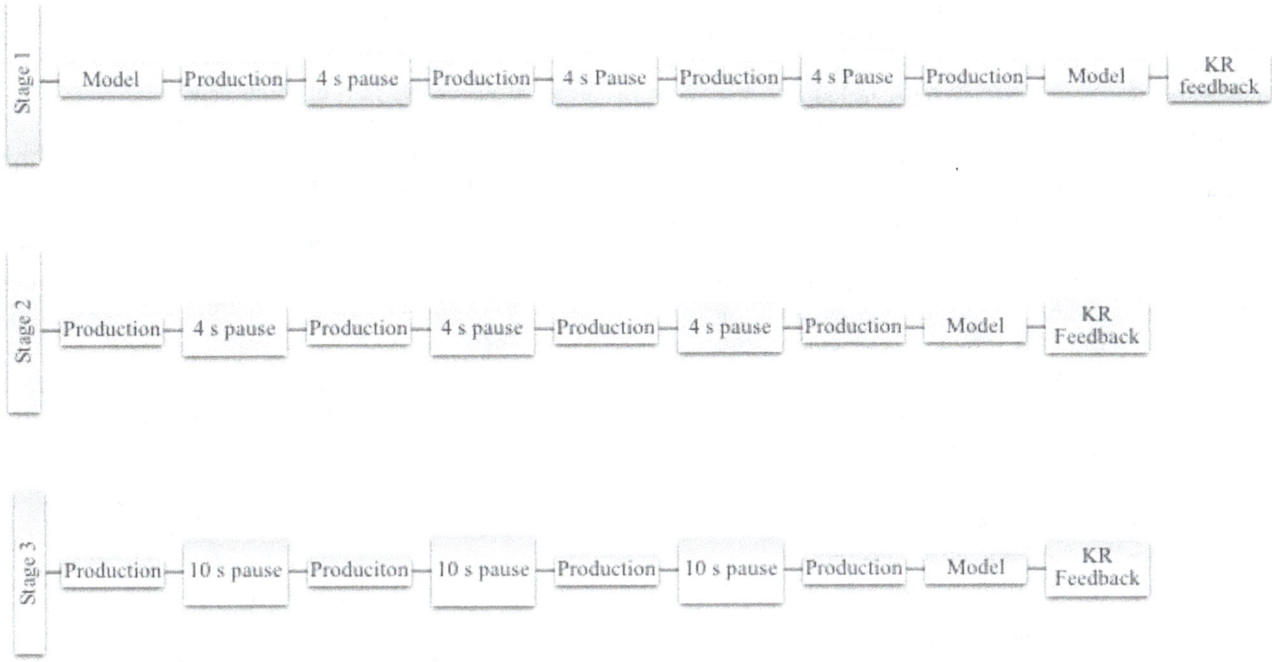

FIGURE 1 | Schematic of practice schedule according to motor learning guided treatment protocol.

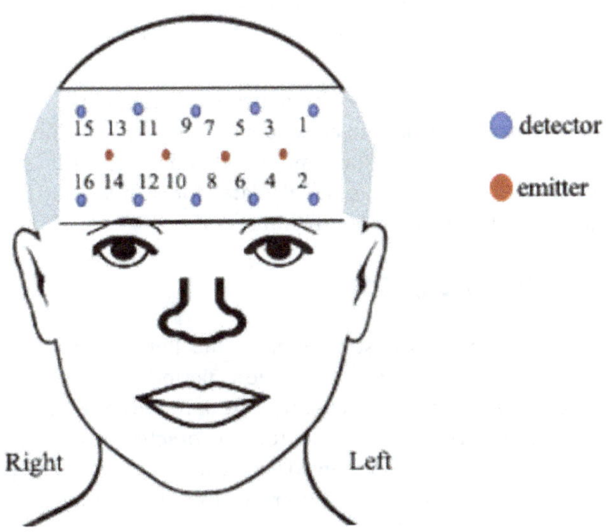

FIGURE 2 | Functional near-infrared configuration.

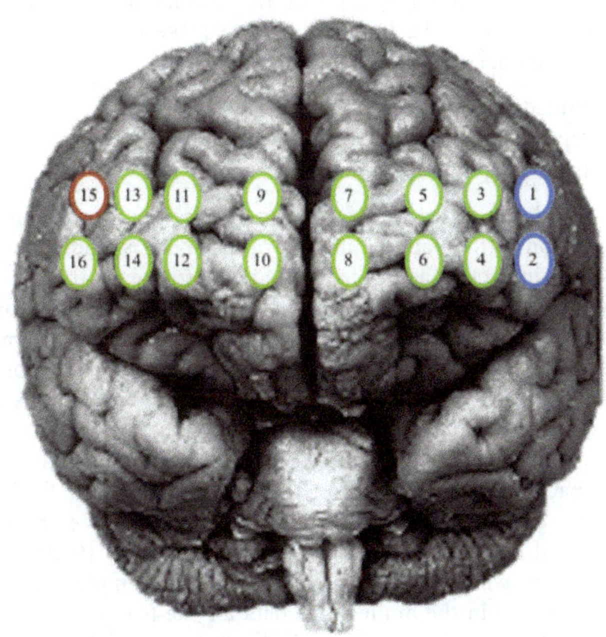

FIGURE 3 | Optode grouping for regions of interest. Red, Broadmann's Area 9; Green, Broadmann's Area 10; Blue, Broadmann's Area 46.

- δ_l represents the effect due to stage, $l = 1, \ldots, L$

The model can be expressed as

- $Y = X\beta + \varepsilon$,

where Y is the vector of measured intensity by location and day, X is the design matrix, β is the vector of regression coefficients and ε is the vector of error terms. Here, the design matrix X includes information about location, day, participant, and stage. The assumptions that the errors are independent and normally distributed will be made and validated in the investigations.

Estimation of the parameter set β is such that $\hat{\beta} = (X'X)^{-1}X'Y$ assuming that the design matrix X is invertible. If not, using the generalized equation will allow us to obtain a solution, even though they will not be unique. Normality is a strong assumption, and there are other techniques that have been proposed to remedy and avoid biases [32]. Huppert [43] mentions that errors in fNIRS are not independent. Tools are currently under consideration to analyze data under region based interest however, no standard recommendation exists at this time [44].

As previously described in section Analysis, two groupings of channels were considered for this analysis and four different comparisons were of interest within the groupings. Groupings and comparisons of interest are:

- Grouping 1: L (channels 1 and 2) and R (channels 15)
- Grouping 2: L (channels 3–8) and R (channels 9–14, 16)

- Case 1: Compare P1 day 1–12 to P2 day 1–12
- Case 2: Compare P1 day 13–18 to P2 day 1–6
- Case 3: Compare P2 day 1–6 to P2 day 13–18
- Case 4: Compare P1 day 1–6 to P1 day 13–18

Spatial Varying Coefficients Model

To assess the spatial connectivity between the channels, a model parcellation of the channel network is adapted. The advantages of this approach include: biological interpretability, reduction in the parameters to estimate, and quantification of the measure of spatial correlation [34, 42]. However, this model does require precise locations and a distance locator. To take advantage of the data, the model should incorporate:

a) Limited assumptions
b) Dependent component analysis
c) Spatio-temporal correlation approach

To reveal the structure of the functioning of the channels, interaction of the patterns in the brain activity is studied such that the impact is associated with the distance of the blood flow between the channels' proximities. A general framework for the spatial model is that the regions of the brain interact after adjustment on the distance between them such that closer regions are more alike. The goal is to identify any regions of the brain that have strong correlation(s). Since the regions are spatial representations (based on MNI coordinates), we adopted the spatial correlation proposed by Moran [45] and further developed by many authors such as Chen et al. [46] We provide additional benefits in the stage of practice and day (temporal) parameters which substantially describe the complex spatial signal control in the brain image data for the two participants. The 16 channels observed were depicted as regions determined by some coordinate system denoted as the pair (x_i, y_i), we rewrite as (x_{it}, y_{it}) to include day/time information for the change in HbO, HbR, and HbT signals.

We define the model of the Moran intensities as follows:

$$M_{ii'}^{t} = \sum_{i,\, i'} w_{ii'}^{t} f(Y_{it},\, Y_{i't}),\ where$$

- $w_{ii'}^{t}$ represents the spatial weight between regions i and i' for the tth day at each stage and will be defined as $w_{ii'} = e^{-d_{ii'}/d}$, the geographical weight, where d is the average of distances between all possible pairs of points i and i'.
- for the distance matrix, $d_{ii'}$, distance is measured as one step (up, down, left, right) along the shortest distance path between regions and when the regions coincide, (i.e., $i = i'$), the distance is taken from a central node (along the brain stem).
- $f(Y_{it},\, Y_{i't})$ can be seen as the absolute value of the difference in observed intensities at regions i and i' for the tth day and for each stage. Generally, the value of this function is smaller for closer regions.

If we ignore the $f(Y_{it},\, Y_{i't})$, then we will have a symmetric matrix function of the locations only, and we can consider it as a design matrix, and call it M:

$$M_{n\times n} = \begin{bmatrix} M_{11} & \cdots & M_{1n} \\ \vdots & \ddots & \vdots \\ M_{n1} & \cdots & M_{nn} \end{bmatrix}$$

In presenting the autocorrelation between the channel locations, we recall the Pearson correlation coefficient. In the spatial setting, the correlation is formulated as:

$$I = \frac{n}{S}\frac{\sum_{i,j=1}^{n} w_{ij}\left(x_i-\bar{x}\right)\left(x_j-\bar{x}\right)^{n}}{\sum_{i=1}\left(x_i-\bar{x}\right)^2},$$

where S is the sum of all weights $w_{i,j}$ between the i and j observations capturing the closeness of those observations from the channel locations to the main blood and oxygenation supply line. These correlations can be calculated for each day and lead to higher correlations for closer regions.

MODEL RESULTS

The analysis provides estimations of the PFC activity for two participants across the intervention period. The significant difference in PFC activity in the regions of interest was associated with parameters of the schedule of practice and time of intervention. The analysis considered two groupings of channels, and five comparisons were of interest within the groupings.

Generalized Linear Model

Grouping 1 looked at hemodynamic activity in BA 9 and 46. The overall models for Grouping 1 were all significant (**Table 1**). Case 1 compared the hemodynamic activity for both participants according to the left and right hemispheres (location) and stages across 12 treatment days. There was a significant difference in oxygenation between the left and right hemispheres across treatment days and between participants. However, the mean HbO was not significantly different for stage (**Figure 4**). The HbR was significantly different across treatment days, between the left and right hemispheres, and stage. The HbT (total blood volume) was significantly different between participants

TABLE 1 | GLM results for Grouping 1.

Grouping 1, Case 1		HbO				HbR				HbT			
Source	**DF**	**SS/Type I SS**	**Mean square**	***F*-value**	**Pr > F**	**SS/Type I SS**	**Mean square**	***F*-value**	**Pr > F**	**SS/Type I SS**	**Mean square**	***F*-value**	**Pr > F**
Model	14	183.5487	13.1106	13.09	< 0.0001	84.4569	6.0326	11.03	< 0.0001	90.0819	6.4344	15.42	< 0.0001
Participant	1	23.0206	23.0206	22.98	< 0.0001	0.3657	0.3657	0.67	0.4137	10.1341	10.1341	24.29	< 0.0001
Day	11	149.5384	13.5944	13.57	< 0.0001	62.0548	5.6413	10.32	< 0.0001	75.6050	6.8732	16.47	< 0.0001
Location	1	9.5965	9.5965	9.58	0.002	5.3082	5.3082	9.71	0.0019	0.3347	0.3347	0.8	0.3706
Stage	1	1.3931	1.3931	1.39	0.2385	16.7282	16.7282	30.59	< 0.0001	4.0081	4.0081	9.61	0.002
Error	1,065	1066.8195	1.0017			582.4305	0.5469			444.3260	0.4172		
Corrected total	1,079	1250.3682				666.8874				534.4079			

Grouping 1, Case 2		HbO				HbR				HbT			
Source	**DF**	**SS/Type I SS**	**Mean square**	***F*-value**	**Pr > F**	**SS/Type I SS**	**Mean square**	***F*-value**	**Pr > F**	**SS/Type I SS**	**Mean square**	***F*-value**	**Pr > F**
Model	13	70.1940	5.3995	8.09	< 0.0001	51.6237	3.9711	13.02	< 0.0001	31.2892	2.4069	8.19	< 0.0001
Participant	1	5.7708	5.7708	8.64	0.0034	9.5154	9.5154	31.21	< 0.0001	0.4657	0.4657	1.58	0.2087
Day	10	43.8210	4.3821	6.56	< 0.0001	25.9361	2.5936	8.51	< 0.0001	30.5549	3.0555	10.39	< 0.0001
Location	1	15.0519	15.0519	22.55	< 0.0001	11.9065	11.9065	39.05	< 0.0001	0.1841	0.1841	0.63	0.429
Stage	1	5.5502	5.5502	8.31	0.0041	4.2657	4.2657	13.99	0.0002	0.0844	0.0844	0.29	0.5923
Error	526	351.1338	0.6676			160.3800	0.3049			154.6135	0.2939		
Corrected total	539	421.3278				212.0036				185.9027			

Grouping 1, Case 3		HbO				HbR				HbT			
Source	**DF**	**SS/Type I SS**	**Mean square**	***F*-value**	**Pr > F**	**SS/Type I SS**	**Mean square**	***F*-value**	**Pr > F**	**SS/Type I SS**	**Mean square**	***F*-value**	**Pr > F**
Model	13	69.6870	5.3605	13.27	< 0.0001	34.6996	2.6692	16.31	< 0.0001	32.6763	2.5136	13.06	< 0.0001
Day	11	33.8503	3.0773	7.62	< 0.0001	20.6024	1.8729	11.44	< 0.0001	26.8968	2.4452	12.7	< 0.0001
Location	1	7.8551	7.8551	19.44	< 0.0001	5.5188	5.5188	33.71	< 0.0001	0.2056	0.2056	1.07	0.3018
Stage	1	27.9816	27.9816	69.25	< 0.0001	8.5783	8.5783	52.4	< 0.0001	5.5738	5.5738	28.96	< 0.0001
Error	526	212.5316	0.4041			86.1068	0.1637			101.2422	0.1925		
Corrected total	539	282.2186				120.8064				133.9185			

Grouping 1, Case 4		HbO				HbR				HbT			
Source	**DF**	**SS/Type I SS**	**Mean square**	***F*-value**	**Pr > F**	**SS/Type I SS**	**Mean square**	***F*-value**	**Pr > F**	**SS/Type I SS**	**Mean square**	***F*-value**	**Pr > F**
Model	13	163.3517	12.5655	8.36	< 0.0001	91.0219	7.0017	8.8	< 0.0001	57.7002	4.4385	6.62	< 0.0001
Day	11	114.1590	10.3781	6.9	< 0.0001	84.4629	7.6784	9.65	< 0.0001	46.6404	4.2400	6.32	< 0.0001
Location	1	22.1112	22.1112	14.7	0.0001	3.2222	3.2222	4.05	0.0447	2.8043	2.8043	4.18	0.0414
Stage	1	27.0815	27.0815	18.01	< 0.0001	3.3368	3.3368	4.19	0.0411	8.2555	8.2555	12.31	0.0005
Error	526	790.9344	1.5037			418.7175	0.7960			352.7480	0.6706		
Corrected total	539	954.2861				509.7394				410.4482			

Grouping 1: L (optodes 1 and 2) and R (optode 15).

across treatment days and stages. For all cases compared, HbT differed significantly across treatment days, remaining steady and consistent between the left and right hemispheres. In case 2, the HbT did not differ between participants for stage only when comparing the first six treatment days on P2's second set of stimuli and P1's first (and only) set of stimuli.

Grouping 2 looked at hemodynamic models of BA 10. The overall models for Grouping 2 were all significant (**Table 2**).

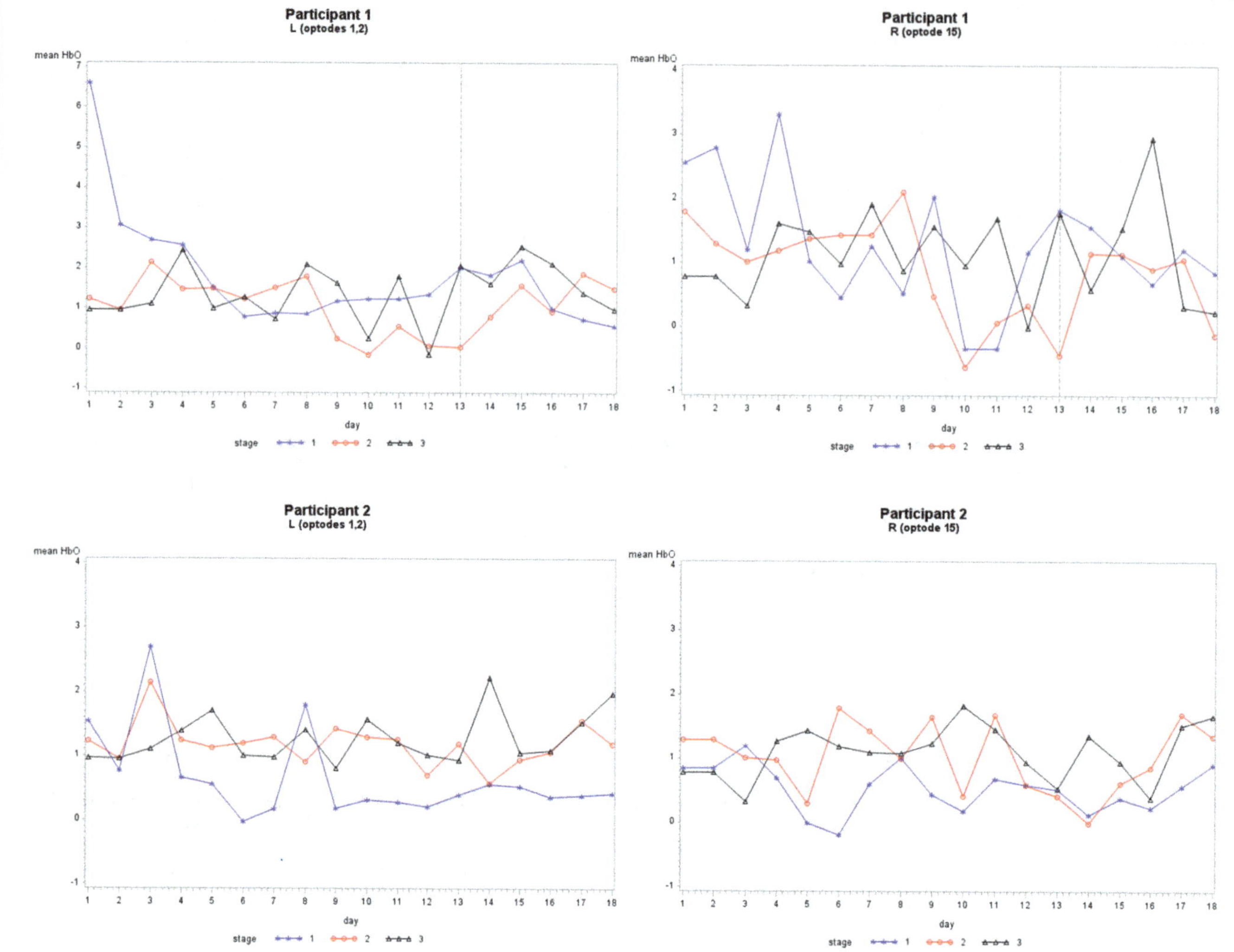

FIGURE 4 | The mean HbO activity for Grouping 1 by stage for each participant. The vertical line at day 13 for participant 1 indicates set 1 to the left and set 2 to the right.

In case 1, HbO was significant for all covariates; however, the HbR for the location was not significant, and the HbR and HbT for stages were not significant. The hemodynamic activity was significant for all covariates during the first six training days for P2 on set 2 and P1. The HbO and HbT were not significant between the left and hemispheres for P1 during the first six treatment days on both sets of stimuli, or P2 during the first and last six treatment days. As illustrated in **Figure 5**, there was bilateral BA 10 activity for all stages of practice. For P1 in Stage 1, there was a higher bilateral BA 10 activity during the initial treatment days with both sets of stimuli. The activity for P2 was similar in each stage; a decrease occurred only in Stage 1 (clinician model) across treatment days.

As illustrated in **Figure 6**, there was a consistent HbO level in channels 1 (inferior frontal gyrus, IFG), 5 (superior frontal gyrus, SFG), 12 (SFG), and 14 (middle frontal gyrus, MFG) for P1 during each stage. The HbO levels during Stage 2 (no model) were higher in channels 12 and 14, while channels 1 and 5 had lower levels. In Stage 3 (no model, 10 s delay), there were distributed HbO levels across multiple channels in the left and right hemispheres. For P2, there was a consistent level of HbO measured in channels 4 (MFG), 10, and 12 in all stages. The HbO concentration became increasingly distributed in the left and right hemispheres across each stage of practice. Active only during Stage 2 and Stage 3 was a cluster of channels 1–4 (IFG and MFG). Channels 13, 14, 15, and 16 (IFG and MFG) recruited during Stage 3, mirrored the left PFC activity pattern.

Figure 7 illustrates the differences in the pattern of PFC activity between the participants across the training days. A similar distribution pattern of PFC activity during the initial days of practice for both sets of stimuli is evident for P1. The activity localized in channels 5, 12, and 14 (bilateral SFG and right MFG) with practice. The activity pattern for P2 remains fairly distributed across the duration of the treatment with consistent activity noted in channels 1, 4, and 12 (left IFG and MFG and right SFG).

TABLE 2 | GLM results for Grouping 2.

Grouping 2, Case 1		**HbO**				**HbR**				**HbT**			
Source	**DF**	**SS/Type I SS**	**Mean square**	***F*-value**	**Pr > F**	**SS/Type I SS**	**Mean square**	***F*-value**	**Pr > F**	**SS/Type I SS**	**Mean square**	***F*-value**	**Pr > F**
Model	14	335.0409	23.9315	35.57	< 0.0001	98.6628	7.0473	22.72	< 0.0001	99.9450	7.1389	26.63	< 0.0001
Participant	1	89.4638	89.4638	132.96	< 0.0001	12.5283	12.5283	40.38	< 0.0001	16.0724	16.0724	59.95	< 0.0001
Day	11	232.5083	21.1371	31.41	< 0.0001	86.1188	7.8290	25.24	< 0.0001	81.4310	7.4028	27.61	< 0.0001
Location	1	2.5590	2.5590	3.8	0.0512	0.0103	0.0103	0.03	0.8553	2.4277	2.4277	9.06	0.0026
Stage	1	10.5098	10.5098	15.62	< 0.0001	0.0054	0.0054	0.02	0.8949	0.0139	0.0139	0.05	0.8197
Error	4,665	3138.9328	0.6729			1447.1880	0.3102			1250.6904	0.2681		
Corrected total	4,679	3473.9737				1545.8507				1350.6354			

Grouping 2, Case 2		**HbO**				**HbR**				**HbT**			
Source	**DF**	**SS/Type I SS**	**Mean square**	***F*-value**	**Pr > F**	**SS/Type I SS**	**Mean square**	***F*-value**	**Pr > F**	**SS/Type I SS**	**Mean square**	***F*-value**	**Pr > F**
Model	13	243.2028	18.7079	25.27	< 0.0001	69.0569	5.3121	17.26	< 0.0001	104.5049	8.0388	35.97	< 0.0001
Participant	1	7.3032	7.3032	9.86	0.0017	2.8725	2.8725	9.34	0.0023	0.9957	0.9957	4.46	0.0349
Day	10	181.5326	18.1533	24.52	< 0.0001	61.9178	6.1918	20.12	< 0.0001	74.1704	7.4170	33.19	< 0.0001
Location	1	4.7764	4.7764	6.45	0.0112	1.3435	1.3435	4.37	0.0368	1.0319	1.0319	4.62	0.0318
Stage	1	49.5906	49.5906	66.97	< 0.0001	2.9230	2.9230	9.50	0.0021	28.3069	28.3069	126.66	< 0.0001
Error	2,326	1722.3117	0.7405			715.7016	0.3077			519.8270	0.2235		
Corrected total	2,339	1965.5145				784.7585				624.3319			

Grouping 2, Case 3		**HbO**				**HbR**				**HbT**			
Source	**DF**	**SS/Type I SS**	**Mean square**	***F*-value**	**Pr > F**	**SS/Type I SS**	**Mean square**	***F*-value**	**Pr > F**	**SS/Type I SS**	**Mean square**	***F*-value**	**Pr > F**
Model	13	270.2793	20.7907	46.6	< 0.0001	75.9453	5.8419	41.67	< 0.0001	88.4753	6.8058	41.26	< 0.0001
Day	11	172.1855	15.6532	35.08	< 0.0001	60.0004	5.4546	38.90	< 0.0001	52.3434	4.7585	28.85	< 0.0001
Location	1	0.5960	0.5960	1.34	0.2479	0.9894	0.9894	7.06	0.008	0.0496	0.0496	0.30	0.5836
Stage	1	97.4978	97.4978	218.52	< 0.0001	14.9556	14.9556	106.67	< 0.0001	36.0823	36.0823	218.75	< 0.0001
Error	2,326	1037.8166	0.4462			326.1254	0.1402			383.6599	0.1649		
Corrected total	2,339	1308.0959				402.0707				472.1352			

Grouping 2, Case 4		**HbO**				**HbR**				**HbT**			
Source	**DF**	**SS/Type I SS**	**Mean square**	***F*-value**	**Pr > F**	**SS/Type I SS**	**Mean square**	***F*-value**	**Pr > F**	**SS/Type I SS**	**Mean square**	***F*-value**	**Pr > F**
Model	13	350.1448	26.9342	27.28	< 0.0001	121.3271	9.3329	22.22	< 0.0001	92.4890	7.1145	19.47	< 0.0001
Day	11	341.2457	31.0223	31.43	< 0.0001	117.9066	10.7188	25.52	< 0.0001	91.9264	8.3569	22.87	< 0.0001
Location	1	0.5921	0.5921	0.6	0.4387	0.5129	0.5129	1.22	0.2692	0.0013	0.0013	0.00	0.9516
Stage	1	8.3070	8.3070	8.41	0.0038	2.9076	2.9076	6.92	0.0086	0.5613	0.5613	1.54	0.2154
Error	2,326	2296.1557	0.9872			976.8735	0.4200			850.1106	0.3655		
Corrected total	2,339	2646.3005				1098.2006				942.5996			

Grouping 2: L (optodes 3–8) and R (optodes 9–14, 16).

Spatial Varying Coefficients Model

For the spatial model, we first focus on the absolute value of the HbO intensity observed in three regions: channels 1, 2, and 15. The resulting distance matrix utilizing the distance as described in section Spatial Varying Coefficients Model is:

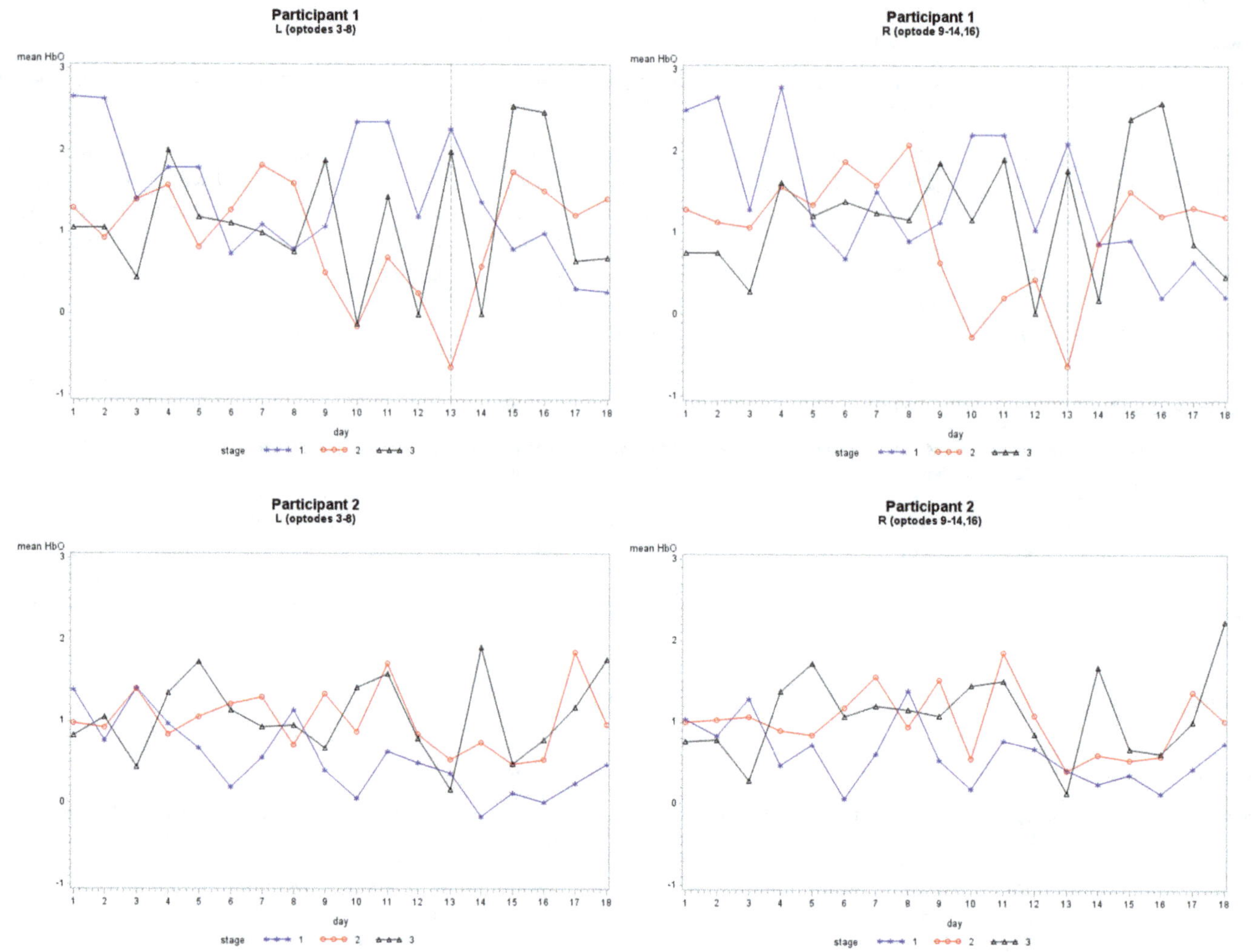

FIGURE 5 | The mean HbO activity for Grouping 2 by stage for each participant. The vertical line at day 13 for participant 1 indicates set 1 to the left and set 2 to the right.

$$d_{ii'} = \begin{bmatrix} 4 & 1 & 7 \\ 1 & 4 & 8 \\ 7 & 8 & 4 \end{bmatrix} \quad (1)$$

Next, we calculate the geographical weights, $w_{ii'}$, defined as $w_{ii'} = e^{-d_{ii'}/\bar{d}}$, where $\bar{d}$ is the average of distances between all possible pairs of points i and i' [47].

$$w_{ii'} = \begin{bmatrix} 0.2765 & 0.7251 & 0.1054 \\ 0.7251 & 0.2765 & 0.0764 \\ 0.1054 & 0.0764 & 0.2765 \end{bmatrix} \quad (2)$$

Then we calculate the Moran statistics for each participant at each stage. **Figure 8** are the graphs of the Moran statistics; a Loess smoothing curve has been added to easily detect trends between participants.

The degree of correlation measured by the Moran statistics between the participants indicated that the correlation between channels 1 and 2 vs. channel 15 was always higher for P1 (red) in Stage 1, whereas, in Stages 2 and 3, the correlations cross. This last remark was especially visible in Stage 2 after day 6. This shows that there was a magnitude of PFC activity (HbO, HbR, and HbT) represented in Stage 1 (clinician model) that lessened in Stage 2 (no model) and 3 (no model, 10 s delay) overtime.

The Pearson correlation of HbO values and differences during each stage and channel for each participant were significant, revealing regional variations in the level of HbO activity in the PFC (**Figure 9**). The darkened main diagonal shape is evidence of the identity autocorrelation. The plot shades close to the main diagonal correspond to short range correlation, while the plot shades distant from the main diagonal correspond to long range correlations. The level of HbO activity in Stage 1 was higher for channels 5, 14, and 16 for P1, while channels 1, 2, 12, and 15 were higher for P2. In stage 2, the level of HbO activity for all channels was consistently higher for P1 compared to P2. In Stage 3, the level of HbO activity was dominant in the left hemisphere for P1; whereas, the right hemisphere was dominant for P2.

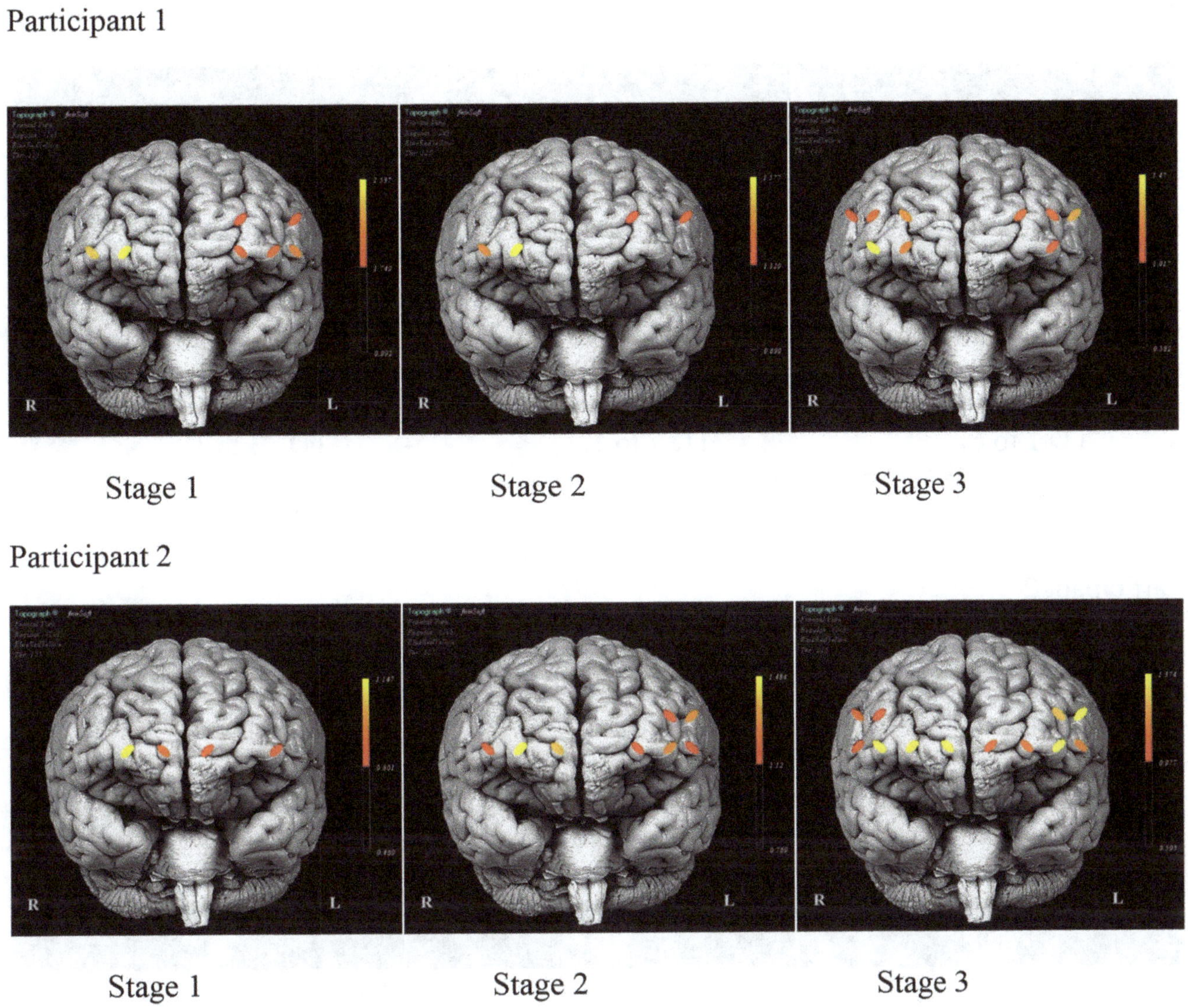

FIGURE 6 | The mean prefrontal cortex activity by optode during each stage of practice during training days 1–12 for each participant.

DISCUSSION

This study investigated the utility of using fNIRS to measure the hemodynamic activity in the PFC during AOS treatment. Of interest was the hemodynamic PFC activity in response to the different methods for eliciting speech and the length of the delay between productions in the treatment protocol. Differences in the PFC activity for regions of interest associated with the schedule of practice in each stage of the treatment protocol and the spatial connectivity of the pattern of activity were determined using two statistical methods. Overall, the results of the GLM model confirm the involvement of cognitive control processes during speech motor learning and the association of cognitive demands and the schedule of practice while learning.

Previous studies reporting a task associated with PFC activity guided the groupings for the comparisons. The GLM model investigated the areas of the PFC associated with higher-order cognitive planning tasks (BA 9 and 46) in the first grouping and working memory tasks (BA 10) in the second grouping. Consistent with prior research, there was a high level of bilateral PFC activity associated with new learning [22–24]. The level of activity in the right hemisphere decreased across treatment days for P1, while the left hemisphere remained steady [20]. This pattern of activity is consistent with previous reports of left hemisphere dominance with speaking tasks [18, 19] and the dual-processing theory [22]. The speech behavior during retention measures for P1 characterized speech productions as consistent decrease occurrences of segmental and intersegmental durations with fewer long durations compared to baseline [27]. However, for P2, the PFC activity remained bilateral, indicating the involvement of cognitive control processes throughout the intervention period. The speech behavior during

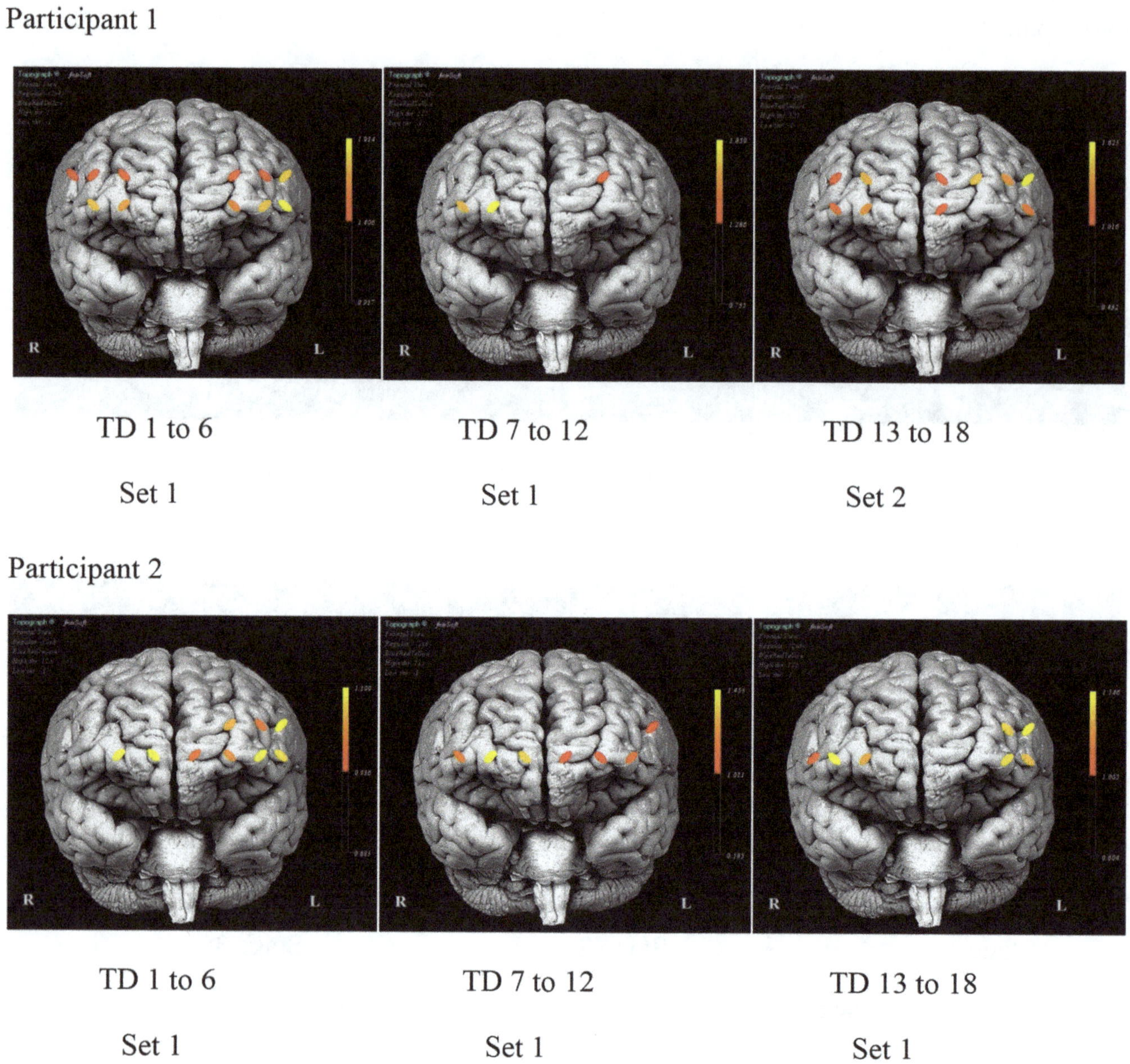

FIGURE 7 | The mean prefrontal cortex activity by optode for selected day ranges for each participant.

retention measures was characterized as inconsistent day to day performance related to distortions and lengthened segmental and intersegmental duration [27]. These findings, combined with behavior outcomes for the retention of learning [27], provide initial evidence of cognitive control processes associated with speech motor learning during the rehabilitation of AOS. Replication of these findings across more participants would assist in the further interpretation of these findings and its potential use as a clinical tool.

The three stages in the MLG treatment protocol differ in the speech elicitation methods and length of time between each production. The verbal model used in Stage 1 is a common prompt for eliciting speech in persons with AOS. Novel to the MLG treatment protocol is eliciting speech with a written prompt only in Stage 2 and Stage 3. The rationale for using this method is the low reliance on external prompts and proximity to independent productions. The GLM, along with the mapping of the functional connectivity interpreted under a spatial-temporal correlation measure using the Spatial Varying Model, presents the activity at the channel level. The regions active during Stage 1 were active during all speech tasks, similar to the activity reported during standard speech production [48]. The pattern of activity, during Stage 2, for P2, and both participants in Stage 3, was similar to the activity pattern in the first six treatment days. The

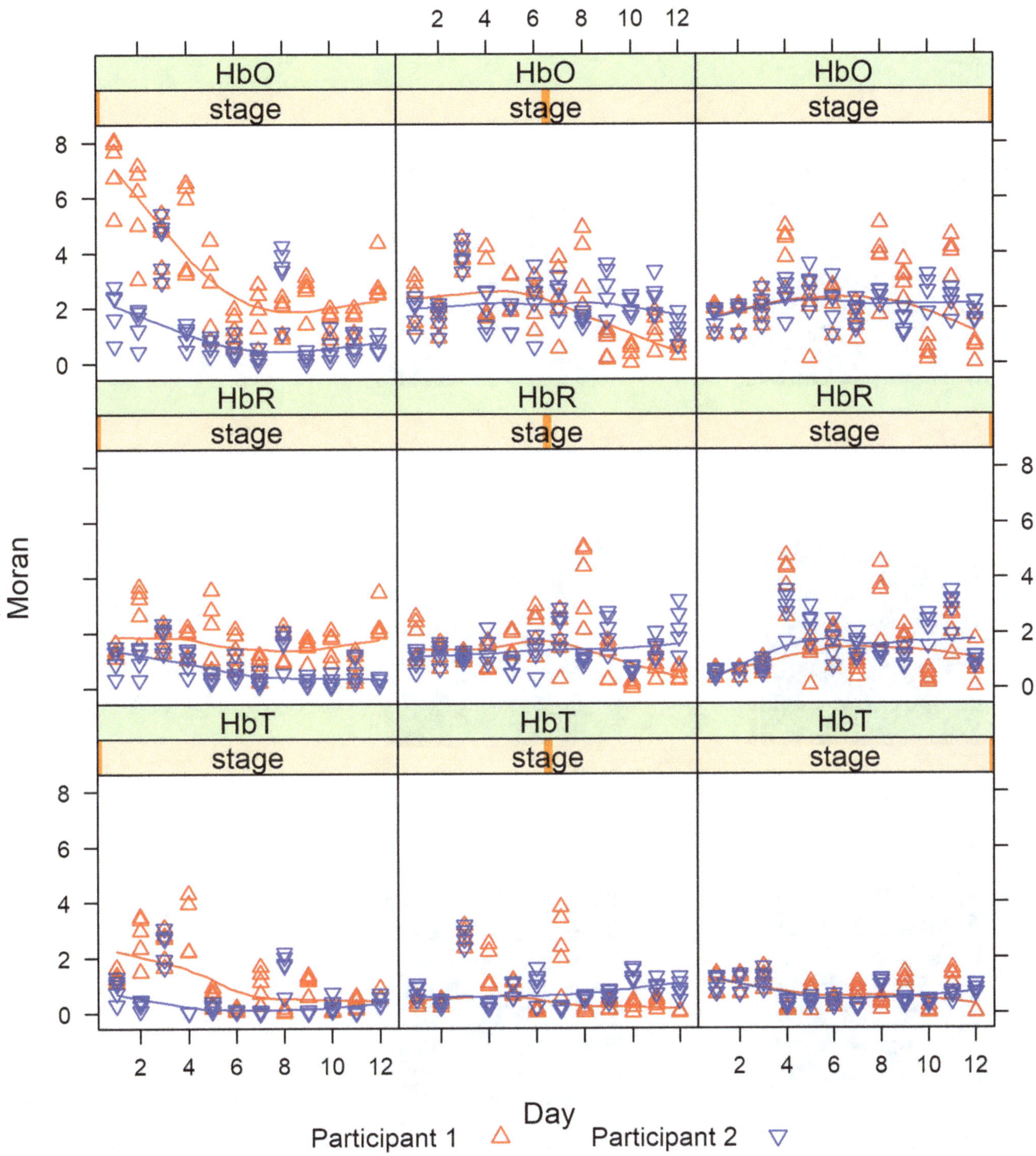

FIGURE 8 | Moran statistic for Grouping 1 for HbO, HbR, and HbT by stage for days 1–12.

decreased magnitude of activity in the IFG after treatment day 6 for P1, paired with the behavior retention outcome characterizing no lengthened segmental or intersegmental speech characteristics [27], provides evidence for the association between speech motor learning and cognitive control processes during the rehabilitation of AOS. Further, the dominant activity for P1 during Stage 2 in left hemisphere activity during Stage 3 compared to P2 are of great interest in understanding the implications the structure of practice has on the cognitive demands during rehabilitation. The model distinguishes a shift in the magnitude of activity in

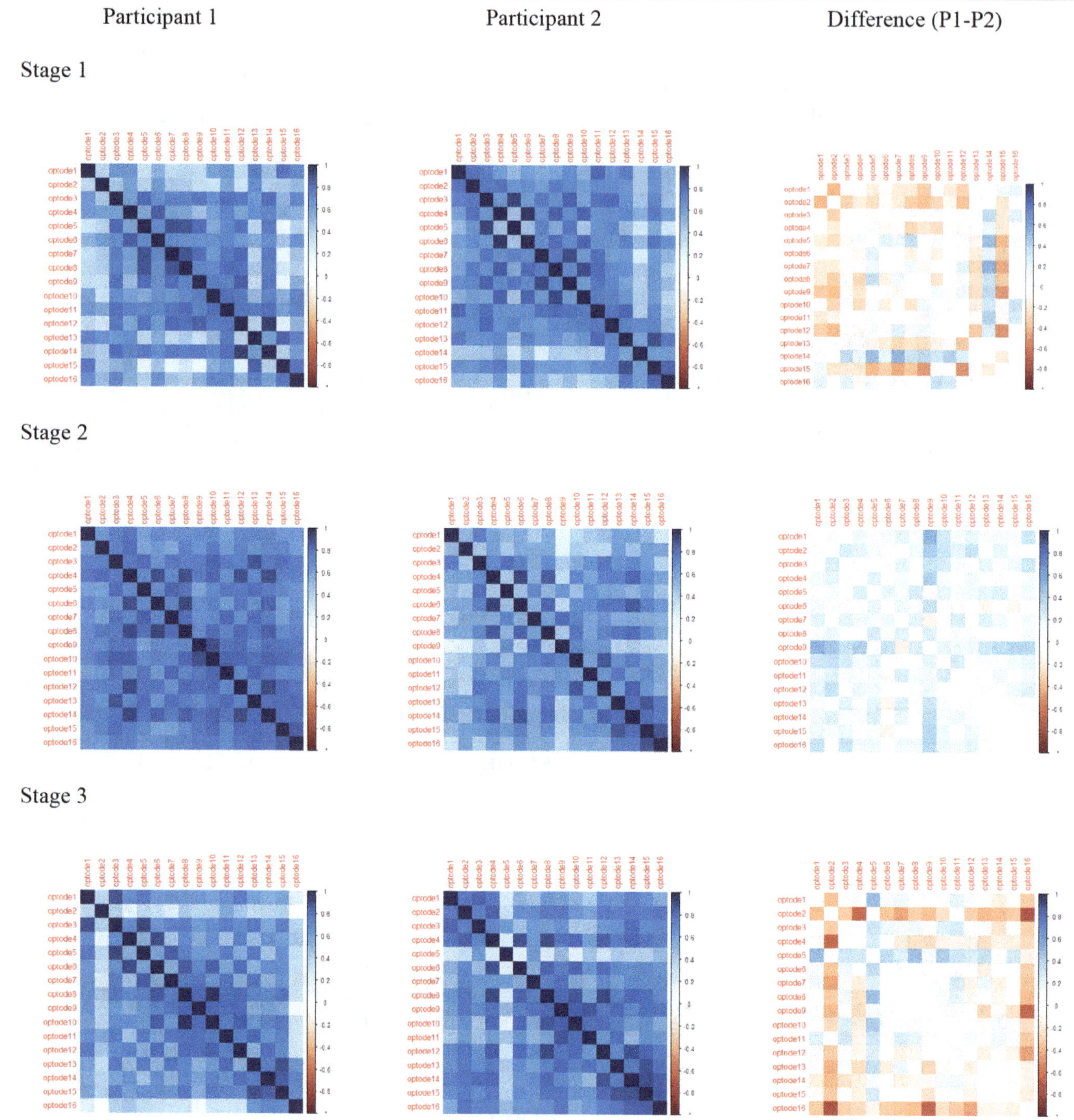

FIGURE 9 | Pearson correlations for HbO: rows are stage 1, stage 2, and stage 3; columns are Participant 1, Participant 2, and Difference (P1–P2).

the left and right IFG and differentiates the regional variation of PFC activity associated with the structure of practice and speech motor learning in these two participants.

The delay between productions and the augmented schedule of feedback in the MLG protocol elicits metacognitive skills for self-reflection, comparing the internal representation of the intended outcome while maintaining the desired outcome representation. The extended length of the delay in Stage 3 compared to Stage 1 and 2, is associated with high bilateral activity in both participants. This neurophysiological evidence explains that the extended delay increases the complexity and cognitive function during the task [49, 50]. The behavior retention outcomes paired with the pattern of neural activity during the training are evidence of the potential use for fNIRS as a tool to guide clinical decisions during rehabilitation in persons with AOS and aphasia.

While this study provides evidence that the length of time between productions engages cognitive functions for learning, without specific characteristics of the speech production during the practice activity, interpretation for clinical application is limited. The lesion size and location for the participants were unavailable and would have contributed information regarding the neural reorganization. One cannot avoid the possibility of individual anatomical differences, or those associated with aging, and consistent placement using anatomical markers [42, 51]. The objective of this study was to examine the utility of using fNIRS and proof of concept for a structural equation to identify individual differences in brain connectivity associated with AOS outcomes during rehabilitation. Probabilistic interpretations in a Spatio-temporal context illustrate common regions of activity associated with the presumed function in healthy subjects for cognitive control processes.

The outcomes of this study provide pilot data to support further investigation of the use of functional imaging to advance clinical practices. The difference in the neural activity of the two participants in this study in response to different aspects of the practice schedule warrants further investigation. The pairing of behavioral and neural functioning data has the potential to advance translational clinical research. While the study is limited in its generalization to the patient population, the evidence presented supports future rigorous investigation of the statistical methods. The trend in neuroimaging research is the use of multimodal imaging techniques. The reality is that studies of this capacity are limited to large scale research institutions. The ability to use statistical methods in lieu of multimodal imaging techniques has the potential to extend contributions to the body of literature at a broader scale from all clinical research institutions.

ETHICS STATEMENT

The studies involving human participants were reviewed and approved by the Old Dominion University Institutional Review Board for Human Subjects. The patients/participants provided their written informed consent to participate in this study.

AUTHOR CONTRIBUTIONS

RJ and ND designed the study. RJ collected data. JM and ND performed the analysis of the data. RJ, JM, and ND wrote the paper. RC added input and organizational flow for the manuscript. All authors contributed to the article and approved the submitted version.

REFERENCES

1. Skipper JI, Devlin JT, Lametti DR. The hearing ear is always found close to the speaking tongue: review of the role of the motor system in speech perception. *Brain Lang.* (2017) **164**:77–105. doi: 10.1016/j.bandl.2016.10.004
2. Duffy JR. *Motor Speech Disorders: Substrates, Differential Diagnosis, and Management.* 3rd ed. St. Louis, MO: Mosby (2013).
3. Basilakos A. Contemporary approaches to the management of post-stroke apraxia of speech. *Semin Speech Lang.* (2018) **39**:25–36. doi: 10.1055/s-0037-1608853
4. Moser D, Basilakos A, Fillmore P, Fridriksson J. Brain damage associated with apraxia of speech: evidence from case studies. *Neurocase.* (2016) **22**:346–56. doi: 10.1080/13554794.2016.1172645
5. Ballard KJ, Tourville JA, Robin DA. Behavioral, computational, and neuroimaging studies of acquired apraxia of speech. *Front Hum Neurosci.* (2014) **8**:892. doi: 10.3389/fnhum.2014.00892
6. Hillis AE, Work M, Barker PB, Jacobs MA, Breese EL, Maurer K. Re-examining the brain regions crucial for orchestrating speech articulation. *Brain.* (2004) **127**(Pt 7):1479–87. doi: 10.1093/brain/awh172
7. Ballard KJ, Wambaugh JL, Duffy JR, Layfield C, Maas E, Mauszycki S, et al. Treatment for acquired apraxia of speech: a systematic review of intervention research between 2004 and 2012. *Am J Speech Lang Pathol.* (2015) **24**:316–37. doi: 10.1044/2015_AJSLP-14-0118
8. Wambaugh JL, Duffy JR, McNeil MR, Robin DA, Rogers MA. Treatment guidelines for acquired apraxia of speech: treatment descriptions and recommendations. *J Med Speech Lang Pathol.* (2006) **14**:35–66.
9. Wambaugh JL, Duffy JR, McNeil MR, Robin DA, Rogers MA. Treatment guidelines for acquired apraxia of speech: a synthesis and evaluation of the evidence. *J Med Speech Lang Pathol.* (2006) 14:Xv–Xxxiii.
10. Bohland JW, Bullock D, Guenther FH. Neural representations and mechanisms for the performance of simple speech sequences. *J Cogn Neurosci.* (2010) **22**:1504–29. doi: 10.1162/jocn.2009.21306
11. Tourville JA, Guenther FH. The DIVA model: a neural theory of speech acquisition and production. *Lang Cogn Process.* (2011) **26**:952–81. doi: 10.1080/01690960903498424
12. Miller HE, Guenther FH. Modelling speech motor programming and apraxia of speech in the DIVA/GODIVA neurocomputational framework. *Aphasiology.* (2020) 1–18. doi: 10.1080/02687038.2020.1765307
13. Ayaz H, Izzetoglu K, Shewokis PA, Onaral B. Sliding-window motion artifact rejection for functional near-infrared spectroscopy. *Conf Proc IEEE Eng Med Biol Soc.* (2010) **2010**:6567–70. doi: 10.1109/IEMBS.2010.5627113
14. Wijeakumar S, Huppert TJ, Magnotta VA, Buss AT, Spencer JP. Validating an image-based fNIRS approach with fMRI and a working memory task. *Neuroimage.* (2017) **147**:204–18. doi: 10.1016/j.neuroimage.2016.12.007
15. Cui X, Bray S, Bryant DM, Glover GH, Reiss AL. A quantitative comparison of NIRS and fMRI across multiple cognitive tasks. *Neuroimage.* (2011) **54**:2808–21. doi: 10.1016/j.neuroimage.2010.10.069
16. Sato H, Yahata N, Funane T, Takizawa R, Katura T, Atsumori H, et al. A NIRS-fMRI investigation of prefrontal cortex activity during a working memory task. *Neuroimage.* (2013) **83**:158–73. doi: 10.1016/j.neuroimage.2013.06.043
17. Scarapicchia V, Brown C, Mayo C, Gawryluk JR. Functional magnetic resonance imaging and functional near-infrared spectroscopy: insights from combined recording studies. *Front Hum Neurosci.* (2017) **11**:419. doi: 10.3389/fnhum.2017.00419
18. Sakatani K, Xie Y, Lichty W, Li S, Zuo H. Language-activated cerebral blood oxygenation and hemodynamic changes of the left prefrontal cortex in poststroke aphasic patients: a near-infrared spectroscopy study. *Stroke.* (1998) **29**:1299–304. doi: 10.1161/01.STR.29.7.1299
19. Chaudhary U, Hall M, DeCerce J, Rey G, Godavarty A. Frontal activation and connectivity using near-infrared spectroscopy: verbal fluency language study. *Brain Res Bull.* (2011) **84**:197–205. doi: 10.1016/j.brainresbull.2011.01.002
20. Kahlaoui K, Di Sante G, Barbeau J, Maheux M, Lesage F, Ska B, et al. Contribution of NIRS to the study of prefrontal cortex for verbal fluency in aging. *Brain Lang.* (2012) **121**:164–73. doi: 10.1016/j.bandl.2011.11.002

21. Matsui M, Tanaka K, Yonezawa M, Kurachi M. Activation of the prefrontal cortex during memory learning: near-infrared spectroscopy study. *Psychiatry Clin Neurosci.* (2007) 61:31–8. doi: 10.1111/j.1440-1819.2007.01607.x
22. Chein JM, Schneider W. Neuroimaging studies of practice-related change: fMRI and meta-analytic evidence of a domain-general control network for learning. *Brain Res Cogn Brain Res.* (2005) 25:607–23. doi: 10.1016/j.cogbrainres.2005.08.013
23. Niendam TA, Laird AR, Ray KL, Dean YM, Glahn DC, Carter CS. Meta-analytic evidence for a superordinate cognitive control network subserving diverse executive functions. *Cogn Affect Behav Neurosci.* (2012) 12:241–68. doi: 10.3758/s13415-011-0083-5
24. Dajani DR, Uddin LQ. Demystifying cognitive flexibility: implications for clinical and developmental neuroscience. *Trends Neurosci.* (2015) **38**:571–8. doi: 10.1016/j.tins.2015.07.003
25. Schmidt RA, Lee T. *Motor Control and Learning: A Behavioral Approach.* 5th ed. Champaign, IL: Human Kinetics (2011).
26. Johnson RK, Lasker JP, Stierwalt JAG, MacPherson MK, LaPointe LL. Motor learning guided treatment for acquired apraxia of speech: a case study investigating factors that influence treatment outcomes. *Speech Lang Hear.* (2017) **21**:213–23. doi: 10.1080/2050571X.2017.1388488
27. Johnson RK, Lott A, Prebor J. A comparison of outcome measures for speech motor learning in acquired apraxia of speech using motor learning guided treatment. *Clin Arch Commun Disord.* (2018) **3**:1–13. doi: 10.21849/cacd.2018.00304
28. Lasker JP, Stierwalt JAG, Hageman CF, LaPointe LL. Using motor learning guided theory and augmentative and alternative communication to improve speech production in profound apraxia: a case example. *J Med Speech Lang Pathol.* (2008) 16:225–33.
29. Lasker JP, Stierwalt JAG, Spence M, Cavin-Root C. Using webCam interactive technology to implement treatment for severe apraxia: a case example. *J Med Speech Lang Pathol.* (2010) 18:71–6.
30. Qureshi NK, Naseer N, Noori FM, Nazeer H, Khan RA, Saleem S. Enhancing classification performance of functional near-infrared spectroscopy- brain-computer interface using adaptive estimation of general linear model coefficients. *Front Neurorobot.* (2017) **11**:33. doi: 10.3389/fnbot.2017.00033
31. Schroeter ML, Bücheler MM, Müller K, Uludag K, Obrig H, Lohmann G, et al. Towards a standard analysis for functional near-infrared imaging. *Neuroimage.* (2004) **21**:283–90. doi: 10.1016/j.neuroimage.2003.09.054
32. Tak S, Ye JC. Statistical analysis of fNIRS data: a comprehensive review. *Neuroimage.* (2014) **85**(Pt 1):72–91. doi: 10.1016/j.neuroimage.2013.06.016
33. Ye JC, Tak S, Jang KE, Jung J, Jang J. NIRS-SPM: statistical parametric mapping for near-infrared spectroscopy. *Neuroimage.* (2009) **44**:428–47. doi: 10.1016/j.neuroimage.2008.08.036
34. Huertas I, Oldehinkel M, van Oort ESB, Garcia-Solis D, Mir P, Beckmann CF, et al. A Bayesian spatial model for neuroimaging data based on biologically informed basis functions. *Neuroimage.* (2017) **161**:134–48. doi: 10.1016/j.neuroimage.2017.08.009
35. Kratochwill TR, Hitchcock JH, Horner RH, Levin JR, Odom SL, Rindskopf DM, et al. Single-Case intervention research design standards. *Rem Spec Educ.* (2013) **34**:26–38. doi: 10.1177/0741932512452794
36. Johnson RK. Motor learning guided treatment for acquired apraxia of speech. *Speech Lang Hear.* (2017) **21**:202–12. doi: 10.1080/2050571X.2017.1379721
37. Ayaz H, Shewokis PA, Curtin A, Izzetoglu M, Izzetoglu K, Onaral B. Using MazeSuite and functional near infrared spectroscopy to study learning in spatial navigation. *J Vis Exp.* (2011). doi: 10.3791/3443
38. Izzetoglu M, Izzetoglu K, Bunce S, Ayaz H, Devaraj A, Onaral B, et al. Functional near-infrared neuroimaging. *IEEE Trans Neural Syst Rehabil Eng.* (2005) **13**:153–9. doi: 10.1109/TNSRE.2005.847377
39. Chance B, Anday E, Nioka S, Zhou S, Hong L, Worden K, et al. A novel method for fast imaging of brain function, non-invasively, with light. *Opt Express.* (1998) **2**:411–23. doi: 10.1364/OE.2.000411
40. Liu Y, Piazza EA, Simony E, Shewokis PA, Onaral B, Hasson U, et al. Measuring speaker-listener neural coupling with functional near infrared spectroscopy. *Sci Rep.* (2017) **7**:43293. doi: 10.1038/srep43293
41. Cui X, Bray S, Reiss AL. Functional near infrared spectroscopy (NIRS) signal improvement based on negative correlation between oxygenated and deoxygenated hemoglobin dynamics. *Neuroimage.* (2010) **49**:3039–46. doi: 10.1016/j.neuroimage.2009.11.050
42. Chen M, Blumen HM, Izzetoglu M, Holtzer R. Spatial coregistration of functional near-Infrared spectroscopy to brain MRI. *J Neuroimaging.* (2017) **27**:453–60. doi: 10.1111/jon.12432
43. Huppert TJ. Commentary on the statistical properties of noise and its implication on general linear models in functional near-infrared spectroscopy. *Neurophotonics.* (2016) **3**:010401. doi: 10.1117/1.NPh.3.1.010401
44. Santosa H, Zhai X, Fishburn F, Huppert TJ. The NIRS brain AnalyzIR toolbox. *Algorithms.* (2018) **11**:73. doi: 10.3390/a11050073
45. Moran PAP. Notes on coinuous stochastic phenomena. *Biometrika.* (1950) **37**:17–23. doi: 10.1093/biomet/37.1-2.17
46. Chen Y. On the four types of weight functions for spatial contiguity matrix. *Lett Spat Resour Sci.* (2012) **5**:65–72. doi: 10.1007/s12076-011-0076-6
47. Murakami D, Yoshida T, Seya H, Griffith DA, Yamagata Y. A Moran coefficient-based mixed effects approach to investigate spatially varying relationships. *Spat Stat.* (2017) **19**:68–89. doi: 10.1016/j.spasta.2016.12.001
48. Hoffman P, Jefferies E, Haffey A, Littlejohns T, Ralph MAL. Domain-specific control of semantic cognition: a dissociation within patients with semantic working memory deficits. *Aphasiology.* (2013) **27**:740–64. doi: 10.1080/02687038.2012.751578
49. Fishburn FA, Norr ME, Medvedev AV, Vaidya CJ. Sensitivity of fNIRS to cognitive state and load. *Front Hum Neurosci.* (2014)8:76. doi: 10.3389/fnhum.2014.00076
50. Baker CM, Burks JD, Briggs RG, Conner AK, Glenn CA, Morgan JP, et al. A connectomic atlas of the human cerebrum-chapter 2: the lateral frontal lobe. *Oper Neurosurg.* (2018) **15**(Suppl. 1):S10–74. doi: 10.1093/ons/opy254
51. Amyot F, Zimmermann T, Riley J, Kainerstorfer JM, Chernomordik V, Mooshagian E, et al. Normative database of judgment of complexity task with functional near infrared spectroscopy–application for TBI. *Neuroimage.* (2012) **60**:879–83. doi: 10.1016/j.neuroimage.2012.01.104

14

Statistical Analysis of Multi-Relational Network Recovery

***Zhi Wang* [1], *Xueying Tang* [2*] *and Jingchen Liu* [1]**

[1] *Department of Statistics, Columbia University, New York, NY, United States,* [2] *Department of Mathematics, University of Arizona, Tucson, AZ, United States*

***Correspondence:**
Xueying Tang
xytang@math.arizona.edu

In this paper, we develop asymptotic theories for a class of latent variable models for large-scale multi-relational networks. In particular, we establish consistency results and asymptotic error bounds for the (penalized) maximum likelihood estimators when the size of the network tends to infinity. The basic technique is to develop a non-asymptotic error bound for the maximum likelihood estimators through large deviations analysis of random fields. We also show that these estimators are nearly optimal in terms of minimax risk.

Keywords: multi-relational network, knowledge graph completion, tail probability, risk, asymptotic analysis, non-asymptotic analysis, maximum likelihood estimation

1. INTRODUCTION

A multi-relational network (MRN) describes multiple relations among a set of entities simultaneously. Our work on MRNs is mainly motivated by its applications to knowledge bases that are repositories of information. Examples of knowledge bases include WordNet [1], Unified Medical Language System [2], and Google Knowledge Graph (https://developers.google.com/knowledge-graph). They have been used as the information source in many natural language processing tasks, such as word-sense disambiguation and machine translation [3–5]. A knowledge base often includes knowledge on a large number of real-world objects or concepts. When a knowledge base is characterized by MRN, the objects and concepts correspond to nodes, and knowledge types are relations. **Figure 1** provides an excerpt from an MRN in which "Earth," "Sun," and "solar system" are three nodes. The knowledge about the orbiting patterns of celestial objects forms a relation "orbit," and the knowledge on classification of the objects forms another relation "belong to" in the MRN.

An important task of network analysis is to recover the unobserved network based on data. In this paper, we consider a latent variable model for MRNs. The presence of an edge from node i to node j of relation type k is a Bernoulli random variable Y_{ijk} with success probability M_{ijk}. Each node is associated with a vector, $\boldsymbol{\theta}$, called the embedding of the node. The probability M_{ijk} is modeled as a function f of the embeddings, $\boldsymbol{\theta}_i$ and $\boldsymbol{\theta}_j$, and a relation-specific parameter vector $\boldsymbol{w}_k$. This is a natural generalization of the latent space model for single-relational networks [6]. Recently, it has been successfully applied to knowledge base analysis [7–14]. Various forms of f are proposed, such as distance models [7], bilinear models [12–14], and neural networks [15]. Computational algorithms are proposed to improve link prediction for knowledge bases [16, 17]. The statistical properties of the embedding-based MRN models have not been rigorously studied. It remains unknown whether and to what extent the underlying distribution of MRN can be recovered, especially when there are a large number of nodes and relations.

The results in this paper fill in the void by studying the error bounds and asymptotic behaviors of the estimators for M_{ijk}'s for a general class of models. This is a challenging problem due to the following facts. Traditional statistical inference of latent variable models often requires a (proper

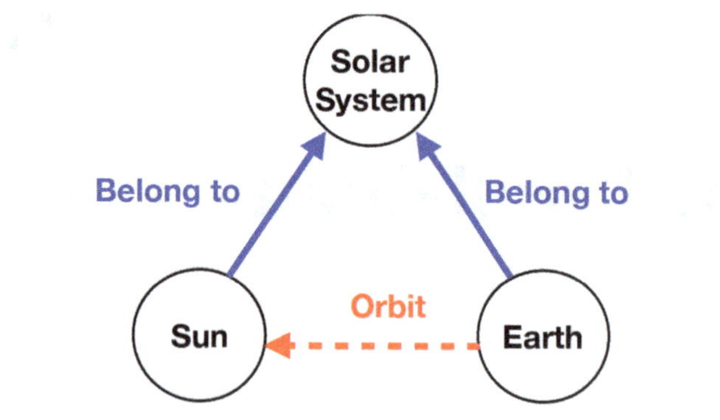

FIGURE 1 | An example of the MRN representation of a knowledge base.

or improper) prior distribution for $\boldsymbol{\theta}_i$. In such settings, one works with the marginalized likelihood with $\boldsymbol{\theta}_i$ integrated out. For the analysis of MRN, the sample size and the latent dimensions are often so large that the above-mentioned inference approaches are computationally infeasible. For instance, a small-scale MRN could have a sample size as large as a few million, and the dimension of the embeddings is as large as several hundred. Therefore, in practice, the prior distribution is often dropped, and the latent variables $\boldsymbol{\theta}_i$'s are considered as additional parameters and estimated via maximizing the likelihood or penalized likelihood functions. The parameter space is thus substantially enlarged due to the addition of $\boldsymbol{\theta}_i$'s whose dimension is proportionate to the number of entities. As a result, in the asymptotic analysis, we face a double-asymptotic regime of both the sample size and the parameter dimension.

In this paper, we develop results for the (penalized) maximum likelihood estimator of such models and show that under regularity conditions the estimator is consistent. In particular, we overcome the difficulty induced by the double-asymptotic regime via non-asymptotic bounds for the error probabilities. Then, we show that the distribution of MRN can be consistently estimated in terms of average Kullback-Leibler (KL) divergence even when the latent dimension increases slowly as the sample size tends to infinity. A probability error bound is also provided together with the upper bound for the risk (expected KL divergence). We further study the lower bound and show the near-optimality of the estimator in terms of minimax risk. Besides the average KL divergence, similar results can be established for other criteria, such as link prediction accuracy.

The outline of the remaining sections is as follows. In section 2, we provide the model specification and formulate the problem. Our main results are presented in section 3. Finite sample performance is examined in section 4 through simulated and real data examples. Concluding remarks are included in section 5.

2. PROBLEM SETUP

2.1. Notation

Let $|\cdot|$ be the cardinality of a set and $\times$ be the Cartesian product. Set $\{1,\ldots,N\}$ is denoted by $[N]$. The sign function $\operatorname{sgn}(x)$ is defined to be 1 for $x \geq 0$ and 0 otherwise. The logistic function is denoted by $\sigma(x) = e^x/(1+e^x)$. Let 1_A be the indicator function on event A. We use $U\left[a,b\right]$ to denote the uniform distribution on $[a,b]$ and $\mathrm{Ber}(p)$ to denote the Bernoulli distribution with probability p. The KL divergence between $\mathrm{Ber}(p)$ and $\mathrm{Ber}(q)$ is written as $D(p||q) = p\log\frac{p}{q} + (1-p)\log\frac{1-p}{1-q}$. We use $\|\cdot\|$ to denote the Euclidean norm for vectors and the Frobenius norm for matrices.

For two real positive sequences $\{a_n\}$ and $\{b_n\}$, we write $a_n = O(b_n)$ if $\limsup_{n\to\infty} a_n/b_n < \infty$. Similarly, we write $a_n = \Omega(b_n)$ if $\limsup_{n\to\infty} b_n/a_n < \infty$ and $a_n = o(b_n)$ if $\lim_{n\to\infty} a_n/b_n = 0$. We denote $a_n \lesssim b_n$ if $\limsup_{n\to\infty} a_n/b_n \leq 1$. When $\{a_n\}$ and $\{b_n\}$ are negative sequences, $a_n \lesssim b_n$ means $\liminf_{n\to\infty} a_n/b_n \geq 1$. In some places, we use $b_n \gtrsim a_n$ as an interchangeable notation of $a_n \lesssim b_n$. Finally, if $\lim_{n\to\infty} a_n/b_n = 1$, we write $a_n \sim b_n$.

2.2. Model

Consider an MRN with N entities and K relations. Given $i,j \in [N]$ and $k \in [K]$, the triple $\lambda = (i,j,k)$ corresponds to the edge from entity i to entity j of relation k. Let $\Lambda = [N] \times [N] \times [K]$ denote the set of all edges. We assume in this paper that an edge can be either present or absent in a network and use $Y_\lambda \in \{0,1\}$ to indicate the presence of edge λ. In some scenarios, the status of an edge may have more than two types. Our analysis can be generalized to accommodate these cases.

We associate each entity i with a vector $\boldsymbol{\theta}_i$ of dimension d_E and each relation k with a vector $\boldsymbol{w}_k$ of dimension d_R. Let $\mathcal{E} \subseteq \mathbb{R}^{d_E}$ be a compact domain where the embeddings $\boldsymbol{\theta}_1,\ldots,\boldsymbol{\theta}_N$ live. We call $\mathcal{E}$ the entity space. Similarly, we define a compact relation space $\mathcal{R} \subseteq \mathbb{R}^{d_R}$ for the relation-specific parameters $\boldsymbol{w}_1,\ldots,\boldsymbol{w}_K$. Let $\boldsymbol{x} = (\boldsymbol{\theta}_1,\ldots,\boldsymbol{\theta}_N,\boldsymbol{w}_1,\ldots,\boldsymbol{w}_K)$ be a vector in the product space $\Theta = \mathcal{E}^N \times \mathcal{R}^K$. The parameters associated with edge $\lambda = (i,j,k)$ is then $\boldsymbol{x}_\lambda = (\boldsymbol{\theta}_i,\boldsymbol{\theta}_j,\boldsymbol{w}_k)$. We assume that given $\boldsymbol{x}$, elements in $\{Y_\lambda \mid \lambda \in \Lambda\}$ are independent with each other and that the log odds of $Y_\lambda = 1$ is

$$\log\frac{P\left(Y_\lambda = 1|\boldsymbol{x}\right)}{P\left(Y_\lambda = 0|\boldsymbol{x}\right)} = \phi\left(\boldsymbol{x}_\lambda\right), \text{for } \lambda \in \Lambda. \tag{1}$$

Here ϕ is defined on $\mathcal{E}^2 \times \mathcal{R}$, and $\phi\left(\boldsymbol{x}_\lambda\right)$ is often called the score of edge λ.

We will use Y to represent the $N \times N \times K$ tensor formed by $\{Y_\lambda \mid \lambda \in \Lambda\}$ and $M(\boldsymbol{x})$ to represent the corresponding probability tensor $\left\{P(Y_\lambda = 1 \mid \boldsymbol{x}) \mid \lambda \in \Lambda\right\}$. Our model is given by

$$Y_\lambda \sim \mathrm{Ber}\left(M_\lambda\left(\boldsymbol{x}^*\right)\right), \tag{2}$$

$$M_\lambda(\boldsymbol{x}) = \sigma\left(\phi\left(\boldsymbol{x}_\lambda\right)\right), \lambda \in \Lambda, \tag{3}$$

where $\boldsymbol{x}^*$ stands for the true value of $\boldsymbol{x}$ and Y_λ's are independent. In the above model, the probability of the presence of an edge is entirely determined by the embeddings of the corresponding entities and the relation-specific parameters. This imposes a low-dimensional latent structure on the probability tensor $M^* = M(\boldsymbol{x}^*)$.

We specify our model using a generic function ϕ. It includes various existing models as special cases. Below are two examples of ϕ.

1. Distance model [7].

$$\phi\left(\theta_i, \theta_j, w_k\right) = b_k - \|\theta_i + a_k - \theta_j\|^2, \tag{4}$$

where $\theta_i, \theta_j, a_k \in \mathbb{R}^d$, $b_k \in \mathbb{R}$ and $w_k = (a_k, b_k)$. In the distance model, relation k from node i to node j is more likely to exist if θ_i shifted by a_k is closer to θ_j under the Euclidean norm.
2. Bilinear model [9].

$$\phi\left(\theta_i, \theta_j, w_k\right) = \theta_i^T \operatorname{diag}(w_k)\theta_j, \tag{5}$$

where $\theta_i, \theta_j, w_k \in \mathbb{R}^d$ and $\operatorname{diag}(w_k)$ is a diagonal matrix with w_k as the diagonal elements. Model (5) is a special case of the more general model $\phi\left(\theta_i, \theta_j, w_k\right) = \theta_i^T W_k \theta_j$, where $W_k \in \mathbb{R}^{d\times d}$ is a matrix parameterized by $w_k \in \mathbb{R}^{d_R}$. Trouillon et al. [12], Nickel et al. [13], and Liu et al. [14] explored different ways of constructing W_k.

Very often, only a small portion of the network is observed [18]. We assume that each edge in the MRN is observed independently with probability γ and that the observation of an edge is independent of Y. Let $\mathcal{S} \subset \Lambda$ be the set of observed edges. Then the elements in $\mathcal{S}$ are independent draws from Λ. For convenience, we use n to represent the expected number of observed edges, namely, $n = E\left[|\mathcal{S}|\right] = \gamma|\Lambda| = \gamma N^2 K$. Our goal is to recover the underlying probability tensor M^* based on the observed edges $\{Y_\lambda \mid \lambda \in \mathcal{S}\}$.

REMARK 1. *Ideally, if there exists x^* such that $Y_\lambda = sgn\left(M_\lambda(x^*) - \frac{1}{2}\right)$ for all $\lambda \in \Lambda$, then Y can be recovered with no error under x^*. This is, however, a rare case in practice, especially for large-scale MRN. A relaxed assumption is that Y can be recovered with some low dimensional x^* and noise $\{\epsilon_\lambda\}$ such that*

$$Y_\lambda = sgn\left(M_\lambda(x^*) + \epsilon_\lambda - \frac{1}{2}\right), \quad \epsilon_\lambda \overset{i.i.d}{\sim} U\left[-\frac{1}{2}, \frac{1}{2}\right], \quad \forall \lambda \in \Lambda. \tag{6}$$

By introducing the noise term, we formulate the deterministic MRN as a random graph. The model described in (2) is an equivalent but simpler form of (6).

2.3. Estimation

According to (2), the log-likelihood function of our model is

$$l\left(x; Y_{\mathcal{S}}\right) = \sum_{\lambda\in\mathcal{S}} Y_\lambda \log M_\lambda(x) + (1 - Y_\lambda)\log\left(1 - M_\lambda(x)\right). \tag{7}$$

We omit the terms $\sum_{\lambda\in\mathcal{S}}\log\gamma + \sum_{\lambda\notin\mathcal{S}}\log(1-\gamma)$ in (7) since γ is not the parameter of interest. To obtain an estimator of M^*, we take the following steps.

1. Obtain the maximum likelihood estimator (MLE) of x^*,

$$\hat{x} = \underset{x\in\Theta}{\operatorname{argmax}}\, l\left(x; Y_{\mathcal{S}}\right). \tag{8}$$

2. Use the plug-in estimator

$$\hat{M} = M(\hat{x}) \tag{9}$$

as an estimator of M^*.

In (8), the estimator $\hat{x}$ is a maximizer over the compact parameter space $\Theta = \mathcal{E}^N \times \mathcal{R}^K$. The dimension of Θ is

$$m = Nd_E + Kd_R,$$

which grows linearly in the number of entities N and the number of relations K.

2.4. Evaluation Criteria

We consider the following criteria to measure the error of the above-mentioned estimator. They will be used in both the main results and numerical studies.

1. Average KL divergence of the predictive distribution from the true distribution

$$L(\hat{M}, M^*) = \frac{1}{|\Lambda|}\sum_{\lambda\in\Lambda} D(M_\lambda^* || \hat{M}_\lambda). \tag{10}$$

2. Mean squared error of the predicted scores

$$MSE_\phi = \frac{1}{|\Lambda|}\sum_{\lambda\in\Lambda}\left(\phi(\hat{x}_\lambda) - \phi(x_\lambda^*)\right)^2. \tag{11}$$

3. Link prediction error

$$\widehat{err} = \frac{1}{|\Lambda|}\sum_{\lambda\in\Lambda} 1_{\hat{Y}_\lambda \neq Y_\lambda^*}, \tag{12}$$

where $\hat{Y}_\lambda = \operatorname{sgn}\left(\hat{M}_\lambda - \frac{1}{2}\right)$ and $Y_\lambda^* = \operatorname{sgn}\left(M_\lambda^* - \frac{1}{2}\right)$.

REMARK 2. *The latent attributes of entities and relations are often not identifiable, so the MLE $\hat{x}$ is not unique. For instance, in (4), the values of ϕ and $M(x)$ remain the same if we replace θ_i and a_k, respectively by $\Gamma\theta_i + t$ and Γa_k, where t is an arbitrary vector in $\mathbb{R}^{d_E}$ and Γ is an orthonormal matrix. Therefore, we consider the mean squared error of scores, which are identifiable.*

3. MAIN RESULTS

We first provide results of the MLE in terms of KL divergence between the estimated and the true model. Specifically, we investigate the tail probability $P(L(\hat{M}, M^*) > t)$ and the expected loss $E[L(\hat{M}, M^*)]$. In section 3.1, we discuss upper bounds for the two quantities. The lower bounds are provided in section 3.2. In section 3.3, we extend the results to penalized maximum likelihood estimators (pMLE) and other loss functions. All proofs are included in the **Supplementary Materials**.

3.1. Upper Bounds

We first present an upper bound for the tail probability $P(L(\hat{M}, M^*) > t)$ in Lemma 1. The result depends on the tensor size, the number of observed edges, the functional form of ϕ, and the geometry of parameter space Θ. The lemma explicitly quantifies the impact of these element on the error probability. It is key to the subsequent analyses. Lemma 2 gives a non-asymptotic upper bound for the expected loss (risk). We then establish the consistency of $\hat{M}$ and the asymptotic error bounds in Theorem 1.

We will make the following assumptions throughout this section.

ASSUMPTION 1. *$x^* \in \Theta = \mathcal{E}^N \times \mathcal{R}^K$, where $\mathcal{E}$ and $\mathcal{R}$ are Euclidean balls of radius U.*

ASSUMPTION 2. *The function ϕ is Lipschitz continuous under the Euclidean norm,*

$$|\phi(\boldsymbol{u}) - \phi(\boldsymbol{v})| \leq \alpha\|\boldsymbol{u} - \boldsymbol{v}\|, \quad \forall \boldsymbol{u}, \boldsymbol{v} \in \mathcal{E}^2 \times \mathcal{R}, \tag{13}$$

where α is a Lipschitz constant.

Assumption 1 is imposed for technical convenience. The results can be easily extended to general compact parameter spaces. Let $C = \sup_{\boldsymbol{u} \in \mathcal{E}^2 \times \mathcal{R}} |\phi(\boldsymbol{u})|$. Without loss of generality, we assume that $C \geq 2$.

LEMMA 1. *Consider $\hat{M}$ defined in (9) and the average KL divergence L in (10). Under Assumptions 1 and 2, for every $t > 0$, $\beta > 0$ and $0 < s < nt$,*

$$P\left(L(\hat{M}, M^*) \geq t\right) \leq \exp\left\{-\frac{nt - s}{C} h\left(\frac{1}{2} - \frac{s}{2nt}\right)\right\} \left(1 + \frac{2\sqrt{3}\alpha U n(1+\beta)}{s}\right)^m + \exp\left\{-n\beta h(\beta)\right\}, \tag{14}$$

where $m = Nd_E + Kd_R$ is the dimension of Θ, $n = \gamma N^2 K$ is the expected number of observations, and $h(u) = (1+\frac{1}{u})\log(1+u) - 1$.

In the proof of Lemma 1, we use Bennett's inequality to develop a uniform bound that does not depend on the true parameters. It is sufficient for the current analysis. If the readers need sharper bounds, they can read through the proof and replace the Bennett's bound by the usual large deviation rate function which provides a sharp exponential bound that depends on the true parameters. We don't pursue this direction in this paper.

Lemma 2 below gives an upper bound of risk $E[L(\hat{M}, M^*)]$, which follows from Lemma 1.

LEMMA 2. *Consider $\hat{M}$ defined in (9) and loss function L in (10). Let $C_1 = 18C$, $C_2 = 8\sqrt{3}\alpha U$ and $C_3 = 2\max\{C_1, C_2\}$. If Assumptions 1 and 2 hold and $\frac{n}{m} \geq C_2 + e$, then*

$$E[L(\hat{M}, M^*)] \leq C_3 \frac{m}{n} \log\frac{n}{m} + \frac{C_1}{n} \exp\left\{-m\log\frac{n}{m}\right\} + \frac{3}{n}\exp\left\{-\frac{1}{3}\left(n + C_3 m \log\frac{n}{m}\right)\right\}. \tag{15}$$

We are interested in the asymptotic behavior of the tail probability in two scenarios: (i) t is a fixed constant and (ii) t decays to zero as the number of entities N tends to infinity. The following theorem gives an asymptotic upper bound for the tail probability and the risk.

THEOREM 1. *Consider $\hat{M}$ defined in (9) and the loss function L in (10). Let the number of entities $N \to \infty$ and $C, K, U, d_E, d_R, \alpha$, and γ be fixed constants. If Assumptions 1 and 2 hold, we have the following asymptotic inequalities.*
When t is a fixed constant,

$$\log P(L(\hat{M}, M^*) \geq t) \lesssim -\frac{t}{5C} n. \tag{16}$$

When $t = 10C\frac{m}{n}\log\frac{n}{m}$,

$$\log P(L(\hat{M}, M^*) \geq t) \lesssim -m\log\frac{n}{m}. \tag{17}$$

Furthermore,

$$E[L(\hat{M}, M^*)] \lesssim 10C\frac{m}{n}\log\frac{n}{m}. \tag{18}$$

The consistency of $\hat{M}$ is implied by (16) and the rate of convergence is $|\log P(L(\hat{M}, M^*) \geq t)| = \Omega(N^2)$ if t is a fixed constant. The rate decreases to $\Omega(N\log N)$ for the choice of t producing (17). It is also implied by (17) that $L(\hat{M}, M^*) = O(\frac{1}{N}\log N)$ with high probability. We show in the next section that this upper bound is reasonably sharp.

The condition that K, U, d_E, d_R, and α are fixed constants can be relaxed. For instance, we can let U, d_E, d_R, and α go to infinity slowly at the rate $O(\log N)$ and K at the rate $O(N)$. We can let γ go to zero provided that $\frac{m}{n}\log\frac{n}{m} = o(1)$.

3.2. Lower Bounds

We show in Theorem 2 that the order of the minimax risk is $\Omega(\frac{m}{n})$, which implies the near optimality of $\hat{M}$ in (9) and the upper bound $O(\frac{m}{n}\log\frac{n}{m})$ in Theorem 1. To begin with, we introduce the following definition and assumption.

DEFINITION 1. For $\boldsymbol{u} = (\boldsymbol{\theta}, \boldsymbol{\theta}', \boldsymbol{w}) \in \mathcal{E}^2 \times \mathcal{R}$, the r-neighborhood of $\boldsymbol{u}$ is

$$\mathcal{N}_r(\boldsymbol{u}) = \left\{(\boldsymbol{\eta}, \boldsymbol{\eta}', \boldsymbol{\zeta}) \in \mathcal{E}^2 \times \mathcal{R} \mid \|\boldsymbol{\eta} - \boldsymbol{\theta}\| \leq r, \|\boldsymbol{\eta}' - \boldsymbol{\theta}'\| \leq r, \|\boldsymbol{\zeta} - \boldsymbol{w}\| \leq r\right\}.$$

Similarly, for $\boldsymbol{x} = (\boldsymbol{\theta}_1, \ldots, \boldsymbol{\theta}_N, \boldsymbol{w}_1, \ldots, \boldsymbol{w}_K) \in \mathcal{E}^N \times \mathcal{R}^K$, the r-neighborhood of $\boldsymbol{x}$ is

$$\mathcal{N}_r(\boldsymbol{x}) = \left\{(\boldsymbol{\eta}_1, \ldots, \boldsymbol{\eta}_N, \boldsymbol{\zeta}_1, \ldots, \boldsymbol{\zeta}_K) \in \mathcal{E}^N \times \mathcal{R}^K \mid \|\boldsymbol{\eta}_i - \boldsymbol{\theta}_i\| \leq r, \|\boldsymbol{\zeta}_k - \boldsymbol{w}_k\| \leq r, \forall i \in [N], k \in [K]\right\}.$$

ASSUMPTION 3. *There exists $\boldsymbol{u}_0 \in \mathcal{E}^2 \times \mathcal{R}$ and $r, \kappa > 0$ such that $\mathcal{N}_r(\boldsymbol{u}_0) \subset \mathcal{E}^2 \times \mathcal{R}$ and*

$$\left|\sigma(\phi(\boldsymbol{u})) - \sigma(\phi(\boldsymbol{v}))\right| \geq \kappa\|\boldsymbol{u} - \boldsymbol{v}\|, \quad \forall \boldsymbol{u}, \boldsymbol{v} \in \mathcal{N}_r(\boldsymbol{u}_0). \tag{19}$$

THEOREM 2. *Let $b = \sup_{u \in \mathcal{N}_r(u_0)} \sigma(\phi(u))$. Under Assumptions 2 and 3, if $r^2 \geq \frac{(m/16-1)b(1-b)}{12\alpha^2 n}$, then for any estimator $\hat{M}$, there exists $x^* \in \Theta$ such that*

$$P\left(L(\hat{M}, M^*) > \tilde{C}\frac{m/16-1}{n}\right) \geq \frac{1}{2}, \tag{20}$$

where $\tilde{C} = \frac{\kappa^2 b(1-b)}{108\alpha^2}$. Consequently, the minimax risk

$$\min_{\hat{M}} \max_{M^*} E[L(\hat{M}, M^*)] \geq \tilde{C}\frac{m/16-1}{2n}. \tag{21}$$

3.3. Extensions

3.3.1. Reguralization

In this section, we extend our asymptotic results in Theorem 1 to regularized estimators. In practice, regularization is often considered to prevent overfitting. We consider a regularization similar to elastic net [19]

$$l_\rho(x; Y_{\mathcal{S}}) = l(x; Y_{\mathcal{S}}) - \rho_1\|x\|_1 - \rho_2\|x\|^2, \tag{22}$$

where $\|\cdot\|_1$ stands for L_1 norm and $\rho_1, \rho_2 \geq 0$ are regularization parameters. The pMLE is

$$\hat{x} = \arg max_{x \in \Theta} l_\rho(x; Y_{\mathcal{S}}). \tag{23}$$

Note that the MLE in (8) is a special case of the pMLE above with $\rho_1 = \rho_2 = 0$. Since $\hat{x}$ is shrunk toward $\mathbf{0}$, without loss of generality, we assume that $\mathcal{E}$ and $\mathcal{R}$ are centered at $\mathbf{0}$. We generalize Theorem 1 to pMLE in the following theorem.

THEOREM 3. *Consider the estimator $\hat{M}$ given by (23) and (9) and the loss function L in (10). Let the number of entities $N \to \infty$ and $C, K, U, d_E, d_R, \alpha, \gamma$ be absolute constants. If Assumptions 1 and 2 hold and $\rho_1 + \rho_2 = o(\log N)$, then asymptotic inequalities (16), (17), and (18) in Theorem 1 hold.*

3.3.2. Other Loss Functions

We present some results for the mean squared error loss MSE_ϕ defined in (11) and the link prediction error $\widehat{err}$ defined in (12). Corollaries 1 and 2 give upper and lower bounds for MSE_ϕ, and Corollary 3 gives an upper bound for $\widehat{err}$ under an additional assumption.

COROLLARY 1. *Under the setting of Theorem 3 with the loss function replaced by MSE_ϕ, we have the following asymptotic results.*
If t is a fixed constant,

$$\log P\left(MSE_\phi \geq t\right) \lesssim -\frac{5\sigma(C)\left(1-\sigma(C)\right)t}{2C}n. \tag{24}$$

If $t = \frac{20C}{\sigma(C)(1-\sigma(C))}\frac{m}{n}\log\frac{n}{m}$,

$$\log P\left(MSE_\phi \geq t\right) \lesssim -m\log\frac{n}{m}. \tag{25}$$

Furthermore,

$$E\left[MSE_\phi\right] \lesssim \frac{20C}{\sigma(C)\left(1-\sigma(C)\right)}\frac{m}{n}\log\frac{n}{m}. \tag{26}$$

COROLLARY 2. *Under the setting of Theorem 2 with the loss function replaced by MSE_ϕ, we have*

$$P\left(MSE_\phi > \tilde{C}\frac{m/16-1}{8n}\right) \geq \frac{1}{2}, \tag{27}$$

and

$$\min_{\hat{M}} \max_{M^*} E\left[MSE_\phi\right] \geq \tilde{C}\frac{m/16-1}{16n}. \tag{28}$$

ASSUMPTION 4. *There exists $\varepsilon > 0$ such that $\left|M^*_\lambda - \frac{1}{2}\right| \geq \varepsilon$ for every $\lambda \in \Lambda$.*

COROLLARY 3. *Under the setting of Theorem 3 with the loss function replaced by $\widehat{err}$ and Assumption 4 added, we have the following asymptotic results.*
If t is a fixed constant,

$$\log P\left(\widehat{err} \geq t\right) \lesssim -\frac{2\varepsilon^2 t}{5C}n. \tag{29}$$

If $t = \frac{5C}{\varepsilon^2}\frac{m}{n}\log\frac{n}{m}$,

$$\log P\left(\widehat{err} \geq t\right) \lesssim -m\log\frac{n}{m}. \tag{30}$$

Furthermore,

$$E\left[\widehat{err}\right] \lesssim \frac{5C}{\varepsilon^2}\frac{m}{n}\log\frac{n}{m}. \tag{31}$$

3.3.3. Sparse Representations

We are interested in sparse entity embeddings and relation parameters. Let $\|\cdot\|_0$ be the number of non-zero elements of a vector and τ be a pre-specified sparsity level of x (i.e., the proportion of non-zero elements). Let $m_\tau = m\tau$ be the upper bound of non-zero parameters, that is, $\|x^*\|_0 \leq m_\tau$. Consider the following estimator

$$\hat{x} = \operatorname*{argmax}_{x \in \Theta} l\left(x; Y_{\mathcal{S}}\right) \quad \text{subject to} \quad \|x\|_0 \leq m_\tau. \tag{32}$$

The estimator defined above maximizes the L_0-penalized log-likelihood.

THEOREM 4. *Consider $\hat{M}$ defined in (32) and (9) and the loss function L in (10). Let the number of entities $N \to \infty$ and $\tau, C, K, U, d_E, d_R, \alpha$ be absolute constants. Under Assumptions 1 and 2, the following asymptotic inequalities hold.*
If t is a fixed constant,

$$\log P(L(\hat{M}, M^*) \geq t) \lesssim -\frac{t}{5C}n. \tag{33}$$

If $t = 10C\frac{m_\tau}{n}\log\frac{n}{m_\tau}$,

$$\log P(L(\hat{M}, M^*) \geq t) \lesssim -m_\tau\log\frac{n}{m_\tau}. \tag{34}$$

Furthermore,

$$E[L(\hat{M}, M^*)] \lesssim 10C\frac{m_\tau}{n}\log\frac{n}{m_\tau}. \tag{35}$$

We omit the results for other loss functions as well as the lower bounds since they can be analogously obtained.

4. NUMERICAL EXAMPLES

In this section, we demonstrate the finite sample performance of $\hat{M}$ through simulated and real data examples. Throughout the numerical experiments, AdaGrad algorithm [20] is used to compute $\hat{x}$ in (8) or (23). It is a first-order optimization method that combines stochastic gradient descent (SGD) [21] with adaptive step sizes for finding the local optima. Since the objective function in (8) is non-convex, a global maximizer is not guaranteed. Our objective function usually has many global maximizers, but, empirically, we found the algorithm works well on MRN recovery and the recovery performance is insensitive to the choice of the starting point of SGD. Computationally, SGD is also more appealing to handle large-scale MRNs than those more expensive global optimization methods.

4.1. Simulated Examples

In the simulated examples, we fix $K = 20$, $d = 20$ and consider various choices of N ranging from 100 to 10,000 to investigate the estimation performance as N grows. The function ϕ we consider is a combination of the distance model (4) and the bilinear model (5),

$$\phi\left(\boldsymbol{\theta}_i, \boldsymbol{\theta}_j, \boldsymbol{w}_k\right) = \left(\boldsymbol{\theta}_i + \boldsymbol{a}_k - \boldsymbol{\theta}_j\right)^T \operatorname{diag}\left(\boldsymbol{b}_k\right)\left(\boldsymbol{\theta}_i + \boldsymbol{a}_k - \boldsymbol{\theta}_j\right), \quad (36)$$

where $\boldsymbol{\theta}_i, \boldsymbol{\theta}_j, \boldsymbol{a}_k, \boldsymbol{b}_k \in \mathbb{R}^d$ and $\boldsymbol{w}_k = (\boldsymbol{a}_k, \boldsymbol{b}_k)$. We independently generate the elements of $\boldsymbol{\theta}_i^*$, $\boldsymbol{a}_k^*$, and $\boldsymbol{b}_k^*$ from normal distributions $N(0,1)$, $N(0,1)$, and $N(0,0.25)$, respectively, where $N(\mu, \sigma^2)$ denotes the normal distribution with mean μ and variance σ^2. To guarantee that the parameters are from a compact set, the normal distributions are truncated to the interval $[-20, 20]$. Given the latent attributes, each Y_{ijk} is generated from the Bernoulli distribution with success probability $M_{ijk}^* = \sigma(\phi(\boldsymbol{\theta}_i^*, \boldsymbol{\theta}_j^*, \boldsymbol{w}_k^*))$. The observation probability γ takes value from $\{0.005, 0.01, 0.02\}$. For each combination of γ and N, 100 independent datasets are generated. For each dataset, we compute $\hat{x}$ and $\hat{M}$ in (8) and (9) with AdaGrad algorithm and then calculate $L(\hat{M}, M^*)$ defined in (10) as well as the link prediction error $\widehat{err}$ defined in (12). The two types of losses are averaged over the 100 datasets for each combination of N and γ to approximate the theoretical risks $E[L(\hat{M}, M^*)]$ and $E[\widehat{err}]$. These quantities are plotted against N in log scale in **Figure 2**. As the figure shows, in general, both risks decrease as N increases. When N is small, n/m is not large enough to satisfy the condition $n/m \geq C_2 + e$ in Lemma 2 and the expected KL risk increases at the beginning. After N gets sufficiently large, the trend agrees with our asymptotic analysis.

4.2. Real Data Example: Knowledge Base Completion

WordNet [1] is a large lexical knowledge base for English. It has been used in word sense disambiguation, text classification, question answering, and many other tasks in natural language processing [3, 5]. The basic components of WordNet are groups of words. Each group, called a synset, describes a distinct concept. In WordNet, synsets are linked by conceptual-semantic and lexical relations, such as super-subordinate relation and antonym. We model WordNet as an MRN with the synsets as entities and the links between synsets as relations.

Following Bordes et al. [7], we use a subset of WordNet for analysis. The dataset contains 40,943 synsets and 18 types of relations. A triple (i, j, k) is called valid if relation k from entity i to entity j exists, i.e., $Y_{ijk} = 1$. All the other triples are called invalid triples. Among more than 3.0×10^{10} possible triples in WordNet, only 151,442 triples are valid. We assume that 141,442 valid triples and the same proportion of invalid triples are observed. The goal of our analysis is to recover the unobserved part of the knowledge base. We adopt the ranking procedure, which is commonly used in knowledge graph embedding literature, to evaluate link predictions. Given a valid triple $\lambda = (i, j, k)$, we rank estimated scores for all the invalid triples inside $\Lambda_{\cdot jk} = \left\{(i', j, k) \mid i' \in [N]\right\}$ in descending order and call the rank of $\phi\left(\hat{x}_\lambda\right)$ as the head rank of λ, denoted by H_λ. Similarly, we can define the tail rank T_λ and the relation rank R_λ by ranking $\phi\left(\hat{x}_\lambda\right)$ among the estimated scores of invalid triples in $\Lambda_{ij\cdot}$ and $\Lambda_{i\cdot k}$, respectively. For a set V of valid triples, the prediction performance can be evaluated by rank-based criteria, mean rank (MR), mean reciprocal rank (MRR), and hits at q (Hits@q), which are defined as

$$\mathrm{MR_E} = \frac{1}{2|V|}\sum_{\lambda \in V} H_\lambda + T_\lambda, \mathrm{MR_R} = \frac{1}{|V|}\sum_{\lambda \in V} R_\lambda,$$

$$\mathrm{MRR_E} = \frac{1}{2|V|}\sum_{\lambda \in V} \frac{1}{H_\lambda} + \frac{1}{T_\lambda}, \quad \mathrm{MRR_R} = \frac{1}{|V|}\sum_{\lambda \in V} \frac{1}{R_\lambda},$$

and

$$\mathrm{Hits_E@}q = \frac{1}{2|V|}\sum_{\lambda \in V} \mathbf{1}_{\{H_\lambda \leq q\}} + \mathbf{1}_{\{T_\lambda \leq q\}},$$

$$\mathrm{Hits_R@}q = \frac{1}{|V|}\sum_{\lambda \in V} \mathbf{1}_{\{R_\lambda \leq q\}}.$$

The subscripts E and R represent the criteria for predicting entities and relations, respectively. Models with higher MRRs, Hits@q's or lower MRs are more preferable. In addition, MRR is more robust to outliers than MR.

The three models described in (4), (5), and (36) are considered in our data analysis and we refer to them as Model 1, 2, and 3, respectively. For each model, the latent dimension d takes value from $\{50, 100, 150, 200, 250\}$. Due to the high dimensionality of the parameter space, L_2 penalized MLE is used to obtain the estimated latent attributes $\hat{x}$, with tuning parameters $\rho_1 = 0$ and ρ_2 chosen from $\{0, 10^{-2}, 10^{-3}, 10^{-4}, 10^{-5}\}$ in (22). Since information criteria based dimension and tuning parameter selection is computationally intensive for dataset of this scale, we set aside 5,000 of the unobserved valid triples as a validation set and select the d and ρ_2 that produce the smallest $\mathrm{MRR_E}$ on this validation set. The model with the selected d and ρ_2 is then evaluated on the test set consisting of the rest 5,000 unobserved valid triples.

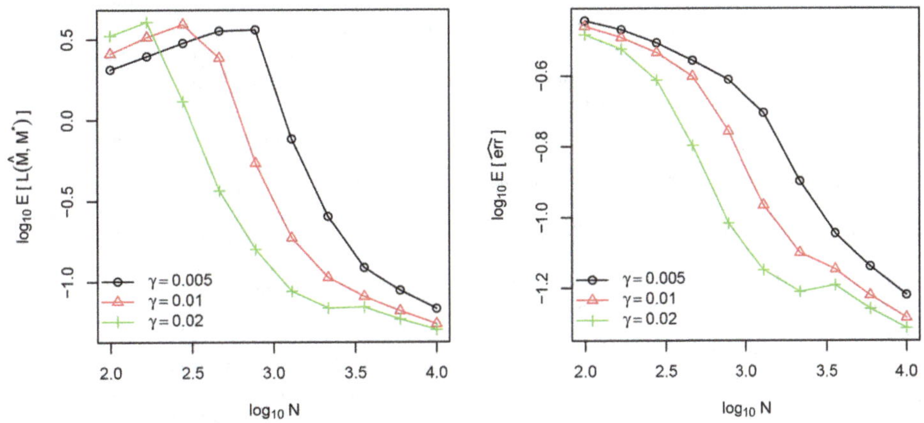

FIGURE 2 | Average Kullback-Leibler divergence (left) and average link prediction error (right) of $\hat{M}$ for different choices of N and γ.

TABLE 1 | Results for WordNet data analysis.

Method	(d, ρ_2)	MR_E	MRR_E	$Hits_E@10$	MR_R	MRR_R	$Hits_R@1$
Model 1	$(100, 10^{-5})$	385	0.64	0.888	1.41	0.896	0.817
Model 2	$(250, 10^{-4})$	769	0.94	0.945	1.31	0.968	0.959
Model 3	$(200, 10^{-4})$	499	0.94	0.947	1.13	0.978	0.967
CP	–	–	0.075	0.125	–	–	–
RESCAL	–	–	0.890	0.928	–	–	–

The results for CP and RESCAL are extracted from Trouillon et al. [12] and Nickel et al. [13].

The computed evaluation criteria on the test set are listed in **Table 1**. The table also includes the selected d and ρ_2 for each of the three score models. Models 2 and 3 generate similar performance. The MRRs for the two models are very close to 1, and the Hits@q's are higher than 90%, suggesting that the two models can identify the valid triples very well. Although Model 1 is inferior to the other two models in terms of most of the criteria, it outperforms them in MR_E. The results imply that Model 2 and Model 3 could perform extremely bad for a few triples.

In addition to Models 1–3, we also display the performance of the Canonical Polyadic (CP) decomposition [22] and a tensor factorization approach, RESCAL [23]. Their MRR_E and $Hits_E@10$ results on the WordNet dataset are extracted from Trouillon et al. [12] and Nickel et al. [13], respectively. Both methods, especially CP, are outperformed by Model 3.

5. CONCLUDING REMARKS

In this article, we focused on the recovery of large-scale MRNs with a small portion of observations. We studied a generalized latent space model where entities and relations are associated with latent attribute vectors and conducted statistical analysis on the error of recovery. MLEs and pMLEs over a compact space are considered to estimate the latent attributes and the edge probabilities. We established non-asymptotic upper bounds for estimation error in terms of tail probability and risk, based on which we then studied the asymptotic properties when the size of MRN and latent dimension go to infinity simultaneously. A matching lower bound up to a log factor is also provided.

We kept ϕ generic for theoretical development. The choice of ϕ is usually problem-specific in practice. How to develop a data-driven method for selecting an appropriate ϕ is an interesting problem to investigate in future works.

Besides the latent space models, sparsity [24] or clustering assumptions [25] have been used to impose low-dimensional structures in single-relational networks. An MRN can be seen as a combination of several heterogeneous single-relational networks. The distribution of edges may vary dramatically across relations. Therefore, it is challenging to impose appropriate sparsity or cluster structures on MRNs. More empirical and theoretical studies are needed to quantify the impact of heterogeneous relations and to incorporate the information for recovering MRNs.

AUTHOR CONTRIBUTIONS

ZW provided the theoretical proofs. ZW and XT conducted the simulation and real data analysis. All authors contributed to manuscript writing and revision.

REFERENCES

1. Miller GA. WordNet: a lexical database for English. *Commun ACM.* (1995) **38**:39–41. doi: 10.1145/219717.219748
2. McCray AT. An upper-level ontology for the biomedical domain. *Compar Funct Genomics.* (2003) **4**:80–4. doi: 10.1002/cfg.255
3. Gabrilovich E, Markovitch S. Wikipedia-based semantic interpretation for natural language processing. *J Artif Intell Res.* (2009) **34**:443–98. doi: 10.1613/jair.2669
4. Scott S, Matwin S. Feature engineering for text classification. In: *ICML* Vol. **99** Bled (1999). p. 379–88.
5. Ferrucci D, Brown E, Chu-Carroll J, Fan J, Gondek D, Kalyanpur AA, et al. Building Watson: an overview of the DeepQA project. *AI Mag.* (2010) **31**:59–79. doi: 10.1609/aimag.v31i3.2303
6. Hoff PD, Raftery AE, Handcock MS. Latent space approaches to social network analysis. *J Am Stat Assoc.* (2002) **97**:1090–8. doi: 10.1198/016214502388618906
7. Bordes A, Usunier N, Garcia-Duran A, Weston J, Yakhnenko O. Translating embeddings for modeling multi-relational data. In: Burges CJC, Bottou L, Welling M, Ghahramani Z, Weinberger KQ, editors. *Advances in Neural Information Processing Systems 26.* Curran Associates, Inc. (2013). p. 2787–95. Available online at: http://papers.nips.cc/paper/5071-translating-embeddings-for-modeling-multi-relational-data.pdf
8. Wang Z, Zhang J, Feng J, Chen Z. Knowledge graph embedding by translating on hyperplanes. In: *AAAI.* Québec City, QC (2014). p. 1112–9.
9. Yang B, Yih SWt, He X, Gao J, Deng L. Embedding entities and relations for learning and inference in knowledge bases. In: *Proceedings of the International Conference on Learning Representations.* San Diego, CA (2015).
10. Lin Y, Liu Z, Sun M, Liu Y, Zhu X. Learning entity and relation embeddings for knowledge graph completion. In: *AAAI.* Austin, TX (2015). p. 2181–7.
11. Garcia-Duran A, Bordes A, Usunier N, Grandvalet Y. Combining two and three-way embedding models for link prediction in knowledge bases. *J Artif Intell Res.* (2016) **55**:715–42. doi: 10.1613/jair.5013
12. Trouillon T, Welbl J, Riedel S, Gaussier É, Bouchard G. Complex embeddings for simple link prediction. In: *International Conference on Machine Learning.* (2016). p. 2071–80.
13. Nickel M, Rosasco L, Poggio TA. Holographic embeddings of knowledge graphs. In: *AAAI.* Phoenix, AZ (2016). p. 1955–61.
14. Liu H, Wu Y, Yang Y. Analogical inference for multi-relational embeddings. In: *Proceedings of the 34th International Conference on Machine Learning.* Vol. **70**. Sydney, NSW (2017). p. 2168–78.
15. Socher R, Chen D, Manning CD, Ng A. Reasoning with neural tensor networks for knowledge base completion. In: Burges CJC, Bottou L, Welling M, Ghahramani Z, Weinberger KQ, editors. *Advances in Neural Information Processing Systems.* Lake Tahoe, NV: Curran Associates, Inc. (2013) p. 926–34.
16. Kotnis B, Nastase V. Analysis of the impact of negative sampling on link prediction in knowledge graphs. *arXiv.* (2017) 170806816.
17. Kanojia V, Maeda H, Togashi R, Fujita S. Enhancing knowledge graph embedding with probabilistic negative sampling. In: *Proceedings of the 26th International Conference on World Wide Web Companion.* Perth, WA (2017) p. 801–2. doi: 10.1145/3041021.3054238
18. Min B, Grishman R, Wan L, Wang C, Gondek D. Distant supervision for relation extraction with an incomplete knowledge base. In: *HLT-NAACL.* Atlanta, GA (2013). p. 777–82.
19. Zou H, Hastie T. Regularization and variable selection via the elastic net. *J R Stat Soc B.* (2005) **67**:301–20. doi: 10.1111/j.1467-9868.2005.00503.x
20. Duchi J, Hazan E, Singer Y. Adaptive subgradient methods for online learning and stochastic optimization. *J Mach Learn Res.* (2011) **12**:2121–59. doi: 10.5555/1953048.2021068
21. Robbins H, Monro S. A stochastic approximation method. *Ann Math Stat.* (1951) **22**:400–7. doi: 10.1214/aoms/1177729586
22. Hitchcock FL. The expression of a tensor or a polyadic as a sum of products. *J Math Phys.* (1927) **6**:164–89. doi: 10.1002/sapm192761164
23. Nickel M, Tresp V, Kriegel HP. A three-way model for collective learning on multi-relational data. In: *Proceedings of the 28th International Conference on International Conference on Machine Learning. ICML'11.* Madison, WI: Omnipress (2011). p. 809–16.
24. Tran N, Abramenko O, Jung A. On the sample complexity of graphical model selection from non-stationary samples. *IEEE Trans Signal Process.* (2020) **68**:17–32. doi: 10.1109/TSP.2019.2956687
25. Jung A, Hero AO III, Mara AC, Jahromi S, Heimowitz A, Eldar YC, et al. Semi-supervised learning in network-structured data via total variation minimization. *IEEE Trans Signal Process.* (2019) **67**:6256–69. doi: 10.1109/TSP.2019.2953593

15

A Mathematical Analysis of the Improving Sequence Effect for Monetary Rewards

Salvador Cruz Rambaud[1], María José Muñoz Torrecillas[1] and Adriana Garcia[2]*

[1] *Departamento de Economía y Empresa, Universidad de Almería, Almería, Spain,* [2] *Faculty of Economics and Business, University of Groningen, Groningen, Netherlands*

Correspondence:
María José Muñoz Torrecillas
mjmtorre@ual.es

In this paper, we mathematically formalize the concept of improving sequence effect, which is one of the main anomalies of the discounted utility model [1]. The improving sequence effect implies a preference for a given sequence of outcomes, which increase over time, and has been empirically demonstrated for both monetary and nonmonetary results (hedonic experiences and health-related outputs). Nevertheless, to date, there is no mathematical treatment of this anomaly in the context of intertemporal choice, which allows us to relate this paradox to other anomalies, such as the delay and magnitude effects. In this way, the present manuscript has filled this gap. More specifically, we have proved that the improving sequence effect for monetary rewards cannot be rationalized by using a separable discount function but only by considering a non-separable discount function. Moreover, under certain conditions, we have proved that the delay and magnitude effects are necessary conditions for the existence of the improving sequence effect.

Keywords: improving sequence effect, intertemporal choice, discounted utility model, delay effect, magnitude effect, discount function

1. INTRODUCTION

Generally speaking, the *improving sequence effect* is an anomaly revealed in the ambit of intertemporal choice by which individuals prefer sequences of outcomes, which increase over time, rather than decrease, or flat sequences with equal mean [2]. In other words, people like improvement in such a way that they would prefer to leave the best outcomes for the last maturities of the sequence [3–5]. Thus, rather than experiencing the best outcome sooner, people usually prefer to postpone a good outcome. Obviously, this is a violation of the discounted utility model [1], since individuals show a negative time preference for sequences of outcomes rather than a positive time preference, as displayed for single outcomes.

Read and Powell [6] define the improving sequence effect for monetary sequences as follows: "Given a choice between two ways of distributing a fixed amount of money over time, either as a falling sequence going from large to small, or a rising sequence going from small to large, a rational decision maker will choose a falling sequence. This is because at a non-zero rate of interest the falling sequence dominates the rising one, since any unspent money can earn interest. Despite this, there is evidence that people often prefer rising to falling sequences."

More specifically, Loewenstein and Sicherman [7] and Loewenstein and Prelec [8] define this anomaly for income sequences in the following way: "Given a positive real rate of interest, a worker presented with alternative income sequences $X = (x_1, \ldots, x_n)$ and $Y = (y_1, \ldots, y_n)$ for otherwise identical jobs where

$\sum_{i=1}^{n} x_i = \sum_{i=1}^{n} y_i$, $x_i > y_i$, for $i = 1, \ldots, j$, and $y_i > x_i$, for $i = j+1, \ldots, n$, should select X over Y. [...] Sequence X dominates Y; by selecting X and saving appropriately, workers could enjoy greater consumption in every period." However, empirical evidence shows that workers actually prefer increasing wage profiles over flat or decreasing wage profiles of greater monetary present value.

Despite this, many researchers have theoretically and empirically studied the improving sequence effect, most are cited in the following section, and none presented a mathematical analysis of this anomaly. In this way, the main contribution of this paper is the rationalization of the improving sequence effect as an anomaly, observed in the context of an intertemporal choice by using a non-separable discount function. Moreover, this manuscript finds an interesting result linking the improving sequence effect to the delay and magnitude effects.

Therefore, the main objective of this paper is to present an analysis of the improving sequence effect for monetary sequences. The organization of this paper is as follows. Section 2 provides three mathematical definitions of the improving sequence effect and the implications between them. On the other hand, section 3 includes a proof of the present value maximization principle for decreasing income sequences by using a separable discount function. Section 4 introduces the concept of a non-separable discount function to explain the improving sequence effect in the context of discrete and continuous distributions of capital. Finally, after demonstrating the relationship between the improving sequence effect and the delay and magnitude effects (section 5), section 6 summarizes and concludes the study.

2. MATHEMATICAL DEFINITIONS OF THE IMPROVING SEQUENCE EFFECT

Although there are several theoretical definitions of the improving sequence effect (seen in the Introduction section of this paper), to the best of our knowledge, no mathematical definitions have been proposed yet. For this reason, we provide three mathematical definitions, which support the ideas previously indicated in the Introduction section. But, before this, we have to highlight two important observations:

- Our definitions restrict the idea provided by Loewenstein and Sicherman [8] because we only consider income sequences variable in (increasing or decreasing) arithmetic progression and not general income sequences. Even a preference for a mixed response (increasing-decreasing-increasing) has been found in the case of nonmonetary outcomes [9].
- However, our third definition allows for a given income sequence to be preferred, not only over all decreasing income sequences but also over the remaining increasing income sequences.

Before discussing the three definitions of the improving sequence effect, let us introduce the following definitions from the past.

DEFINITION 1. *Let R be a set of money amounts and T a set of time instants. The pair (R, T) is said to be a distribution of capital if there is a one-to-one correspondence between the elements of R and the elements of T.*

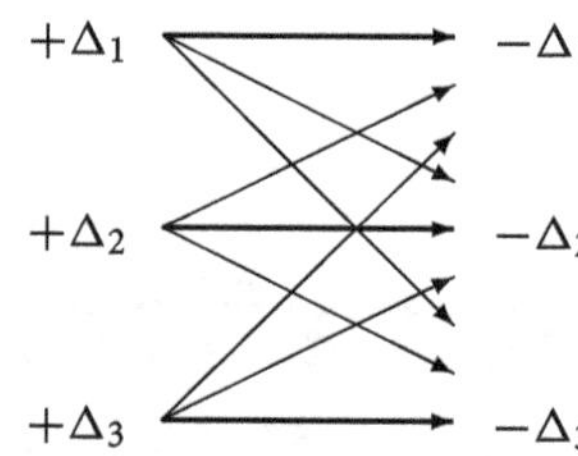

Chart 1 | Scheme corresponding to Definition 2.

In our definitions of the improving sequence effect, we will only consider a special set of money amounts, viz a sequence variable in an arithmetic progression.

DEFINITION 2. *Given an amount S and a period of time n, an intertemporal choice is said to satisfy the improving sequence effect if, for every $\Delta > 0$, the sequence*

$$(a, 1), (a + \Delta, 2), (a + 2\Delta, 3), \ldots, (a + (n-1)\Delta, n), \qquad (1)$$

where $a = \frac{S}{n} - \frac{n-1}{2}\Delta > 0$, is preferred over the rest of the decreasing sequences variable in an arithmetic progression whose positive terms mature at $1, 2, \ldots, n$, all terms summing up to S.

Definition 2 can be schematically viewed in **Chart 1**, where the arrow $+\Delta_i \longrightarrow -\Delta_j$ $(i, j = 1, 2, 3)$ means that the sequence

$$(a, 1), (a + \Delta_i, 2), (a + 2\Delta_i, 3), \ldots, (a + (n-1)\Delta_i, n),$$

is preferred to

$$(a, 1), (a - \Delta_j, 2), (a - 2\Delta_j, 3), \ldots, (a - (n-1)\Delta_j, n),$$

both satisfying the conditions displayed in definition 2.

Definition 2, which is called the *strong definition* of the improving sequence effect, does not completely fit the idea underlying this anomaly since, in the empirical studies on monetary sequences [2, 4, 6, 10–14], there are some decreasing sequences, which are preferred over other improving sequences, whose terms sum up to the same amount. Therefore, we provide a new definition, called the *semi-strong definition* of the improving sequence effect.

DEFINITION 3. *Given an amount S and a period of time n, an intertemporal choice is said to satisfy the improving sequence effect if, for every $\Delta > 0$, the sequence (1) is preferred over the following decreasing sequence*

$$(a + (n-1)\Delta, 1), (a + (n-2)\Delta, 2), \ldots, (a + \Delta, n-1), (a, n),$$

all terms summing up to S.

Chart 2 schematizes definition 3.

Despite introducing the former two definitions, the following one fits the idea of the improving sequence effect provided by Loewenstein and Sicherman [8] better. In other words, the conditions imposed on definition 2 and 3, need to be relaxed, giving rise to the *weak definition* of the improving sequence effect.

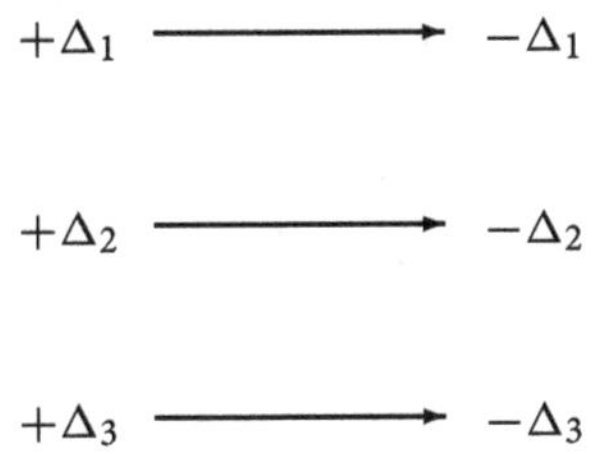

Chart 2 | Scheme corresponding to Definition 3.

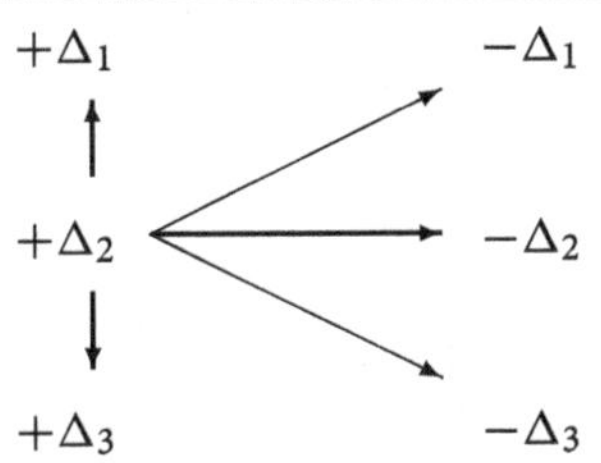

Chart 3 | Scheme corresponding to Definition 4.

DEFINITION 4. *Given an amount S and a period of time n, the intertemporal choice is said to satisfy the improving sequence effect if there is a* $\Delta > 0$ *such that sequence (1) is preferred over the rest of the sequences whose positive terms mature at* $1, 2, \ldots, n$ *and are constant or variable (increasing or decreasing) in an arithmetic progression, all terms summing up to S.*

Chart 3 schematizes definition 4.

These three definitions fit the idea of the improving sequence effect (and the negative time preference). However, as indicated, taking into account the results obtained from the empirical studies on this effect, definition 4 fits the empirically observed behavior better than the other behaviors and also allows a computational treatment, when analytically solving the equation, leading to the preferred sequence.

PROPOSITION 1. *Definition 2 implies definition 3, and definition 3 implies definition 4.*

Proof. See **Appendix** in Supplementary Material. □

Observe that definitions 2, 3, and 4 involve all values of the common difference Δ such that the terms of the corresponding sequences are positive. However, the set of differences could be restricted to a neighborhood of zero of the form $]-k,k[$, where $k > 0$, and incremental times to an interval $[0,h]$, where $h > 0$, giving rise to the respective "local" concepts of improving sequence effect. Finally, in the rest of this paper, we only considered definitions 3 and 4.

3. PRESENT VALUE MAXIMIZATION PRINCIPLE FOR INCOME SEQUENCES

Loewenstein and Sicherman [7] proved through a numerical example that the present value of a decreasing sequence is greater than the present value of the constant and the improving sequences, provided that their terms total the same amount ($150,000). This statement can be generalized to mean that the present value of a sequence variable in an arithmetic progression, whose terms total a fixed amount, is decreasing with respect to the difference of the progression. This result is consistent with the financial principles: the smaller the difference, the higher the amount of the first term, which consequently is discounted using a shorter period of time than the others. Moreover, it is relevant to note that Loewenstein and Sicherman [7] used the exponential discount, which is characterized by the use of a constant instantaneous discount rate. If the instantaneous discount rate [15] is constant (exponential discounting), it is intuitive to see that the discount function cannot explain the improving sequence effect. Now, we wonder if this result can be generalized to any separable discount function, regardless of whether its instantaneous discount rate is increasing or decreasing [16]. The answer to this question is affirmative, but before that, we will introduce some formal nomenclature and notation.

Let $F(t)$ be a (separable) discount function and let $a, a+\Delta, a+2\Delta, \ldots, a+(n-1)\Delta$ be an ordinary annuity variable in an arithmetic progression, where $a > 0$ is the first term and Δ is the common difference, such that the sum of all terms is constant and equal to S:

$$a + (a+\Delta) + (a+2\Delta) + \cdots + [a+(n-1)\Delta] = S.$$

Next, the following result can be stated.

PROPOSITION 2. *The present value of the ordinary annuity whose terms are variable in arithmetic progression:*

$$a, a+\Delta, a+2\Delta, \ldots, a+(n-1)\Delta$$

such that

$$a + (a+\Delta) + (a+2\Delta) + \cdots + [a+(n-1)\Delta] = S,$$

using any separable discount function, is decreasing with respect to Δ.

Proof. See **Appendix** in Supplementary Material. □

Proposition 2 can also be shown if $a, a+\Delta, a+2\Delta, \ldots, a+(n-1)\Delta$ is an annuity due variable in arithmetic progression.

COROLLARY 1. *The present value of the ordinary annuity variable in arithmetic progression*

$$a, a+\Delta, a+2\Delta, \ldots, a+(n-1)\Delta$$

such that

$$a + (a+\Delta) + (a+2\Delta) + \cdots + [a+(n-1)\Delta] = S,$$

using the exponential discounting $F(t) = (1+i)^{-t}$, *with* $i > 0$, *is decreasing with respect to* Δ. □

4. THE CASE OF NON-SEPARABLE DISCOUNT FUNCTIONS

Proposition 2 shows that, in order to explain the improving sequence effect, it is not possible to consider a separable discount function [17]. So, our first task was to define a non-separable discount function. See, for example, Lisei's paper [18].

DEFINITION 5. *A non-separable discount function is a real-valued function*

$$F : \mathbb{R} \times \mathbb{R}^+ \to \mathbb{R}$$

such that

$$(x, t) \mapsto F(x, t),$$

satisfying the following conditions:

1. *$F(x, 0) = x$, for every $x \in \mathbb{R}$.*
2. *$F(0, t) = 0$, for every $t \in \mathbb{R}^+$.*
3. *F is increasing with respect to x. If F is differentiable, $\frac{\partial F(x,t)}{\partial x} > 0$.*
4. *F is decreasing (respectively, increasing) with respect to t, if $x > 0$ (respectively, if $x < 0$). If F is differentiable, $\frac{\partial F(x,t)}{\partial t} < 0$ (respectively, $\frac{\partial F(x,t)}{\partial t} > 0$), if $x > 0$ (respectively, if $x < 0$).*

As an immediate consequence of conditions 1 and 3, it can be deduced that

$$F(x, t) < x, \text{ if } x > 0, \text{ and } F(x, t) > x, \text{ if } x < 0.$$

$F(x, t)$ represents the amount at time 0, which is indifferent to the amount x at time t.

4.1. Improving Sequence Effect With Discrete Distributions of Capital

In order to verify the existence of the improving sequence effect by using a given non-separable discount function, we had to solve the following problem: to maximize the present value

$$PV = \sum_{t=0}^{n-1} F(a + \Delta t, t + 1).$$

This optimization problem is equivalent to solving the following equation in Δ:

$$\frac{dPV}{d\Delta} = 0,$$

that is to say,

$$\sum_{t=0}^{n-1} \left(t - \frac{n-1}{2}\right) \left.\frac{\partial F(x, t+1)}{\partial x}\right|_{x=a+\Delta t} = 0, \qquad (2)$$

where

$$a = \frac{S}{n} - \frac{n-1}{2}\Delta.$$

Among all solutions of equation (2), we had to dismiss those such that some $a + \Delta t$ $(t = 0, 1, \ldots, n-1)$ was negative. Taking into account that all solutions are relative extremes, we had to choose the value of Δ corresponding to the greatest present value. To do this, if Δ_0 is a solution of equation (2), we had to verify that

$$\left.\frac{d^2PV}{d\Delta^2}\right|_{x_0=a_0+\Delta_0 t} < 0, \qquad (3)$$

where

$$a_0 = \frac{S}{n} - \frac{n-1}{2}\Delta_0,$$

that is to say

$$\sum_{t=0}^{n-1} \left(t - \frac{n-1}{2}\right)^2 \left.\frac{\partial^2 F(x, t+1)}{\partial x^2}\right|_{x_0=a_0+\Delta_0 t} < 0.$$

EXAMPLE 1. *The non-separable discount function $F(x, t) = x \exp\{-kt/x^\alpha\}$, with $k > 0$ and $\alpha > 1$ [19], satisfies the improving sequence effect for $k = 0.10$, $n = 3$, $\alpha = 2$, and $S = 6$. In effect, in this case,*

$$\frac{\partial F(x, t)}{\partial x} = \left(1 + \frac{kt\alpha}{x^\alpha}\right) \exp\{-kt/x^\alpha\}$$

and

$$\frac{\partial^2 F(x, t)}{\partial x^2} = \frac{kt\alpha}{x^{\alpha+1}} \left(1 + \frac{kt}{x^\alpha} - \alpha\right) \exp\{-kt/x^\alpha\}.$$

To solve equation (2), we obtained the value $\Delta_0 = 0.536$. In other words, this value of Δ maximizes the present value, as represented in **Figure 1**. Proposition 2 proved that if the discount function depends on time t and linearly on the reward x

$$F(x, t) = xF(t),$$

it is not rational that the improving sequence effect can hold. In other words, the improving sequence effect can occur when the discount function depends on time and is not linear with respect to the discounted outcome

$$F = F(x, t),$$

as confirmed by example 1.

Next, we wondered if alternatively this calculation could be easier if we worked with continuous distributions of capital. Before continuing with the development of the next section, we are going to introduce the following definition.

DEFINITION 6. *A distribution of capital $D = (R, T)$ is said to be continuous if T is a real interval and $R(t)$ is a continuous function [20], where $R(t)dt$ is the elemental amount corresponding to time t.*

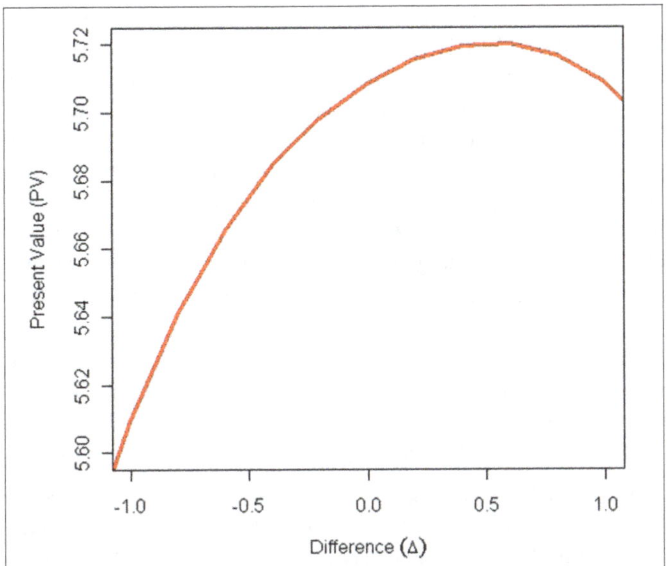

FIGURE 1 | Present value as a function of the common difference Δ (Example 1).

4.2. Improving Sequence Effect With Continuous Distributions of Capital

The objective of this subsection is to use continuous annuities for the analysis and description of the improving sequence effect. In this case, as indicated, we can not use separable discount functions. Therefore, we have to introduce the expression of the present value of a continuous annuity by using a non-separable discount function, $F(x,t)$:

$$PV = \lim_{n\to\infty} \sum_{k=1}^{n} F\left[R\left(\frac{k}{n}l\right), \frac{k}{n}l\right]\frac{l}{n}$$

or, equivalently,

$$\int_0^l F(R(t), t)\mathrm{d}t.$$

Obviously, in the treatment of the improving sequence effect, we will work with linear continuous annuities. More specifically, we will work with the family of continuous annuities defined by a straight line, $R(t) = mt + n$, such that $R(t) > 0$, for every t in $[0, l]$, and where the area defined by $R(t)$, the x-axis, and the vertical lines $t = 0$ and $t = l$ is constant and equal to S. From this, there are two important remarks:

1. The slope m can be positive or negative as long as the condition $R(t) > 0$, for every t in $[0, l]$, holds.
2. The values of l and S are given as constants.

In this context, the expression of the present value of the continuous annuity $R(t) = mt+n$, valued with the non-separable discount function $F(x, t)$, is

$$PV = \int_0^l F(mt + n, t)\mathrm{d}t. \tag{4}$$

Before going further, it is necessary to show the relationship between m and n. To do this, we took the following condition into account:

$$\int_0^l (mt + n)\mathrm{d}t = S,$$

from where

$$\left(m\frac{t^2}{2} + nt\right)\Bigg|_0^l = S,$$

or, equivalently,

$$m\frac{l^2}{2} + nl = S,$$

and finally,

$$n = \frac{S}{l} - \frac{ml}{2}.$$

Therefore,

$$R(t) = m\left(t - \frac{l}{2}\right) + \frac{S}{l}.$$

Now, we formulate the following question: Is there any value of m which maximizes the present value? If this value m_0 is positive, the improving sequence effect would hold. To do this, let us calculate the first derivative of the present value with respect to m:

$$\frac{\mathrm{d}PV}{\mathrm{d}m} = \int_0^l \frac{\partial F(x,t)}{\partial x}\Bigg|_{x=mt+n} \frac{\mathrm{d}R(t)}{\mathrm{d}m}\mathrm{d}t.$$

Hence,

$$\frac{\mathrm{d}R(t)}{\mathrm{d}m} = t - \frac{l}{2}$$

and consequently

$$\frac{\mathrm{d}PV}{\mathrm{d}m} = \int_0^l \frac{\partial F(x,t)}{\partial x}\Bigg|_{x=\left(t-\frac{l}{2}\right)m+\frac{S}{l}} \left(t - \frac{l}{2}\right)\mathrm{d}t. \tag{5}$$

Observe that, in equation (5), it is satisfied that

- $\frac{\partial F(x,t)}{\partial x}\Big|_{x=\left(t-\frac{l}{2}\right)m+\frac{S}{l}} > 0$, by the definition of $F(x, t)$.
- $t - \frac{l}{2} \leq 0$, for every t in $\left[0, \frac{l}{2}\right]$, and $t - \frac{l}{2} \geq 0$, for every t in $\left[\frac{l}{2}, l\right]$.

Therefore, the sign of the integral depends on the increase or the decrease of the factor $\frac{\partial F(x,t)}{\partial x}\Big|_{x=\left(t-\frac{l}{2}\right)m+\frac{S}{l}}$ with respect to t.

Finally, in order to prove that the obtained value of m, denoted by m_0, is a maximum, it is necessary that

$$\left.\frac{d^2PV}{dm^2}\right|_{m=m_0} < 0.$$

Therefore, the use of continuous distributions of capital can facilitate the resolution of equation (2), which arises in the case of a discrete distribution of capital, and it is presented as an alternative of easy calculation.

5. IMPROVING SEQUENCE EFFECT AND THE DELAY AND MAGNITUDE EFFECTS

Before analyzing the relationship between the improving sequence effect and the delay and magnitude effects, the following statement can be enunciated:

PROPOSITION 3. *A non-separable discount function $F(x,t)$ describes the semi-strong improving sequence effect, for every period length $h \leq h_0$, where $h_0 > 0$, and every amount difference $-\Delta_0 \leq \Delta \leq \Delta_0$, with $\Delta_0 > 0$, if and only if $\frac{\partial^2 F(x,t)}{\partial x \partial t} \geq 0$ and, moreover, the set*

$$M := \left\{(x,t) \in \mathbb{R} \times \mathbb{R}^+ : \frac{\partial^2 F(x,t)}{\partial x \partial t} > 0\right\}$$

is dense in the set

$$N := \left\{(x,t) \in \mathbb{R} \times \mathbb{R}^+ : \frac{\partial^2 F(x,t)}{\partial x \partial t} \geq 0\right\}$$

in the usual topology of $\mathbb{R}^2$.

Proof. See **Appendix** in Supplementary Material. □

In what follows, our aim was to mathematically relate the increasing sequence effect to the delay and magnitude effects but, before this, we had to define them. *Delay effect* consists of using higher discount rates for short intervals than for large ones. This way, discount rates decrease as waiting time (to obtain the reward) increases.

In the experiment conducted by Benzion et al. [21], the inferred (mean) discount rates for postponing the receipt of a 200-dollar amount were 42.8, 25.5, 23.0, and 19.5% for delays of 6 months, 1, 2, and 4 years, respectively. In other words, they obtained higher average discount rates for shorter time intervals.

More specifically, Prelec and Loewenstein [22] formalized this effect in the following way: $(x, t_1) \sim (y, t_2)$ implies $(x, t_1 + h) \prec (y, t_2 + h)$, for $x < y$ and $h > 0$.

On the other hand, some researchers found that the rates used to discount large amounts were lower than the rates used to discount small ones, which was labeled as the *magnitude effect* [5]. Thaler [23] carried out an experiment where the participants were indifferent between receiving \$15 right away and \$60 in a year, \$250 now and \$350 in a year, and \$3,000 immediately and \$4,000 in a year, implying discount rates of 139, 34, and 29%, respectively.

This effect was expressed formally by Prelec and Loewenstein [22] as follows: $(x, t_1) \sim (y, t_2)$ implies $(\alpha x, t_1) \prec (\alpha y, t_2)$, for $\alpha > 1$ and $0 < x < y$ (which implies $t_1 < t_2$). Moreover, Gerber and Rohde [24] proposed this definition: preferences exhibit the magnitude effect at date t if the outcomes x, y, X, and Y are given with $0 < x < X$ and $0 < y < Y$, then $(x, 0) \sim (y, t)$ and $(X, 0) \sim (Y, t)$ implies $X/Y > x/y$. In the same way, Read [25] mathematically defined the magnitude effect: imagine that a small outcome (x_1^s) and a large outcome (x_1^l) are respectively equated with outcomes x_2^s and x_2^l available at different times in the future, such as $u(x_1^s) = u(x_2^s)$ and $u(x_1^l) = u(x_2^l)$, then $x_1^l/x_2^l > x_1^s/x_2^s$. However, we will use a stronger definition of the magnitude effect: $(x, t_1) \sim (y, t_2)$ implies $(x + k, t_1) \prec (y + k, t_2)$, for $k > 0$ and $0 < x < y$.

There are two behavioral explanations for this anomaly:

1. Individuals perceive and are influenced not only by relative differences but also by absolute differences in monetary amounts. The difference between \$100 now and \$150 in a year seems to be greater than the difference between \$10 now and \$15 in a year; for this reason, a lot of people are willing to wait for \$50 but not for \$5 [22].
2. Mental accounting [26] affects the amount entered into a mental checking. Large amounts are considered as savings and small amounts as consumption. Thus, the cost of waiting for a small reward may be perceived as a foregone consumption, whilst the opportunity cost of waiting for a large amount is perceived as a foregone interest [27].

Delay effect is equivalent to expecting that the instantaneous discount rate

$$\delta(x,t) = -\frac{\frac{\partial F(x,t)}{\partial t}}{x\frac{\partial F(x,t)}{\partial x}}$$

is decreasing with respect to the time t, that is to say,

$$\frac{\partial \delta(x,t)}{\partial t} < 0.$$

Analogously, magnitude effect is equivalent to expecting that the instantaneous discount rate is decreasing with respect to the amount x, that is to say,

$$\frac{\partial \delta(x,t)}{\partial x} < 0.$$

Once the delay and magnitude effects are characterized, the following two corollaries demonstrate that they are necessary conditions for the existence of the semi-strong improving sequence effect in terms of proposition 3.

COROLLARY 2. *If a non-separable discount function $F(x,t)$ describes the semi-strong improving sequence effect for every period length $h \leq h_0$ and every amount difference $-\Delta_0 \leq \Delta \leq \Delta_0$ and $\frac{\partial^2 F(x,t)}{\partial t^2} \geq 0$, then it satisfies the delay effect.*

Proof. See **Appendix** in Supplementary Material. □

EXAMPLE 2. *Let us consider the so-called hyperbolic discount function deformed by the amount [28], that is to say, the following non-separable discount function:*

$$F(x,t) = \frac{x}{1+it/x} = \frac{x^2}{x+it},$$

defined for every $x \in \mathbb{R}^+$, *where* $i > 0$. *Thus, one has*

- $\frac{\partial F(x,t)}{\partial t} = -\frac{ix^2}{(x+it)^2}$.
- $\frac{\partial F(x,t)}{\partial x} = \frac{x^2+2ixt}{(x+it)^2}$.

Therefore, as

$$\frac{\partial^2 F(x,t)}{\partial t^2} = \frac{2i^2x^2}{(x+it)^3} > 0$$

and

$$\delta(x,t) = \frac{ix^2}{x^3+2ix^2t} = \frac{i}{x+2it},$$

$F(x,t)$ *satisfies the condition involved in the statement of corollary 2 and the delay effect (observe that* $\delta(x,t)$ *is strictly decreasing with respect to* t*).*

However, as

$$\frac{\partial^2 F(x,t)}{\partial x \partial t} = -\frac{2i^2xt}{(x+it)^3} < 0,$$

$F(x,t)$ *does not satisfy the local semi-strong definition of the improving sequence effect.*

COROLLARY 3. *If a non-separable discount function* $F(x,t)$ *describes the semi-strong improving sequence effect for every period length* $h \leq h_0$ *and every amount difference* $-\Delta_0 \leq \Delta \leq \Delta_0$ *and* $\frac{\partial^2 F(x,t)}{\partial x^2} \geq 0$, *then it satisfies the magnitude effect.*

Proof. See **Appendix** in Supplementary Material. □

These results fit the experimental findings from Duffy and Smith [14] who found a positive relationship between the preference for increasing payments (sequence effect) and the size of those payments (magnitude effect). Nevertheless, the converse statement of corollary 3 is not true, as shown in example 3.

EXAMPLE 3. *With the same non-separable discount function of example 2, one has*

$$\frac{\partial^2 F(x,t)}{\partial x^2} = \frac{2i^2t^2}{(x+it)^3} > 0.$$

Thus, $F(x,t)$ *satisfies the condition involved in the statement of corollary 3 and the magnitude effect (observe that* $\delta(x,t)$ *is strictly decreasing with respect to* x*). However, as indicated in example 2,* $F(x,t)$ *does not satisfy the local semi-strong definition of the improving sequence effect.*

Summarizing, under the same hypothesis of corollaries 2 and 3 some additional conditions are required in the instantaneous discount rate so that the converse statements holds. This task will be left for further research.

6. CONCLUSION

This paper contributes to the existing literature on the improving sequence effect by presenting a mathematical analysis of this anomaly. First of all, in order to relate this effect with other effects, we proposed three mathematical definitions of this paradox. These three definitions fit the idea of the improving sequence effect (negative time preference); however, based on the results from the empirical studies, definition 4 fits the empirically observed behavior better than the others and allows a computational treatment to find the best sequence.

Moreover, it has been mathematically proven that the present value of a sequence variable in an arithmetic progression, whose terms total a fixed amount, decreases with respect to the difference of the progression either using the exponential discount function or any other discount function. Additionally, we introduced a new methodology to detect and explain the improving sequence effect with discrete distributions of capital by using non-separable discount functions (see Example 1).

Finally, a relationship between the improving sequence effect and the delay and magnitude effects was presented. More specifically, we proved that, under certain hypotheses, the semi-strong improving sequence effect is a sufficient (but not a necessary) condition for both the delay and magnitude effects.

AUTHOR CONTRIBUTIONS

All authors listed have made a substantial, direct and intellectual contribution to the work, and approved it for publication.

FUNDING

The authors gratefully acknowledge the financial support from the Spanish Ministry of Economy and Competitiveness [National R&D Project DER2016-76053-R] and from the Spanish Ministry of Economy and Science and the European Regional Development Fund-ERDF/FEDER [National R&D Project ECO2015-66504-P].

ACKNOWLEDGMENTS

We are very grateful for the comments and suggestions offered by two referees.

REFERENCES

1. Samuelson PA. A note on measurement of utility. *Rev Econ Stud.* (1937) **4**:155–61. doi: 10.2307/2967612
2. Loewenstein G, Prelec D. Preferences for sequences of outcomes. *Psychol Rev.* (1993) **100**:91–108. doi: 10.1037/0033-295X.100.1.91
3. Chapman GB. Temporal discounting and utility for health and money. *J Exp Psychol Learn Mem Cogn.* (1996) **22**:771–91. doi: 10.1037/0278-7393.22.3.771
4. Guyse J, Keller L, Eppel T. Valuing environmental outcomes: preferences for constant or improving sequences. *Organ Behav Hum Decis Process.* (2002) **87**:253–77. doi: 10.1006/obhd.2001.2965
5. Cruz Rambaud S, Muñoz Torrecillas MJ. An analysis of the anomalies in traditional discounting models. *Int J Psychol Psychol Ther.* (2004) **4**:105–28.
6. Read D, Powell M. Reasons for sequence preferences. *J Behav Decis Making* (2002) **15**:433–60. doi: 10.1002/bdm.429
7. Loewenstein G, Sicherman N. Do workers prefer increasing wage profiles? *J Labor Econ.* (1991) **9**:67–84. doi: 10.1086/298259
8. Loewenstein G, Prelec D. Negative time preference. *Am Ec Rev.* (1991) **81**:347–52.
9. Baucells M, Smith D, Weber M. Preferences over constructed sequences: Empirical evidence from music. *Darden Business School Working Paper* (2016).
10. Chapman GB. Expectations and preferences for sequences of health and money. *Organ Behav Hum Decis Process.* (1996) **67**:59–75.
11. Gigliotti G, Sopher B. Analysis of intertemporal choice: a new framework and experimental results. *Theor Decis.* (2003) **55**:209–33. doi: 10.1023/B:THEO.0000044601.83386.7d
12. Manzini P, Mariotti M, Mittone L. Choosing monetary sequences: theory and experimental evidence. *Theor. Decis.* (2010) **69**:327–54. doi: 10.1007/s11238-010-9214-7
13. Matsumoto D, Peecher ME, Rich JS. Evaluations of outcome sequences. *Organ Behav Hum Decis Process.*(2000) **83**:331–52. doi: 10.1006/obhd.2000.2913
14. Duffy S, Smith J. Preference for increasing wages: how do people value various streams of income? *Judgm Decis Making* (2013) **8**:74–90.
15. Cruz Rambaud S, Valls Martínez MC. *Introducción a las Matemáticas Financieras.* 3rd ed. Madrid: Pirámide (2014).
16. Cruz Rambaud S, Muñoz Torrecillas MJ. A generalization of the q-exponential discounting function. *Physica A Stat Mech Appl.* (2013) **392**:3045–50. doi: 10.1016/j.physa.2013.03.009
17. Musau A. Hyperbolic discount curves: a reply to Ainslie. *Theor Decis.* (2014) **76**:9–30. doi: 10.1007/s11238-013-9361-8
18. Lisei G. Su un'equazione funzionale collegata alla scindibilità delle leggi finanziarie. *Giornalle dell'Istituto Italiano degli Attuari* (1979) anno XLII:19–24.
19. Noor J. Intertemporal choice and the magnitude effect. *Games Econ Behav.* (2011) **72**:255–70. doi: 10.1016/j.geb.2010.06.006
20. Gil Peláez L. *Matemática de la Operaciones Financieras.* Madrid: Editorial AC (1989).
21. Benzion U, Rapaport A, Yagil J. Discount rates inferred from decisions: an experimental study. *Manage Sci.* (1989) **35**:270–84. doi: 10.1287/mnsc.35.3.270
22. Prelec D, Loewenstein G. Decision making over time and under uncertainty: a common approach. *Manage Sci.* (1991) **37**:770–86. doi: 10.1287/mnsc.37.7.770
23. Thaler R. Some empirical evidence on dynamic inconsistency. *Econ Lett.* (1981) **1**:201–7. doi: 10.1016/0165-1765(81)90067-7
24. Gerber A, Rohde K. Anomalies in intertemporal choice? *Swiss Finance Institute Research Paper.* Geneva (2007).
25. Read D. Intertemporal choice. *Working Paper LSEOR 0358.* London (2003).
26. Shefrin HM, Thaler RH. The behavioral life-cycle hypothesis. *Econ Inquiry* (1988) **26**:609–43. doi: 10.1111/j.1465-7295.1988.tb01520.x
27. Loewenstein G, Thaler R. Anomalies: intertemporal choice. *J Econ Perspect.* (1989) **3**:181–93. doi: 10.1257/jep.3.4.181
28. Cruz Rambaud S, Parra Oller IM, Valls Martínez MC. The amount-based deformation of the q-exponential discount function: a joint analysis of delay and magnitude effects. *Physica A Stat Mech Appl.* (2018) **508**:788–96. doi: 10.1016/j.physa.2018.05.152

16

CPD-Structured Multivariate Polynomial Optimization

Muzaffer Ayvaz[1,2*] *and Lieven De Lathauwer*[1,2]

[1] *Department of Electrical Engineering (ESAT), KU Leuven, Leuven, Belgium,* [2] *Group Science, Engineering and Technology, KU Leuven Kulak, Kortrijk, Belgium*

Correspondence:
Muzaffer Ayvaz
muzaffer.ayvaz@kuleuven.be

We introduce the Tensor-Based Multivariate Optimization (TeMPO) framework for use in nonlinear optimization problems commonly encountered in signal processing, machine learning, and artificial intelligence. Within our framework, we model nonlinear relations by a multivariate polynomial that can be represented by low-rank symmetric tensors (multi-indexed arrays), making a compromise between model generality and efficiency of computation. Put the other way around, our approach both breaks the curse of dimensionality in the system parameters and captures the nonlinear relations with a good accuracy. Moreover, by taking advantage of the symmetric CPD format, we develop an efficient second-order Gauss–Newton algorithm for multivariate polynomial optimization. The presented algorithm has a quadratic per-iteration complexity in the number of optimization variables in the worst case scenario, and a linear per-iteration complexity in practice. We demonstrate the efficiency of our algorithm with some illustrative examples, apply it to the blind deconvolution of constant modulus signals, and the classification problem in supervised learning. We show that TeMPO achieves similar or better accuracy than multilayer perceptrons (MLPs), tensor networks with tensor trains (TT) and projected entangled pair states (PEPS) architectures for the classification of the MNIST and Fashion MNIST datasets while at the same time optimizing for fewer parameters and using less memory. Last but not least, our framework can be interpreted as an advancement of higher-order factorization machines: we introduce an efficient second-order algorithm for higher-order factorization machines.

Keywords: multivariate polynomial, numerical optimization, tensor decomposition, Gauss-Newton algorithm, factorization machines, higher order factorization machines, tensor network, image classification

1. INTRODUCTION

Many problems in data science, signal processing, machine learning and artificial intelligence (AI) can be thought of determining the nonlinear relationship between input and output data. Several strategies have been developed to efficiently model these nonlinear interactions. However, due to the higher-order nature of input and output data, developing scalable algorithms to model these nonlinear interactions is a challenging research direction. Another major issue is the large number of system parameters needed to model the physical phenomena under consideration. For example, large numbers of layers and neurons are needed in deep neural networks (DNNs). Multivariate polynomials are also utilized to model nonlinear continuous functions. However, this approach suffers from an exponential increase in the number of coefficients with the degree of the polynomial. This is known as the curse of dimensionality and is a major drawback that inhibits the development of efficient algorithms.

Tensor decompositions such as canonical polyadic decomposition (CPD) and tensor trains (TT) are promising tools for breaking the curse of dimensionality. Tensors are multi-indexed arrays. They preserve the higher-order structure which is inherent in data, are able to model nonlinear interactions, and can be decomposed uniquely under mild conditions [1–3]. Efficient numerical optimization algorithms have been developed for tensor decompositions. In the context of CPD, the Gauss–Newton algorithm using both line search and trust-region frameworks have been effectively implemented by exploiting the CPD structure [4–6]. A low complexity damped Gauss-Newton algorithm has also been proposed [7]. Moreover, a randomized block sampling approach has been proposed which achieves linear time complexity for the CPD of large tensors by utilizing the Gauss–Newton algorithm [8]. Many data science problems such as latent factor analysis have been solved by reformulating them as tensor decomposition problems [9–12]. An inexact Gauss–Newton algorithm has been proposed for scaling the CPD of large tensors with non-least-squares cost functions [13]. Moreover, generalized Gauss–Newton algorithm with its efficient parallel implementation has been proposed for tensor completion with generalized loss functions [14]. Our aim in this work is to extend the efficient numerical approaches to a broader class of problems that includes not only tensor decompositions but also the optimization of multilinear/polynomial cost functions. Examples include, but are not limited to matrix and tensor eigenvalue problems, nonlinear dimensionality reduction, nonlinear blind source separation, multivariate polynomial regression, and classification problems.

In this study, we develop a framework called Tensor-Based Multivariate Polynomial Optimization (TeMPO) to deal with nonlinear optimization problems commonly encountered in signal processing, machine learning and artificial intelligence. A preliminary version, where only rank-1 CPD is considered with application in blind identification, appeared as the conference paper [15]. In the TeMPO framework, these nonlinear functions are approximated or modeled by multivariate polynomials. Then, low-rank tensors are used to represent the polynomial under consideration. This approach reduces the number of parameters that define the system, and hence enables us to develop efficient numerical optimization algorithms. To further elaborate on the proposed methodology, let us consider the optimization problem

$$\min_{p} \quad l(p(\mathbf{z}), \boldsymbol{\theta}), \tag{1}$$

where $l: \mathcal{R} \times \mathcal{R}^M \rightarrow \mathcal{R}^+$ denotes a loss function such as the mean squared error, $p: \mathcal{R}^I \rightarrow \mathcal{R}$ denotes an unknown multivariate polynomial, $\mathbf{z} \in \mathcal{R}^I$ denotes input data, and $\boldsymbol{\theta} \in \mathcal{R}^M$ denotes output data. We compactly represent the polynomial $p(\mathbf{z})$ through low-rank tensors. One possible way to do this is to write the polynomial as a sum of homogeneous polynomials as follows:

$$p(\mathbf{z}) := \sum_{j=0}^{d} \mathcal{T}_j \mathbf{z}^j, \tag{2}$$

where $\mathcal{T}_j$ denotes a low-rank tensor of order j, and $\mathcal{T}_j \mathbf{z}^j$ denotes the mode-n product (see Section 2.1) of a tensor $\mathcal{T}_j$ and the vector $\mathbf{z}$ for all modes. As by convention, $\mathcal{T}_0$ is assumed to be scalar and $\mathbf{z}^0$ is assumed to be scalar 1. From now on, we call (2) a type I model. We can represent a multivariate polynomial with a single tensor by utilizing a process called homogenization, and augmenting the independent variable $\mathbf{z}$ by a constant 1 as

$$p(\widetilde{\mathbf{z}}) := \mathcal{W}\widetilde{\mathbf{z}}^d, \tag{3}$$

where $\mathcal{W}$ is a tensor of order d, and $\widetilde{\mathbf{z}} = [1; \mathbf{z}]$. Hereafter, we call (3) a type II model.

An n-variate polynomial of degree d has $\mathcal{O}\left(n^d\right)$ coefficients. This exponential dependence on d is the so-called curse of dimensionality. In the TeMPO framework, we break the curse of dimensionality by assuming low-rank structure in the coefficient tensors. For example, when rank- R symmetric CPD structure is used, the number of parameters needed to represent the n-variate polynomial of degree d is ndR which is linear in the number of variables. Several low-rank structures for tensors have been introduced in the literature [1, 2, 16], e.g., canonical polyadic decomposition (CPD), Tucker decomposition, hierarchical Tucker decomposition (HT) [17], tensor train decomposition (TT) [18]. All of these structures can be incorporated into the TEMPO framework; however, in this paper we restrict ourselves to symmetric CPDs. Note that different types of low-rank structure allow us to represent different sub-classes of polynomials. Of course, different representations differ in storage space, and computational complexity. A more detailed exposition will be given in Section 3.2. Note also that the type I model allows us to constrain each term separately while the type II model does not. Therefore, the type I model is a more general representation of multivariate polynomials which may provide better results depending on the applications.

Besides breaking the curse of dimensionality, exploiting low-rank representations of tensors enables us to derive efficient expressions for objective function and gradient evaluations. These then lead us to develop scalable algorithms. We apply our framework for image classification by adapting the second-order Gauss–Newton algorithm and exploiting the symmetric CPD structure in two different tensor representations of multivariate polynomials. We show that the TeMPO framework with symmetric CPD structure achieves similar or better accuracy than various methods such as MLPs, and tensor networks with different structures for the classical MNIST and Fashion MNIST datasets while using fewer parameters and therefore less memory.

Related Work

Several tensor-based methods have been reported in the literature for regression and classification, two problems that are in the class of problems (1). In most of these approaches, a linear model

$$y = \langle \mathcal{W}, \mathcal{X} \rangle + b, \tag{4}$$

is used where $\mathcal{W}$ denotes a weight tensor and $\mathcal{X}$ represents nonlinear features of the input data. This model corresponds to the type II model when a symmetric CPD structure is imposed on the weight tensor $\mathcal{W}$ and $\mathcal{X}$ is composed of polynomial features of input data. Clearly, imposing different structures to the weight tensor $\mathcal{W}$ and using different nonlinear features in the tensor $\mathcal{X}$ leads to a different representation of the nonlinear interaction between input data and output data. For example, *exponential machines* utilize the tensor train format in the weight tensor with a norm regularization term in the optimization [19]. In this approach, the Riemannian gradient descent algorithm is used for solving the optimization problem. In a similar approach, tensor trains is used with the feature map $\phi(x_j) = \left[\cos\left(\frac{\pi x_j}{2}\right), \sin\left(\frac{\pi x_j}{2}\right)\right]$, by using the density matrix renormalization group (DMRG) algorithm and the first-order ADAM algorithm for the optimization of different cost functions [20, 21]. The same feature map is also used for the linear model (4) by imposing projected entangled pair states (PEPS) structure on the weight tensor $\mathcal{W}$ [22]. The CPD format in model (4) has also been studied in the realm of tensor regression with the Frobenius norm and group sparsity norm regularization terms while using a coordinate-descent approach [23]. A similar model is also considered by utilizing the symmetric CPD format and the second-order Gauss–Newton algorithm with algebraic initialization for multivariate polynomial regression [24]. Several approaches have been proposed that utilize CPD or Tucker formats in tensor regression that use different regularization strategies to prevent the overfitting [25, 26]. Also, the hierarchical Tucker (HT) format has been used in the tensor regression context for the generalized linear model (GLM) $y = \boldsymbol{\alpha}^{\mathrm{T}}\mathbf{x} + \langle \mathcal{W}, \mathcal{X} \rangle$. This approach was successfully applied to brain imaging data sets and uses a block relaxation algorithm, which solves a sequence of lower dimensional optimization problems [27].

Similarly, several models related to the type I model are considered in various settings. For example, Kar and Karnick use random polynomial features and parameterize the coefficients of the polynomial under consideration [28]. The parameterization used in this approach has been shown to be equivalent to imposing the CPD format to the weight tensor $\mathcal{W}$ [29]. Another approach is *factorization machines* which use a multivariate polynomial kernel in the realm of support vector machines (SVM) [30]. For second-order factorization machines a first-order stochastic gradient descent algorithm has been proposed. This approach has a linear time complexity. Higher-order factorization machines use the ANOVA kernel to achieve a linear time complexity and have been successfully applied to link prediction models using stochastic gradient descent [31]. The ANOVA kernel does not use symmetric tensors in the representation and instead only considers combinations of distinct features [31]. Also, factorization machines in the symmetric CPD format have been considered using first-order and BFGS type algorithms [32]. *Tensor machines* generalize both the Kar-Karnick random features approach and factorization machines. It has been shown that these approaches correspond to specific types of tensor machines in the CPD format. Further, it has been shown that empirical risk minimization is an efficient method for finding locally optimal tensor machines if the optimization algorithm avoids saddle points [29].

As can be seen from the literature summary above, one of the differences between our approach and the above methods is the model used. The type I model (2) has not been examined with the symmetric CPD structure in the weight tensors, to the best of our knowledge. Another difference of our approach from the above methods is the algorithm used. While first-order algorithms are used in most of these approaches, we utilize the second-order batch Gauss–Newton (GN) algorithm. Although first-order methods have the advantage of lower per-iteration complexity, second-order GN algorithms generally require fewer iterations to converge and fewer hyperparameters to be optimized. Moreover, the GN algorithm using trust-region is more robust in the sense that it converges to a (local) minimum for any starting point under mild conditions and it is less prone to swamps (many iterations with little to no improvement) [5, 6, 33].

We summarize our contributions as follows:

- We develop a TeMPO framework that is able to solve many nonlinear problems with ubiquitous applications in signal processing, machine learning and artificial intelligence. Moreover, we develop an efficient second-order Gauss–Newton algorithm for optimizing multivariate polynomials in the CPD format.
- We determine the conditions where the tensorized linear model (4) with polynomial features and the multivariate polynomial model (2) coincide when the symmetric CPD format is used in their representations.
- We show that TeMPO achieves similar or better accuracy than various methods such as multilayer perceptrons (MLPs), tensor networks with different architectures including tensor trains (TT), tree tensor networks, and projected entangled pair states (PEPS). We also show that TeMPO requires the optimization for fewer parameters and less memory than these methods for the classification of the MNIST and Fashion MNIST datasets.
- Last but not least, our framework can be interpreted as an advancement of higher-order factorization machines; we introduce an efficient second-order Gauss–Newton algorithm for higher-order factorization machines.

The remaining part of this article is organized as follows. In Section 2, we describe notation and background information concerning tensors. In Section 3, we describe the TeMPO framework in a more detailed manner. Section 3 also covers the details of representation of polynomials by symmetric CPD structured tensors. In Section 3, we also show how to exploit the symmetric CPD structure to obtain efficient expressions for the gradient and Jacobian-vector products which are necessary for the Gauss–Newton algorithm. The formulation of the image classification problem in the context of TeMPO, numerical experiments and related discussions will be covered in Section 4. We conclude our paper with future remarks in the last section.

2. PRELIMINARIES

2.1. Notation

A tensor is a higher-order generalization of a vector (first-order) and a matrix (second-order). Following established conventions, we denote scalars, vectors, matrices, and tensors by a, $\mathbf{a}$, $\mathbf{A}$, and $\mathcal{A}$, respectively. The transpose of a matrix $\mathbf{A}$ is denoted as $\mathbf{A}^{\mathrm{T}}$. The ith column vector of a matrix $\mathbf{A}$ is denoted as $\mathbf{a}_i$, i.e., $\mathbf{A} = (\mathbf{a}_1, \mathbf{a}_2 \ldots)$. The entry with row index i and column index j in a matrix $\mathbf{A}$, i.e., $(\mathbf{A})_{ij}$, is denoted by a_{ij}. Similarly, $(\mathcal{A})_{i_1 i_2 \ldots i_N}$ is denoted by $a_{i_1 i_2 \ldots i_N}$. Diag($\mathbf{a}$) denotes the diagonal matrix whose entries are composed from the vector $\mathbf{a}$. On the other hand, diag($\mathbf{A}$) denotes a vector composed from the diagonal elements of $\mathbf{A}$. The vectorization operator vec($\mathbf{A}$) for $\mathbf{A} \in \mathbb{K}^{I \times J}$ stacks all the columns of $\mathbf{A}$ into a column vector $\mathbf{a} \in \mathbb{K}^{IJ}$. The reverse operation unvec($\mathbf{a}$) reshapes a vector $\mathbf{a}$ into a matrix $\mathbf{A} \in \mathbb{K}^{I \times J}$. The identity matrix of size $(K \times K)$ is denoted by $\mathbf{I}_K$. A vector of length K with all entries equal to 1 is denoted by $\mathbf{1}_K$. The l_2 norm of a vector $\mathbf{a}$ is denoted by $\|\mathbf{a}\|_2$. The row-wise and column-wise concatenation of two vectors $\mathbf{a}$ and $\mathbf{b}$ is denoted by $[\mathbf{a}, \mathbf{b}]$ and $[\mathbf{a}; \mathbf{b}]$, respectively. The outer product, Kronecker product, Khatri–Rao product, and Hadamard product are denoted by $\circledast$, $\otimes$, $\odot$, and $*$, respectively. The nth power of a vector $\mathbf{x}$ with respect to Kronecker product is defined as $\mathbf{x}^{\otimes n} = \mathbf{x} \otimes \mathbf{x}^{\otimes(n-1)}$, with $\mathbf{x}^{\otimes 0} = 1$. Similarly, $\mathbf{x}^{\odot n}$ and $\mathbf{x}^{* n}$ denotes the nth power of vector $\mathbf{x}$ with respect to Khatri–Rao product and Hadamard product, respectively. The mode-n product of a tensor $\mathcal{A} \in \mathbb{K}^{I_1 \times I_2 \times \ldots \times I_N}$ (with $\mathbb{K}$ meaning either $\mathcal{R}$ or $\mathbb{C}$) and a vector $\mathbf{x} \in \mathbb{K}^{I_n}$, denoted by $\mathcal{A} \bullet_n \mathbf{x}^{\mathrm{T}}$, is defined element-wise as $\left(\mathcal{A} \bullet_n \mathbf{x}^{\mathrm{T}}\right)_{i_1 i_2 \cdots i_{n-1} i_{n+1} \cdots i_N} = \sum_{i_n=1}^{I_N} a_{i_1 i_2 \cdots i_n \cdots i_N} x_{i_n}$. The mode-$n$ product of a tensor $\mathcal{A} \in \mathbb{K}^{I \times I \times \ldots \times I}$ of order k and a vector $\mathbf{x} \in \mathbb{K}^{I}$ for all modes is defined as

$$\mathcal{A}\mathbf{x}^k \stackrel{\text{def}}{=} \mathcal{A} \bullet_1 \mathbf{x}^{\mathrm{T}} \bullet_2 \mathbf{x}^{\mathrm{T}} \ldots \bullet_k \mathbf{x}^{\mathrm{T}}.$$

A mode-n vector or mode-n fiber of a tensor $\mathcal{A} \in \mathbb{K}^{I_1 \times I_2 \times \ldots \times I_N}$ is a vector obtained by fixing every index except the nth. The mode-n matricization of $\mathcal{A}$ is a matrix $\mathbf{A}_{[n;N,N-1,\ldots,n+1,n-1,\ldots,1]}$ collecting all the mode-n vectors as its columns. For example, an entry $a_{i_1 i_2 i_3}$ of a tensor $\mathcal{A} \in \mathbb{K}^{I \times J \times K}$ is mapped to the (i_2, q) entry of the matrix $\mathbf{A}_{[2;3,1]}$ with $q = i_1 + (i_3 - 1)I$. The binomial coefficient is denoted by $C_n^k = \frac{n!}{(n-k)!k!}$. Some useful definitions are listed below.

Definition 1 (Symmetric Tensor). *A tensor $\mathcal{A} \in \mathbb{K}^{I \times I \times \ldots \times I}$ of order k is called symmetric if its entries are invariant under the permutation of its indices.*

As a consequence of this definition, the matrix representations of symmetric tensors in different modes are all equal.

Definition 2 (Rank of a Tensor). *A rank-1 tensor of order N is the outer product of N nonzero vectors. The rank of a tensor is equal to the minimal number of rank-1 terms that yield the tensor as their sum.*

Definition 3 (Kronecker Product). *Given two matrices $\boldsymbol{A} \in \mathbb{K}^{I \times J}$ and $\boldsymbol{B} \in \mathbb{K}^{K \times L}$, their Kronecker product is*

$$\boldsymbol{A} \otimes \boldsymbol{B} = \begin{bmatrix} a_{1,1}\boldsymbol{B} & \cdots & a_{1,J}\boldsymbol{B} \\ \vdots & \ddots & \vdots \\ a_{I,1}\boldsymbol{B} & \cdots & a_{I,J}\boldsymbol{B} \end{bmatrix} \in \mathbb{K}^{IK \times JL}.$$

Definition 4 (Khatri–Rao Product). *Given two matrices $\boldsymbol{A} \in \mathbb{K}^{I \times K}$ and $\boldsymbol{B} \in \mathbb{K}^{J \times K}$ with the same number of columns, their Khatri–Rao product, also known as columnwise Kronecker product, is*

$$\boldsymbol{A} \odot \boldsymbol{B} = [\boldsymbol{a}_1 \otimes \boldsymbol{b}_1, \boldsymbol{a}_2 \otimes \boldsymbol{b}_2, \ldots, \boldsymbol{a}_K \otimes \boldsymbol{b}_K] \in \mathbb{K}^{IJ \times K},$$

where $\boldsymbol{a}_i$ and $\boldsymbol{b}_i$ denote the ith column of the matrices $\boldsymbol{A}$ and $\boldsymbol{B}$, respectively.

Definition 5 (Hadamard Product). *Given two matrices $\boldsymbol{A} \in \mathbb{K}^{I \times J}$ and $\boldsymbol{B} \in \mathbb{K}^{I \times J}$ with the same size, their Hadamard product is the elementwise product, i.e.,*

$$\boldsymbol{A} * \boldsymbol{B} = \begin{bmatrix} a_{1,1}b_{1,1} & \cdots & a_{1,J}b_{1,J} \\ \vdots & \ddots & \vdots \\ a_{I,1}b_{I,1} & \cdots & a_{I,J}b_{I,J} \end{bmatrix} \in \mathbb{K}^{I \times J}.$$

The following properties will be useful for our derivations.

Property 1. *Let $\boldsymbol{A} \in \mathbb{K}^{I \times J}, \boldsymbol{X} \in \mathbb{K}^{J \times K}, \boldsymbol{B} \in \mathbb{K}^{K \times L}$. Then*

$$vec(\boldsymbol{AXB}) = \left(\boldsymbol{B}^T \otimes \boldsymbol{A}\right) vec(\boldsymbol{X}) \in \mathbb{K}^{IL}.$$

Moreover, if $\boldsymbol{X} \in \mathbb{K}^{J \times J}$ is a diagonal matrix and $\boldsymbol{B} \in \mathbb{K}^{J \times L}$, then

$$vec(\boldsymbol{AXB}) = \left(\boldsymbol{B}^T \odot \boldsymbol{A}\right) diag(\boldsymbol{X}) \in \mathbb{K}^{IL}.$$

Property 2. *Let $\boldsymbol{A} \in \mathbb{K}^{I \times J}, \boldsymbol{B} \in \mathbb{K}^{K \times J}, \boldsymbol{C} \in \mathbb{K}^{I \times L}$, and $\boldsymbol{D} \in \mathbb{K}^{K \times L}$. Then*

$$(\boldsymbol{A} \odot \boldsymbol{B})^T (\boldsymbol{C} \odot \boldsymbol{D}) = \left(\boldsymbol{A}^T\boldsymbol{C}\right) * \left(\boldsymbol{B}^T\boldsymbol{D}\right) \in \mathbb{K}^{J \times L}.$$

Property 3. *For matrices $\boldsymbol{A} \in \mathcal{R}^{I \times J}$ and $\boldsymbol{B} \in \mathcal{R}^{J \times K}$, and for the function $f(\boldsymbol{A}, \boldsymbol{B}) = \boldsymbol{AB}$, the following equations hold:*

$$\frac{\partial vec(f(\boldsymbol{A}, \boldsymbol{B}))}{\partial vec(\boldsymbol{A})} = \boldsymbol{B}^T \otimes \boldsymbol{I}_I, \qquad \frac{\partial vec(f(\boldsymbol{A}, \boldsymbol{B}))}{\partial vec(\boldsymbol{B})} = \boldsymbol{I}_K \otimes \boldsymbol{A}.$$

2.2. Canonical Polyadic Decomposition

Here, we will briefly describe the canonical polyadic decomposition. A more detailed description of CPD can be found in [1] and references therein. The CPD writes a tensor $\mathcal{T} \in \mathbb{K}^{I_1 \times I_2 \times \ldots \times I_N}$ as a sum of R rank-1 tensors and is denoted by $\left[\!\left[\mathbf{U}^{(1)}, \ldots, \mathbf{U}^{(N)}\right]\!\right]$, with its factor matrices $\mathbf{U}^{(n)} \in \mathbb{K}^{I_n \times R}$, where R equals the rank of the tensor. This is a shortcut notation for

$$\mathcal{T} = \sum_{r=1}^{R} \mathbf{u}_r^{(1)} \circledast \mathbf{u}_r^{(2)} \circledast \ldots \circledast \mathbf{u}_r^{(N)},$$

where $\mathbf{u}_r^{(n)}$ denotes the rth column of the factor matrix $\mathbf{U}^{(n)}$. CPD is essentially unique under mild conditions [34–37], and has found many applications in signal processing and machine learning [1].

For symmetric tensors, all the factor matrices are equal, i.e.,

$$\mathcal{T} = [\![\mathbf{U}, \mathbf{U} \ldots, \mathbf{U}; \mathbf{c}^{\mathrm{T}}]\!] = \sum_{r=1}^{R} c_r (\mathbf{u}_r \circledast \mathbf{u}_r \circledast \ldots \circledast \mathbf{u}_r) \in \mathbb{K}^{I \times I \times \ldots \times I},$$

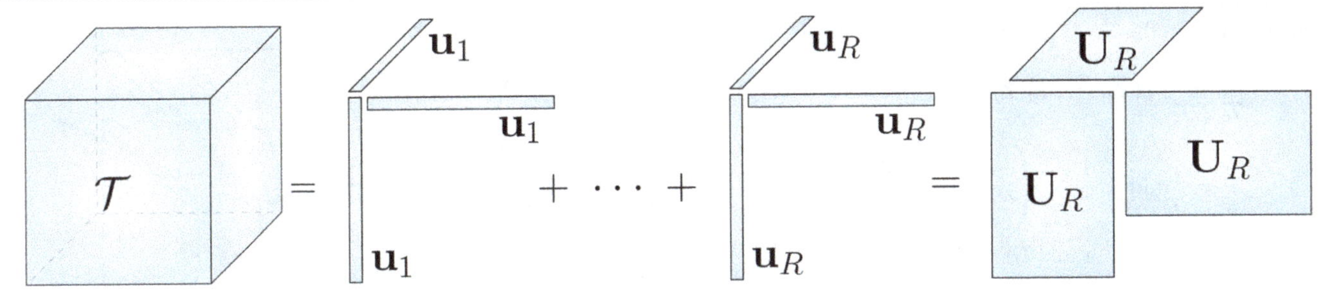

FIGURE 1 | Polyadic decomposition of a third order symmetric tensor $\mathcal{T}$. It is called canonical (CPD) if R is equal to the rank of $\mathcal{T}$, i.e., R is minimal. It allows compact representation of polynomials.

where $\mathbf{U} \in \mathbb{K}^{I\times R}$, and $\mathbf{c} \in \mathbb{K}^{R}$ is a vector of weights which allows us to give minus signs to the factors for even-degree symmetric tensors, see **Figure 1**. The matrix unfolding of a symmetric CPD is given by

$$\mathbf{T} = \mathbf{U}\mathrm{Diag}(\mathbf{c})(\mathbf{U}\odot\mathbf{U}\odot\cdots\odot\mathbf{U})^{\mathrm{T}}.$$

3. TENSOR-BASED MULTIVARIATE POLYNOMIAL OPTIMIZATION

The primary aim of the TeMPO framework is to develop efficient algorithms for modeling nonlinear phenomena commonly encountered in the areas of signal processing, machine learning, and artificial intelligence [15]. To achieve this, we assume structure in the nonlinear function $f:\mathcal{R}^I \to \mathcal{R}^N$ that maps the input data to output data. In our framework, we first assume smoothness in f and approximate it as multivariate polynomial $p:\mathcal{R}^I \to \mathcal{R}^N$. Then, we approximate p with low-rank tensors. This allows us to achieve efficiency both in storing the coefficients of the approximation and in performing computations with those coefficients. Although any continuous function on a compact domain can be approximated by polynomials arbitrarily well according to the Stone–Weierstrass theorem, polynomial approximations used in practice can pose several numerical issues such as the Runge phenomenon. Several strategies have been proposed to overcome these numerical issues, such as using different polynomial bases and Tikhonov regularization [38, 39]. In this work, we will focus more on computational issues of the TeMPO framework; however, it is possible to incorporate these strategies with TeMPO using slight modifications. In the remaining part of this section, we describe the scope of TeMPO. Then we will describe two types of tensor representations of multivariate polynomials where the symmetric CPD structure is imposed on the coefficient tensors. Next we will briefly describe the Gauss–Newton algorithm using the dogleg trust-region method and show how to exploit the symmetric CPD structure in the computation of Jacobian and Jacobian-vector products that are necessary for the Gauss–Newton algorithm.

3.1. Scope of the TeMPO Framework

The TeMPO framework concerns optimization problems with continuous cost functions on compact domains, namely multilinear/polynomial cost functions with or without additional constraints, which is a more general setting than tensor decomposition or retrieval of a tensor factorization. To better describe the scope, let us consider the following class of objective functions:

$$l(\boldsymbol{\theta}, p(\mathbf{z})), \tag{5}$$

where $l:\mathcal{R}\times\mathcal{R}^M \to \mathcal{R}^+$ denotes the performance measure of the model to be optimized, $p:\mathcal{R}^I \to \mathcal{R}$ denotes a multivariate polynomial represented by low-rank tensors, $\mathbf{z} \in \mathcal{R}^I$ denotes input data, and $\boldsymbol{\theta} \in \mathcal{R}^M$ denotes output data. A broad range of objective functions are in the class of (5). For example, the objective function for the estimation of the CPD of a third-order tensor $\mathcal{T}$ can be written as

$$\frac{1}{2}\left\|\boldsymbol{\theta} - p(\mathbf{A},\mathbf{B},\mathbf{C})\right\|_2^2, \quad \text{where} \quad p(\mathbf{A},\mathbf{B},\mathbf{C}) = \mathrm{vec}\left(\llbracket\mathbf{A},\mathbf{B},\mathbf{C}\rrbracket\right).$$

Other tensor decomposition problems, such as block term decomposition (BTD), also fit into TeMPO. The symmetric best rank-1 approximation problem [40], which can also be formulated as

$$\max_{\mathbf{z}\in\mathcal{R}^I} \quad \mathcal{T}\mathbf{z}^d, \quad \text{subject to} \quad \|\mathbf{z}\| = 1, \tag{6}$$

is another example problem that fits into the framework. Note that (6) is expressed as the maximization of an objective function, rather than as the decomposition of a tensor; indeed TeMPO allows one to address more general problems. For the symmetric best rank-1 approximation problem, several approaches such as higher-order power method [40], generalized Rayleigh–Newton iteration and the alternating least squares methods [41], SVD-based algorithms [42], semi-definite relaxations [43] have been proposed. Problems from unsupervised learning such as nonlinear dimensionality reduction, manifold learning, nonlinear blind source separation, and nonlinear independent component analysis also fit into TeMPO. Similarly, problems from supervised learning fit into TeMPO as well. In this work, we will focus on the regression and classification problem and derive expressions for Jacobian and Jacobian-vector products, which are necessary for the Gauss–Newton algorithm. However, the derivations here can be extended to the other problems without much effort.

Given data points $(y_k, \mathbf{z}_k)$, the regression problem can be formulated within the TeMPO framework for the type I model as

$$\begin{aligned} &\min_{\mathcal{T}_0,\ldots,\mathcal{T}_d} \ p(\mathcal{T}_0,\ldots,\mathcal{T}_d,\mathbf{Z}) \\ &= \min_{\mathcal{T}_0,\ldots,\mathcal{T}_d} \ \frac{1}{2}\sum_{k=1}^{K}\left(y_k - \mathcal{T}_0 - \sum_{j=1}^{d}\mathcal{T}_j\mathbf{z}_k^j\right)^2, \\ &\text{subject to} \quad \operatorname{rank}(\mathcal{T}_j) = R_j \end{aligned} \tag{7}$$

where $R_j \in \mathbb{N}^+$ is a small integer, $\mathcal{T}_j \in \mathcal{R}^{I\times I\times\ldots\times I}$ denotes the low-rank structured coefficient tensor of order j to be optimized, $\mathcal{T}_0 \in \mathcal{R}$ denotes a scalar, $\mathbf{Z} \in \mathcal{R}^{I\times K}$ denotes the data matrix, $\mathbf{z}_k$ denotes the kth column of $\mathbf{Z}$ and K is the number of available data points. For the type II model, the regression problem takes the form

$$\begin{aligned} &\min_{\mathcal{T}} \ p(\mathcal{T}, \widetilde{\mathbf{Z}}) = \min_{\mathcal{T}} \ \frac{1}{2}\sum_{k=1}^{K}\left(y_k - \mathcal{T}\widetilde{\mathbf{z}}_i^d\right)^2, \\ &\text{subject to} \quad \operatorname{rank}(\mathcal{T}) = R \end{aligned} \tag{8}$$

where $\mathcal{T}$ denotes the low-rank structured coefficient tensor of order d to be optimized, $\widetilde{\mathbf{Z}} \in \mathcal{R}^{(I+1)\times K}$ denotes the augmented input data matrix, and $\widetilde{\mathbf{z}}_k$ denotes the kth column of $\widetilde{\mathbf{Z}}$, i.e., $\widetilde{\mathbf{z}}_k = [1; \mathbf{z}_k]$.

3.2. Tensor Representation of Polynomials

In this subsection, we examine the type I and type II model in detail. A (symmetric) tensor $\mathcal{T}$ of order d and dimension n can be associated with a homogeneous n-variate polynomial $p(\mathbf{z})$ of degree d [44], as shown in Equation (3).

Type I: Since any polynomial can be written as a sum of homogeneous polynomials of increasing degrees, any polynomial of degree d can be written by using tensors of order up to d, as shown in Equation (2). Note that in the tensor representation of polynomials, any tensor can be assumed to be symmetric without loss of generality. Indeed, any homogeneous polynomial $p(\mathbf{z})$ of degree $d \in \mathbb{N}$ can be represented by a multilinear form $\mathcal{T}\mathbf{z}^d$, where $\mathcal{T} \in \mathbb{K}^{I\times I\times\ldots\times I}$ is a symmetric tensor of order d and $\mathbf{z} \in \mathbb{K}^I$.

To see this, suppose a homogeneous polynomial $p(\mathbf{z})$ is represented as

$$p(\mathbf{z}) = \widetilde{\mathcal{T}}\mathbf{z}^d = \sum_{i_1,i_2,\ldots,i_d=1}^{I} \widetilde{t}_{i_1 i_2\ldots i_d} z_{i_1} z_{i_2}\ldots z_{i_d},$$

where $\widetilde{\mathcal{T}} \in \mathbb{K}^{I\times I\times\ldots\times I}$ is a tensor of order d. Since the terms $z_{i_1} z_{i_2}\ldots z_{i_d}$ are invariant under the permutation of indices, we may write

$$p(\mathbf{x}) = \sum_{i_1,i_2,\ldots,i_d=1}^{I} t_{i_1 i_2\ldots i_d} z_{i_1} z_{i_2}\ldots z_{i_d},$$

$$\text{where} \quad t_{i_1 i_2\ldots i_d} = \frac{1}{d!}\sum_{(i_1,i_2,\ldots,i_d)\in\Pi(i_1 i_2\ldots i_d)} \widetilde{t}_{i_1 i_2\ldots i_d},$$

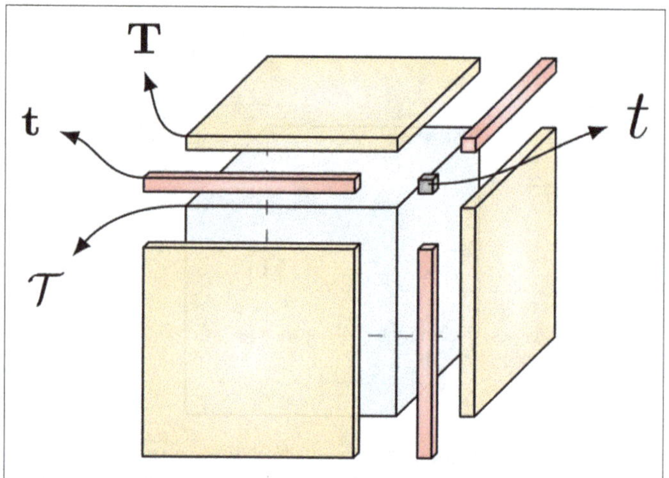

FIGURE 2 | By applying the homogenization process, symmetric tensors can represent the coefficients of non-homogeneous polynomials. For example, by stacking the coefficients t, $\mathbf{t}$, $\mathbf{T}$, and $\mathcal{T}$ of the third degree polynomial into a tensor as shown above, we can represent it with a symmetric third-order tensor. Image reproduced from Debals [46].

here $\Pi(i_1 i_2\ldots i_d)$ denotes the collection of all permutation of indices $(i_1, i_2, \ldots, i_d)$. Since the entries of $\mathcal{T}$ are invariant under the permutation of indices, we can conclude that $\mathcal{T}$ is symmetric.

The above discussion reveals the fact that there are infinitely many representations of a given polynomial. Indeed two representations with tensors $\mathcal{T}$ and $\mathcal{W}$ are equal so long as the summation of the corresponding entries over the permutation of indices remains the same, i.e.,

$$\sum_{(i_1,i_2,\ldots,i_d)\in\Pi(i_1 i_2\ldots i_d)} t_{i_1 i_2\ldots i_d} = \sum_{(i_1,i_2,\ldots,i_d)\in\Pi(i_1 i_2\ldots i_d)} w_{i_1 i_2\ldots i_d}$$

In the ANOVA kernel used in higher-order factorization machines, all $t_{\Pi(i_1 i_2\ldots i_d)}$ are set to zero except $t_{(i_1<i_2<\ldots<i_d)}$ [31], which leads to a sparse representation. In this paper, we use symmetric tensors for two reasons. The first reason is that the CPD of a symmetric tensor can be expressed by a single factor matrix. Therefore, the symmetric CPD representation of multivariate polynomial requires fewer number of parameters in comparison with a non-symmetric representation. The second reason is that there is a rich history of the representation of polynomials with symmetric tensors in the field of algebraic geometry under the name of the Waring problem [45].

Type II: Augmenting the independent variable vector $\mathbf{z}$, by a constant 1[1], i.e., $\widetilde{\mathbf{z}} = [1; \mathbf{z}^{\mathrm{T}}]$ leads to a different representation of non-homogeneous polynomials that uses a single dth order symmetric tensor for the inhomogeneous multivariate polynomial of degree d, as shown in Equation (3). This process is called homogenization [46] and is graphically illustrated in **Figure 2**. If we just use full tensors, the type I and II

[1]Since the weight vector $\mathbf{c}$ is used in the parametrization of tensors, different choices of constant in $\widetilde{\mathbf{z}}$ lead to mathematically equivalent cost functions in the optimization problems. On the other hand, the choice of the constant may imply numerical differences—in situations of this type, one should generally choose a constant that "makes sense for the application."

models are interchangeable. However, it is important to note that *when low-rank structure is imposed on the coefficient tensors,* both representations yield different classes of low-rank multivariate polynomial. Hence, these approaches may lead to different results depending on the application. The former approach requires more parameters since it uses more factor matrices. The difference in the number of parameters should be taken into account to prevent underfitting and overfitting. A more detailed description for storage complexity is given in Section 3.5. Moreover, the type I model allows us to constrain each term in the representation separately. In modeling multivariate polynomials, one might not wish the terms of different order to have some shared structure, in which case one should choose the type I model to work with. Similarly, the type II model should be chosen, if some shared structure is desired in the terms of different order. To further elaborate on the effects of homogenization on the rank of a tensor, let us consider the following proposition.

Proposition 1. *Let $p(\mathbf{z}) : \mathcal{R}^I \to \mathcal{R}$ be a multivariate polynomial of order d defined as in equation (2) by a scalar $\mathcal{T}_0$ and symmetric tensors $\mathcal{T}_j \in \mathcal{R}^{I\times I\times\ldots\times I}$ for $j = 1, 2, \ldots, d$. Moreover, let $\mathcal{W} \in \mathcal{R}^{(I+1)\times\ldots\times(I+1)}$ be the corresponding tensor obtained from the homogenization process. The tensors $\mathcal{W}$ and $\mathcal{T}_j$ have the same rank R if and only if the tensors $\mathcal{T}_j$ admit unique CPDs with shared factor matrices and a weight vector $\mathbf{c}$, i.e.,*

$$\mathcal{T}_j = \left[\!\left[\mathbf{U}, \ldots, \mathbf{U}; C_d^j (\mathbf{c}^T)^{\odot(d-j)}\right]\!\right], \quad and \quad \mathcal{T}_0 = \sum_{r=1}^{R} \left((\mathbf{c}^T)^{\odot d}\right)_r .$$

Proof 1. *Let the CPD of the tensor $\mathcal{W}$ be defined as $[\![\mathbf{V}, \ldots, \mathbf{V}]\!]$, where, for convenience but without loss of generality, the weights of the rank-1 terms are assumed to be 1. Since $\mathcal{W}$ is obtained by the homogenization process, partitioning $\mathbf{V}$ as $[\mathbf{v}^T; \mathbf{Q}]$ and using the definition of CPD, we obtain*

$$\mathcal{T}_j = \left[\!\left[\mathbf{Q}, \ldots, \mathbf{Q}; C_d^j (\mathbf{v}^T)^{\odot(d-j)}\right]\!\right], \quad and \quad \mathcal{T}_0 = \sum_{r=1}^{R} \left((\mathbf{v}^T)^{\odot d}\right)_r . \tag{9}$$

Since the CPDs of the tensors $\mathcal{T}_j$ are unique, the equality (9) holds if and only if the equalities $\mathbf{Q} = \mathbf{U}$ and $\mathbf{v} = \mathbf{c}$ also hold.

Remark 1. *In the above proof, we assumed that the vector $\mathbf{v}$ does not contain any zero elements. Note that if the vector $\mathbf{v}$ does contain zero elements, it cancels the corresponding rank-1 terms. Therefore, in that case rank($\mathcal{W}$) > rank($\mathcal{T}_j$), for $j = 1, \ldots, d-1$. Moreover, the uniqueness of the CPDs of $\mathcal{T}_j$ implies that rank($\mathcal{W}$) $\geq$ rank($\mathcal{T}_j$). Since the equality rank($\mathcal{W}$) = rank($\mathcal{T}_j$) holds only when the tensors $\mathcal{T}_j$ have shared factor matrices as described above, we can conclude that in all other cases rank($\mathcal{W}$) > rank($\mathcal{T}_j$).*

Proposition 1 together with Remark 1 reveals the fact that if $\mathcal{W}$ admits a rank-R CPD, there exists tensors $\mathcal{T}_j$ that admit rank-R_j CPDs with shared factors and $R_j \leq R$. Hence, the expressive power of the type II model is weaker than the type I model, i.e., the type II model requires higher rank values than the type I model to be able to model functions of the same complexity. In other words, the set of polynomials represented by the type II model is a strict subset of the set of polynomials represented by the type I model for the same rank values.

Although we focus in this study on the type I and type II models in the symmetric CPD format, the TeMPO framework is not limited to these. TeMPO collects low-rank tensor representations of multivariate polynomials under a roof by utilizing various other tensor decompositions such as TT, HT, and non-symmetric and partially symmetric CPD formats[2]. In this way, TeMPO breaks the curse of dimensionality and makes it possible to develop second-order efficient algorithms for the optimization of a more general class of multivariate polynomials. Moreover, use of structured tensors and multilinear algebra makes it easy to incorporate other polynomial bases and, more generally, other nonlinear feature maps rather than the standard polynomial bases to the TeMPO framework. From this point of view, TeMPO can be interpreted as a generalization of higher-order factorization machines that use particular types of multivariate polynomials with the standard polynomial bases and utilize first-order and BFGS type algorithms [30–32, 47].

3.3. Gauss–Newton Algorithm

Most standard first-order and second-order numerical optimization algorithms can be used for solving problem (8). Since the objective function under consideration is a least-squares function, we will utilize the second-order batch Gauss–Newton (GN) algorithm using a trust-region to take advantage of its attractive properties such as quadratic convergence near a local optimum point, resistant to swamps, suitable to incorporate constraints easily and eligible to exploit multilinear structure. In the case the objective function is not least squares, the inexact GN algorithm can also be utilized. Below, we briefly describe the GN algorithm using a trust-region, and then derive the expressions for Jacobian and Jacobian-vector products for tensors in the symmetric CPD format. In nonlinear least-squares problems, the objective function is the squared error between a data vector $\mathbf{y}$ and a nonlinear model $m(\mathbf{z})$ [6, 33]:

$$f(\mathbf{z}) = \frac{1}{2} \left\| m(\mathbf{z}) - \mathbf{y} \right\|_2^2 = \frac{1}{2} \mathbf{r}^T \mathbf{r}, \tag{10}$$

where $\mathbf{z} \in \mathcal{R}^I$. The algorithm updates the initial guess iteratively by taking a step length α_k in the direction $\mathbf{p}_k$ at the iteration k, i.e.,

$$\mathbf{z}_k = \mathbf{z}_{k-1} + \alpha_k \mathbf{p}_k,$$

until some stopping criteria are satisfied. Line search and trust-region are the two main approaches used to determine α_k and $\mathbf{p}_k$. Here, we focus on the dogleg trust-region approach. In this approach, one sets $\alpha_k = 1$. Then, given a trust-region of radius δ_k, the GN step $\mathbf{p}_k^{gn}$ and the steepest descent step $\mathbf{p}_k^{sd}$ for the current iteration, the step direction $\mathbf{p}_k$ is determined by the following procedure:

[2]Note that the non-symmetric and partially symmetric CPD formats are fairly straightforward variants of the symmetric CPD format, and derivations presented in Section 3.4 can be generalized to these formats with slight modifications.

Algorithm 1: GN algorithm using dogleg trust-region for the type II model.

Input : $\mathbf{Z}$ – Input data matrix
$\mathbf{y}$ – Vector of values (labels in the classification case) for each data point in $\mathbf{Z}$
$\mathbf{U}, \mathbf{c}$ – Initial factor matrix and weight vector
$\mathcal{T}_0$ – Initial scalar
Output: $\mathbf{U}, \mathbf{c}$ – Optimized factor matrix and weight vector
$\mathcal{T}_0$ – Optimized scalar

while *not converged* **do**
  $\mathbf{r}_k \leftarrow$ Compute residual vector using equations (21) and (22)
  $\mathbf{g}_k \leftarrow$ Compute gradient using equation (27)
  $\mathbf{p}_k \leftarrow$ Solve linear systems of equations (12) using CG method
  $\mathbf{U}, \mathbf{c}, \mathcal{T}_0 \leftarrow$ Update via dogleg trust-region explained in Subsection 3.3
end

- If $\|\mathbf{p}_k^{gn}\|_2 \leq \delta_k$, then $\mathbf{p}_k = \mathbf{p}_k^{gn}$.
- If $\|\mathbf{p}_k^{gn}\|_2 > \delta_k$ and $\left\|\mathbf{p}_k^{sd}\right\|_2 > \delta_k$, then $\mathbf{p}_k = \delta_k \mathbf{p}_k^{sd} / \left\|\mathbf{p}_k^{sd}\right\|_2$.
- If $\|\mathbf{p}_k^{gn}\|_2 > \delta_k$ and $\left\|\mathbf{p}_k^{sd}\right\|_2 \leq \delta_k$, then $\mathbf{p}_k = \tau_k \mathbf{p}_k^{sd} + \beta_k(\mathbf{p}_k^{gn} - \tau_k \mathbf{p}_k^{sd})$, where $\tau_k = -\left\|\mathbf{p}_k^{sd}\right\|_2^2 / \left\|\mathbf{J}_k \mathbf{p}_k^{sd}\right\|_2^2$, and β_k is selected such that $\|\mathbf{p}_k\|_2 = \delta_k$.

The steepest descent step $\mathbf{p}_k^{sd}$ is given by $-\mathbf{J}_k^{\mathrm{T}}\mathbf{r}_k$. To compute the GN step, a second order Taylor series approximation for the objective function is used. The optimal direction for the GN step $\mathbf{p}_k^{gn}$ can be obtained by solving the optimization problem,

$$\mathbf{p}_k^{gn} = \arg\min_{\mathbf{p}} \widetilde{f}(\mathbf{p}), \quad \text{with} \quad \widetilde{f}(\mathbf{p}) = f(\mathbf{z}_k) + \mathbf{p}^{\mathrm{T}}\mathbf{g}_k + \frac{1}{2}\mathbf{p}^{\mathrm{T}}\mathbf{H}_k\mathbf{p}, \tag{11}$$

where $\mathbf{g}_k$ denotes the gradient and $\mathbf{H}_k$ denotes the Hessian at the current iteration. Setting $\partial\widetilde{f}(\mathbf{p})/\partial\mathbf{p}$ to zero, the solution of (11) can be obtained by solving the linear system of equations

$$\mathbf{H}_k\mathbf{p}_k^{gn} = -\mathbf{g}_k, \quad \text{with} \quad \mathbf{g}_k = \mathbf{J}_k^{\mathrm{T}}\mathbf{r}_k, \tag{12}$$

where $\mathbf{J}_k$ denotes the Jacobian of $f(\mathbf{z}_k)$ at iteration k, and $\mathbf{r}_k = m(\mathbf{z}_k) - \mathbf{y}$. However, in real-life applications, explicit computation of the Hessian is often expensive. To overcome this, GN approximates the Hessian by the Grammian matrix as

$$\mathbf{H}_k \approx \mathbf{J}_k^{\mathrm{T}}\mathbf{J}_k.$$

In this study, we used the conjugate gradient (CG) algorithm for solving (12) together with the dogleg trust-region approach which is implemented in Tensorlab [11]. The overall algorithm is summarized in **Algorithm 1**.

3.4. Exploiting the Symmetric CPD Format

As described above, the GN algorithm minimizes a cost function in the form of Equation (10). The gradient of this objective function can be written as $\mathbf{J}^{\mathrm{T}}\mathbf{r}$, and the Hessian is approximated by $\mathbf{J}^{\mathrm{T}}\mathbf{J}$, where $\mathbf{J}$ is the Jacobian matrix composed of partial derivatives of the residual vector $\mathbf{r}$. Hence, it is sufficient to derive expressions for the Jacobian and Jacobian-vector products. We begin with the first-order derivatives of the multilinear form $\mathcal{T}\mathbf{z}^d$, where $\mathcal{T}$ is in the symmetric CPD format, with respect to its factors and then proceed to the derivation of Jacobian and Jacobian-vector products for problems (7) and (8). The derivations made here can be used in other TeMPO problems with slight modifications.

3.4.1. Derivatives of the Multilinear Form in the Symmetric CPD Format

By using the matrix unfolding of the tensor in the symmetric CPD format and Property 2 of Khatri–Rao product, the multilinear form $\mathcal{T}\mathbf{z}^d$ can be written as

$$\mathcal{T}\mathbf{z}^d = \mathbf{c}^{\mathrm{T}}\left(\mathbf{U}^{\mathrm{T}}\mathbf{z}\right)^{*d}, \tag{13}$$

which will be useful for our derivations below.

Lemma 1. *Let $\mathcal{T} \in \mathbb{K}^{I\times I\times\ldots\times I}$ be a symmetric tensor of order d and its CPD given as $\mathcal{T} = [\![U, \ldots, U; c^T]\!]$. Then the derivative of the multilinear form $\mathcal{T}z^d$ with respect to vec(U) can be obtained as*

$$\frac{\partial \mathcal{T}z^d}{\partial vec(U)} = \left((c * w) \otimes z\right)^T,$$

where $w = d\left(U^T z\right)^{*(d-1)}$.

Proof 2. *The proof immediately follows from Equation (13) and successive application of Property 3.*

Lemma 2. *Let $\mathcal{T} \in \mathbb{K}^{I\times I\times\ldots\times I}$ be symmetric tensor of order d and its CPD is given as $\mathcal{T} = [\![U, \ldots, U; c^T]\!]$. Then the derivative of multilinear form $\mathcal{T}z^d$ with respect to vector* **c** *can be obtained as*

$$\frac{\partial \mathcal{T}z^d}{\partial c} = \left(z^T U\right)^{*d}.$$

Proof 3. *The proof immediately follows from Property 3 and Equation (13).*

3.4.2. Exploiting Structure in the Type I Model

Objective Function: The construction of the residual vector $\mathbf{r}$ and the computation of its l_2 norm is sufficient for computing the objective function in (7). By utilizing Property 2 and Equation (13), the residual vector can be expressed as $\mathbf{r} = \mathbf{y} - \boldsymbol{\mu}$, where each entry of the vector $\boldsymbol{\mu} \in \mathcal{R}^K$ is defined as

$$\mu_k = \mathcal{T}_0 + \sum_{j=1}^{d} \mathbf{c}_j^{\mathrm{T}}\mathbf{w}_{j,k}^{*j},$$

in which $\mathbf{w}_{j,k} = \mathbf{U}_j^{\mathrm{T}}\mathbf{z}_k$ with $\mathbf{U}_j \in \mathcal{R}^{I\times R}$, and $\mathbf{w}_{j,k}^{*j}$ denotes the jth elementwise power of the vector $\mathbf{w}_{j,k}$. By defining $\mathbf{W}_j =$

$[\mathbf{w}_{j,1}, \mathbf{w}_{j,2}, \dots, \mathbf{w}_{j,K}]$, we can write the residual vector $\mathbf{r}$ in a compact form as

$$\mathbf{r} = \mathbf{y} - \mathcal{T}_0 \cdot \mathbf{1}_K - \sum_{j=1}^{d} \left(\mathbf{c}_j^{\mathrm{T}} \left(\mathbf{W}_j^{*j} \right) \right)^{\mathrm{T}}, \tag{14}$$

Using the above Equation (14), the objective function can be computed as the l_2 norm of the residual vector $\mathbf{r}$.

Jacobian: The Jacobian matrix for problem (7), with the tensors $\mathcal{T}_j$ in their symmetric CPD format, can be written in a compact form as

$$\mathbf{J} = [\mathbf{J}_1; \dots; \mathbf{J}_K], \quad \text{where}$$
$$\mathbf{J}_k = \left[1, \frac{\partial \mathbf{r}_k}{\partial \mathrm{vec}(\mathbf{U}_1)}, \dots, \frac{\partial \mathbf{r}_k}{\partial \mathrm{vec}(\mathbf{U}_d)}, \frac{\partial \mathbf{r}_k}{\partial \mathbf{c}_1}, \dots, \frac{\partial \mathbf{r}_k}{\partial \mathbf{c}_d} \right]. \tag{15}$$

Note that we used the fact $\partial \mathbf{r}_k / \partial \mathcal{T}_0 = 1$ in the above equation. By utilizing Lemma 1 and Lemma 2, the derivative of each term of the residual vector with respect to $\mathbf{U}_j$ and $\mathbf{c}_j$ can be expressed as

$$\frac{\partial \mathbf{r}_k}{\partial \mathrm{vec}(\mathbf{U}_j)} = -j \left[\left(\mathbf{c}_j * \mathbf{w}_{j,k}^{*(j-1)} \right) \otimes \mathbf{z}_k \right]^{\mathrm{T}}, \quad \text{and} \quad \frac{\partial \mathbf{r}_k}{\partial \mathbf{c}_j} = \left(\mathbf{w}_{j,k}^{*j} \right)^{\mathrm{T}}. \tag{16}$$

By defining $\widetilde{\mathbf{W}}_j = -j \left(\mathbf{W}_j^{*(j-1)} \right)$ for $j = 1, \dots, d$, and $\mathbf{Z} = [\mathbf{z}_1, \mathbf{z}_2, \dots, \mathbf{z}_K]$, the Jacobian matrix $\mathbf{J}$ in (15) can be obtained in the following compact block form:

$$\mathbf{J} = \left[\mathbf{1}_K, ((\mathbf{C}_1 \widetilde{\mathbf{W}}_1) \odot \mathbf{Z})^{\mathrm{T}}, \dots, ((\mathbf{C}_d \widetilde{\mathbf{W}}_d) \odot \mathbf{Z})^{\mathrm{T}}, \mathbf{V} \right], \tag{17}$$

where $\mathbf{V}$ is a $K \times d$ block matrix in which each block is defined as $\mathbf{V}_{k,j} = \left(\mathbf{w}_{j,k}^{*j} \right)^{\mathrm{T}}$, $\mathbf{C}_j = \mathrm{Diag}(\mathbf{c}_j)$, and d is the degree of the polynomial under consideration. Since we only need the Jacobian-vector products for the GN algorithm, the explicit construction of the Jacobian matrix is not required. The Jacobian-vector products can be obtained in a more memory-efficient way as described below.

Jacobian-Vector Product: The product of Jacobian $\mathbf{J}$ by a vector $\mathbf{x}$ can be obtained using block matrix operations. The product of each block term by a vector $\mathrm{vec}(\mathbf{X}_j) = \mathbf{x}_j$ can be obtained by utilizing properties 1 and 2 as

$$((\mathbf{C}_j \widetilde{\mathbf{W}}_j) \odot \mathbf{Z})^{\mathrm{T}} \mathbf{x}_j = \left[\left(\mathbf{X}_j^{\mathrm{T}} \mathbf{Z} \right) * (\mathbf{C}_j \widetilde{\mathbf{W}}_j) \right]^{\mathrm{T}} \mathbf{1}_R. \tag{18}$$

Note that the multiplication of a matrix by $\mathbf{1}_R$ from the right is equivalent to summing the columns of the matrix under consideration. Therefore, neither the multiplication by $\mathbf{1}_R$ nor the transposition of the matrix $\left(\mathbf{X}_j^{\mathrm{T}} \mathbf{Z} \right) * (\mathbf{C}_j \widetilde{\mathbf{W}}_j)$ in Equation (18) is necessary to obtain the Jacobian-vector product. Note also that, since the matrices $\mathbf{C}_j$ are diagonal, the product $\mathbf{C}_j \widetilde{\mathbf{W}}_j$ can be obtained in a memory efficient way by multiplying the rows of $\widetilde{\mathbf{W}}_j$ by the corresponding diagonal elements of $\mathbf{C}_j$ without explicitly forming the matrices $\mathbf{C}_j$. Overall, the product of the Jacobian $\mathbf{J}$ and the vector $\mathbf{x}$ can be obtained by partitioning the vector $\mathbf{x}$, i.e., $\mathbf{x} = [x_1; \mathbf{x}_1; \mathbf{x}_2; \dots; \mathbf{x}_d; \mathbf{x}_v]$, and by using the Equations (17) and (18) as

$$\mathbf{J}\mathbf{x} = x_1 \cdot \mathbf{1}_K + \sum_{j=1}^{d} \left[\left(\mathbf{X}_j^{\mathrm{T}} \mathbf{Z} \right) * (\mathbf{C}_j \widetilde{\mathbf{W}}_j) \right]^{\mathrm{T}} \mathbf{1}_R + \mathbf{V}\mathbf{x}_v,$$

where $\mathbf{X}_j = \mathrm{unvec}(\mathbf{x}_j)$.

Jacobian Transpose -Vector Product and Gradient: In a similar way, block-wise multiplication of the Jacobian transpose $\mathbf{J}^{\mathrm{T}}$ by a vector can be obtained from the expression

$$((\mathbf{C}_j \widetilde{\mathbf{W}}_j) \odot \mathbf{Z})\mathbf{x} = \mathrm{vec} \left(\mathbf{Z} \mathrm{Diag}(\mathbf{x}) (\mathbf{C}_j \widetilde{\mathbf{W}}_j)^{\mathrm{T}} \right). \tag{19}$$

Note that right multiplication by a diagonal matrix can be done efficiently by only multiplying the columns of the matrix with the corresponding diagonal elements without explicitly forming the diagonal matrix. Overall, by defining $\boldsymbol{\xi}_j = \mathrm{vec} \left(\mathbf{Z} \mathrm{Diag}(\mathbf{x}) (\mathbf{C}_j \widetilde{\mathbf{W}}_j)^{\mathrm{T}} \right)$, we can obtain the product of the Jacobian transpose $\mathbf{J}^{\mathrm{T}}$ and a vector $\mathbf{x}$ in the following form:

$$\mathbf{J}^{\mathrm{T}} \mathbf{x} = \left[\sum_{k=1}^{K} x_k; \boldsymbol{\xi}_1; \boldsymbol{\xi}_2; \dots; \boldsymbol{\xi}_d; \mathbf{V}^{\mathrm{T}} \mathbf{x} \right]. \tag{20}$$

The gradient can be obtained by the product of the Jacobian transpose $\mathbf{J}^{\mathrm{T}}$ and the residual vector $\mathbf{r}$. Defining $\boldsymbol{\eta}_j = \mathrm{vec} \left(\mathbf{Z} \mathrm{Diag}(\mathbf{r}) (\mathbf{C}_j \widetilde{\mathbf{W}}_j)^{\mathrm{T}} \right)$ and utilizing the Equations (19) and (20), we can obtain the gradient as

$$\mathbf{g} = \left[\sum_{k=1}^{K} r_k; \boldsymbol{\eta}_1; \boldsymbol{\eta}_2; \dots; \boldsymbol{\eta}_d; \mathbf{V}^{\mathrm{T}} \mathbf{r} \right].$$

3.4.3. Exploiting Structure in the Type II Model

Objective Function: The computation of the objective function for the type II model is similar to that of the type I model. Utilizing Property 2 and Equation (13), the residual vector for problem (8) can be obtained as $\mathbf{r} = \mathbf{y} - \boldsymbol{\mu}$ with

$$\boldsymbol{\mu} = \left[\mathbf{c}^{\mathrm{T}} \mathbf{w}_1^{*d}; \ \mathbf{c}^{\mathrm{T}} \mathbf{w}_2^{*d}; \ \dots; \ \mathbf{c}^{\mathrm{T}} \mathbf{w}_K^{*d} \right], \tag{21}$$

where $\mathbf{w}_k = \mathbf{U}^{\mathrm{T}} \mathbf{z}_k$. By defining $\mathbf{W} = [\mathbf{w}_1, \mathbf{w}_2, \dots, \mathbf{w}_K]$, we can write the residual vector $\mathbf{r}$ in a compact form as

$$\mathbf{r} = \mathbf{y} - \left(\mathbf{c}^{\mathrm{T}} \left(\mathbf{W}^{*d} \right) \right)^{\mathrm{T}}, \tag{22}$$

Using the above Equation (22), the objective function can be computed as the l_2 norm of the residual vector $\mathbf{r}$.

Jacobian: The Jacobian matrix of the cost function in (8) can be defined in a compact form as

$$\mathbf{J} = [\mathbf{J}_1; \mathbf{J}_2; \dots; \mathbf{J}_K], \quad \text{where} \quad \mathbf{J}_k = \left[\frac{\partial \mathbf{r}_k}{\partial \mathrm{vec}(\mathbf{U})}, \frac{\partial \mathbf{r}_k}{\partial \mathbf{c}} \right]. \tag{23}$$

Utilizing Lemma 1 and Lemma 2 and using the equations in (16), the parts of $\mathbf{J}_k$ in Equation (23) can be written as

$$\frac{\partial \mathbf{r}_k}{\partial \text{vec}(\mathbf{U})} = -d\Big[\left(\mathbf{c} * \mathbf{w}_k^{*(d-1)}\right) \otimes \widetilde{\mathbf{z}}_k\Big]^{\mathrm{T}}, \quad \frac{\partial \mathbf{r}_k}{\partial \mathbf{c}} = \left(\mathbf{w}_k^{*d}\right)^{\mathrm{T}}.$$

By defining $\widetilde{\mathbf{W}} = -d\left(\mathbf{W}^{*(d-1)}\right)$, $\mathbf{V} = \left[\frac{\partial \mathbf{r}_1}{\partial \mathbf{c}}; \frac{\partial \mathbf{r}_2}{\partial \mathbf{c}}; \ldots; \frac{\partial \mathbf{r}_K}{\partial \mathbf{c}}\right]$, and $\mathbf{Z} = [\widetilde{\mathbf{z}}_1, \widetilde{\mathbf{z}}_2, \ldots, \widetilde{\mathbf{z}}_K]$, the Jacobian matrix can be obtained in the following compact form:

$$\mathbf{J} = \left[((\mathbf{C}\widetilde{\mathbf{W}}) \odot \mathbf{Z})^{\mathrm{T}}, \mathbf{V}\right]. \tag{24}$$

As mentioned earlier, explicit construction of the Jacobian matrix $\mathbf{J}$ is not required. We only require the Jacobian-vector and Jacobian transpose-vector products and derive efficient expressions for these products below.

Jacobian-Vector Product: The product of the Jacobian matrix $\mathbf{J}$ and a vector $\mathbf{x}$ can be obtained in a similar way as for the type I model, by partitioning the vector $\mathbf{x}$, i.e., $\mathbf{x} = [\mathbf{x}_u; \mathbf{x}_c]$ and utilizing properties 1 and 2 and Equation (24), as

$$\mathbf{J}\mathbf{x} = \left[(\mathbf{X}_u^{\mathrm{T}}\mathbf{Z}) * (\mathbf{C}\widetilde{\mathbf{W}})\right]^{\mathrm{T}} \mathbf{1}_R + \mathbf{V}\mathbf{x}_c, \tag{25}$$

where $\mathbf{X}_u = \text{unvec}(\mathbf{x}_u)$. As mentioned earlier for Equation (18), explicit construction of the diagonal matrix $\mathbf{C}$ is not required. The product $\mathbf{C}\widetilde{\mathbf{W}}$ can be obtained in a memory efficient way by multiplying the rows of $\widetilde{\mathbf{W}}$ by the corresponding diagonal elements of $\mathbf{C}$.

Jacobian Transpose -Vector Product and Gradient: In similar way, utilizing properties 1 and 2 and Equation (24), the product of Jacobian transpose $\mathbf{J}^{\mathrm{T}}$ and a vector $\mathbf{x}$ can be written as

$$\mathbf{J}^{\mathrm{T}}\mathbf{x} = \left[\text{vec}\left(\mathbf{Z}\text{Diag}(\mathbf{x})(\mathbf{C}\widetilde{\mathbf{W}})^{\mathrm{T}}\right); \mathbf{V}^{\mathrm{T}}\mathbf{x}\right]. \tag{26}$$

Since the gradient is the product of the Jacobian transpose $\mathbf{J}^{\mathrm{T}}$ and the residual vector $\mathbf{r}$, it directly follows from the above Equation (26) as

$$\mathbf{g} = \left[\text{vec}\left(\mathbf{Z}\text{Diag}(\mathbf{r})(\mathbf{C}\widetilde{\mathbf{W}})^{\mathrm{T}}\right); \mathbf{V}^{\mathrm{T}}\mathbf{r}\right]. \tag{27}$$

3.5. Complexity Analysis

We now analyze the storage and computational complexity of TeMPO where we are optimizing over symmetric rank-R CPD structured tensors $\mathcal{T} \in \mathbb{K}^{I \times I \times \ldots \times I}$ of order d. The analysis is presented here for the type II model. However, since the number of optimization parameters of the type I and type II models (see Equations 2, 3) for an I-variate polynomial of degree d are proportional to each other, the analysis also applies to the type I model. Indeed, the computational complexity of the type I model is d times the computational complexity of the type II model. We also compare with the storage and computational complexity of TT and PEPS tensor networks.

Representing a multivariate polynomial with I independent variables and of degree d in dense format requires storing $C^d_{(I+d)}$ elements. Using Stirling's approximation, it can be shown that the storage complexity for a multivariate polynomial represented in dense format is $\mathcal{O}\left(I^d\right)$ for $d \ll I$. In the symmetric CPD format, we need to store only the factor matrix $\mathbf{U} \in \mathcal{R}^{I \times R}$ and the vector of weights $\mathbf{c} \in \mathcal{R}^R$, where R is the rank of the symmetric CPD. Therefore, the storage complexity for the type II model using the symmetric CPD format is $\mathcal{O}(IR)$. This shows that the symmetric CPD format breaks the curse of dimensionality, since the storage complexity in this format is linear in terms of rank R and dimension I.

As is clear from Equation (22), the construction of the matrix $\mathbf{W}$ and its dth Hadamard (elementwise) power dominates the computational complexity of the objective function. The construction of a single column of the matrix $\mathbf{W}$ requires the multiplication of $\mathbf{U}^{\mathrm{T}} \in \mathcal{R}^{R \times I}$ and $\widetilde{\mathbf{z}}_k \in \mathcal{R}^I$. Thus, the computational complexity of constructing the matrix $\mathbf{W}$ is $\mathcal{O}(IKR)$. The dth Hadamard power of the matrix $\mathbf{W}$ can be computed recursively by using the relation $\mathbf{W}^{*(2m)} = \left(\mathbf{W}^{*m}\right)^{*2}$. Thus, the computational complexity of the dth Hadamard power of the matrix $\mathbf{W} \in \mathcal{R}^{R \times K}$ is $\mathcal{O}\left(\log(d)RK\right)$. Therefore, the total computational complexity for computing the objective function for a batch of size K is $\mathcal{O}\left((I + \log(d))KR\right)$. Since $\log(d) \ll I$, the computational complexity for the objective function in Equation (8) is $\mathcal{O}(IKR)$.

The gradient of the objective function in Equation (8) can be obtained by multiplying the Jacobian transpose $\mathbf{J}^{\mathrm{T}}$ by the residual vector $\mathbf{r}$. As shown in Equation (27), this operation requires multiplication of a matrix $\mathbf{Z} \in \mathcal{R}^{I \times K}$ by a diagonal matrix $\text{Diag}(\mathbf{r})$, and the multiplication of the matrices $\mathbf{Z}\text{Diag}(\mathbf{r})$ and $(\mathbf{C}\widetilde{\mathbf{W}})^{\mathrm{T}}$ with sizes $(I \times K)$ and $(K \times R)$, respectively. Note that the entries of the product $\mathbf{C}\widetilde{\mathbf{W}}$ were already obtained in the computation of the objective function. Further, the computational complexity for the product $\mathbf{Z}\text{Diag}(\mathbf{r})$ is $\mathcal{O}(IK)$. Consequently, the computational complexity for the multiplication of $\mathbf{Z}\text{Diag}(\mathbf{r})$ and $(\mathbf{C}\widetilde{\mathbf{W}})^{\mathrm{T}}$ is $\mathcal{O}(IKR)$. In addition, the computation of $\mathbf{V}^{\mathrm{T}}\mathbf{r}$ in Equation (27) requires $\mathcal{O}(KR)$ operations. However $KR \ll IKR$. Therefore, the total computational complexity for computing the gradient is $\mathcal{O}(IKR)$ for $R \gg 1$.

In addition, TeMPO uses the GN algorithm for the optimization. However, this is not a requirement and first-order methods can also be utilized within TeMPO as well. GN requires solving the linear system of equations in (12). Tensorlab's implementation of GN uses the conjugate-gradient (CG) method which requires only the Grammian-vector product for solving (12). This operation requires multiplication of the Jacobian and its transpose by a vector. The computational complexity of multiplying the transpose of Jacobian by a vector is $\mathcal{O}(IKR)$ as described above. The computationally most expensive operations in the multiplication of Jacobian by a vector are the multiplication of matrices $\mathbf{X}_u^{\mathrm{T}}$ and $\mathbf{Z}$ with sizes $(R \times I)$ and $(I \times K)$, and the Hadamard product of two matrices of size $(R \times K)$ as shown in Equation (25). Hence, the computational complexity of computing $\mathbf{J}\mathbf{x}$ is $\mathcal{O}(IKR)$. Note that the entries of the product $\mathbf{C}\widetilde{\mathbf{W}}$ were already obtained in the computation of the objective function. Therefore, the total computational complexity for a single CG iteration is $\mathcal{O}(2IKR)$. Note that a large number of

TABLE 1 | The comparison of the computational complexity of TEMPO with TT and PEPS tensor networks for a batch size of K.

	Calls (per iter)	TEMPO (Type I model)	PEPS	TT-N
Storage		$\mathcal{O}(dIR)$		$\mathcal{O}(nIR_{TT}^2)$
Objective func.	1	$\mathcal{O}((dIKR)$	$\mathcal{O}(KR_{BT}^3R_{PS}^6)$	$\mathcal{O}(nIR_{TT}^2 + R_{TT}^3 \log(I))$
Gradient	1	$\mathcal{O}(dIKR)$	$\mathcal{O}(\alpha KR_{BT}^3R_{PS}^6)$	$\mathcal{O}(\alpha(nIKR_{TT}^2 + KR_{TT}^3 \log(I)))$
Gramian-vector	it_{CG}	$\mathcal{O}(2dIKR)$	–	–

CG iterations in the solution of linear equations for the GN algorithm might increase the computation time compared to first-order algorithms. In fact, the number of CG iterations scales with the number of optimization variables (IR), if the exact solution is desired in the solution of the normal equations. This may lead to an quadratic complexity of $\mathcal{O}\left(2(IR)^2K\right)$. However, we observed in our experiments that a small number of CG iterations were sufficient to obtain accurate results. For example, we set the maximum number CG iterations to 10 for the classification of the MNIST and Fashion MNIST datasets, where the number of unknowns is $784 \times R$ with R ranging from 10 to 150.

The storage complexity of a tensor network with TT architecture is bounded by $\mathcal{O}\left(nIR_{TT}^2\right)$ for a tensor of order I with dimensions $(n \times n \times \ldots \times n)$, where R_{TT} denotes the TT-rank [48]. n is equal to 2 and I is the size of a single image in the image classification applications presented in [20, 21]. Note that the storage complexity of TT increases with powers of the TT-rank R_{TT}. The total computational complexity of TT for computing the objective function has been reported as $\mathcal{O}\left(nIR_{TT}^2 + R_{TT}^3 \log(I)\right)$, when the contraction order defined in [21] is used. When the sweeping algorithm described in [20] is used, the computational complexity of the objective function for TT is $\mathcal{O}\left(n^3R_{TT}^3I\right)$ for a single data point of size I. Similar to the storage complexity, the computational complexity of the objective function for TT increases with powers of the TT-rank of the tensor under consideration. On the other hand, automatic differentiation (AD) is one of methods used to compute the gradient of TT. The computational complexity of automatic differentiation is linear in the complexity of the evaluation of the objective function [49]. Therefore, the computational complexity of the gradient for TT tensor network presented in [21] is $\mathcal{O}\left(\alpha\left(nIKR_{TT}^2 + KR_{TT}^3 \log(I)\right)\right)$, for a batch size of K with $\alpha > 1$. The total computational complexity of TT tensor network for a batch size of K has been reported as $\mathcal{O}\left(mR_{TT}^2(R_{TT} + K)\right)$ for a single iteration of the stochastic Riemannian gradient descent algorithm [19]. As it is clear from the above discussion, both the storage and the computational complexity of TT increases with a power of the TT-rank regardless of the algorithm used, while for TeMPO it increases linearly with the symmetric CPD rank in the symmetric CPD case.

The computational complexity of a single forward pass of PEPS for a batch size of K is $\mathcal{O}\left(KR_{BT}^3R_{PS}^6\right)$, when the boundary matrix product state method is used. Here R_{BT} is the bond dimension (rank) of the boundary matrix product state of PEPS and R_{PS} is the bond dimension of PEPS. In addition, the backward pass for PEPS requires $\mathcal{O}\left(\alpha KR_{BT}^3R_{PS}^6\right)$ operations (with $\alpha > 1$), when automatic differentiation is used [22].

The above analysis shows that TeMPO is computationally less expensive than TT and PEPS, even though it uses a second-order algorithm. All these results are summarized in **Table 1**. The fundamental reason for this is the linear storage complexity of the symmetric CPD format. Both TT and PEPS involve third and higher-order tensors, which makes their computational complexity increase with powers of the bond dimension. On the other hand, the CPD format is known to be numerically less stable than the TT format, which relies on orthogonal matrices.

4. NUMERICAL EXPERIMENTS

We conducted an experiment on the regression problem using synthetic data to illustrate the TeMPO framework and compared TeMPO with different implementations of SVMs in Section 4.1. Next, we applied our framework to the blind deconvolution of constant modulus (CM) signals and compared with the analytical CM algorithm (ACMA) [50], the optimal step-size CM algorithm (OSCMA) [51], and the LS-CPD framework [52] in Section 4.2. In Section 4.3, we further illustrate TeMPO with the image classification problem. We performed experiments on the MNIST and Fashion MNIST datasets and compared the accuracy and number of optimization parameters with MLPs, and TT and PEPS tensor networks. We performed experiments on a computer with an Intel Core i7-8850H CPU at 2.60 GHz with 6 cores and 32 GB of RAM using MATLAB R2021b and Tensorlab 3.0 [11].

In our blind deconvolution experiments, we used the complex GN algorithm with the conjugate gradient Steihaug method. We used the second-order batch Gauss–Newton algorithm for the regression and classification, following the same intuition as in [53]. In each epoch of the algorithm, we randomly shuffle the data points in the training set and process all data points by dividing them into batches. In the regression and binary classification case, we optimize a single cost function. In the multi-label classification case, for each batch, we randomly select a cost function f_l defined for each label to optimize. Thus our algorithm does not guarantee that each f_l will be trained by all training images in each epoch in the multi-label classification case. To guarantee this, the algorithm can be modified such that for each batch all cost functions f_l are optimized at the cost of increasing CPU time by a factor of the number of classes L. However, in that case the algorithm might need fewer epochs to converge. The overall algorithm is summarized in **Algorithm 2**. **Algorithm 2**

Algorithm 2: Batched GN algorithm using dogleg trust-region for regression and classification for the type II model.

Input : $\mathbf{Z}$ – Input data matrix
$\mathbf{y}$ – Vector of values (labels in the classification case) for each data point in $\mathbf{Z}$
$\mathbf{U}_1, \ldots, \mathbf{U}_L$ – Initial factor matrices for each label (single in the regression case)
$\mathbf{c}_1, \ldots, \mathbf{c}_L$ – Initial weight vectors for each label (single in the regression case)
$\mathcal{T}_{0,l}$ – Initial scalar for each label (single in the regression case)
epoch – Number of epochs
batchsize – Batch size

Output: $\mathbf{U}_1, \ldots, \mathbf{U}_L$ – Optimized factor matrices for each label (single in the regression case)
$\mathbf{c}_1, \ldots, \mathbf{c}_L$ – Optimized weight vectors for each label (single in the regression case)

```
for each epoch do
    Shuffle input data
    for each batch do
        l ← 1
        if multi-label classification then
            l ←
            Randomly select label l to optimize f_l , 0 < l ≤ L
        end
        U_l, c_l, T_{0,l} ← Optimize f_l using Algorithm 1
    end
end
```

is given for the type II model for the ease of explanation. Slight modifications are sufficient to obtain an algorithm for the type I model.

We define the relative error as the relative difference in l_2 norm $\|\mathbf{f} - \widehat{\mathbf{f}}\|_2 / \|\mathbf{f}\|_2$ with $\widehat{\mathbf{f}}$ an estimate for a vector $\mathbf{f}$, and the signal-to-noise ratio (SNR) as $20\log_{10}(\|\mathbf{f}\|_2/\|\boldsymbol{\eta}\|_2)$, where $\boldsymbol{\eta} = \widehat{\mathbf{f}} - \mathbf{f}$.

4.1. Regression

In this experiment, we considered a low-rank smooth function $f(\mathbf{x}) : \mathcal{R}^N \to \mathcal{R}$, namely

$$f(\mathbf{x}) = \sum_{r=1}^{R_f} \alpha_r e^{(\mathbf{a}_r^{\mathrm{T}}\mathbf{x})}, \tag{28}$$

where $\mathbf{x} \in [-1,1]^N$, R_f is the rank of the function $f(\mathbf{x})$, and the coefficients α_r are scalars randomly chosen from the standard normal distribution. We generated 5,000 test samples and 1,000 training samples for $N = 50$ and $R_f = 8$. Each vector $\mathbf{a}_r \in \mathcal{R}^N$ was a unit norm vector drawn from the standard normal distribution. Each of the samples of $\mathbf{x}$ was drawn from the uniform distribution. We initialized each factor matrix with a matrix whose elements were randomly drawn from the standard normal distribution, and scaled it to unit norm. We initialized each weight vector in the same way as the factor matrices. We approximated $f(\mathbf{x})$ by the type I and type II model of degree 5 whose coefficient tensors were represented in the rank-R symmetric CPD format. We set the batch size to 500 and the maximum number GN iterations to 5 for each batch. In **Figure 3**, we show the median relative test and training errors for $R = \{2,4,8,16\}$ as a function of the number of epochs for 100 trials. Each epoch corresponds to optimization over the full training set. It is clear from **Figure 3** that TeMPO produces more accurate results and generalizes better when using higher rank values, for both the type I and type II model. Good performance is also observed for $R = 16 > R_f = 8$, meaning that TeMPO is robust to over-estimation of the number of parameters. For low rank values, i.e., $R < R_f$, the type I model produces better results than the type II model because it involves more parameters that can be tuned, cf. the discussion of Proposition 1.

In the second stage of the experiment, we trained the type I and type II model for a multivariate polynomial of degree 5 with noisy measurements. We added Gaussian noise to the function values for a given SNR, i.e.,

$$\widetilde{f}(\mathbf{x}) = \sum_{r=1}^{R} \alpha_r e^{(\mathbf{a}_r^{\mathrm{T}}\mathbf{x})} + \eta, \tag{29}$$

where η denotes the noise. We run our algorithm with the same settings as in the noiseless case for an SNR ranging from 10 dB to 50 dB. In **Figure 4**, we show the median errors for 100 trials as a function of SNR. We have similar observations as in the noiseless case. Apart from these observations; although the accuracy of our algorithm decreases for SNR ≤ 20 (dB), it still maintains good accuracy for SNR > 20 (dB), as shown in **Figure 4**. Moreover, as can be observed from the **Figure 4** (left), the type I model overfits for $R = \{8, 16\}$ and SNR ≤ 20 (dB) in agreement with the result of Proposition 1.

In our next experiment, we trained the type I and type II model with larger-size samples, i.e., $N = 250$ and $R = \{8, 16, 32, 64\}$, to assess how the CPU time depends on the rank. In **Figure 5**, we show the median CPU time per epoch as a function of the rank. It is evident from the figure that the computational complexity of the type I model is d times the computational complexity of the type II model (cf. Section 3.5). Moreover, **Figure 5** confirms that the computational complexity of our algorithm is linear in the rank (cf. Section 3.5).

In our next experiment, we examined the generalization abilities of the Gauss–Newton and ADAM [54] algorithms in our framework. We trained the type I model for a multivariate polynomial of degree 5 with both of these algorithms for different number of training samples to fit the rank-8 function given as in Equation (29). We set $R = 8$, $N = 50$, and SNR $= 20$(dB). For the ADAM algorithm, we set the step size, the exponential decay rate for the first momentum (β_1), and the exponential decay rate for the second momentum (β_2) to 0.01, 0.9, and 0.99, respectively. In **Figure 6**, we show the median training and test accuracies of these algorithms for the number of training samples ranging from 500 to 5,000 as a function of the number of epochs for 100 trials. It is evident from **Figure 6** that the presented Gauss–Newton algorithm produces more accurate results than

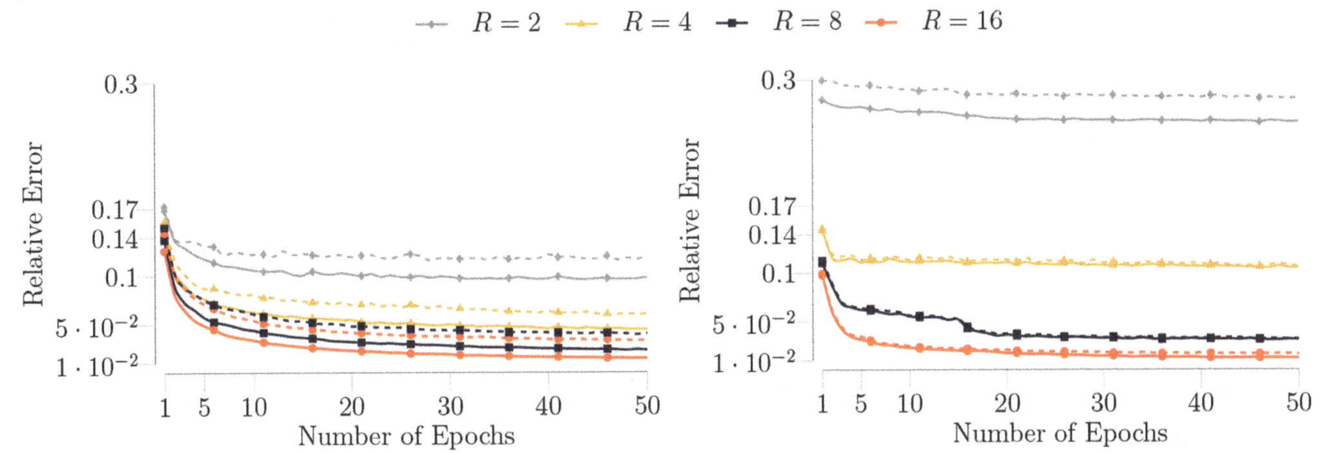

FIGURE 3 | (Left) The median test (dashed lines) and training (solid lines) errors of the type I model for 100 trials on the synthetic data for a rank-8 function given as in Equation (28). The number of samples for the training dataset is set to 5, 000 and for the test dataset it is set to 1, 000. The batch size is set to 500 and the maximum number of GN iterations is set to 5. **(Right)** The median test (dashed lines) and training (solid lines) errors of the type II model with the same algorithm settings. TeMPO produces more accurate results and generalizes better for higher rank values for both the type I and type II model. The performance is robust to overparameterization ($R > R_f$). The type I model produces better results for low rank values, i.e., $R < R_f$.

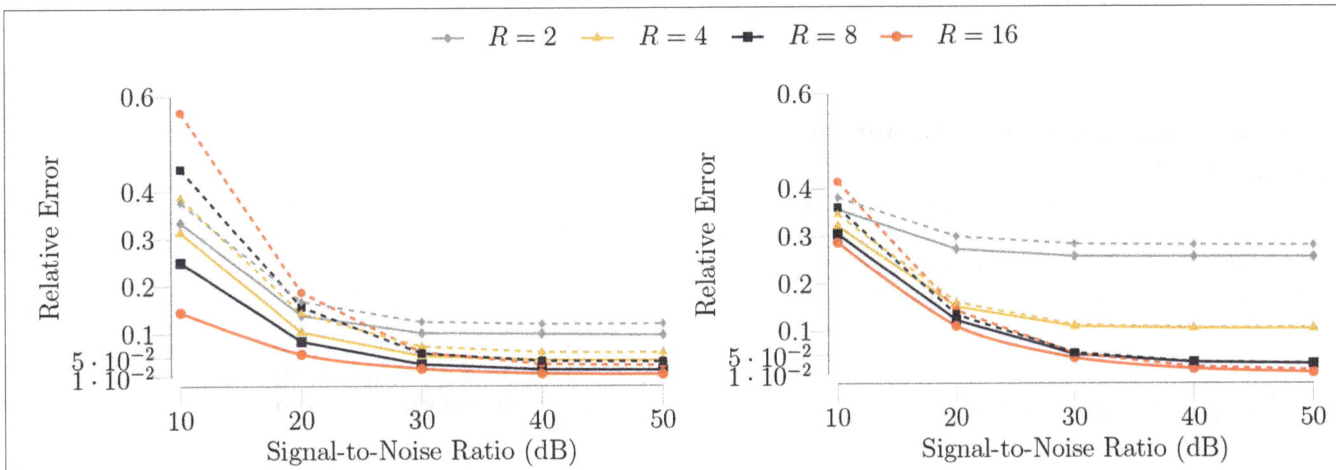

FIGURE 4 | (Left) The median test (dashed lines) and training (solid lines) errors of the type I model for 100 trials on the synthetic noisy data for a rank-8 function given as in Equation (29). The number of samples for the training dataset is set to 5, 000 and for test dataset is set to 1, 000. The batch size is set to 500 and the maximum number of GN iterations is set to 5. **(Right)** The median test (dashed lines) and training (solid lines) errors of the type II model with the same algorithm settings. TeMPO produces more accurate results and generalizes better for higher rank values for both the type I and type II model in the presence of noise as well. Again, the type I model produces better results for low rank values, i.e., $R < R_f$, because it involves more parameters than the type II model.

the ADAM algorithm and also requires fewer number of epochs to converge in these experimental settings.

We also compared TeMPO with SVMs using a polynomial kernel. We run the same experiment for a number of training samples ranging from 500 to 5, 000. We set the rank to 8, i.e., $R = R_f$ for TeMPO. We used the built-in Matlab routine `fitrsvm` and LS-SVMlab toolbox [55, 56]. We set the degree of polynomial kernel to 5, i.e., equal to the degree of the type I and type II model for `fitrsvm`. LS-SVMlab automatically tunes the degree to 3 to find the best fit. In **Figure 7** (left), we show the median test and training errors for SVM, the type I and type II model. It is clear from **Figure 7** (left) that the type I and type II model generalize better than `fitrsvm`. A possible reason is the dense parameterization of SVMs, while TeMPO uses low-rank parameterization. Moreover, as shown in **Figure 7** (right), our algorithm is faster than SVMs for numbers of training samples above 1, 000. This is due to the higher memory requirement of SVMs. Typically, kernel based methods such as LS-SVM have a storage and computational complexity of $\mathcal{O}(N^2)$ [55], with N the number of training samples. In contrast, **Figure 7** (right) confirms that the computational complexity of TeMPO is linear in the number of training samples (cf. Section 3.5).

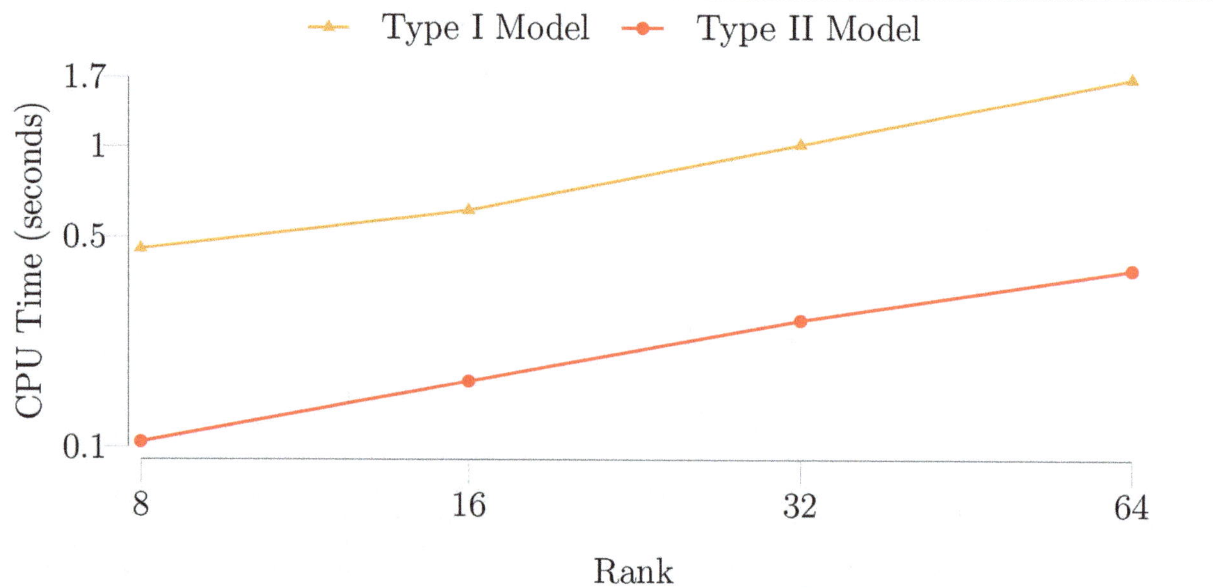

FIGURE 5 | The median CPU time (seconds) per epoch for the type I and type II model as a function of the rank for a rank-8 function given as in Equation (28) for 100 trials. The number of samples for the training dataset is set to 5, 000 and for the test dataset it is set to 1, 000. The batch size is set to 500 and the maximum number of GN iterations is set to 5. The figure confirms that the computational complexity of the type I model is d times the computational complexity of the type II model (cf. Section 3.5). Moreover, the computational complexity of the algorithm is linear in the rank (cf. Section 3.5). The figure is in a logarithmic scale on the horizontal axis.

4.2. Blind Deconvolution of Constant Modulus Signals

Blind deconvolution can be formulated as a multivariate polynomial optimization (MPO) problem and hence it fits into the TeMPO framework [15]. In this illustrative example, we limit ourselves to an autoregressive single-input single-output (SISO) system [57], given by

$$\sum_{l=0}^{L} w_l \cdot y[k-l] = s[k] + n[k], \text{ for } k = 1, \dots, K, \tag{30}$$

where $y[k]$, $s[k]$, and $n[k]$ are the measured output signal, the input signal and the noise at the kth measurement, respectively, and w_l denotes the lth filter coefficient. Ignoring the noise for the ease of derivation, (30) can be written as:

$$\mathbf{Y}^{\mathrm{T}}\mathbf{w} = \mathbf{s}, \tag{31}$$

where $\mathbf{Y} \in \mathbb{C}^{L\times K}$ is a Toeplitz matrix and its rows are the subsequent observations under the assumption that we have $K + L - 1$ samples $y[-L+1], \dots, y[K]$. The vector $\mathbf{w} \in \mathbb{K}^L$ contains the filter coefficients and the kth entry of the source vector $\mathbf{s} \in \mathbb{C}^K$ is the input signal at the kth time instance, i.e., $s_k = s[k]$. In blind deconvolution, one attempts to find the original input signal $\mathbf{s}$ and the filter coefficients $\mathbf{w}$ by only observing the output signal $\mathbf{Y}$. Thus, constraints on signals and/or channel have to be imposed to obtain interpretable results. The constant modulus (CM) criterion is a widely used input constraint [58]. The CM property, which holds for phase- or frequency-modulated signals [50, 59] can be written as:

$$|s_k|^2 = c, \text{ for } k = 1, 2, \dots, K. \tag{32}$$

Here, c is a constant scalar. By substituting s_k defined in (31) into (32), we obtain

$$(\mathbf{Y} \odot \overline{\mathbf{Y}})^{\mathrm{T}}(\mathbf{w} \otimes \overline{\mathbf{w}}) = c \cdot \mathbf{1}_K. \tag{33}$$

Following the same intuition as in [60], by multiplying (33) from the left with a Householder reflector $\mathbf{Q}$ [61], generated for $c \cdot \mathbf{1}_K$, and removing the first equation[3], we obtain

$$\mathbf{M}(\mathbf{w} \otimes \overline{\mathbf{w}}) = \mathbf{0}. \tag{34}$$

Here, $\mathbf{M} = \widetilde{\mathbf{Q}}(\mathbf{Y} \odot \overline{\mathbf{Y}})^{\mathrm{T}}$, and $\widetilde{\mathbf{Q}}$ is obtained by removing the first row of the Householder reflector $\mathbf{Q}$. In applications, $\mathbf{M}(\mathbf{w} \otimes \overline{\mathbf{w}})$ will not vanish exactly due to the presence of noise. Hence, we look for the solution which minimizes its l_2 norm as

$$\min_{\mathbf{w},\overline{\mathbf{w}}} f(\mathbf{w}, \overline{\mathbf{w}}) = \min_{\mathbf{w},\overline{\mathbf{w}}} \frac{1}{2} \left\| \mathbf{M}(\mathbf{w} \otimes \overline{\mathbf{w}}) \right\|_2^2, \quad \text{subject to } \|\mathbf{w}\| = 1. \tag{35}$$

The objective function in (35) is a homogeneous multivariate polynomial of degree 4 in which the coefficient tensor $\mathcal{W}$ is given as a rank-1 Hermitian symmetric CPD, i.e.,

$$\mathcal{W} = \mathbf{w} \otimes \overline{\mathbf{w}} \otimes \overline{\mathbf{w}} \otimes \mathbf{w} := [\![\mathbf{w}, \overline{\mathbf{w}}, \overline{\mathbf{w}}, \mathbf{w}]\!]. \tag{36}$$

Exploiting the rank-1 Hermitian symmetric CPD structure in (36) and the structure of $\mathbf{M}$, which is a special case of Lemma 1 and Lemma 2, efficient expressions for the computation

[3] The first equation is only a normalization constraint.

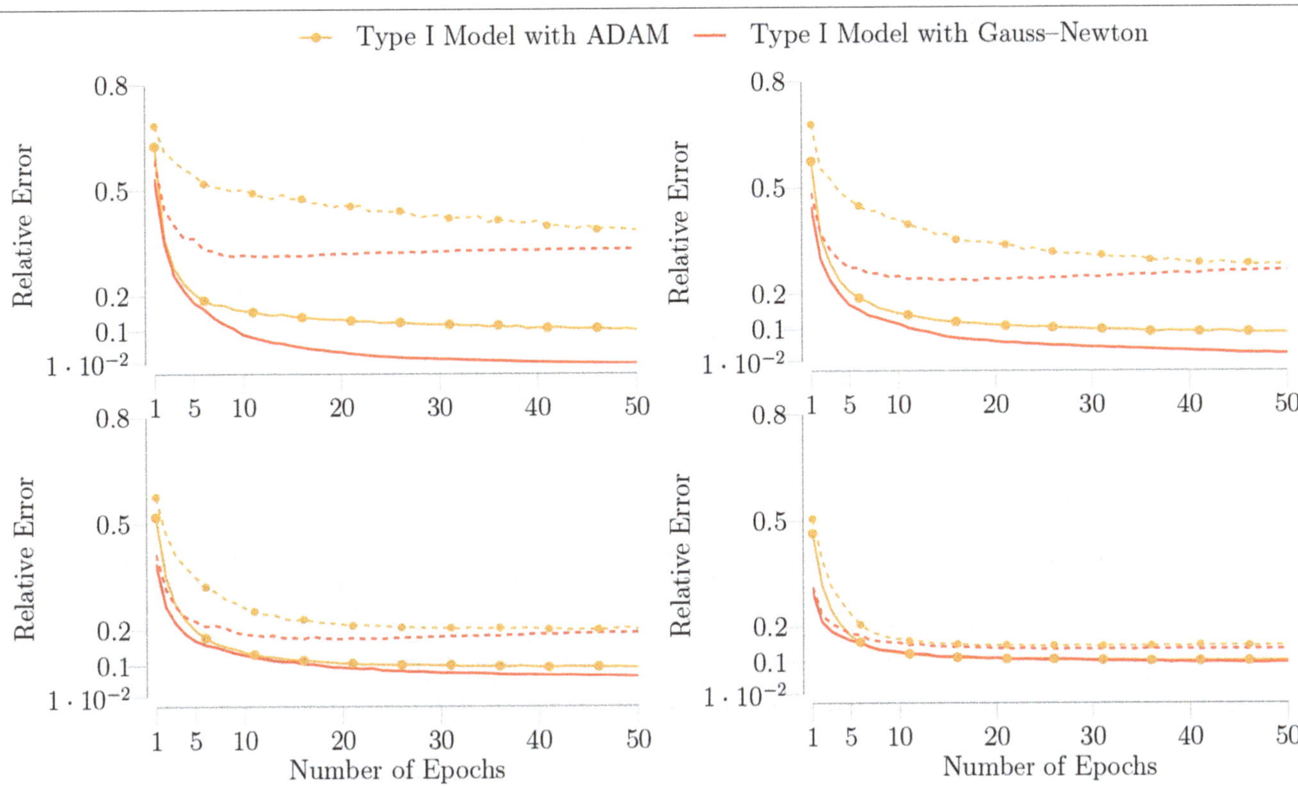

FIGURE 6 | Comparison of the median test (dashed lines) and training (solid lines) errors of the Gauss–Newton and the ADAM algorithms as a function of the number of epochs for 100 trials. The type I model for a rank-8 function given as in Equation (29) in the presence of SNR 20 dB Gaussian noise is used to generate the training and the test sets. The batch size is set to 10% of the training set size. For the Gauss–Newton algorithm, the maximum number of GN iterations and CG iterations is set to 1 and 5, respectively. For the ADAM algorithm, the step size, β_1 and β_2 are set to 0.01, 0.9, and 0.99, respectively. The number of training samples is set to 500 **(top-left)**, 1,000 **(top-right)**, 2,000 **(bottom-left)**, and 5,000 **(bottom-right)**. The presented Gauss–Newton algorithm produces more accurate results than the ADAM algorithm and also requires fewer number of epochs to converge in these experimental settings.

of Jacobian-vector products for the problem (35) have been presented in [15].

A number of algorithms have been developed to solve (33) and (34). The analytical CM algorithm (ACMA) [50] writes (34) as a generalized matrix eigenvalue problem in the absence of noise, and under the assumption that the null space of **M** is one dimensional, which makes ACMA more restrictive than TeMPO. In the presence of noise, ACMA writes (34) as the simultaneous diagonalization of a number of matrices and solves it by extended QZ iteration. Gradient descent and stochastic gradient descent algorithms have also been proposed for the minimization of the expected value of $\{(|\mathbf{y}_\mathbf{n}^\mathrm{T}\mathbf{w}| - c)^2\}$. The optimal step-size CMA (OSCMA) [51] algorithm uses a gradient descent algorithm, which computes the step size algebraically. The problem in (35) can also be interpreted as a linear system with a rank-1 constrained solution, which fits the LS-CPD framework in [52]. LS-CPD solves (33) by relaxing the complex conjugate $\overline{\mathbf{w}}$ to a possibly different vector $\mathbf{v} \in \mathbb{C}^L$ and utilizing the second-order GN algorithm using dogleg trust-region method. We solve (35) by utilizing the complex GN algorithm using the conjugate gradient Steihaug method implemented in TensorLab 3.0 [11]. We compare with these algorithms in terms of computation time and accuracy.

We consider an autoregressive model of degree $L = 10$ with coefficients uniformly distributed on $[0, 1]$, sample length $K = 600$, and $c = 1$. We add scaled Gaussian noise to the measurements to obtain a particular SNR. We run 50 experiments starting from the algebraic solution presented in [52] for LS-CPD, OSCMA, and TeMPO. In **Figure 8** (left), we show the median relative error on **w** as a function of SNR. It is clear from **Figure 8** (left) that TeMPO achieves similar accuracy as LS-CPD and OS-CMA, which are more accurate than ACMA. In **Figure 8** (right), we show the median CPU time in seconds as a function of SNR. Clearly, TeMPO is faster than ACMA, OS-CMA, and LS-CPD for SNR $\geq$ 10(dB) by exploiting the structure of the data.

4.3. Image Classification

Multi-class image classification amounts to the determination of a possibly nonlinear function f that maps input images $\mathbf{Z}_k$ to integer scalar labels y_k, which are known for a training set. In this study, we represent f by a multivariate polynomial p. Following the one-versus-all strategy, we define a cost function f_l for each label that maps the input image $\mathbf{Z}_k$ to a scalar value as

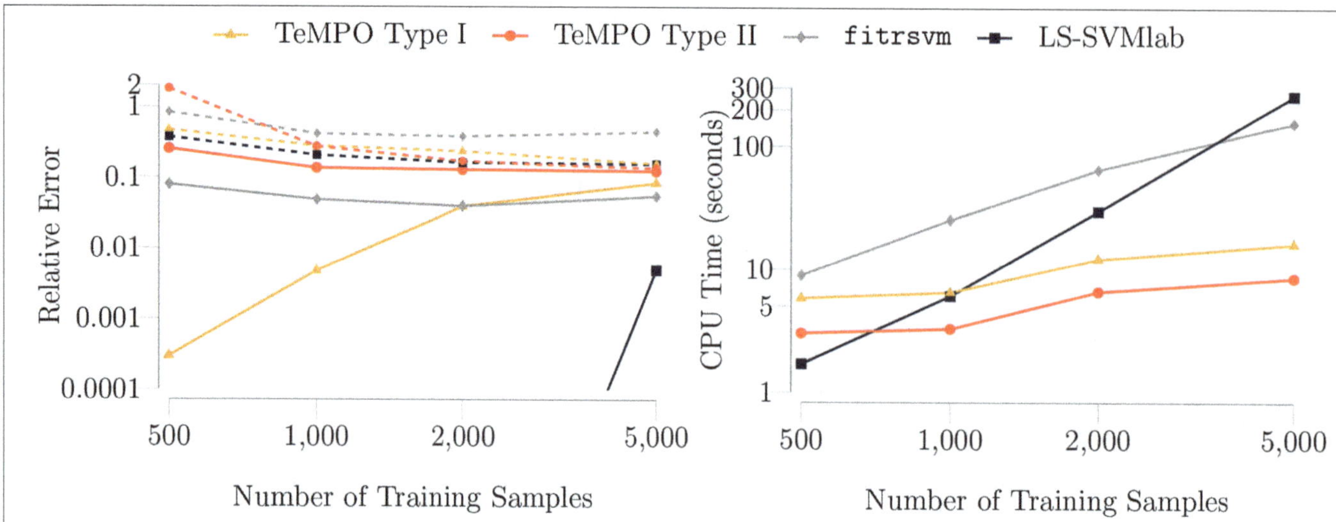

FIGURE 7 | (Left) The median test (dashed lines) and training (solid lines) errors of SVMs with polynomial kernel, the type I and type II model for a rank-8 function given as in Equation (29) in the presence of SNR 20 dB Gaussian noise as a function of the number of training samples for 100 trials. The batch size is equal to 10% of the training set size. The maximum number of GN iterations is set to 5 for the type I and type II model. Specifically, for the SVMs, the built-in Matlab routine `fitrsvm` and LS-SVMlab toolbox were used to obtain the results. The relative errors of LS-SVMLab for the sample sizes 500, 1,000, and 2,000 are 1.6e − 6, 2.2e − 6 and 3.3e − 6, respectively. The presented algorithm generalizes better than `fitrsvm` in these experimental settings. **(Right)** The median CPU times (seconds) with the same setting. The computational complexity of our algorithm is linear in the problem size as expected, and it is faster than SVMs for numbers of training samples above 1,000. The figures are in a logarithmic scale on both the horizontal and vertical axes.

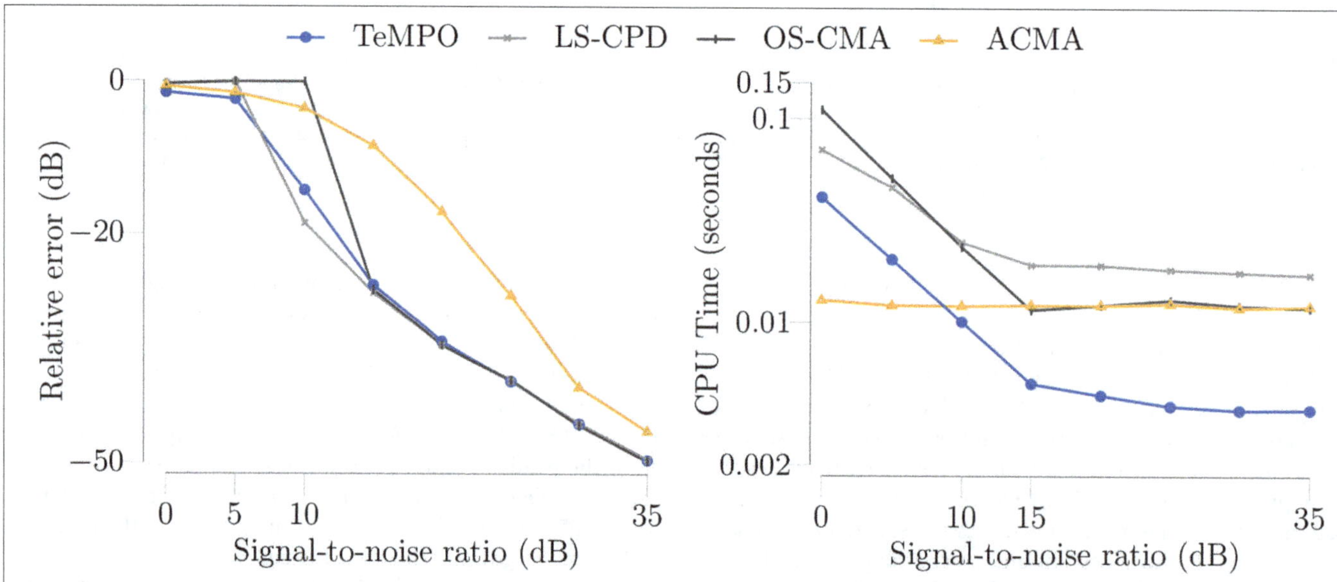

FIGURE 8 | (Left) The median relative errors (dB) of LS-CPD, OS-ACMA, ACMA, and TeMPO with respect to SNR (dB) for an autoregressive model of degree $L = 10$ with uniformly distributed coefficients between zero and one, sample length $K = 600$ for 50 trials. TeMPO obtains similar accuracy to LS-CPD, OS-CMA, while obtaining more accurate results than ACMA. **(Right)** The median CPU times (seconds) with the same settings. TeMPO is faster than other algorithms for SNR > 10 (dB).

$$f_l(p_l, \mathbf{z}_1, \ldots, \mathbf{z}_K) = \frac{1}{2}\sum_{k=1}^{K}\left(y_k - p_l(\mathbf{z}_k)\right)^2,$$

where $\mathbf{z}_k = \mathrm{vec}(\mathbf{Z}_k)$ and where $y_k = 1$ if $\mathbf{z}_k$ is labeled as l and $y_k = 0$ otherwise. The polynomial p_l can be chosen within the type I or the type II model class. For the type I model, the optimization problem can be written as

$$\min_{p_l} f_l(p_l, \mathbf{z}_1, \ldots, \mathbf{z}_K), \quad \text{subject to} \quad p_l(\mathbf{z}_k) = \mathcal{T}_{l,0} + \sum_{j=1}^{d} \mathcal{T}_{l,j}\mathbf{z}_k^j,$$
$$\text{and} \quad \mathcal{T}_{l,j} = \left[\!\left[\mathbf{U}_{l,j}, \ldots, \mathbf{U}_{l,j}; \mathbf{c}_{l,j}^{\mathrm{T}}\right]\!\right], \tag{37}$$

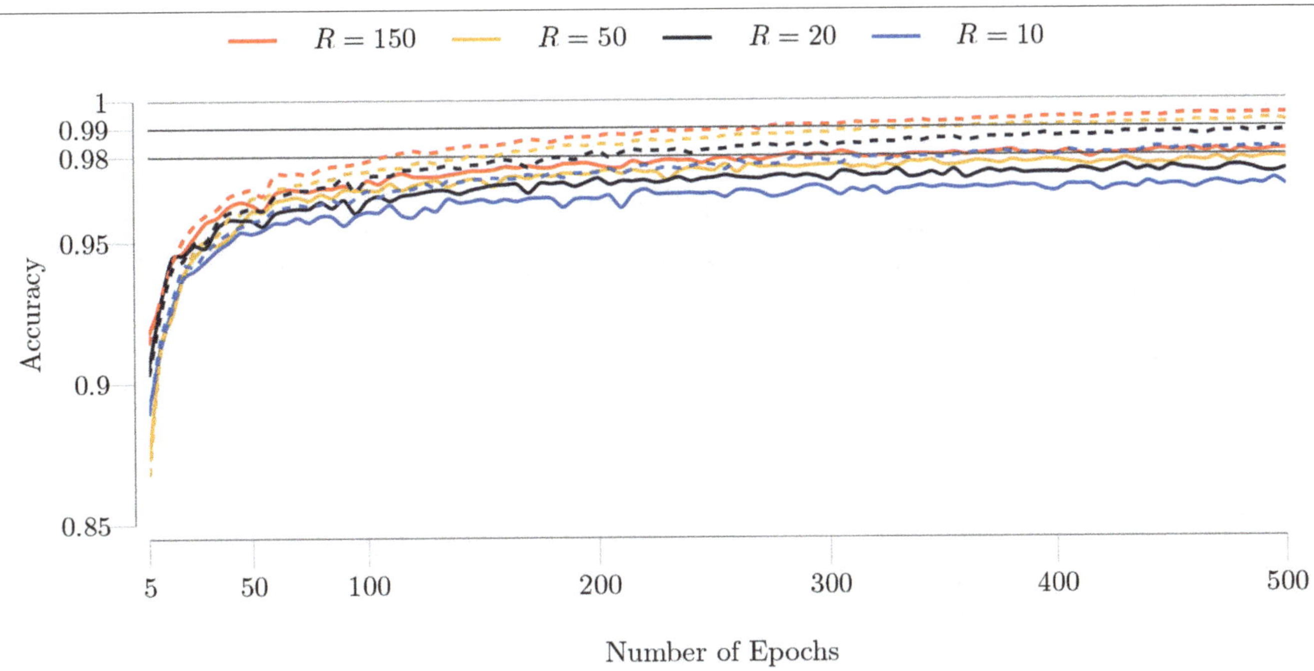

FIGURE 9 | Test (solid lines) and training (dashed lines) accuracies of the type I model for the MNIST dataset with respect to the number of epochs. The full training set (60, 000 images) and test set (10, 000 images) are used. The batch size is set to 100 and the maximum number of GN iterations is set to 1. TeMPO achieves high accuracy even for low rank values, i.e., $R = \{10, 20\}$. Both the test and training accuracy increase mildly as the rank increases.

where d is the degree of the polynomial under consideration. Note that we substitute the symmetric CPD structure given as a constraint into the objective function, and hence obtain and solve an unconstrained optimization problem. For the type II model, the optimization problem can be written as

$$\min_{p_l} f_l(p_l, \mathbf{z}_1, \ldots, \mathbf{z}_K), \quad \text{subject to} \quad p_l(\mathbf{z}_k) = \mathcal{T}_l \mathbf{z}_k^d,$$
$$\text{and} \quad \mathcal{T}_l = [\![\mathbf{U}_l, \ldots, \mathbf{U}_l; \mathbf{c}_l^{\mathrm{T}}]\!] .$$

After the optimization of f_l for each label l, the classification is done by computing each $p_l(\mathbf{s})$ for the data point $\mathbf{s}$ to be classified and selecting the value of l for which $|p_l(\mathbf{s})|$ is largest.

4.3.1. Experiments

We performed several experiments by varying the parameters rank and maximum number of GN iterations to illustrate the TeMPO framework for the classification of the MNIST and Fashion MNIST datasets. We kept the maximum number of CG iterations equal to 10, the degree of the multivariate polynomial to 3, the tolerance for the objective function and optimization variables equal to $1e - 10$, the inner solver tolerance equal to $1e - 10$, and the trust-region radius equal to 0.1, throughout the experiments.

We initialized each factor matrix with a matrix whose elements were randomly drawn from the standard normal distribution, and scaled it to unit norm. Similarly, we initialized each weight vector $\mathbf{c}_l$ with a vector whose elements were randomly drawn from the standard normal distribution and scaled it to unit norm.

Datasets

Modified National Institute of Standards and Technology (MNIST) handwritten digit database [62] and the Fashion MNIST database [63] are used for this study. Both datasets contain gray scale images of size (28×28). The training sets of both datasets are composed of 60, 000 images and test sets are composed of 10, 000 images. The images have been size-normalized and centered in a fixed-size image. We rescale images such that every pixel value is in the interval $[0, 1]$ and the mean of each image is zero. Then, we vectorize, i.e., stack each column vertically in a vector, each image to a vector of size 784. For the type II model, we augment the resulting vector by the scalar 1. Similar pre-processing steps are necessary for also tensor networks. Additionally, they may require the encoding input data which increases the storage and the computational resource requirement.

Results and Comparisons

Results of the Type I Model

We first trained the type I model on the total MNIST training set for various rank values ranging from 10 to 150 to illustrate the effect of rank on the accuracy. We set the batch size to 100 and the maximum number of GN iterations to 1. We show the training history in **Figure 9**. It is evident from **Figure 9** that TeMPO achieves high accuracy even for low rank values, i.e., $R = \{10, 20\}$. Increasing the rank mildly improves both the test and training accuracy, with the improvement getting smaller as the rank increases.

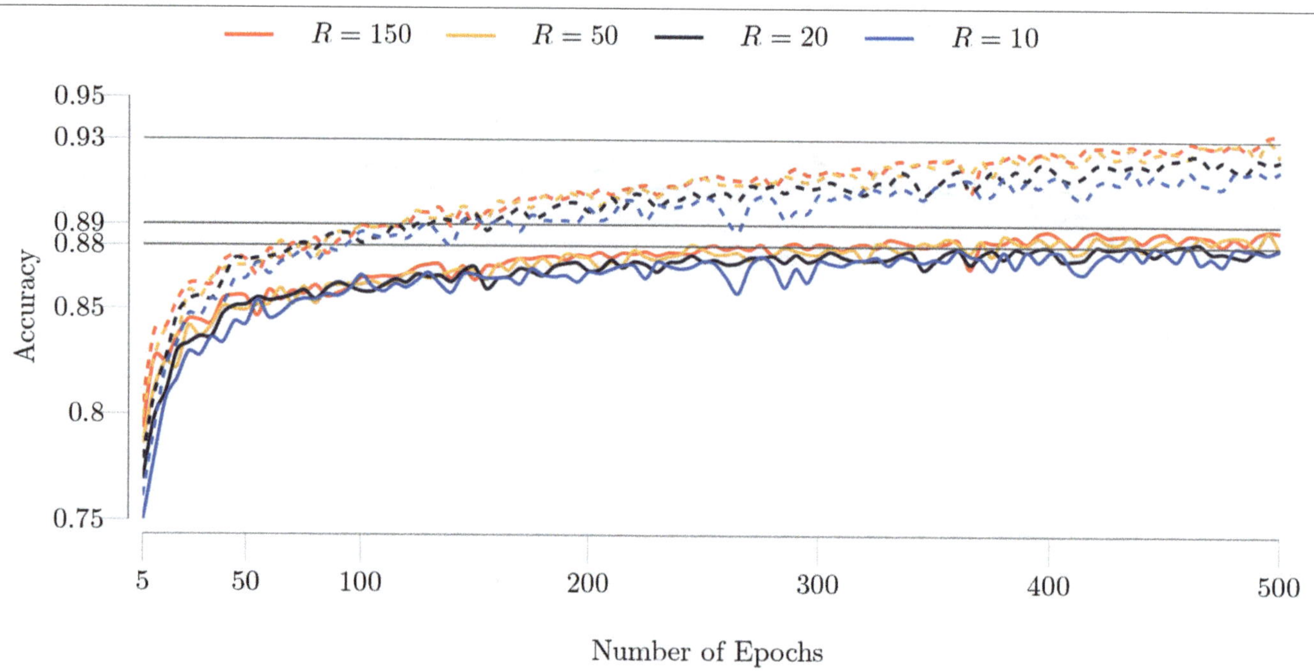

FIGURE 10 | Test (solid lines) and training (dashed lines) accuracies of the type I model for the Fashion MNIST dataset with respect to the number of epochs. The full training set (60, 000 images) and test set (10, 000 images) are used. The batch size is set to 100 and the maximum number of GN iterations is set to 1. Similar to the MNIST dataset, TeMPO achieves good accuracy even for low rank values and both the test and training accuracy mildly increase as the rank increases.

We repeated the same experiments for the Fashion MNIST dataset, which is harder to classify. We show the training history in **Figure 10**. The observations made for the MNIST dataset also apply to the Fashion MNIST dataset. However, the test and training accuracy are lower for the Fashion MNIST dataset in agreement with previous works. Also, our algorithm requires more epochs to converge for the Fashion MNIST dataset.

In our next experiment, we set the maximum number of GN iterations to 5. We observed that our algorithm needs fewer epochs to converge and produces more accurate results with this setting. The comparison for the MNIST and Fashion MNIST dataset is shown in **Figures 11**, **12**, respectively. The improvement in the test accuracy for the Fashion MNIST dataset is around 1% and more pronounced than the improvement in the test accuracy for the MNIST dataset. TeMPO achieves around 98.30% test accuracy for the MNIST dataset and around 90% test accuracy for the Fashion MNIST dataset with $R = 150$.

Results of the Type II Model

We repeated the same experiments for the type II model. We used the same settings as in the type I model. However, we set the batch size to 200 to obtain an accuracy similar to that of the type I model. We show the training history in **Figure 13**. Similar to previous experiments, our algorithm performs well even for low rank values, and produces more accurate results for higher rank values. TeMPO achieves around 98% test accuracy and 100% training accuracy after 200 epochs with $R = 150$ for the MNIST dataset.

In **Figure 14**, we show the training history for the Fashion MNIST dataset. Similar to the type I model, the test and training accuracy is lower than the MNIST dataset. The algorithm converges around 100 epochs and achieves around 89.30% test accuracy with $R = 150$. Moreover, our algorithm achieves around 99% training accuracy after 400 epochs.

We repeated the same experiments with the maximum number of GN iterations set to 5. The comparisons for the MNIST and Fashion MNIST datasets are shown in **Figure 15**. Contrary to our observation for the type I model, the test accuracy now decreases for both datasets. A possible reason is that when the residuals are big, doing more GN iterations may not lead a better direction for minimizing (37). A similar observation has been made in [53], for training DNNs. It is experimentally shown that higher number of CG iterations might not produce more accurate results if the Hessian obtained by mini-batch is not reliable due to non-representative batches and/or big residuals. On the other hand, if the residuals are small, higher number of CG iterations can produce more accurate results thanks to the curvature information [53].

Comparisons

We now compare TeMPO with different models, namely: TT tensor networks [21], TT structured tree tensor networks (TTN) [64], multi-layer perceptron (MLP) with 784−1000−10 neurons, MLP with a convolution layer (CNN-MLP), PEPS, and PEPS with

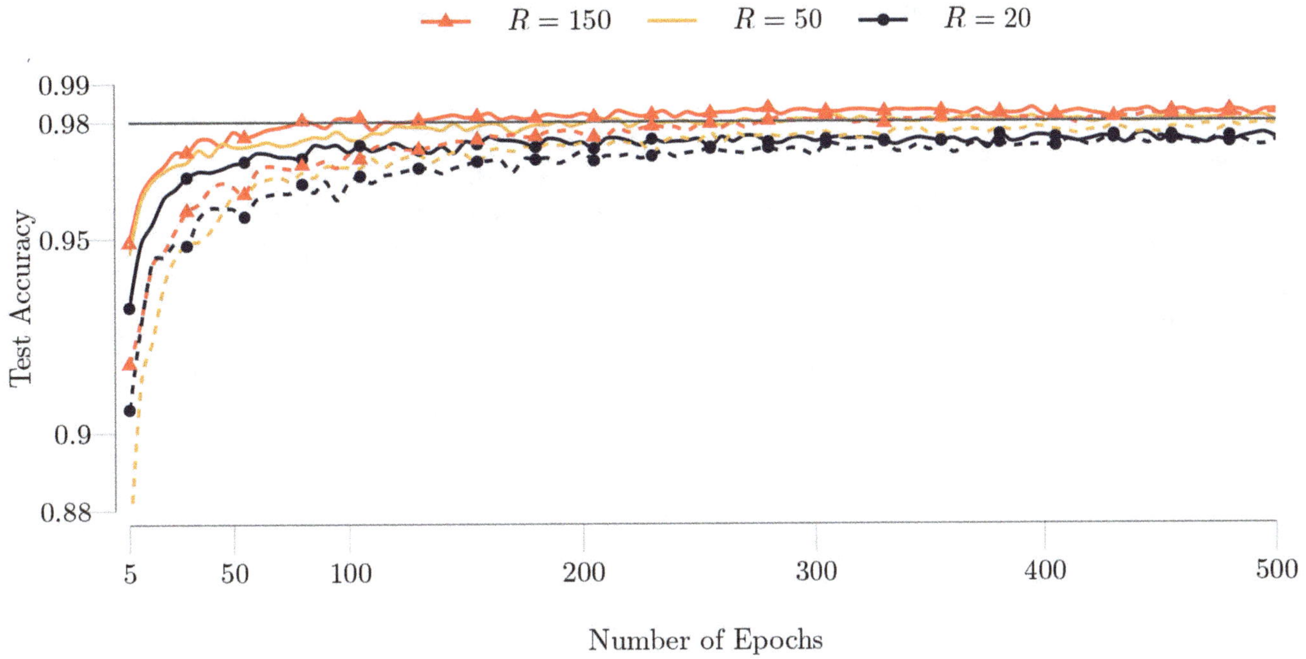

FIGURE 11 | Comparison of test accuracies of the type I model on the MNIST dataset for different maximum number of GN iterations as a function of the number of epochs. The full training set (60, 000 images) and test set (10, 000 images) are used. The batch size is set to 100 and the maximum number of GN iterations is set to 1 (dashed lines) and to 5 (solid lines).

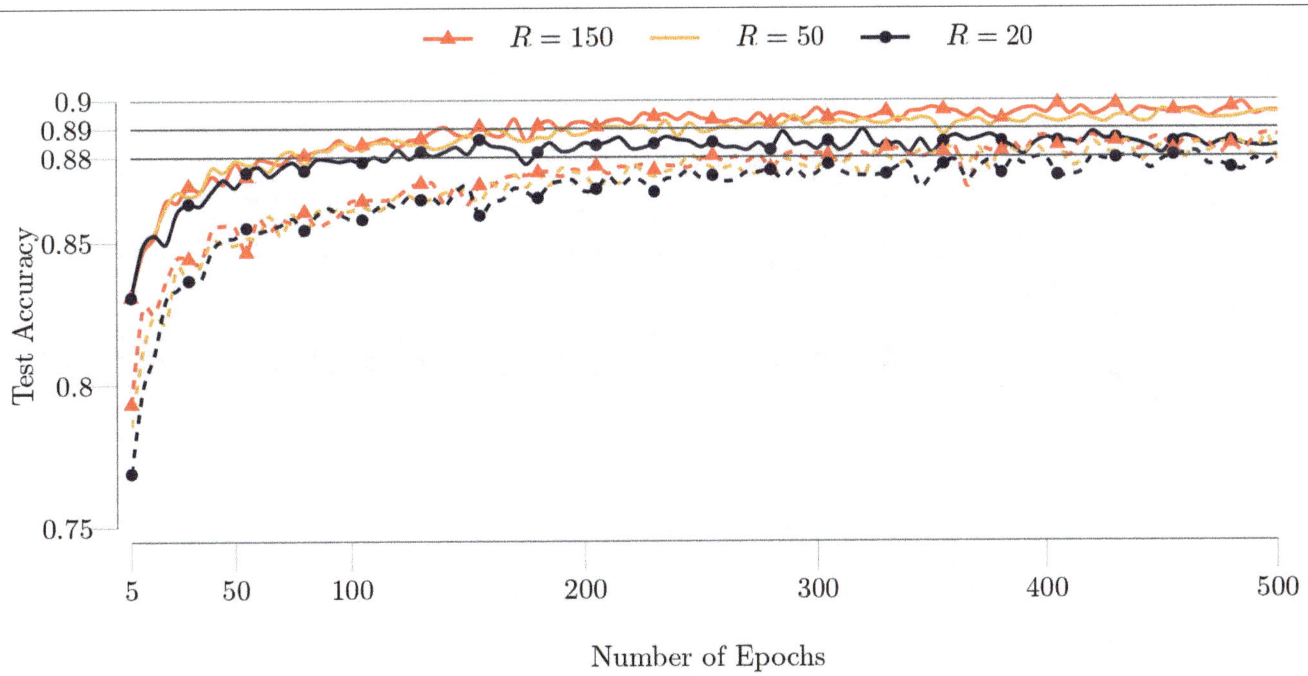

FIGURE 12 | Comparison of test accuracies of the type I model on the Fashion MNIST dataset for different maximum number of GN iterations as a function of the number of epochs. The full training set (60, 000 images) and test set (10, 000 images) are used. The batch size is set to 100 and the maximum number of GN iterations is set to 1 (dashed lines) and to 5 (solid lines).

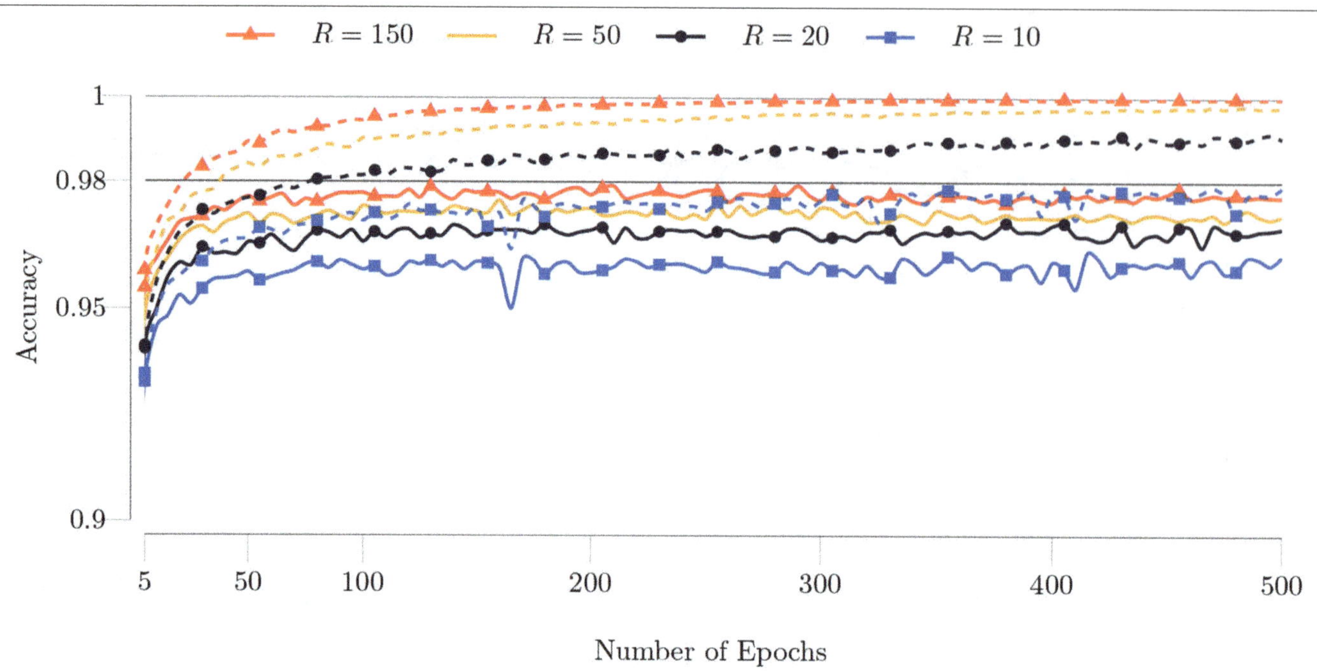

FIGURE 13 | Test (solid lines) and training (dashed lines) accuracies of the type II model for the MNIST dataset with respect to the number of epochs. The full training set (60, 000 images) and test set (10, 000 images) are used. The batch size is set to 200 and the maximum number of GN iterations is set to 1. Both the test and training accuracy increase as the rank increases. The improvement in the accuracy gets smaller as the rank increases. The algorithm achieves around 100% training accuracy after 200 epochs.

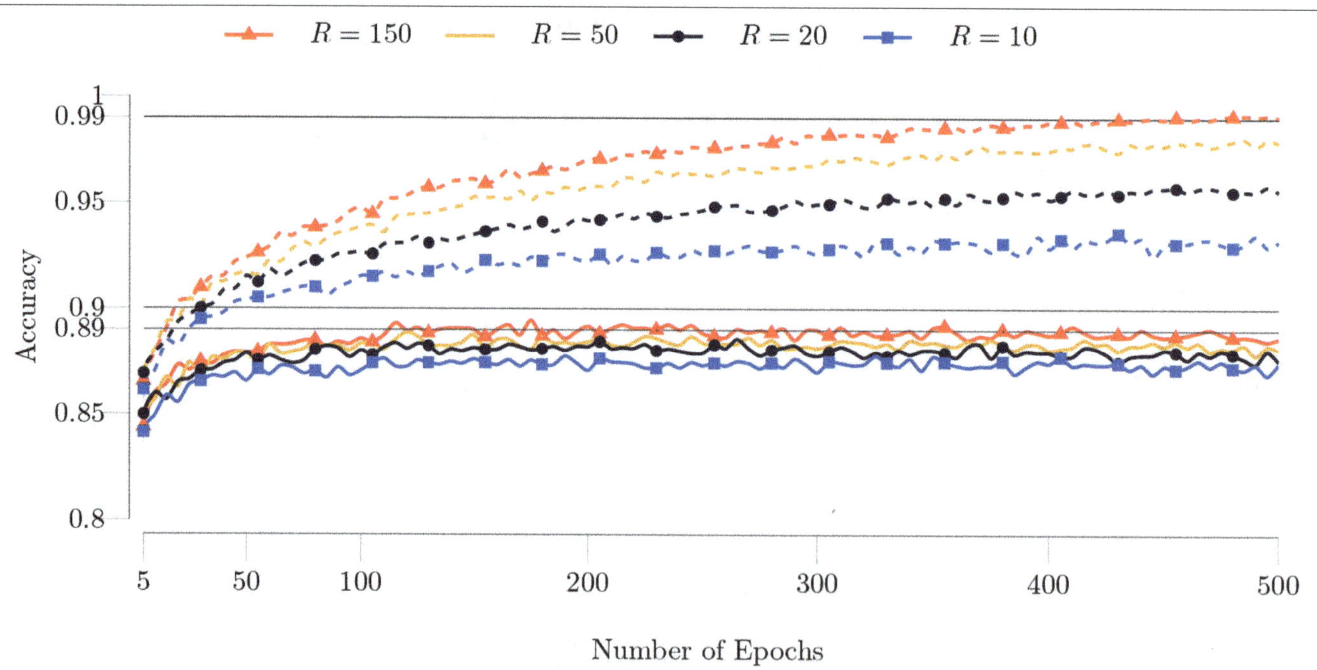

FIGURE 14 | Test (solid lines) and training (dashed lines) accuracies of the type I model for the MNIST dataset with respect to the number of epochs. The full training set (60, 000 images) and test set (10, 000 images) are used. The batch size is set to 200 and the maximum number of GN iterations is set to 1. Both the test and training accuracy increase as the rank increases. Also the improvement in the accuracy gets smaller as the rank increases. The algorithm achieves around 99% training accuracy after 400 epochs.

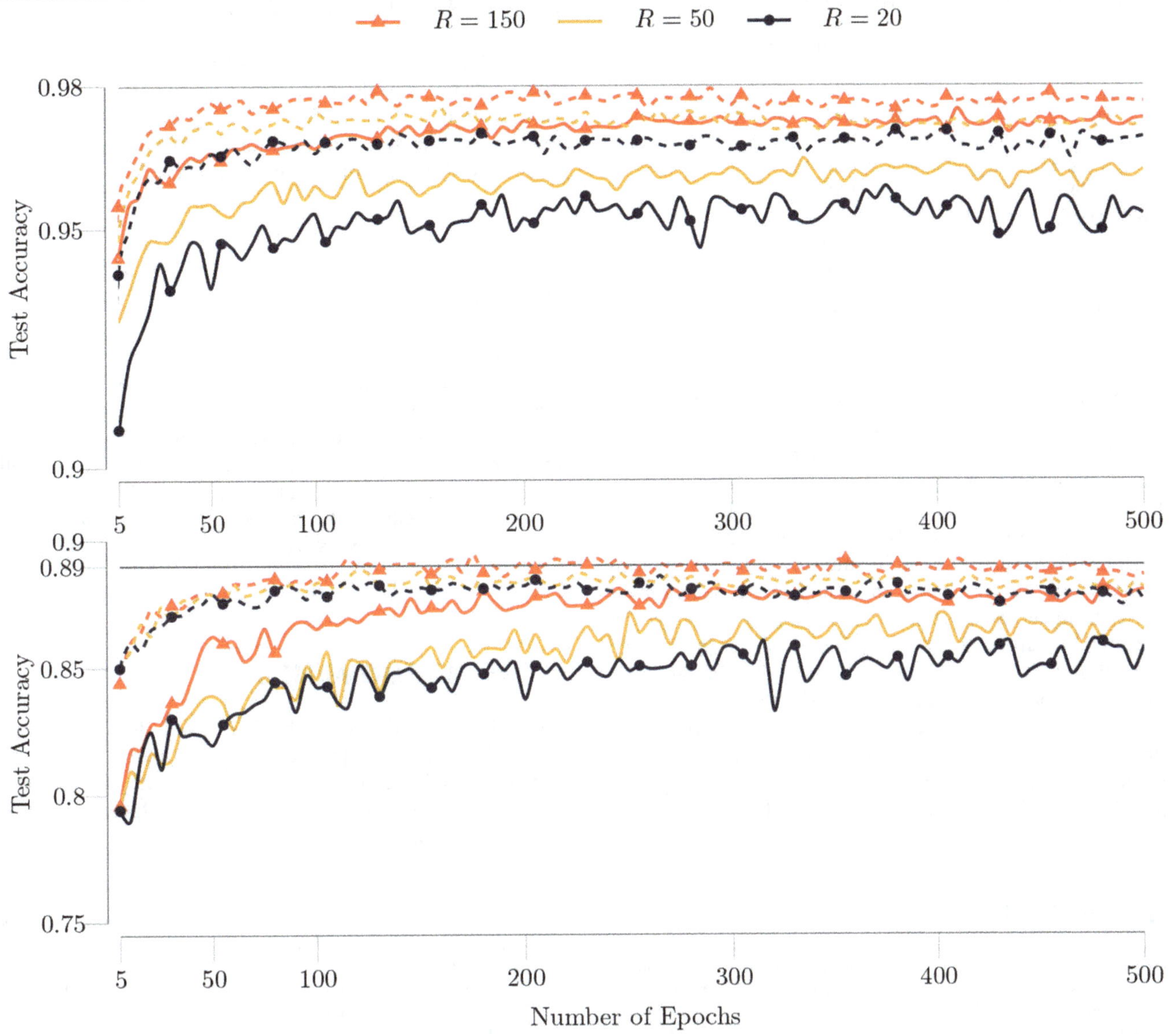

FIGURE 15 | Comparison of test accuracies of the type II model on the MNIST **(top)** and Fashion MNIST **(bottom)** datasets for different maximum number of GN iterations as a function of the number of epochs. The full training set (60, 000 images) and test set (10, 000 images) are used. The batch size is set to 200 and the maximum number of GN iterations is set to 1 (dashed lines) and to 5 (solid lines).

a convolution layer (CNN-PEPS) [22]. We compare in terms of the test accuracy for the Fashion MNIST dataset. We summarize the test accuracy of different models in **Table 2**. TeMPO achieves better accuracy than TT, PEPS and MLP, while optimizing for fewer parameters and using less memory (cf. **Table 1**). The accuracy of TeMPO is lower than CNN-MLP and CNN-PEPS as expected, since it does not use a convolution layer. Note that the accuracy of TeMPO can further be improved by tuning the parameters such as the rank, the number of CG iterations, the trust-region radius, the batch size and the degree of the multivariate polynomial.

5. CONCLUSION AND FUTURE WORK

We presented the TeMPO framework for use in nonlinear optimization problems arising in signal processing, machine learning, and artificial intelligence. We modeled the nonlinearities in these problems by multivariate polynomials represented by low rank tensors. In particular, we investigated the symmetric CPD format in this study. By taking the advantage of low rank symmetric CPD structure, we developed an efficient second-order batch Gauss–Newton algorithm. We demonstrated the efficiency of TeMPO with some illustrative examples, and

TABLE 2 | The test accuracy of different models for the Fashion MNIST dataset.

Model	Test accuracy (%)
TT	88.0
MLP	88.3
PEPS	88.3
TTN	89.0
TeMPO (Type II)	**89.3**
TeMPO (Type I)	**89.9**
CNN–MLP	91.0
CNN–PEPS	91.2

The bold values indicate the results from the proposed methods.

with the blind deconvolution of constant modulus signals. We showed that TeMPO achieves similar or better classification rates than MLPs, TT and PEPS tensor networks on the MNIST and Fashion MNIST datasets while optimizing for fewer parameters and using less memory space.

The non-symmetric and partially symmetric CPD formats are fairly straightforward variants of the symmetric CPD format in which the factor matrices can be mutually different. Efficient algorithms can be developed for multivariate polynomials in these formats by utilizing the derivations presented in this study. We are investigating other tensor formats such as HT and TT in our framework as well. HT and TT require more parameters than the CPD format. However, they break the curse of dimensionality in a numerically stable way. We are also exploring other polynomial bases, and more generally other nonlinear feature maps to further improve the accuracy and numerical stability of our framework.

AUTHOR CONTRIBUTIONS

MA developed the theory and Matlab implementation. He is the main contributor to the numerical experiments and also wrote the first draft of the manuscript. LD conceived the idea and supervised the project. Both authors contributed to manuscript revision, read, and approved the submitted version.

FUNDING

Research supported by: (1) Flemish Government: This research received funding from the Flemish Government (AI Research Program). LD and MA are affiliated to Leuven. AI-KU Leuven institute for AI, B-3000, Leuven, Belgium. This work was supported by the Fonds de la Recherche Scientifique – FNRS and the Fonds Wetenschappelijk Onderzoek – Vlaanderen under EOS Project no G0F6718N (SeLMA). (2) KU Leuven Internal Funds: C16/15/059, IDN/19/014.

ACKNOWLEDGMENTS

The authors would like to thank E. Evert, N. Govindarajan, and S. Hendrikx for proofreading the manuscript and N. Vervliet for valuable discussions. The authors also thank the two referees whose comments/suggestions helped improve and clarify this manuscript.

REFERENCES

1. Sidiropoulos N, De Lathauwer L, Fu X, Huang K, Papalexakis EE, Faloutsos C. Tensor decomposition for signal processing and machine learning. *IEEE Trans Signal Process.* (2017) 65:3551–82. doi: 10.1109/TSP.2017.2690524
2. Cichocki A, Mandic DP, De Lathauwer L, Zhou G, Zhao Q, Caiafa CF, et al. Tensor decompositions for signal processing applications: from two-way to multiway component analysis. *IEEE Signal Process Mag.* (2015) 32:145–63. doi: 10.1109/MSP.2013.2297439
3. Kolda TG, Bader BW. Tensor decompositions and applications. *SIAM Rev.* (2009) 51:455–500. doi: 10.1137/07070111X
4. Sorber L, Van Barel M, De Lathauwer L. Optimization-based algorithms for tensor decompositions: Canonical polyadic decomposition, decomposition in rank-$(L_r, L_r, 1)$ terms, and a new generalization. *SIAM J Optim.* (2013) 23:695–720. doi: 10.1137/120868323
5. Sorber L, Van Barel M, De Lathauwer L. Unconstrained optimization of real functions in complex variables. *SIAM J Optim.* (2012) 22:879–98. doi: 10.1137/110832124
6. Vervliet N, De Lathauwer L. Numerical optimization based algorithms for data fusion. In: Cocchi M, editor. *Data Fusion Methodology and Applications.* Vol. 31. Amsterdam; Oxford; Cambridge: Elsevier (2019). p. 81–128. doi: 10.1016/B978-0-444-63984-4.00004-1
7. Phan AH, Tichavský P, Cichocki A. Low Complexity Damped Gauss-Newton Algorithms for CANDECOMP/PARAFAC. *arXiv:1205.2584.* (2013) 34:126–47. doi: 10.1137/100808034
8. Vervliet N, De Lathauwer L. A randomized block sampling approach to canonical polyadic decomposition of large-scale tensors. *IEEE J Selec Top Sign Process.* (2016) 10:284–95. doi: 10.1109/JSTSP.2015.2503260
9. Comon P, Jutten C. *Handbook of Blind Source Separation: Independent Component Analysis and Applications.* Oxford; Burlington: Academic Press; Elsevier (2009).
10. Vervliet N, Debals O, Sorber L, De Lathauwer L. Breaking the curse of dimensionality using decompositions of incomplete tensors: tensor-based scientific computing in big data analysis. *IEEE Signal Process Mag.* (2014) 31:71–9. doi: 10.1109/MSP.2014.2329429
11. Vervliet N, Debals O, Sorber L, Van Barel M, De Lathauwer L. *Tensorlab 3.0.* (2016). Available online at https://www.tensorlab.net (accessed December, 2021).
12. Vervliet N. *Compressed Sensing Approaches to Large-Scale Tensor Decompositions.* Leuven: KU Leuven (2018).
13. Vandecappelle M, Vervliet N, Lathauwer LD. Inexact generalized gauss-newton for scaling the canonical polyadic decomposition with non-least-squares cost functions. *IEEE J Selec Top Sign Process.* (2021) 15:491–505. doi: 10.1109/JSTSP.2020.3045911
14. Singh N, Zhang Z, Wu X, Zhang N, Zhang S, Solomonik E. Distributed-memory tensor completion for generalized loss functions in python using new sparse tensor kernels. *arXiv:191002371.* (2021). doi: 10.48550/arXiv.1910.02371
15. Ayvaz M, De Lathauwer L. Tensor-based multivariate polynomial optimization with application in blind identification. In: *(2021) 29th Europian Signal Processing Conference, (EUSIPCO).* Dublin (2021). p. 1080–4. doi: 10.23919/EUSIPCO54536.2021.9616070
16. Grasedyck L, Kressner D, Tobler C. A literature survey of low-rank tensor approximation techniques. *GAMM-Mitteil.* (2013) 36:53–78. doi: 10.1002/gamm.201310004

17. Grasedyck L. Hierarchical singular value decomposition of tensors. *SIAM J Matrix Anal Appl.* (2010) 31:2029–54. doi: 10.1137/090764189
18. Oseledets IV, Tyrtyshnikov EE. Breaking the curse of dimensionality, or how to use SVD in many dimensions. *SIAM J Sci Comput.* (2009) 31:3744–59. doi: 10.1137/090748330
19. Novikov A, Trofimov M, Oseledets IV. Exponential machines. In: *5th International Conference on Learning Representations, ICLR 2017.* Toulon (2017). Available online at: https://openreview.net/forum?id=rkm1sE4tg
20. Stoudenmire EM, Schwab DJ. Supervised learning with tensor networks. In: Lee D, Sugiyama M, Luxburg U, Guyon I, Garnett R, editors. *Advances in Neural Information Processing Systems.* Vol. 29. Barcelona: Curran Associates, Inc. (2016). Available online at: https://proceedings.neurips.cc/paper/2016/file/5314b9674c86e3f9d1ba25ef9bb32895-Paper.pdf
21. Efthymiou S, Hidary J, Leichenauer S. TensorNetwork for machine learning. *arXiv: 190606329.* (2019). doi: 10.48550/arXiv.1906.06329
22. Cheng S, Wang L, Zhang P. Supervised learning with projected entangled pair states. *Phys Rev B.* (2021) 103:125117. doi: 10.1103/PhysRevB.103.125117
23. Guo W, Kotsia I, Patras I. Tensor learning for regression. *IEEE Trans Image Process.* (2012) 21:816–27. doi: 10.1109/TIP.2011.2165291
24. Hendrikx S, Boussé M, Vervliet N, De Lathauwer L. Algebraic and optimization based algorithms for multivariate regression using symmetric tensor decomposition. In: *Proceedings of the (2019) IEEE International Workshop on Computational Advances in Multi-Sensor Adaptive Processing (CAMSAP).* Guadeloupe (2019). p. 475–9. doi: 10.1109/CAMSAP45676.2019.9022662
25. Rabusseau G, Kadri H. Low-rank regression with tensor responses. In: Lee D, Sugiyama M, Luxburg U, Guyon I, Garnett R, editors. *Advances in Neural Information Processing Systems.* Vol. 29. Barcelona: Curran Associates, Inc. (2016). Available online at: https://proceedings.neurips.cc/paper/2016/file/3806734b256c27e41ec2c6bffa26d9e7-Paper.pdf
26. Yu R, Liu Y. Learning from multiway data: simple and efficient tensor regression. In: Balcan MF, Weinberger KQ, editors. *Proceedings of the 33rd International Conference on Machine Learning, Vol. 48 of Proceedings of Machine Learning Research.* New York, NY (2016). p. 373–81. Available online at: https://proceedings.mlr.press/v48/yu16.html
27. Hou M, Chaib-Draa B. Hierarchical Tucker tensor regression: application to brain imaging data analysis. In: *Proceedings of the (2015) IEEE International Conference on Image Processing (ICIP 2015).* Québec, QC (2015). p. 1344–8. doi: 10.1109/ICIP.2015.7351019
28. Kar P, Karnick H. Random feature maps for dot product kernels. In: Lawrence ND, Girolami M, editors. *Proceedings of the Fifteenth International Conference on Artificial Intelligence and Statistics, Vol. 22 of Proceedings of Machine Learning Research.* La Palma (2012). p. 583–91. Available online at: https://proceedings.mlr.press/v22/kar12.html
29. Yang J, Gittens A. Tensor machines for learning target-specific polynomial features. *arxiv: 150401697.* (2015). doi: 10.48550/arXiv.1504.01697
30. Rendle S. Factorization machines. In: *(2010) IEEE International Conference on Data Mining.* Sydney (2010). p. 995–1000. doi: 10.1109/ICDM.2010.127
31. Blondel M, Fujino A, Ueda N, Ishihata M. Higher-order factorization machines. In: *Proceedings of the 30th International Conference on Neural Information Processing Systems, NIPS'16.* Red Hook, NY: Curran Associates Inc. (2016). p. 3359–67.
32. Blondel M, Ishihata M, Fujino A, Ueda N. Polynomial networks and factorization machines: new insights and efficient training algorithms. In: *Proceedings of the 33rd International Conference on International Conference on Machine Learning.* Vol. 48. New York, NY (2016). p. 850–8.
33. Nocedal J, Wright S. *Numerical Optimization.* New York, NY: Springer (2006).
34. Kruskal JB. Three-way arrays: rank and uniqueness of trilinear decompositions, with application to arithmetic complexity and statistics. *Linear Algeb Appl.* (1977) 18:95–138. doi: 10.1016/0024-3795(77)90069-6
35. Sidiropoulos ND, Bro R. On the uniqueness of multilinear decomposition of N-way arrays. *J Chemometr.* (2000) 14:229–39. doi: 10.1002/1099-128X(200005/06)14:3<229::AID-CEM587>3.0.CO;2-N
36. Domanov I, De Lathauwer L. On the uniqueness of the canonical polyadic decomposition of third-order tensors – Part ii: uniqueness of the overall decomposition. *SIAM J Matrix Anal Appl.* (2013) 34:876–903. doi: 10.1137/120877258
37. Domanov I, De Lathauwer L. Canonical polyadic decomposition of third-order tensors: relaxed uniqueness conditions and algebraic algorithm. *arXiv:1501.07251.* (2017) 513:342–75. doi: 10.1016/j.laa.2016.10.019
38. Boyd JP, Ong JR. Exponentially-convergent strategies for defeating the Runge phenomenon for the approximation of non-periodic functions, part I: single-interval schemes. *Commun Comput Phys.* (2009) 5:484–97.
39. Trefethen LN. *Approximation Theory and Approximation Practice, Extended Edition.* Philadelphia, PA: SIAM (2019). doi: 10.1137/1.9781611975949
40. De Lathauwer L, De Moor B, Vandewalle J. On the best rank-1 and rank-$(R_1, R_2, \cdots, R_N)$ approximation of higher-order tensors. *SIAM J Matrix Anal Appl.* (2000) 21:1324–42. doi: 10.1137/S0895479898346995
41. Zhang T, Golub G. Rank-one approximation to high order tensors. *SIAM J Matrix Anal Appl.* (2001) 23:534–50. doi: 10.1137/S0895479899352045
42. Guan Y, Chu MT, Chu D. SVD-based algorithms for the best rank-1 approximation of a symmetric tensor. *SIAM J Matrix Anal Appl.* (2018) 39:1095–115. doi: 10.1137/17M1136699
43. Nie J, Wang L. Semidefinite relaxations for best rank-1 tensor approximations. *SIAM J Matrix Anal Appl.* (2013) 35:1155–79. doi: 10.1137/130935112
44. Brachat J, Comon P, Mourrain B, Tsigaridas E. Symmetric tensor decomposition. *Linear Algeb Appl.* (2010) 433:1851–72. doi: 10.1016/j.laa.2010.06.046
45. Alexander J, Hirschowitz A. Polynomial interpolation in several variables. *Adv Comput Math.* (1995) 4:201–22.
46. Debals O. *Tensorization and Applications in Blind Source Separation.* Leuven: KU Leuven (2017).
47. Blondel M, Niculae V, Otsuka T, Ueda N. Multi-output Polynomial Networks and Factorization Machines. In: *Advances in Neural Information Processing Systems 30: Annual Conference on Neural Information Processing Systems 2017.* Long Beach, CA (2017). p. 3349–59.
48. Khoromskij BN. *Tensor Numerical Methods in Scientific Computing.* Berlin; Boston: De Gruyter (2018). doi: 10.1515/9783110365917
49. Margossian CC. A review of automatic differentiation and its efficient implementation. *WIREs Data Mining Knowl Discov.* (2019) 9:e1305. doi: 10.1002/widm.1305
50. van der Veen AJ, Paulraj A. An analytical constant modulus algorithm. *IEEE Trans Signal Process.* (1996) 44:1136–55. doi: 10.1109/78.502327
51. Zarzoso V, Comon P. Optimal step-size constant modulus algorithm. *IEEE Trans Commun.* (2008) 56:10–3. doi: 10.1109/TCOMM.2008.050484
52. Boussé M, Vervliet N, Domanov I, Debals O, De Lathauwer L. Linear systems with a canonical polyadic decomposition constrained solution: algorithms and applications. *Numer Linear Algeb Appl.* (2018) 25:e2190. doi: 10.1002/nla.2190
53. Gargiani M, Zanelli A, Diehl M, Hutter F. On the promise of the stochastic generalized Gauss-Newton method for training DNNs. *arXiv: 200602409.* (2020). doi: 10.48550/arXiv.2006.02409
54. Kingma DP, Ba J. Adam: a method for stochastic optimization. In: Bengio Y, LeCun Y, editors. *International Conference on Learning Representations, ICLR 2015.* 3rd Edn. San Diego, CA (2015). Available online at: http://arxiv.org/abs/1412.6980
55. De Brabanter K, Karsmakers P, Ojeda F, Alzate C, De Brabanter J, Pelckmans K, et al. *LS-SVMlab Toolbox User's Guide Version 1.8.* Leuven: ESAT-STADIUS (2010). p. 10–46.
56. Suykens JAK, Van Gestel T, De Brabanter J, De Moor B, Vandewalle J. *Least Squares Support Vector Machines.* Singapore: World Scientific (2002). doi: 10.1142/5089
57. Ljung L. *System Identification: Theory for the User.* 2nd ed. Upper Saddle River, NJ: Prentice Hall (1999). doi: 10.1002/047134608X.W1046
58. Johnson R, Schniter P, Endres TJ, Behm JD, Brown DR, Casas RA. Blind equalization using the constant modulus criterion: a review. *Proc IEEE.* (1998) 86:1927–50. doi: 10.1109/5.720246
59. van der Veen AJ. Algebraic methods for deterministic blind beamforming. *Proc IEEE.* (1998) 86:1987–2008. doi: 10.1109/5.720249
60. De Lathauwer L. Algebraic techniques for the blind deconvolution of Constant Modulus signals. In: *Proceedings of the 12th European Signal Processing Conference (EUSIPCO 2004).* Vienna (2004). p. 225–8.
61. Householder AS. Unitary triangularization of a nonsymmetric matrix. *J ACM.* (1958) 5:339–42. doi: 10.1145/320941.320947

62. Deng L. The MNIST database of handwritten digit images for machine learning research. *IEEE Sign Process Mag.* (2012) 29:141–2. doi: 10.1109/MSP.2012.2211477
63. Xiao H, Rasul K, Vollgraf R. Fashion-MNIST: a novel image dataset for benchmarking machine learning algorithms *arXiv:1708.07747.* (2017). doi: 10.48550/arXiv.1708.07747
64. Stoudenmire EM. Learning relevant features of data with multi-scale tensor networks. *Quant Sci Technol.* (2018) 3:034003. doi: 10.1088/2058-9565/aaba1a

17

Geostatistical Learning: Challenges and Opportunities

Júlio Hoffimann *†, *Maciel Zortea, Breno de Carvalho and Bianca Zadrozny*

IBM Research, Rio de Janeiro, Brazil

***Correspondence:**
Júlio Hoffimann
julio.hoffimann@impa.br

†Present address:
Instituto de Matemática Pura e Aplicada, Rio de Janeiro, Brazil

Statistical learning theory provides the foundation to applied machine learning, and its various successful applications in computer vision, natural language processing and other scientific domains. The theory, however, does not take into account the unique challenges of performing statistical learning in geospatial settings. For instance, it is well known that model errors cannot be assumed to be independent and identically distributed in geospatial (a.k.a. regionalized) variables due to spatial correlation; and trends caused by geophysical processes lead to covariate shifts between the domain where the model was trained and the domain where it will be applied, which in turn harm the use of classical learning methodologies that rely on random samples of the data. In this work, we introduce the *geostatistical (transfer) learning* problem, and illustrate the challenges of learning from geospatial data by assessing widely-used methods for estimating generalization error of learning models, under covariate shift and spatial correlation. Experiments with synthetic Gaussian process data as well as with real data from geophysical surveys in New Zealand indicate that none of the methods are adequate for model selection in a geospatial context. We provide general guidelines regarding the choice of these methods in practice while new methods are being actively researched.

Keywords: geostatistical learning, transfer learning, covariate shift, geospatial, density ratio estimation, importance weighted cross-validation

1 INTRODUCTION

Classical learning theory [1–3] and its applied machine learning methods have been popularized in the geosciences after various technological advances, leading initiatives in open-source software [4–7], and intense marketing from a diverse portfolio of industries. In spite of its popularity, learning theory cannot be applied straightforwardly to solve problems in the geosciences as the characteristics of these problems violate fundamental assumptions used to derive the theory and related methods (e.g., i. i.d. samples).

Among these methods derived under classical assumptions (more on this later), those for estimating the generalization (or prediction) error of learned models in unseen samples are crucial in practice [2]. In fact, estimates of generalization error are widely used for selecting the best performing model for a problem, out of a collection of available models [8], and statistical learning can be posed broadly as minimization of generalization error. If estimates of error are inaccurate because of violated assumptions, then there is great chance that models will be selected inappropriately [9]. The issue is aggravated when models of great expressiveness (i.e., many learning parameters) are considered in the collection since they are quite capable of overfitting the available data [10, 11].

The literature on generalization error estimation methods is vast [8, 12], and we do not intend to review it extensively here. Nevertheless, some methods have gained popularity since their introduction in the mid 70s because of their generality, ease of use, and availability in open-source software:

Leave-one-out (1974). The leave-one-out method for assessing and selecting learning models was based on the idea that to estimate the prediction error on an unseen sample one only needs to hide a seen sample from a dataset and learn the model. Because the hidden sample has a known label, the method can compare the model prediction with the true label for the sample. By repeating the process over the entire dataset, one gets an estimate of the expected generalization error [13]. Leave-one-out has been investigated in parallel by many statisticians, including Nicholson (1960) and Stone (1974), and is also known as ordinary cross-validation.

k-fold cross-validation (1975). The term k-fold cross-validation refers to a family of error estimation methods that split a dataset into non-overlapping "folds" for model evaluation. Similar to leave-one-out, each fold is hidden while the model is learned using the remaining folds. It can be thought of as a generalization of leave-one-out where folds may have more than a single sample [14, 15]. Cross-validation is less computationally expensive than leave-one-out depending on the size and number of folds, but can introduce bias in the error estimates if the number of samples in the folds used for learning is much smaller than the original number of samples in the dataset.

Major assumptions are involved in the derivation of the estimation methods listed above. The first of them is the assumption that samples come from independent and identically distributed (i.i.d.) random variables. It is well-known that spatial samples are not i.i.d., and that spatial correlation needs to be modeled explicitly with geostatistical theory. Even though the sample mean of the empirical error used in those methods is an unbiased estimator of the prediction error regardless of the i.i.d. assumption, the precision of the estimator can be degraded considerably with non-i.i.d. samples.

Motivated by the necessity to leverage non-i.i.d. samples in practical applications, and evidence that model's performance is affected by spatial correlation [16, 17], the statistical community devised new error estimation methods using the spatial coordinates of the samples:

h-block leave-one-out (1995). Developed for time-series data (i.e., data showing temporal dependency), the h-block leave-one-out method is based on the principle that stationary processes achieve a correlation length (the "h") after which the samples are not correlated. The time-series data is then split such that samples used for error evaluation are at least "h steps" distant from the samples used to learn the model [18]. Burman (1994) showed how the method outperformed traditional leave-one-out in time-series prediction by selecting the hyperparameter "h" as a fraction of the data, and correcting the error estimates accordingly to avoid bias.

Spatial leave-one-out (2014). Spatial leave-one-out is a generalization of h-block leave-one-out from time-series to spatial data [19]. The principle is the same, except that the blocks have multiple dimensions (e.g., norm-balls).

Block cross-validation (2016). Similarly to k-fold cross-validation for non-spatial data, block cross-validation was proposed as a faster alternative to spatial leave-one-out. The method creates folds using blocks of size equal to the spatial correlation length, and separates samples for error evaluation from samples used to learn the model. The method introduces the concept of "dead zones", which are regions near the evaluation block that are discarded to avoid over-optimistic error estimates [20, 21].

Unlike the estimation methods proposed in the 70s, which use random splits of the data, these methods split the data based on spatial coordinates and what the authors called "dead zones". This set of heuristics for creating data splits avoids configurations in which the model is evaluated on samples that are too near (< spatial correlation length) other samples used for learning the model. Consequently, these estimation methods tend to produce error estimates that are higher on average than their non-spatial counterparts, which are known to be over-optimistic in the presence of spatial correlation. However, systematic splits of the data introduce bias, which have not been emphasized enough in the literature.

All methods for estimating generalization error in classical learning theory, including the methods listed above, rely on a second major assumption. The assumption that the distribution of unseen samples to which the model will be applied is equal to the distribution of samples over which the model was trained. This assumption is very unrealistic for various applications in geosciences, which involve quite heterogeneous (i.e., variable), heteroscedastic (i.e., with different variability) processes [22].

Very recently, an alternative to classical learning theory has been proposed, known as transfer learning theory, to deal with the more difficult problem of learning under shifts in distribution, and learning tasks [23–25]. The theory introduces methods that are more amenable for geoscientific work [26–28], yet these same methods were not derived for geospatial data (e.g. climate data, Earth observation data, field measurements).

Of particular interest in this work, the covariate shift problem is a type of transfer learning problem where the samples on which the model is applied have a distribution of covariates that differs from the distribution of covariates over which the model was trained [29]. It is relevant in geoscientific applications in which a list of explanatory features is known to predict a response via a set of physical laws that hold everywhere. Under covariate shift, a generalization error estimation method has been proposed:

Importance-weighted cross-validation (2007). Under covariate shift, and assuming that learning models may be misspecified, classical cross-validation is not unbiased. Importance weights can be considered for each sample to recover the unbiasedness property of the method, and this is the core idea of importance-weighted cross-validation [30, 31]. The method is unbiased under covariate shift for the two most common supervised learning tasks: regression and classification.

The importance weights used in importance-weighted cross-validation are ratios between the target (or test) probability density and the source (or train) probability density of covariates. Density ratios are useful in a much broader set of applications including two-sample tests, outlier detection, and distribution comparison. For that reason, the problem of density ratio estimation has become a general statistical problem [32]. Various density ratio estimators have been proposed with

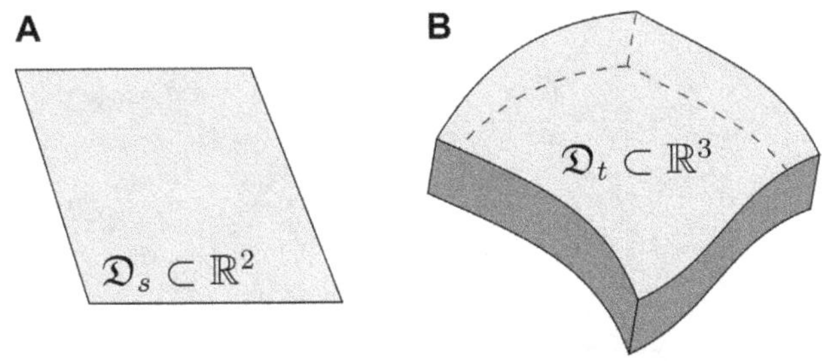

FIGURE 1 | Examples of source and target spatial domains. **(A)** Source domain $\mathcal{D}_s \subset \mathbb{R}^2$, which may represent a satellite view of an area of interest. **(B)** Target domain $\mathcal{D}_t \subset \mathbb{R}^3$, which may represent a body of rock in the subsurface of the Earth.

increasing performance [33–36], yet an investigation is missing that contemplates importance-weighted cross-validation and other existing error estimation methods in geospatial settings.

In this work, we introduce *geostatistical (transfer) learning*, and discuss how most prior work in spatial statistics fits in a specific type of learning from geospatial data that we term *pointwise learning*. In order to illustrate the challenges of learning from geospatial data, we assess existing estimators of generalization error from the literature using synthetic Gaussian process data and real data from geophysical well logs in New Zealand that we made publicly available [37].

The paper is organized as follows. In **Section 2**, we introduce *geostatistical (transfer) learning*, which contains all the elements involved in learning from geospatial data. We define covariate shift in the geospatial setting and briefly review the concept of spatial correlation. In **Section 3**, we define generalization error in geostatistical learning, discuss how it generalizes the classical definition of error in non-spatial settings, and review estimators of generalization error from the literature devised for *pointwise learning*. In **Section 4**, we present our experiments with geospatial data, and discuss the results of different error estimation methods. In **Section 5**, we conclude the work and share a few remarks regarding the choice of error estimation methods in practice.

2 GEOSTATISTICAL LEARNING

In this section, we define the elements of statistical learning in geospatial settings. We discuss the covariate shift and spatial correlation properties of the problem, and illustrate how they affect the involved feature spaces.

Consider a sample space $\mathbf{\Omega}$, a source spatial domain $\mathcal{D}_s \subset \mathbb{R}^{d_s}$, and a target spatial domain $\mathcal{D}_t \subset \mathbb{R}^{d_t}$ on which stochastic processes (i.e., spatial random variables) are defined:

$$\begin{aligned} Z_{s_j} &: \mathcal{D}_s \times \mathbf{\Omega} \to \mathbb{R},\ j = 1, 2, \ldots, n_s \text{ on source domain } \mathcal{D}_s \\ Z_{t_j} &: \mathcal{D}_t \times \mathbf{\Omega} \to \mathbb{R},\ j = 1, 2, \ldots, n_t \text{ on target domain } \mathcal{D}_t \end{aligned} \quad (1)$$

For example, $(Z_{s_j})_{j=1,2,\ldots,n_s}$ may represent a collection of processes observed remotely from satellite on a 2D surface $\mathcal{D}_s \subset \mathbb{R}^2$, whereas $(Z_{t_j})_{j=1,2,\ldots,n_t}$ may represent processes that occur within the 3D subsurface of the Earth $\mathcal{D}_t \subset \mathbb{R}^3$ (see **Figure 1**). Any process Z in these collections can be viewed in two distinct ways:

Geostatistical theory. From the viewpoint of geostatistical theory, samples $z(\cdot, \omega)$ of the process $Z(\mathbf{u}, w)$ are obtained by fixing $\omega \in \mathbf{\Omega}$. These samples are spatial maps that assign a real number to each location $\mathbf{u} \in \mathcal{D}$.

Learning theory. From the viewpoint of statistical learning theory, scalar samples $z(\mathbf{u}, \cdot)$ are obtained by fixing $\mathbf{u} \in \mathcal{D}$. These scalar samples are ordered into a "feature vector" $\mathbf{x_u} = (z_1, z_2, \ldots, z_n)$ for a collection of processes $(Z_j)_{j=1,2,\ldots,n}$, and for a specific location $\mathbf{u} \in \mathcal{D}$. In this case, $\mathbf{X_u} : \mathbf{\Omega} \to \mathbb{R}^n$ denotes the corresponding random vector of features such that $\mathbf{x_u} \sim \mathbf{X_u}$.

In order to define the geostatistical learning problem, we need to understand the joint probability distribution of features for all locations in a spatial domain $\Pr(\{\mathbf{X_u}\}_{\mathbf{u}\in\mathcal{D}})$. This distribution is very complex in general as feature vectors $\mathbf{X_u}$ and $\mathbf{X_v}$ for two different locations $\mathbf{u} \neq \mathbf{v}$ are not independent. The closer the locations $\mathbf{u}, \mathbf{v} \in \mathcal{D}$ in the spatial domain, the more similar are their features $\mathbf{x_u}, \mathbf{x_v} \in \mathbb{R}^n$ in the feature space. Moreover, given that only one realization $z^{obs} = z(\cdot, \omega) \sim Z$ of the process is available at any given time, one must introduce stationarity assumptions inside $\mathcal{D}$ to pool together different scalar samples $z(\mathbf{u}, \cdot)$ from different locations $\mathbf{u} \in \mathcal{D}$ in the spatial domain, and be able to estimate the distribution.

Regardless of the stationarity assumptions involved in modeling these processes, we can assume that inside $\mathcal{D}$ the probability $\Pr_{\mathcal{D}}(\mathbf{X}) = \Pr(\{\mathbf{X_u}\}_{\mathbf{u}\in\mathcal{D}})$ is well-defined. For example, most prior art in statistical learning with geospatial data assume that the pointwise probability of features $\Pr_{\mathbf{u}}(\mathbf{X}) = \Pr(\mathbf{X_u})$ is not a function of location, that is $\Pr_{\mathbf{u}}(\mathbf{X}) = \Pr(\mathbf{X}), \forall \mathbf{u} \in \mathcal{D}$. Under this assumption, samples from everywhere in $\mathcal{D}$ are used to estimate $\Pr(\mathbf{X}) = \Pr(Z_1, Z_2, \ldots, Z_n)$. With the additional assumption that the feature vectors $\mathbf{X_u}$ and $\mathbf{X_v}$ are independent, the joint distribution of features for all locations can be written as $\Pr_{\mathcal{D}}(\mathbf{X}) = \prod_{\mathbf{u}\in\mathcal{D}} \Pr_{\mathbf{u}}(\mathbf{X})$.

Whereas the pointwise stationarity assumption may be reasonable inside a given spatial domain, the assumption of spatial independence of features is rarely defensible in practice. Additionally, pointwise stationarity often does not transfer from a source domain $\mathcal{D}_s$ where the model is learned to a target domain $\mathcal{D}_t$ where the model is applied, and consequently the joint distributions of features differ $\Pr_{\mathcal{D}_s} \neq \Pr_{\mathcal{D}_t}$. Before we can illustrate these two issues in more detail, we need to complete the definition of geostatistical learning problems by introducing the notion of spatial learning tasks.

We have introduced the notion of spatial domain $\mathcal{D}$, and the notion of joint probability of features $\Pr_{\mathcal{D}}(\mathbf{X})$ for all locations in the domain. Now we introduce the notion of spatial learning tasks, which are similar to classical learning tasks, but with the main difference that they can leverage properties of the underlying spatial domain. Classically, a learning task describes an action in terms of available features to produce new data. For example, "predict feature Z_{j_0} from features (Z_{j_1}, Z_{j_2})", or "cluster the samples using features $(Z_{j_1}, Z_{j_2}, Z_{j_3})$" are classical learning tasks. Differently, a spatial learning task T involves the spatial domain $\mathcal{D}$ besides the features, and is therefore more complex. Practical examples from the industry include:

- Mining: The task of segmenting a mineral deposit from drillhole samples using a set of features is a spatial learning task. It assumes the segmentation result to be a *contiguous volume* of rock, which is an additional constraint in terms of spatial coordinates.
- Agriculture: The task of identifying crops from satellite images is a spatial learning task. Locations that have the same crop type *appear together* in the images despite possible noise in image layers (e.g. presence of clouds, animals).
- Petroleum: The task of segmenting formations from seismic data is a spatial learning task because these formations are large-scale *near-horizontal* layers of stacked rock.

Many more examples of spatial learning tasks exist, and others are yet to be proposed. Given the concepts introduced above, we are now ready for the main definition of this section:

Definition (Geostatistical Learning). Let $\mathcal{D}_s$ be a source spatial domain, and $\mathcal{D}_t$ be a target spatial domain. Let $\Pr_{\mathcal{D}_s}(\mathbf{X}_s)$ and $\Pr_{\mathcal{D}_t}(\mathbf{X}_t)$ be the joint distributions of features for all locations in these domains, and let $\boldsymbol{T}_s$ and $\boldsymbol{T}_t$ be two spatial learning tasks. Geostatistical (transfer) learning consists of learning $\boldsymbol{T}_t$ over $\mathcal{D}_t$ using the knowledge acquired while learning $\boldsymbol{T}_s$ over $\mathcal{D}_s$, and assuming that the observed spatial data in $\mathcal{D}_s$ and $\mathcal{D}_t$ are both a single spatial sample of $\Pr_{\mathcal{D}_s}(\mathbf{X}_s)$ and $\Pr_{\mathcal{D}_t}(\mathbf{X}_t)$, respectively.

There are considerable differences between the classical definition of transfer learning [23, 24], and the proposed definition above. First, the distribution we have denoted by $\Pr_{\mathcal{D}}(\mathbf{X})$ is spatial and involves all the locations $\mathbf{u} \in \mathcal{D}$, whereas the distribution in classical transfer learning is the marginal for any specific location, obtained from the assumption of pointwise stationarity $\Pr(\mathbf{X_u}) = \Pr(\mathbf{X})$. Second, we use the term domain to refer to spatial domains $\mathcal{D}$, whereas the non-spatial literature uses the same term for the pair $(\mathbf{X_u}, \Pr(\mathbf{X_u})) = (\mathbf{X}, \Pr(\mathbf{X}))$. Third, the spatial learning task T we have introduced may be described in terms of properties of the spatial domain, which are not available in generic transfer learning problems.

Having understood the main differences between classical and geostatistical learning, we now focus our attention to a specific type of geostatistical transfer learning problem, and illustrate some of the unique challenges caused by spatial dependence.

2.1 Covariate Shift

Assume that the two spatial domains are different $\mathcal{D}_s \neq \mathcal{D}_t$, but that they share a set of processes $(Z_1, Z_2, \ldots, Z_n)$. Additionally, assume that pointwise stationarity holds. Let $Z_o = f(Z_1, Z_2, \ldots, Z_n)$ be a new process obtained as a function of the shared processes, and assume that it has only been observed in $\mathcal{D}_s$ via a measuring device and/or manual labeling. That is, $z_o^{obs}(\cdot, \omega) \sim Z_o$ is a spatial sample of the process Z_o over $\mathcal{D}_s$. Under these assumptions, we can introduce the shared vector of features $\mathbf{X}_s = \mathbf{X}_t = \mathbf{X} = (Z_1, Z_2, \ldots, Z_n, Z_o)$, and the supervised learning task $T_s = T_t = T$ of predicting the process Z_o regardless of location $\mathbf{u} \in \mathcal{D}_s \cup \mathcal{D}_t$.

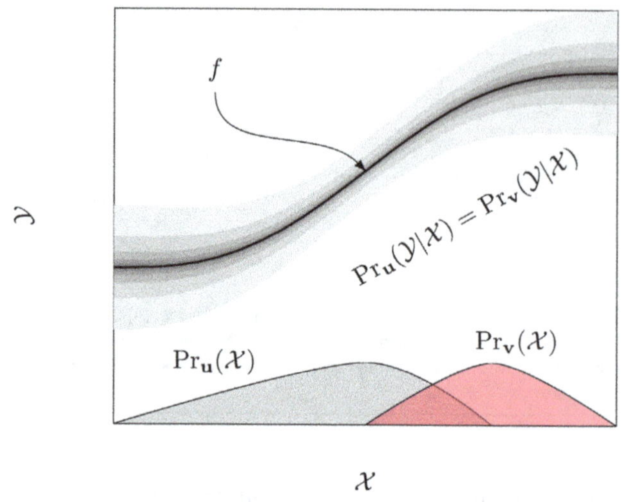

FIGURE 2 | Covariate shift. The true underlying function $\dagger = f(\mathcal{X})$ is approximated equally as $\Pr_{\mathbf{u}}(\dagger|\mathcal{X}) = \Pr_{\mathbf{v}}(\dagger|\mathcal{X})$, $\forall \mathbf{u} \in \mathcal{D}_s$, $\forall \mathbf{v} \in \mathcal{D}_t$. The only difference between source and target spatial domains lies in the distributions of explanatory variables $\Pr_{\mathbf{u}}(\mathcal{X}) \neq \Pr_{\mathbf{v}}(\mathcal{X})$.

Let $\mathcal{X} = \mathbf{X}_{1:n}$ be the explanatory features, and $\mathcal{Y} = \mathbf{X}_{n+1}$ be the response feature. For any $\mathbf{u} \in \mathcal{D}_s$, we can write $\Pr_{\mathbf{u}}(\mathcal{X}, \mathcal{Y}) = \Pr_{\mathbf{u}}(\mathcal{Y}|\mathcal{X})\Pr_{\mathbf{u}}(\mathcal{X})$. Likewise, for any $\mathbf{v} \in \mathcal{D}_t$ we can write $\Pr_{\mathbf{v}}(\mathcal{X}, \mathcal{Y}) = \Pr_{\mathbf{v}}(\mathcal{Y}|\mathcal{X})\Pr_{\mathbf{v}}(\mathcal{X})$. The covariate shift property is defined as follows:

Definition (Covariate Shift). A geostatistical learning problem has the covariate shift property when for any $\mathbf{u} \in \mathcal{D}_s$ and for any $\mathbf{v} \in \mathcal{D}_t$ the distributions $\Pr_{\mathbf{u}}(\mathcal{X}, \mathcal{Y})$ and $\Pr_{\mathbf{v}}(\mathcal{X}, \mathcal{Y})$ differ by $\Pr_{\mathbf{u}}(\mathcal{X}) \neq \Pr_{\mathbf{v}}(\mathcal{X})$ while $\Pr_{\mathbf{u}}(\mathcal{Y}|\mathcal{X}) = \Pr_{\mathbf{v}}(\mathcal{Y}|\mathcal{X})$.

The property is based on the idea that the underlying true function f that created the process $\mathcal{Y} = f(\mathcal{X})$ is the same for all $\mathbf{u} \in \mathcal{D}_s$ and all $\mathbf{v} \in \mathcal{D}_t$. In this case, the function is approximated by the conditional distribution $\Pr_{\mathbf{u}}(\mathcal{Y}|\mathcal{X}) = \Pr_{\mathbf{v}}(\mathcal{Y}|\mathcal{X})$ for each and every location (see **Figure 2**).

In the geosciences, it is very common to encounter problems with covariate shift due to the great variability of natural processes. Whenever a model is 1) learned using labels provided by experts on a spatial domain $\mathcal{D}_s$, is 2) validated with classical train-validation-test methodologies (meaning that it satisfies some performance threshold), and yet 3) performs poorly on a new spatial domain $\mathcal{D}_t$ where the labeling function is expected to be the same, we can conclude that there are shifts in distribution. In **Section 4** we illustrate covariate shifts in real data that we prepared in-house from geophysical surveys in New Zealand.

2.2 Spatial Correlation

Another important issue with geospatial data that is often ignored is spatial dependence, which we illustrate next. As mentioned earlier, the closer are two locations $\mathbf{u}, \mathbf{v} \in \mathcal{D}$ in a spatial domain, the more similar are their features $\mathbf{x_u}, \mathbf{x_v} \in \mathbb{R}^n$ in the feature space. Different statistics are available to quantify this spatial

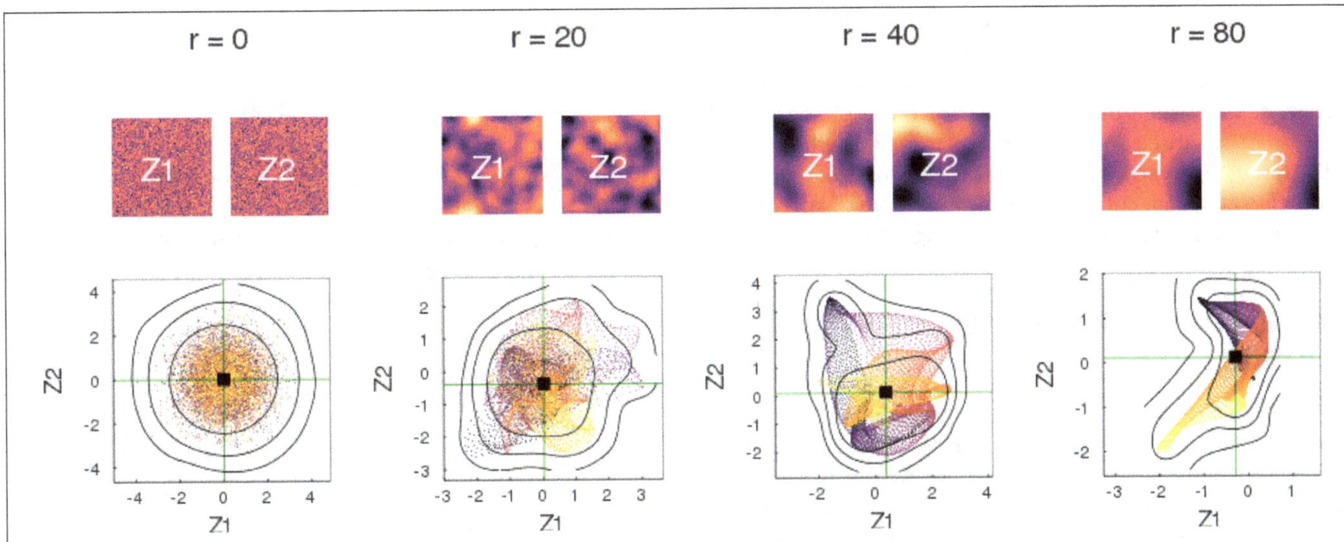

FIGURE 3 | Impact of spatial correlation in feature space. Two Gaussian processes Z_1 and Z_2 are simulated over a domain $\mathcal{D}$ with 100 × 100 pixels. As the variogram range r increases from $r = 0$ to $r = 80$ pixels, the distribution of features $\Pr(\boldsymbol{\mathcal{X}}) = \Pr(Z_1, Z_2)$ is gradually deformed away from the standard Gaussian $\mathcal{N}(0, I)$. Marker colors in scatter plots represent the pixel number in column-major order.

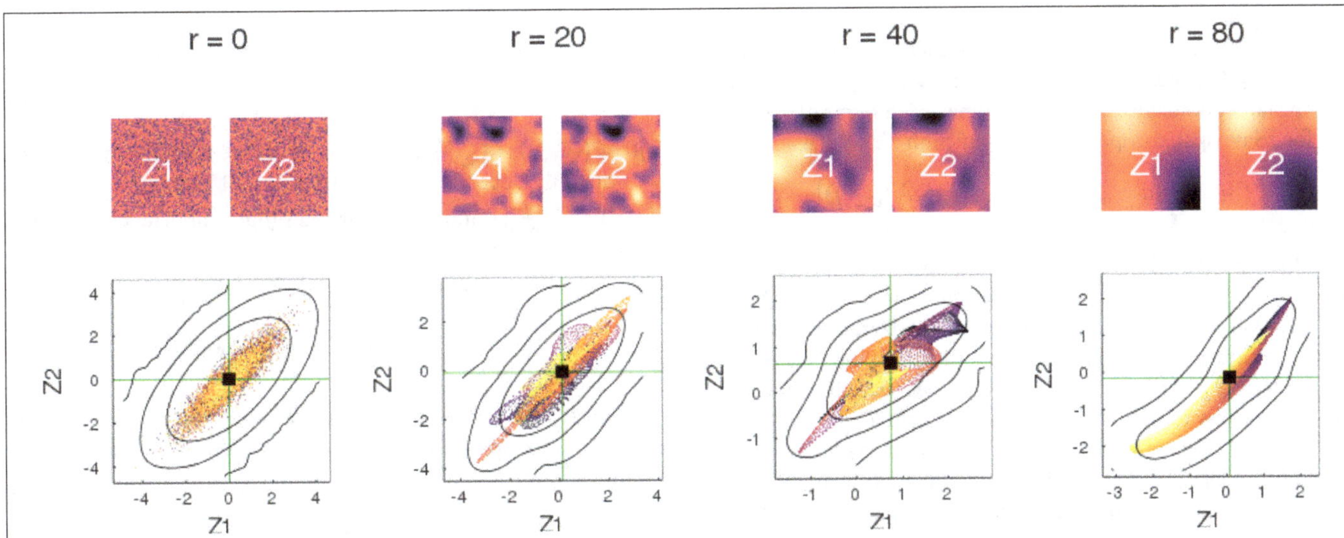

FIGURE 4 | Impact of spatial correlation in feature space with correlated processes. Similar to **Figure 3** with the difference that the processes Z_1 and Z_2 are strongly correlated with correlation $\rho(Z_1, Z_2) = 0.9$.

dependence in a collection of samples, and a popular choice from geostatistics is the variogram $\gamma(h)$, which estimates for each spatial lag $h = \|\mathbf{u} - \mathbf{v}\| \in \mathbb{R}_0^+$ a correlation $\sigma^2 - \gamma(h)$, where σ^2 is the total sill in the samples [22]. Parallel algorithms for efficient variogram estimation exist in the literature [38], and can be useful tools for fast diagnosis of the spatial correlation property:

Definition (Spatial Correlation). A geostatistical learning problem has the spatial correlation property when the variogram of any of the stochastic processes $(Z_{s_j})_{j=1,2,\ldots,n_s}$ and $(Z_{t_j})_{j=1,2,\ldots,n_t}$ defined over $\mathcal{D}_s$ and $\mathcal{D}_t$ has a non-negligible positive range (or correlation length)

Besides serving as a tool for diagnosing spatial correlation in geostatistical learning problems, variograms can also be used to simulate spatial processes with theoretical correlation structure. In **Figure 3**, we illustrate the impact of spatial correlation in the feature space of two independent spatial processes Z_1 and Z_2 simulated with direct (a.k.a. LU) Gaussian simulation [39]. As we increase the variogram range r in a spatial domain $\mathcal{D}$ with 100 × 100 pixels, we observe that the distribution of features $\Pr(\boldsymbol{\mathcal{X}}) = \Pr(Z_1, Z_2)$ is gradually deformed from a standard Gaussian ($r = 0$) to a "boomerang"-shaped distribution ($r = 80$).

Similar deformations are observed when the two processes Z_1 and Z_2 are correlated. In **Figure 4**, we illustrate the impact of spatial correlation for an inter-process correlation of $\rho(Z_1, Z_2) = 0.9$.

Spatial correlation may have different impact in source and target domains $\mathcal{D}_s$ and $\mathcal{D}_t$, and can certainly affect the generalization error of learning models. In our experiments of **Section 4**, we assume that the variogram range of source and target processes are equal (i.e. $r_s = r_t = r$) to facilitate the analysis of the results. In practice, source and target processes may also have different spatial correlation, which is a type of shift that is not considered in classical transfer learning problems.

3 GENERALIZATION ERROR OF LEARNING MODELS

Having defined geostatistical learning problems, and their covariate shift and spatial correlation properties, we now turn into a general definition of generalization error of learning models in geospatial settings. We review an importance-weighted approximation of a related generalization error based on pointwise stationarity assumptions, and the use of an efficient importance-weighted cross-validation method for error estimation.

Consider a geostatistical learning problem $\mathcal{P} = \{(\mathcal{D}_s, \Pr_{\mathcal{D}_s}, T_s), (\mathcal{D}_t, \Pr_{\mathcal{D}_t}, T_t)\}$ with a single supervised spatial learning task $T_s = T_t = T$ (e.g. regression), and assume that a set of response features $\mathcal{Y}_\mathbf{u}$ are created by a function f, based on a set of explanatory features $\mathcal{X}_\mathbf{u}$ for each and every location $\mathbf{u} \in \mathcal{D}_s \cup \mathcal{D}_t$. Our goal is to learn a model $\{\mathcal{Y}_\mathbf{u}\}_{\mathbf{u}\in\mathcal{D}_t} \approx \hat{f}(\{\mathcal{X}\}_{\mathbf{u}\in\mathcal{D}_t})$ over the target domain $\mathcal{D}_t$ that approximates f in terms of expected risk for some spatial supervised loss function $\mathcal{L}$:

$$\hat{f} = \arg\min_{g} \mathbb{E}_{\Pr_{\mathcal{D}_t}}\left[\mathcal{L}\left(\{\mathcal{Y}_\mathbf{u}\}_{\mathbf{u}\in\mathcal{D}_t}, g(\{\mathcal{X}\}_{\mathbf{u}\in\mathcal{D}_t})\right)\right] \tag{2}$$

In the expected value of **Eq 2**, spatial samples of the processes are drawn from $\Pr_{\mathcal{D}_t}$ and rearranged into feature vectors $\mathcal{X}_\mathbf{u}$ and $\mathcal{Y}_\mathbf{u}$ for every location $\mathbf{u} \in \mathcal{D}_t$. The spatial loss function $\mathcal{L}$ compares the spatial map of features from the sample $\{\mathcal{Y}_\mathbf{u}\}_{\mathbf{u}\in\mathcal{D}_t}$ with the approximated map from the model $g(\{\mathcal{X}\}_{\mathbf{u}\in\mathcal{D}_t})$. The model $\hat{f}$ is the model that minimizes the expected loss (or risk) over the target domain $\mathcal{D}_t$.

Definition (Generalization Error). The generalization error of a learning model $\hat{f}$ in a geostatistical learning problem $\mathcal{P}$ is the expected risk attained by the model when spatial samples are drawn from $\Pr_{\mathcal{D}_t}$ over the target domain $\mathcal{D}_t$ (see **Eq 2**).

Unlike the classical definition of generalization error, the definition above for geostatistical learning problems relies on a spatial loss function $\mathcal{L}$, and on spatial samples like those produced via geostatistical simulation [40, 41]. For truly spatial learning models $\hat{f}$ that use multiple locations in the spatial domain to make predictions, this generalization error is more appropriate. In this present work, however; we do not target spatial learning models, and only consider pointwise learning:

Definition (Pointwise Learning). Given a family of classical (non-spatial) learning models $\{\hat{f}_\mathbf{u}\}_{\mathbf{u}\in\mathcal{D}}$, pointwise learning consists of learning the model $\hat{f}(\{\mathcal{X}\}_{\mathbf{u}\in\mathcal{D}}) = \{\hat{f}_\mathbf{u}(\mathcal{X}_\mathbf{u})\}_{\mathbf{u}\in\mathcal{D}}$ that assigns for each location $\mathbf{u} \in \mathcal{D}$ the value $\hat{f}_\mathbf{u}(\mathcal{X}_\mathbf{u})$ independently of the explanatory features at other locations.

More specifically, we consider pointwise learning with families that are made of a single learning model $\{\hat{f}_\mathbf{u}\}_{\mathbf{u}\in\mathcal{D}} = \{\hat{f}\}$. In this case, the model $\dot{f}$ is often learned based on pointwise stationarity assumptions, for some pointwise loss $\dot{\mathcal{L}}$:

$$\dot{f} = \arg\min_{\dot{g}} \mathbb{E}_{\Pr}\left[\dot{\mathcal{L}}(\mathcal{Y}, \dot{g}(\mathcal{X}))\right] \tag{3}$$

Although pointwise learning with a single model is a very simple type of geostatistical learning, it is by far the most widely used approach in the geospatial literature. We acknowledge this fact, and consider an empirical approximation of the pointwise expected risk in **Eq 3** as opposed to the spatial expected risk in **Eq 2**.

An empirical approximation of the pointwise expected risk of a model $\dot{g}$ can be obtained via discretization of the target spatial domain $\mathcal{D}_t$:

$$\mathcal{R}_t(\dot{g}) = \mathbb{E}_{\Pr_{\mathbf{u}\in\mathcal{D}_t}}\left[\dot{\mathcal{L}}(\mathcal{Y}, \dot{g}(\mathcal{X}))\right] \approx \frac{1}{|\mathcal{D}_t|}\sum_{\mathbf{u}\in\mathcal{D}_t} \dot{\mathcal{L}}(\mathcal{Y}_\mathbf{u}, \dot{g}(\mathcal{X}_\mathbf{u})) \tag{4}$$

with $|\mathcal{D}_t|$ the number of locations in the discretization. The problem with this empirical approximation is that the response features $\mathcal{Y}_\mathbf{u}$ are not available in the target domain where the model will be applied. However, it is easy to show that the pointwise expected risk in **Eq 4** can be rewritten with importance weights $\dot{w}(\mathcal{X}, \mathcal{Y})$ when samples from $\mathcal{D}_s$ are drawn instead [30]:

$$\mathcal{R}_t(\dot{g}) = \mathbb{E}_{\Pr_{\mathbf{v}\in\mathcal{D}_t}}\left[\dot{\mathcal{L}}(\mathcal{Y}, \dot{g}(\mathcal{X}))\right] = \mathbb{E}_{\Pr_{\mathbf{u}\in\mathcal{D}_s}}\left[\dot{w}(\mathcal{X}, \mathcal{Y})\dot{\mathcal{L}}(\mathcal{Y}, \dot{g}(\mathcal{X}))\right] \tag{5}$$

with $\dot{w}(\mathcal{X}, \mathcal{Y}) = \frac{\Pr_{\mathbf{v}\in\mathcal{D}_t}(\mathcal{X},\mathcal{Y})}{\Pr_{\mathbf{u}\in\mathcal{D}_s}(\mathcal{X},\mathcal{Y})}$. Under covariate shift, the importance weights only depend on the distribution of explanatory features $\dot{w}(\mathcal{X}) = \frac{\Pr_{\mathbf{v}\in\mathcal{D}_t}(\mathcal{X})}{\Pr_{\mathbf{u}\in\mathcal{D}_s}(\mathcal{X})}$, and we can write a simple importance-weighted empirical approximation:

$$\begin{aligned}\mathcal{R}_t(\dot{g}) &= \mathbb{E}_{\Pr_{\mathbf{u}\in\mathcal{D}_s}}\left[\dot{w}(\mathcal{X})\dot{\mathcal{L}}(\mathcal{Y}, \dot{g}(\mathcal{X}))\right] \\ &\approx \frac{1}{|\mathcal{D}_s|}\sum_{\mathbf{u}\in\mathcal{D}_s} \dot{w}(\mathcal{X}_\mathbf{u})\dot{\mathcal{L}}(\mathcal{Y}_\mathbf{u}, \dot{g}(\mathcal{X}_\mathbf{u}))\end{aligned} \tag{6}$$

Our goal is to find the pointwise model that minimizes the empirical risk approximation $\hat{\mathcal{R}}_t(\dot{g})$ introduced in **Eq 6**:

$$\dot{f} = \arg\min_{\dot{g}} \hat{\mathcal{R}}_t(\dot{g}) \tag{7}$$

Alternatively, our goal is to rank a collection of models $\{\dot{g}_i\}_{i=1,2,\ldots,k}$ based on their empirical risk $\{\hat{\mathcal{R}}_t(\dot{g}_i)\}_{i=1,2,\ldots,k}$ in a geostatistical learning problem to aid model selection.

In order to achieve the stated goals, we need to 1) estimate the importance weights in the empirical risk approximation, and 2) remove the dependence of the approximation on a specific dataset. These two issues are addressed in the following sections.

3.1 Density Ratio Estimation

The empirical approximation of the risk $\hat{\mathcal{R}}_t(\dot{g})$, depends on estimates of the weights $\dot{w}(\mathcal{X}_\mathbf{u})$, which are ratios of probabilities in the target and source domains. The following problem can be posed [32]:

Definition (Density Ratio Estimation). Given two collections of samples $\{\mathcal{X}_\mathbf{u}\}_{\mathbf{u}\in\mathcal{D}_s}$ and $\{\mathcal{X}_\mathbf{v}\}_{\mathbf{v}\in\mathcal{D}_t}$ from source and target domains, estimate the density ratio $\frac{\Pr_{\mathbf{v}\in\mathcal{D}_t}(\mathcal{X})}{\Pr_{\mathbf{u}\in\mathcal{D}_s}(\mathcal{X})}$ at any new sample $\mathcal{X}$. In particular, estimate the ratio at all samples $\{\mathcal{X}_\mathbf{u}\}_{\mathbf{u}\in\mathcal{D}_s}$ from the source.

Efficient methods for density ratio estimation that perform well with high-dimensional features have been proposed in the literature. In this work we consider a fast method named Least Squares Importance Fitting (LSIF) [35, 36]. The LSIF method assumes that the weights are a linear combination of basis functions $\dot{w}(\mathcal{X}_{\mathbf{u}}) = \alpha^\top \varphi(\mathcal{X}_{\mathbf{u}})$ with coefficients to be learned $\alpha = (\alpha_1, \alpha_2, \ldots, \alpha_b)$ and fixed basis $\varphi(\mathcal{X}_{\mathbf{u}}) = (\varphi_1(\mathcal{X}_{\mathbf{u}}), \varphi_2(\mathcal{X}_{\mathbf{u}}), \ldots, \varphi_b(\mathcal{X}_{\mathbf{u}}))$. The LSIF estimator is derived by minimizing the squared error with the true density ratio:

$$\underset{\alpha\in\mathbb{R}^b}{\text{minimize}}\frac{1}{2}\int\left(\dot{w}(\mathcal{X}_u) - \frac{\Pr_{v\in\mathcal{D}t}(\mathcal{X}_u)}{\Pr_{u\in\mathcal{D}s}(\mathcal{X}_u)}\right)^2 \Pr_{u\in\mathcal{D}s}(\mathcal{X}_u)d\mathcal{X}_u \quad (8)$$
$$\text{s.t.} \quad \boldsymbol{\alpha}\succeq\mathbf{0}$$

under the constraint that densities are always positive. By choosing b center features randomly from the target domain $\{\mathcal{X}_i\}_{i=1,2,\ldots,b}$, the method introduces a Gaussian kernel basis $\varphi_i(\mathcal{X}_{\mathbf{u}}) = k(\mathcal{X}_{\mathbf{u}}, \mathcal{X}_i)$ that simplifies the objective function to a matrix form:

$$\underset{\alpha\in\mathbb{R}^b}{\text{minimize}}\ \frac{1}{2}\boldsymbol{\alpha}^\top\mathbf{H}\boldsymbol{\alpha} - \mathbf{h}^\top\boldsymbol{\alpha} + \boldsymbol{\lambda}\mathbf{1}^\top\boldsymbol{\alpha} \quad (9)$$
$$\text{s.t.} \quad \alpha\succeq 0$$

with $\mathbf{H} = \int \varphi(\mathcal{X}_{\mathbf{u}})\varphi(\mathcal{X}_{\mathbf{u}})^\top \Pr_{\mathbf{u}\in\mathcal{D}_s}(\mathcal{X}_{\mathbf{u}})d\mathcal{X}_{\mathbf{u}}$ and $\mathrm{h} = \int \varphi(\mathcal{X}_{\mathbf{v}})\Pr_{\mathbf{v}\in\mathcal{D}_t}(\mathcal{X}_{\mathbf{v}})d\mathcal{X}_{\mathbf{v}}$. The regularization parameter $\lambda \geq 0$ on the coefficients α avoids overfitting, and empirical estimates of both $\mathbf{H}$ and $\mathbf{h}$ are easily obtained with sample averages:

$$\widehat{\mathbf{H}} = \frac{1}{|\mathcal{D}_s|}\sum_{\mathbf{u}\in\mathcal{D}_s}\varphi(\mathcal{X}_{\mathbf{u}})\varphi(\mathcal{X}_{\mathbf{u}})^\top \quad (10)$$
$$\widehat{\mathbf{h}} = \frac{1}{|\mathcal{D}_t|}\sum_{\mathbf{v}\in\mathcal{D}_t}\varphi(\mathcal{X}_{\mathbf{v}})$$

This quadratic optimization problem with linear inequality constraints can be solved very efficiently with modern optimization software [42, 43]. In the end, the optimal coefficients $\alpha^\star$ are plugged back into the basis expansion for optimal estimates of the weights on new samples $\dot{w}(\mathcal{X}) = \alpha^{\star\top}\varphi(\mathcal{X})$.

3.2 Weighted Cross-Validation

In order to remove the dependence of the empirical risk approximation on the dataset, we use importance-weighted cross-validation (IWCV) [30, 31]. As with the traditional cross-validation procedure, the source domain is split into k folds $\mathcal{D}_s = \cup_{j=1}^k \mathcal{D}_s^{(j)}$, and each fold $\mathcal{D}_s^{(j)}$ is hidden for error evaluation while the model $\dot{g}^{(j)}$ is learned on the remaining folds:

$$\widehat{\mathcal{R}}_t^{IWCV}(\dot{g}) = \frac{1}{k}\sum_{j=1}^k \frac{1}{\left|\mathcal{D}_s^{(j)}\right|}\sum_{\mathbf{u}\in\mathcal{D}_s^{(j)}}(\dot{w}(\mathcal{X}_{\mathbf{u}}))^l \mathcal{L}\left(\mathcal{Y}_{\mathbf{u}}, \dot{g}^{(j)}(\mathcal{X}_{\mathbf{u}})\right) \quad (11)$$

The main difference in the IWCV procedure are the weights that multiply each sample. The regularization exponent $l \in [0, 1]$ can be set to zero to recover the traditional estimator, or to a positive value to account for covariate shift. An optimal value for l can be found via hyperparameter search by considering another layer of cross-validation. In this work, we simply set default values for l such as $l = 1$ or $l = 0.5$.

In the rest of the paper, we combine IWCV with LSIF into a method for estimating generalization error that we term *Density Ratio Validation*. Although IWCV is known to outperform classical cross-validation methods in non-spatial settings, little is known about its performance with geospatial data. Moreover, like all prior art, IWCV approximates the pointwise generalization error of **Eq 3** as opposed to the geostatistical generalization error of **Eq 2**, and therefore is limited by design to non-spatial learning models.

4 EXPERIMENTS

In this section, we perform experiments to assess estimators of generalization error under varying covariate shifts and spatial correlation lengths. We consider Cross-Validation (CV), Block Cross-Validation (BCV) and Density Ratio Validation (DRV), which all rely on the same cross-validatory mechanism of splitting data into folds.

First, we use synthetic Gaussian process data and simple labeling functions to construct geostatistical learning problems for which learning models have a known (via geostatistical simulation) generalization error. In this case, we assess the estimators in terms of how well they estimate the actual error under various spatial distributions. Second, we demonstrate how the estimators are used for model selection in a real application with well logs from New Zealand, which can be considered to be a dataset of moderate size in this field.

4.1 Gaussian Processes

Let Z_{s_1}, Z_{s_2} be two Gaussian processes with constant mean μ_s and variogram γ_s defined over $\mathcal{D}_s$, and likewise let Z_{t_1}, Z_{t_2} be two Gaussian processes with constant mean μ_t and variogram γ_t defined over $\mathcal{D}_t$. Denote by r_s the variogram range (or correlation length) and by σ_s^2 the variogram sill (or total variance) of the processes in the source domain. Likewise, denote by r_t and σ_t^2 the range and sill of the variogram in the target domain. It is clear that pointwise stationarity holds inside each of these domains. The feature vector $\mathcal{X}_{\mathbf{u}} \in \mathbb{R}^2$ for any location $\mathbf{u} \in \mathcal{D}_s$ in the source domain has a bivariate Gaussian distribution $\mathcal{N}(\mu_s\mathbf{1}, \sigma_s^2\mathbf{I})$, whereas the feature vector $\mathcal{X}_{\mathbf{v}} \in \mathbb{R}^2$ for any location $\mathbf{v} \in \mathcal{D}_t$ in the target domain has a bivariate Gaussian distribution $\mathcal{N}(\mu_t\mathbf{1}, \sigma_t^2\mathbf{I})$. By constraining the variogram ranges to be equal in source and target domains, that is $r_s = r_t = r$, and by requiring that both variograms pass through the origin (i.e. no nugget effect), we can investigate two types of covariate shift for various ranges r:

Mean shift. Define the shift in the mean as $\delta = c\left|\mu_t - \mu_s\right| \in [0, \infty)$ for some normalization constant $c > 0$. In this experiment, we set the value of this constant to $c = \frac{1}{3\sqrt{2}\sigma_s}$ for convenience so that a $\delta = 1$ becomes equivalent to $\left|\mu_t - \mu_s\right| = 3\sqrt{2}\sigma_s$, which in turn is equivalent to two circles (i.e. bivariate Gaussians) of radii $3\sigma_s$ touching each other along the identity line, see **Figure 5**.

Variance shift. Define the shift in the variance as $\tau = \sigma_t/\sigma_s \in (0, \infty)$. Here, $\tau = 1$ means absence of variance shift, $\tau < 1$ means that the variance where the model is applied is smaller than the variance where the model was trained, and $\tau > 1$ means the exact opposite of $\tau < 1$.

Geostatistical learning problems with $\tau > 1$ are very challenging to solve, and usually require additional extrapolation models, beyond the pointwise learning models discussed in this work. Therefore, we only consider cases with $\tau \leq 1$ in this experiment. More specifically, we consider all

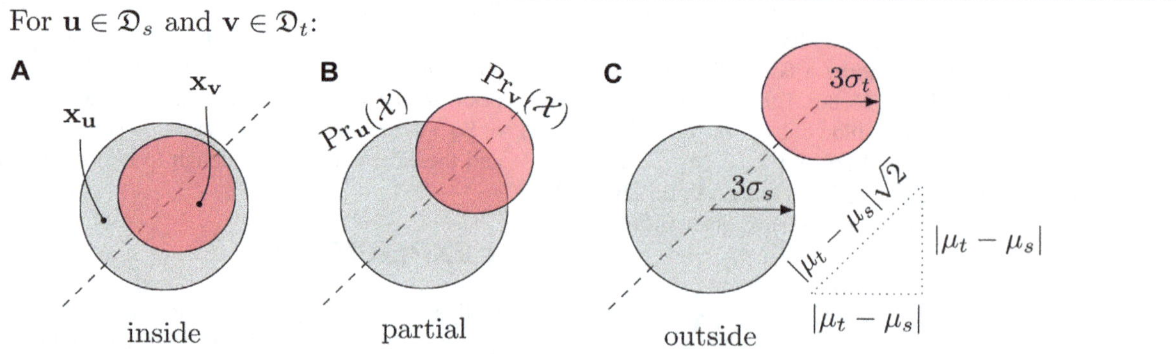

FIGURE 5 | Three possible shift configurations. The target distribution is "inside" the source distribution **(A)**, "outside" **(C)**, or they show "partial" overlap **(B)**. The processes Z_{s_1} and Z_{s_2} are observed at any location $\mathbf{u} \in \mathcal{D}_s$ forming a feature vector $x_\mathbf{u} \in \mathbb{R}^2$. Similarly, the processes Z_{t_1} and Z_{t_2} (which are shifted versions of the previous two) are observed at any location $\mathbf{v} \in \mathcal{D}_t$ forming a feature vector $x_\mathbf{v} \in \mathbb{R}^2$. In this case the features $x_\mathbf{u} \sim \Pr_\mathbf{u}(\mathcal{X})$ and $x_\mathbf{v} \sim \Pr_\mathbf{v}(\mathcal{X})$ are samples of bivariate Gaussians illustrated with circles of radii $3\sigma_s$ and $3\sigma_t$ centered at $\mu_s\mathbf{1}$ and $\mu_t\mathbf{1}$.

combinations of shift in the mean and variance of Gaussian features by varying (δ, τ) in the unit square $\mathcal{B} = [0,1] \times (0,1]$.

Given a shift parameterized by $(\delta, \tau) \in \mathcal{B}$, we can classify it into one of three possible configurations depending on how the source and target distributions of features overlap:

$$\text{config}(\delta,\tau) = \begin{cases} \text{inside}, & 2\delta \le 1-\tau \\ \text{outside}, & 2\delta \ge 1+\tau \\ \text{partial}, & \text{otherwise} \end{cases} \tag{12}$$

The first configuration in **Eq. 12** refers to the case in which the target distribution $\mathcal{N}(\mu_t\mathbf{1}, \sigma_t^2\mathbf{I})$ is "inside" the source distribution $\mathcal{N}(\mu_s\mathbf{1}, \sigma_s^2\mathbf{I})$ meaning that the circle of radius $3\sigma_t$ centered at $\mu_t\mathbf{1}$ is contained in the circle of radius $3\sigma_s$ centered at $\mu_s\mathbf{1}$. Similarly, the second configuration refers to the case in which the target distribution is "outside" the source distribution. Finally, the third configuration refers to a "partial" overlap when the two distributions share a common set of samples but are not entirely one inside of the other. We note, however; that the illustration with circles provided in **Figure 5** is only representative in the absence of spatial correlation (i.e., $r = 0$), see **Figure 3**.

To efficiently simulate multiple spatial samples of the processes over a regular grid domain with 100×100 locations (or pixels), we use spectral Gaussian simulation [44]. We fix the parameters of the source distribution at $\mu_s = 0$ and $\sigma_s = 1$ without loss of generality, and assume no inter-process correlation (i.e., $\rho = 0$) like we did in **Figure 3**. Under these modeling assumptions, we are able to investigate the spatial distribution of features as a function of shift parameters $(\delta, \tau) \in \mathcal{B}$ and variogram ranges $r \in \mathcal{C} = \{0, 10, 20\}$.

To fully specify the geostatistical learning problem, we need to specify a learning task. The task consists of predicting a binary variable $\mathcal{Y}_\mathbf{v}$ at locations $\mathbf{v}$ in the target grid $\mathcal{D}_t$ based on observations $y_\mathbf{u}$ of the variable at locations $\mathbf{u}$ in the source grid $\mathcal{D}_s$. These observations (or labels) are synthesized using known labeling functions such as $y_\mathbf{u} = \text{sgn}(\sin(w\|x\|_{\mathbf{u}p}))$, where $\|\cdot\|_p$ is the p-norm, w is the angular frequency, and sgn is the modified sign function that assigns $+1$ to $x \ge 0$ and -1 otherwise. The observations produced by these functions form alternating patterns in the feature space, which are not trivial to predict with simple learning models, see **Figure 6**. In this experiment, we fix $p = 1$ and $w = 4$ to save computational time. Other norms and angular frequencies produce similar results.

Having defined the problem, we proceed and specify learning models in order to investigate the different estimators of generalization error. We choose two models that are based on different prediction mechanisms [2]:

Decision tree. A pointwise decision tree model $\dot{f}_T$ makes predictions solely based on the features of the sample, without exploiting nearby features in the feature space.

K-nearest neighbors. A pointwise k-nearest neighbors model $\dot{f}_N$ makes predictions based on nearby features, and is sometimes called a "spatial model".

These two models $\dot{f}_T$ and $\dot{f}_N$ are simply representative models from the "non-spatial" and "spatial" families of models. We emphasise, however; that the term "spatial model" can be misleading in the spatial statistics literature. It is important to distinguish "spatial models" such as k-nearest neighbors that exploit the notion of proximity of features in the *feature space* from "geospatial models" that also exploit the proximity of samples in the *physical space* (or spatial domain as we have been calling it) besides their features.

The experiment proceeds as follows. For each shift $(\delta, \tau) \in \mathcal{B}$, each correlation length $r \in \mathcal{C}$, and each pointwise learning model $\dot{f} \in \{\dot{f}_T, \dot{f}_N\}$, we sample a problem $\mathcal{P}_{\delta,\tau,r}$ and estimate the generalization error of the model $\dot{f}$ on the problem with CV, BCV and DRV. We set the hyperparameters of the CV and BCV estimators based on the fact that the correlation length never exceeds 20. For instance, we set the block side in BCV to $s = 20$, and use the equivalent number of folds in CV, i.e., $k = (100/20)^2 = 25$ for a domain with 100×100 pixels. We set the kernel width of the LSIF estimator in DRV to $\sigma = 2$ based on the synthetic Gaussian distributions, and use 10 kernels in the basis expansion. Additionally, we approximate the true generalization error of the models with Monte Carlo simulation over the target domain (e.g. 100 spatial samples).

To facilitate the visualization of the results, we introduce shift functions $\mathcal{S} : \mathcal{B} \to [0, \infty)$ that map the shift parameters (δ, τ) to a single covariate shift value, which can be interpreted loosely as the "difficulty" of the problem:

Kullback-Leibler divergence. The divergence or relative entropy between two distributions p and q is defined as

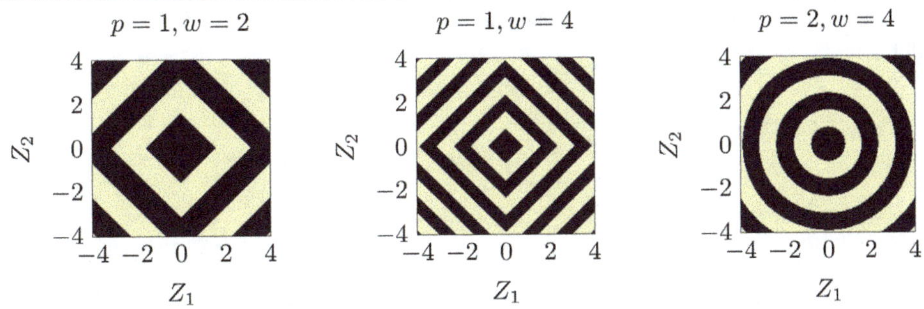

FIGURE 6 | Labeling function $f_{p,w}(x) = \text{sgn}(\sin(w\|x\|_p))$. Labels form alternating patterns in the feature space for different p-norm and angular frequencies w.

$\mathcal{S}_{KL} = (p \parallel q) \int p(x)\log\frac{p(x)}{q(x)}dx$, and can be derived analytically for two (2D) Gaussian distributions $p = \mathcal{N}(\mu_t\mathbf{1}, \sigma_t^2\mathbf{I})$ and $q = \mathcal{N}(\mu_s\mathbf{1}, \sigma_s^2\mathbf{I})$. We derive a formula in terms of δ and τ by fixing $\sigma_s = 1$:

$$\mathcal{S}_{KL}(\delta, \tau) = \delta^2 + \tau^2 - \log(\tau^4) - 1 \tag{13}$$

Jaccard distance. The Jaccard index between two sets A and B is defined as $J(A,B) = \frac{|A\cap B|}{|A\cup B|}$, and the corresponding distance as $\mathcal{S}_J(A,B) = 1 - J(A,B)$. For two (2D) Gaussian distributions, we consider A and B to be circles of radii $3\sigma_s$ and $3\sigma_t$ centered at $\mu_s\mathbf{1}$ and $\mu_t\mathbf{1}$. The distance is then expressed in terms of areas, which can be derived analytically in terms of δ and τ by fixing $\sigma_s = 1$:

$$\begin{aligned}
|A| &= 9\pi \\
|B| &= 9\pi\tau^2 \\
C_1 &= 9\arccos\left(\frac{2\delta^2 + 9(1-\tau^2)}{6\sqrt{2}\delta}\right) \\
C_2 &= 9\tau^2\arccos\left(\frac{2\delta^2 + 9(\tau^2-1)}{6\sqrt{2}\delta\tau}\right) \\
C_3 &= \frac{1}{2}\sqrt{(9(1+\tau)^2 - 2\delta^2)(2\delta^2 - 9(1-\tau)^2)} \\
|A\cap B| &= C_1 + C_2 - C_3 \\
|A\cup B| &= |A| + |B| - |A\cap B|
\end{aligned} \tag{14}$$

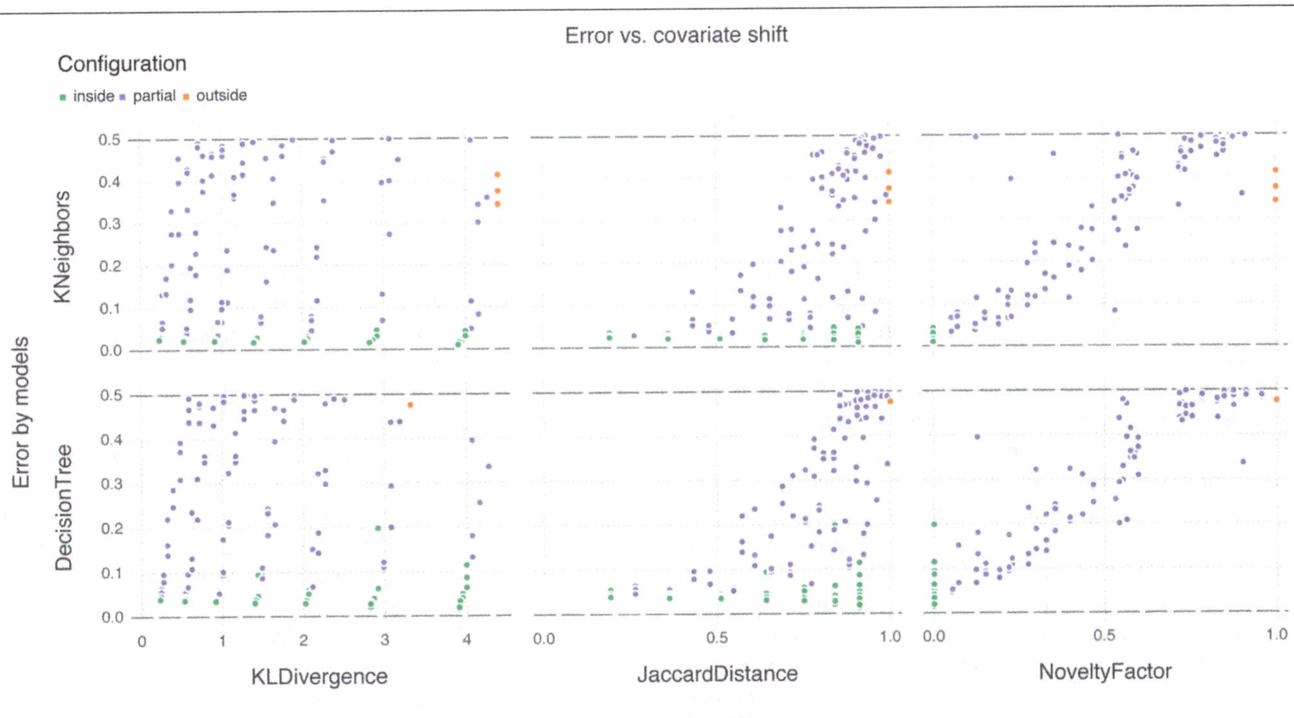

FIGURE 7 | Generalization error of learning models versus covariate shift functions. Among all shift functions, the novelty factor is the only function that groups shift configurations along the horizontal axis. Models behave similarly in terms of generalization error for the given dataset size (100 × 100 pixels), and difficult shift configurations lead to errors that approach the theoretical Bernoulli limit of 0.5.

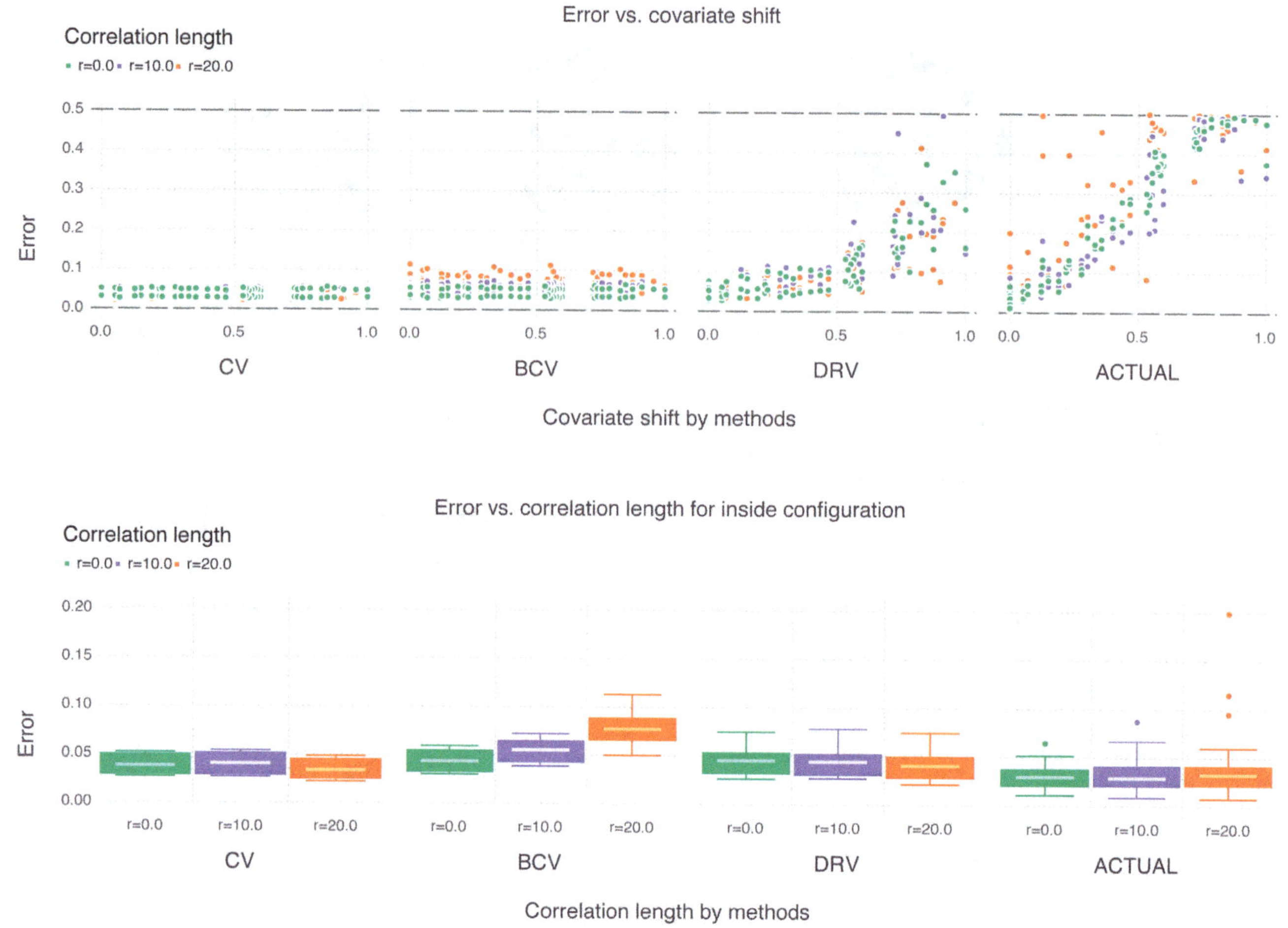

FIGURE 8 | Estimates of generalization error for various shifts (i.e., novelty factor) and various correlations lengths. The box plots for the *inside* configuration illustrate how the estimators behave differently for increasing correlation lengths.

Novelty factor. We propose a new shift function termed the *novelty factor* inspired by the geometric view of Jaccard. First, we define the novelty of B with respect to A as $N(B/A) = \frac{|B-A\cap B|-|A\cap B|}{|B|}$, and notice that it is the fraction of B that is outside of A minus the fraction of B that is inside of A. Second, we restrict the definition to cases with $|B| \leq |A|$ (e.g. Gaussian case with $\tau \leq 1$), and notice that the novelty $N(B/A)$ lies in the interval $[-1, 1]$. Finally, we define the novelty factor $\mathcal{S}_N(A, B) = \frac{N(B/A)+1}{2}$ in the interval $[0, 1]$, which can be easily computed for the Gaussian case using the formulas derived in **Eq 14**.

We plot the true generalization error of the models as a function of the different covariate shifts in **Figure 7**, and color the points according to their shift configuration (see **Eq 12**). In this plot, the horizontal dashed line intercepting the vertical axis at 0.5 represents a model that assigns positive and negative labels to samples at random with equal weight (i.e., Bernoulli variable).

Among the three shift functions, the novelty factor is the only function that groups shift configurations along the horizontal axis. In this case, configurations deemed easy (i.e., where the target distribution is *inside* the source distribution) appear first, then configurations of moderate difficulty (i.e., *partial* overlap) appear next, and finally difficult configurations (i.e., target is *outside* the source) appear near the theoretical Bernoulli limit. The Kullback-Leibler divergence and the Jaccard distance fail to summarize the shift parameters into a one-dimensional visualization, and are therefore omitted in the next plots.

The two models behave similarly in terms of generalization error for the given dataset size (i.e., 100×100 pixels), and therefore can be aggregated into a single scatter to increase the confidence in the observed trends.

We plot the CV, BCV and DRV estimates of generalization error versus covariate shift (i.e., novelty factor) in the top row of **Figure 8**, and color the points according to their correlation length. We omit a few DRV estimates that suffered from numerical instability in challenging *partial* or *outside* configurations, and show the box plot of error estimates in the bottom row of the figure.

First, we emphasize that the CV and BCV estimates remain constant as a function of covariate shift. This is expected given that these estimators do not make use of the target distribution. The DRV estimates increase with covariate shift as expected, but do not follow the same rate of increase of the true (or actual) generalization error obtained with Monte Carlo simulation. Second, we emphasize in the box plots for the *inside* configuration that the correlation length affects the estimators

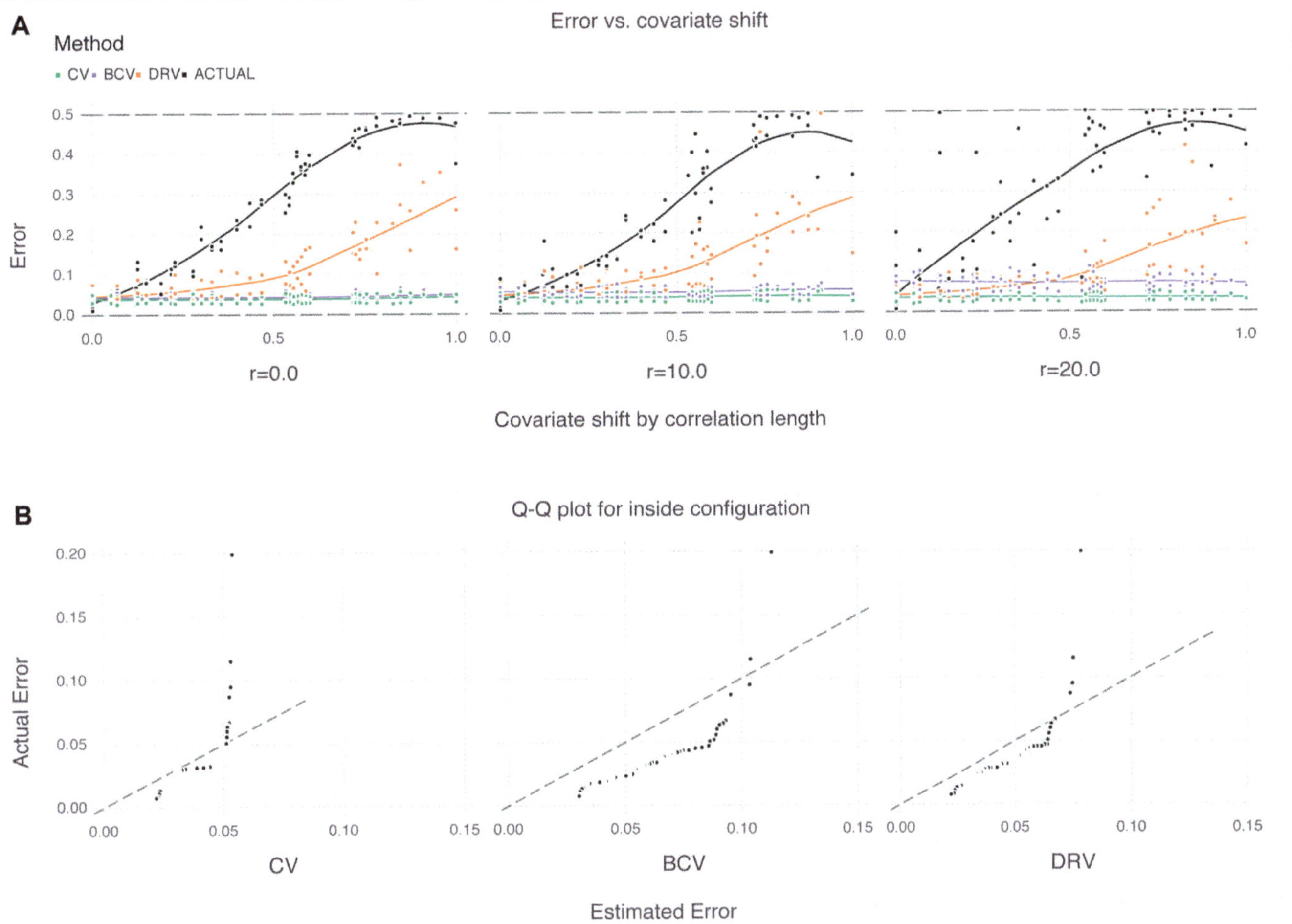

FIGURE 9 | Trends of generalization error for different estimators **(A)** and Q-Q plots of estimated versus actual error for *inside* configuration **(B)**.

differently. The CV estimator becomes more optimistic with increasing correlation length, whereas the BCV estimator becomes less optimistic, a result that is also expected from prior art. Additionally, the interquartile range of the BCV estimator increases with correlation length. It is not clear from the box plots that a trend exists for the DRV estimator. The actual generalization error behaves erratically in the presence of large correlation lengths as indicated by the scatter and box plots.

In order to better visualize the trends in the estimates, we smooth the scatter plots with locally weighted regression per correlation length in the top row of **Figure 9**, and show in the bottom row of the figure the Q-Q plots of the different estimates against the actual generalization error for the *inside* configuration where all estimators are supposed to perform well.

From the figure, there exists a gap between the DRV estimates and the actual generalization error of the models for all covariate shifts. This gap is expected given that the target distribution may be very different from the source distribution, particularly in *partial* or *outside* shift configurations. On the other hand, the gap also seems to be affected by the correlation length, and is largest with 20 pixels of correlation. Additionally, we emphasize in the Q-Q plots that the BCV estimates are biased due to the systematic selection of folds. The BCV estimates are less optimistic than the CV estimates, which is a desired property in practice, however there is no guarantee that the former estimates will approximate well the actual generalization error of the models.

4.2 New Zealand Dataset

Unlike the previous experiment with synthetic Gaussian process data and known generalization error, this experiment consists of applying the CV, BCV and DRV estimators to a real dataset of well logs prepared in-house [37]. We quickly describe the dataset, introduce the related geostatistical learning problems, and use error estimates to rank learning models. Finally, we compare these ranks with an ideal rank obtained with additional label information that is not available during the learning process.

The dataset consists of 407 wells in the Taranaki basin, including the main geophysical logs and reported geological formations. The basin comprises an area of about 330.000 km^2, located broadly onshore and offshore the New Zealand west coast (see **Figure 10**). Well trajectories are georeferenced in UTM coordinates (**X** and **Y**) and true vertical depth (Z).

We split the wells into onshore and offshore locations in order to introduce a geostatistical learning problem with covariate shift.

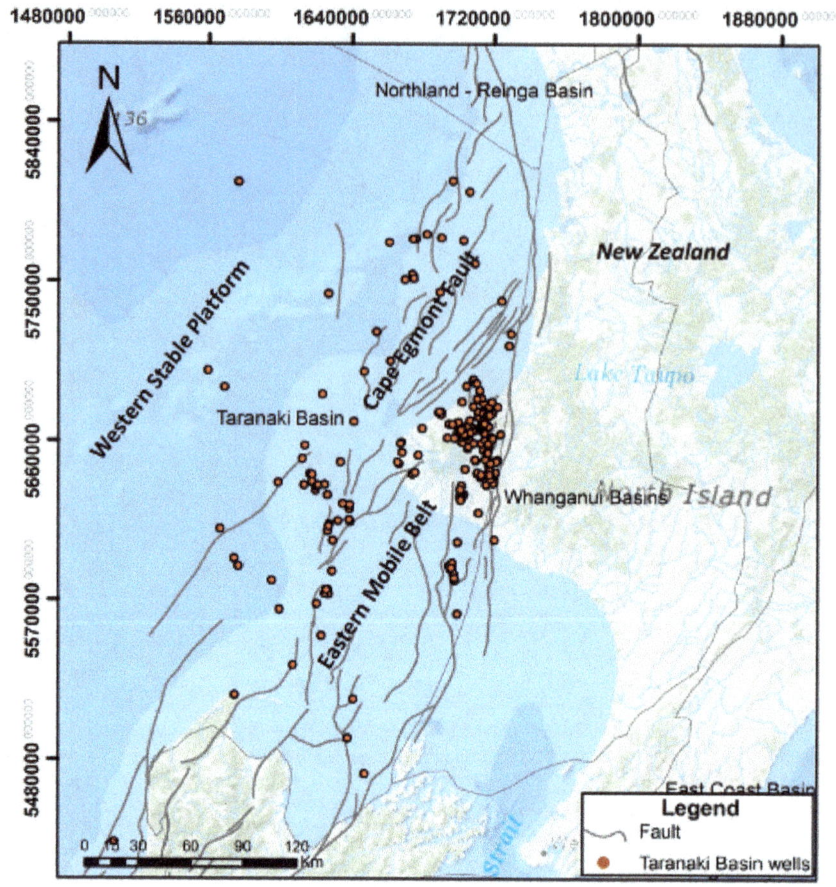

FIGURE 10 | Curated dataset with 407 wells in the Taranaki basin, New Zealand. The basin comprises an area of about 330.000 km^2, located bradly onshore and offshore the New Zealand west coast.

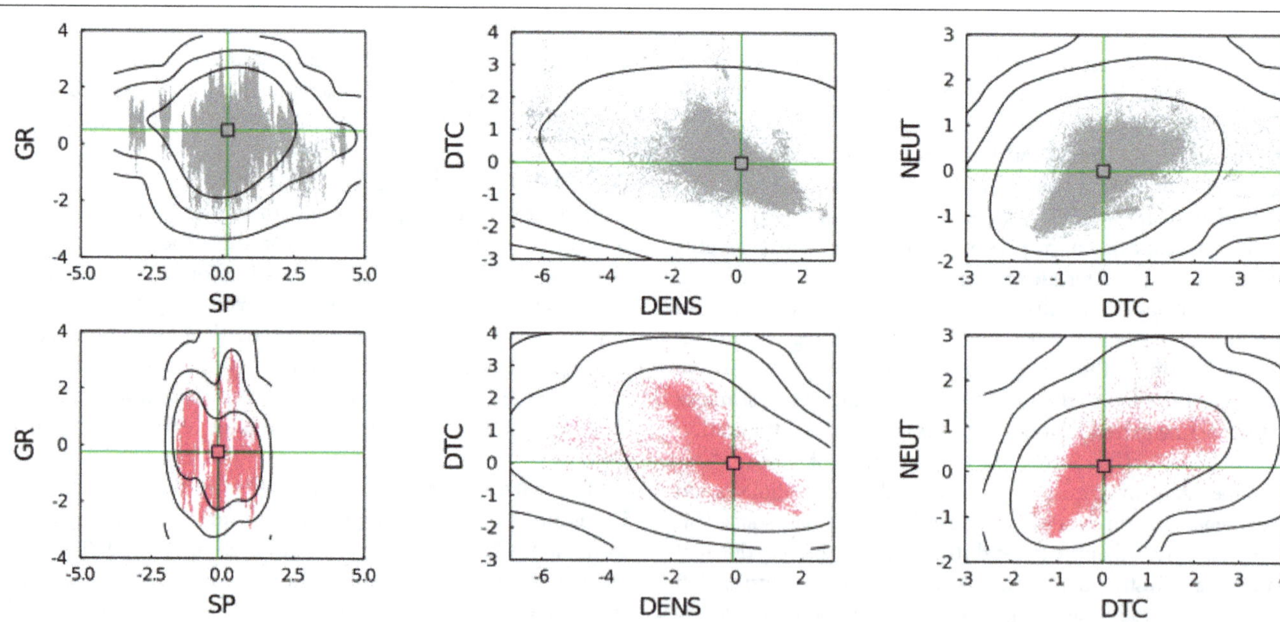

FIGURE 11 | Distribution of main geophysical logs onshore (gray) and offshore (purple) centered by the mean and divided by the standard deviation. Visible covariate shift in the scatter and contour plots.

TABLE 1 | Estimates of generalization error with different estimators for the onshore-to-offshore problem. The CV estimator produces estimates that are the most distant to the actual target error due to covariate shift and spatial correlation. None of the estimators is capable of ranking the models correctly. They all select complex models with low generalization ability.

MODEL	SOURCE	TARGET	CV	BCV	DRV
Ridge	0.262	**0.291**	0.263	**0.326**	**0.278**
Logistic	0.262	**0.294**	**0.262**	0.322	**0.275**
KNeighbors	0.023	**0.355**	**0.034**	**0.227**	0.043
GaussianNB	0.256	**0.321**	**0.256**	**0.288**	0.284
BayesianLDA	0.262	**0.291**	0.261	**0.326**	**0.294**
Perceptron	0.375	**0.395**	**0.344**	0.369	**0.392**
DecisionTree	0.0	**0.378**	**0.034**	**0.249**	0.043
Dummy	0.499	**0.497**	**0.499**	0.502	**0.524**

TARGET RANK	CV RANK	BCV RANK	DRV RANK
BayesianLDA	KNeighbors	KNeighbors	KNeighbors
Ridge	DecisionTree	DecisionTree	DecisionTree
Logistic	GaussianNB	GaussianNB	Logistic
GaussianNB	BayesianLDA	Logistic	Ridge
KNeighbors	Logistic	Ridge	GaussianNB
DecisionTree	Ridge	BayesianLDA	BayesianLDA
Perceptron	Perceptron	Perceptron	Perceptron
Dummy	Dummy	Dummy	Dummy

TABLE 2 | Estimates of generalization error with different estimators for the problem without covariate shift. The BCV estimator produces estimates that are the most distant to the actual target error due to bias from its systematic selection of folds. All estimators are capable of ranking the models in the absence of covariate shift.

MODEL	SOURCE	TARGET	CV	BCV	DRV
Ridge	0.267	**0.273**	**0.269**	**0.329**	0.259
Logistic	0.266	**0.271**	**0.265**	**0.324**	0.254
KNeighbors	0.026	**0.04**	**0.04**	**0.232**	0.042
GaussianNB	0.265	**0.268**	**0.265**	**0.308**	0.262
BayesianLDA	0.267	**0.273**	**0.266**	**0.329**	0.264
Perceptron	0.445	**0.442**	**0.362**	**0.402**	0.392
DecisionTree	0.0	**0.039**	**0.039**	**0.244**	0.041
Dummy	0.499	**0.498**	0.499	**0.497**	**0.476**

TARGET RANK	CV RANK	BCV RANK	DRV RANK
DecisionTree	DecisionTree	KNeighbors	DecisionTree
KNeighbors	KNeighbors	DecisionTree	KNeighbors
GaussianNB	GaussianNB	GaussianNB	Logistic
Logistic	Logistic	Logistic	Ridge
Ridge	BayesianLDA	BayesianLDA	GaussianNB
BayesianLDA	Ridge	Ridge	BayesianLDA
Perceptron	Perceptron	Perceptron	Perceptron
Dummy	Dummy	Dummy	Dummy

The problem consists of predicting the rock formation from well logs offshore after learning a model with well logs and reported (i.e., manually labeled) formations onshore. The well logs considered are gamma ray (GR), spontaneous potential (SP), density (DENS), compressional sonic (DTC) and neutron porosity (NEUT). We eliminate locations with missing values for these logs and investigate a balanced dataset with the two most frequent formations—Urenui and Manganui. We normalize the logs and illustrate the covariate shift property by comparing the scatter plots of onshore and offshore locations in **Figure 11**. Additionally, we define a second geostatistical learning problem without covariate shift. In this case, we join all locations filtered in the previous problem and sample two new sets of locations with sizes respecting the same source-to-target proportion (e.g., 300000 : 50000).

We set the hyperparameters of the error estimators based on variography and according to available computational resources. In particular, we set blocks for the BCV estimators with sides $10000 \times 10000 \times 500$ that are much greater than the vertical and horizontal correlation lengths estimated from empirical variograms. We obtain the corresponding number of folds $k = 99$ for the CV estimator by partitioning the bounding box of onshore wells into blocks with the given sides. Similarly to the previous experiment with synthetic Gaussian process data, we set the kernel width in DRV to $\sigma = 2$ given that the well logs were normalized to have unit variance. Finally, we select a list of learning models to rank including Ridge classification

(Ridge), logistic regression (Logistic), k-nearest neighbors (KNeighbors), naive Bayes (GaussianNB), linear discriminant analysis (BayesianLDA), perceptron (Perceptron), decision tree (DecisionTree), and a dummy model that reproduces the marginal distribution of formations in the source domain (Dummy).

In **Table 1**, we report the results for the onshore-to-offshore problem. In the upper part of the table we compare side-by-side the error estimates of the different methods. We highlight the closest estimates to the target error in blue color, and the most distant in red color. We emphasize that the target error is the error of the model in one single realization of the process, and is *not* the generalization error averaged over multiple spatial realizations. In spite of this important distinction, we still think it is valuable to compare it with the estimates of generalization error given by CV, BCV and DRV since these methods were all derived under pointwise learning assumptions, and are therefore smooth averages over multiple points exactly like the error estimated from the single realization of the target. In the bottom part of the table, we report the model ranks derived from the error estimates as well as the ideal rank derived from the target error.

Among the three estimators of generalization error, the CV estimator produces estimates that are the most distant from the target error, with a tendency to underestimate the error. The BCV estimator produces estimates that are higher than the CV estimates, and consequently closer to the target error in this case. The DRV estimator produces the closest estimates for most models, however; like the CV estimator it fails to approximate the error for models like KNeighbors and DecisionTree that are over-fitted to the source distribution. The three estimators fail to rank the models under covariate shift and spatial correlation. Over-fitted models with low generalization ability are incorrectly ranked at the top of the list, and the best models, which are simple "linear" models, appear at the bottom. We compare these results with the results obtained for the problem without covariate shift in **Table 2**.

From **Table 2**, the CV estimator produces estimates that are the closest to the target error. The BCV estimator produces estimates that are higher than the CV estimates as before, however this time this means that the BCV estimates are the most distant to the target error. The DRV estimator produces estimates that are not the closest nor the most distant to the target error. The three estimators successfully rank the models from simple linear models at the bottom of the list to more complex learning models at the top. Unlike the previous problem with covariate shift, this time complex models like KNeighbors and DecisionTree show high generalization ability.

5 CONCLUSION

In this work, we introduce *geostatistical (transfer) learning*, and demonstrate how most prior art in statistical learning with geospatial data fits into a category we term *pointwise learning*. We define geostatistical generalization error and demonstrate how existing estimators from the spatial statistics literature such as block cross-validation are derived for that specific category of learning, and are therefore unable to account for general spatial errors.

We propose experiments with spatial data to compare estimators of generalization error, and illustrate how these estimators fail to rank models under covariate shift and spatial correlation. Based on the results of these experiments, we share a few remarks related to the choice of estimators in practice:

- The apparent quality of the BCV estimator is falsified in the Q-Q plots of **Figure 9** and in **Table 2**. The systematic bias produced by the blocking mechanism only guarantees that the error estimates are higher than the CV estimates. When the CV estimates are good (i.e. no covariate shift), the BCV estimates are unnecessarily pessimistic.
- The CV estimator is not adequate for geostatistical learning problems that show various forms of covariate shift. Situations without covariate shift are rare in geoscientific settings, and since the DRV estimator works reasonably well for both situations (i.e. with and without shift), it is recommended instead.
- Nevertheless, both the CV and DRV estimators suffer from a serious issue with over-fitted models in which case they largely underestimate the generalization error. For risk-averse applications where one needs to be careful about the generalization error of the model, the BCV estimator can provide more conservative results.
- None of the three estimators were capable of ranking models correctly under covariate shift and spatial correlation. This is an indication that one needs to be skeptical about interpreting similar rankings available in the literature.

Finally, we believe that this work can motivate methodological advances in learning from geospatial data, including research on new estimators of geostatistical generalization error as opposed to pointwise generalization error, and more explicit treatments of spatial coordinates of samples in learning models.

Computer Code Availability

All concepts and methods developed in this paper are made available in the GeoStats.jl project [45]. The project is hosted on GitHub under the MIT[1] open-source license: https://github.com/JuliaEarth/GeoStats.jl.

Experiments of this specific work can be reproduced with the following scripts: https://github.com/IBM/geostats-gen-error.

AUTHOR CONTRIBUTIONS

JH: Conceptualization, methodology, software, formal analysis, investigation, visualization, writing—original draft; MZ: Methodology, validation; BC: Data curation, validation; BZ: Methodology, validation, supervision.

[1]https://opensource.org/licenses/MIT

REFERENCES

1. Vapnik V. *Statistical Learning Theory*. New York: Wiley (1998).
2. Hastie T, Tibshirani R, and Friedman J. *Elements of Statistical Learning*. 2nd ed. (2009). doi:10.1007/978-0-387-84858-7
3. Mitchell TM. *Machine Learning*. New York: McGraw-Hill (1997).
4. Pedregosa F, Varoquaux G, Gramfort A, Michel V, Thirion B, Grisel O, et al. Scikit-learn: Machine Learning in Python. *J Machine Learn Res* (2011) 12:2825–2830.
5. Abadi M, Barham P, Chen J, Chen Z, Davis A, Dean J, Devin M, Ghemawat S, Irving G, Isard M, Kudlur M, Levenberg J, Monga R, Moore S, Murray DG, Steiner B, Tucker P, Vasudevan V, Warden P, Wicke M, Yu Y, and Zheng X. TensorFlow: A System for Large-Scale Machine Learning. In: Proceedings of the 12th USENIX Symposium on Operating Systems Design and Implementation. OSDI (2016). arXiv:1605.08695.
6. Paszke A, Gross S, Massa F, Lerer A, Bradbury J, Chanan G, et al. Pytorch: An Imperative Style, High-Performance Deep Learning Library. In: H Wallach, H Larochelle, A Beygelzimer, F d' Alché-Buc, E Fox, and R Garnett, editors. *Advances in Neural Information Processing Systems 32*. Curran Associates, Inc. (2019). p. 8024–35.
7. Innes M. Flux: Elegant Machine Learning with Julia. *Joss* (2018) 3:602. doi:10.21105/joss.00602
8. Arlot S, and Celisse A. A Survey of Cross-Validation Procedures for Model Selection. *Statist Surv*. (2010). 4: 4728, 2010 . arXiv:0907. doi:10.1214/09-SS054
9. Ferraciolli MA, Bocca FF, and Rodrigues LHA. Neglecting Spatial Autocorrelation Causes Underestimation of the Error of Sugarcane Yield Models. *Comput Electron Agric* (2019) 161:233–40. doi:10.1016/j.compag.2018.09.003
10. Jiang Y, Krishnan D, Mobahi H, and Bengio S. Predicting the Generalization Gap in Deep Networks with Margin Distributions. 7th International Conference on Learning Representations. ICLR (2019).arXiv:1810.00113.
11. Zhang C, Recht B, Bengio S, Hardt M, and Vinyals O. Understanding Deep Learning Requires Rethinking Generalization. In: 5th International Conference on Learning Representations, ICLR 2017 - Conference Track Proceedings (2019).
12. Vehtari A, and Ojanen J. A Survey of Bayesian Predictive Methods for Model Assessment, Selection and Comparison. *Statist Surv* 6. (2012) 03530, 2012 . arXiv:1611. doi:10.1214/12-ss102
13. Stone M. Cross-Validatory Choice and Assessment of Statistical Predictions. *J R Stat Soc Ser B (Methodological)* (1974) 36(2):111–33. doi:10.1111/j.2517-6161.1974.tb00994.x
14. Geisser S. The Predictive Sample Reuse Method with Applications. *J Am Stat Assoc* (1975) 70:320–8. doi:10.1080/01621459.1975.10479865
15. Burman P. A Comparative Study of Ordinary Cross-Validation, V-fold Cross-Validation and the Repeated Learning-Testing Methods. *Biometrika* (1989) 76: 503–14. doi:10.1093/biomet/76.3.503
16. Cracknell MJ, and Reading AM. Geological Mapping Using Remote Sensing Data: A Comparison of Five Machine Learning Algorithms, Their Response to Variations in the Spatial Distribution of Training Data and the Use of Explicit Spatial Information. *Comput Geosciences* (2014) 63:22–33. doi:10.1016/j.cageo.2013.10.008
17. Baglaeva EM, Sergeev AP, Shichkin AV, and Buevich AG. The Effect of Splitting of Raw Data into Training and Test Subsets on the Accuracy of Predicting Spatial Distribution by a Multilayer Perceptron. *Math Geosci* (2020) 52:111–21. doi:10.1007/s11004-019-09813-9
18. Burman P, Chow E, and Nolan D. A Cross-Validatory Method for Dependent Data. *Biometrika* (1994) 81:351–8. doi:10.1093/biomet/81.2.351
19. Le Rest K, Pinaud D, Monestiez P, Chadoeuf J, and Bretagnolle V. Spatial Leave-One-Out Cross-Validation for Variable Selection in the Presence of Spatial Autocorrelation. *Glob Ecol Biogeogr* (2014) 23:811–20. doi:10.1111/geb.12161
20. Roberts DR, Bahn V, Ciuti S, Boyce MS, Elith J, Guillera-Arroita G, et al. Cross-validation Strategies for Data with Temporal, Spatial, Hierarchical, or Phylogenetic Structure *Ecography* (2017) 40:913–29. doi:10.1111/ecog.02881
21. Pohjankukka J, Pahikkala T, Nevalainen P, and Heikkonen J. Estimating the Prediction Performance of Spatial Models via Spatial K-fold Cross Validation. *Int J Geographical Inf Sci* (2017) 31:2001–19. doi:10.1080/13658816.2017.1346255
22. Chilès J-P, and Delfiner P. *Geostatistics: Modeling Spatial Uncertainty*. Hoboken, NJ: Wiley (2012).
23. Pan SJ, and Yang Q. A Survey on Transfer Learning. *IEEE Trans Knowl Data Eng* (2010) 22:1345–59. doi:10.1109/TKDE.2009.191
24. Weiss K, Khoshgoftaar TM, and Wang D. A Survey of Transfer Learning. *J Big Data* (2016) 3. doi:10.1186/s40537-016-0043-6
25. Silver DL, and Bennett KP. Guest Editor's Introduction: Special Issue on Inductive Transfer Learning. *Mach Learn* (2008) 73:215–20. doi:10.1007/s10994-008-5087-1
26. Zadrozny B, Langford J, and Abe N. Cost-sensitive Learning by Cost-Proportionate Example Weighting. In: Proceedings - IEEE International Conference on Data Mining . ICDM (2003). doi:10.1109/icdm.2003.1250950
27. Zadrozny B. Learning and Evaluating Classifiers under Sample Selection Bias. In: Proceedings, Twenty-First International Conference on Machine Learning. ICML (2004). doi:10.1145/1015330.1015425
28. Wei Fan W, Davidson I, Zadrozny B, and Philip S. Yu PS. An Improved Categorization of Classifiers Sensitivity on Sample Selection Bias. In: Proceedings - IEEE International Conference on Data Mining. ICDM (2005). doi:10.1109/ICDM.2005.24
29. Joaquin Q-C, Sugiyama M, Schwaighofer A, and Lawrence ND. *Dataset Shift in Machine Learning*. Cambridge, MA: The MIT Press (2008).
30. Sugiyama M, Blankertz B, Krauledat M, Dornhege G, and Müller K-R. Importance-weighted Cross-Validation for Covariate Shift. In Lecture Notes in Computer Science (Including Subseries Lecture Notes in Artificial Intelligence and Lecture Notes in Bioinformatics(2006). p. 354–63. doi:10.1007/11861898_36
31. Sugiyama M, Krauledat M, and Müller KR. Covariate Shift Adaptation by Importance Weighted Cross Validation. *J Machine Learn Res* (2007) 8:985–1005.
32. Sugiyama M, Suzuki T, and Kanamori T. Density Ratio Estimation in Machine Learning (2012). doi:10.1017/CBO9781139035613
33. Huang J, Smola AJ, Gretton A, Borgwardt KM, and Schölkopf B. Correcting Sample Selection Bias by Unlabeled Data. In Advances in Neural Information Processing Systems (2007). doi:10.7551/mitpress/7503.003.0080
34. Sugiyama M, Nakajima S, Kashima H, Von Bünau P, and Kawanabe M. Direct Importance Estimation with Model Selection and its Application to Covariate Shift Adaptation. In Advances in Neural Information Processing Systems (2009).
35. Kanamori T, Hido S, and Sugiyama M. Efficient Direct Density Ratio Estimation for Non-stationarity Adaptation and Outlier Detection. In Advances in Neural Information Processing Systems (2009).
36. Kanamori T, Hido S, and Sugiyama M. A Least-Squares Approach to Direct Importance Estimation. *J Machine Learn Res* (2009) 10:1391–1445.
37. Carvalho BW, Oliveira M, Avalone M, Hoffimann J, Szwarcman D, Guevara Diaz J, et al. Taranaki basin Curated Well Logs. *Zenodo* (2020). doi:10.5281/zenodo.383295510.5281/zenodo.3832955
38. Hoffimann J, and Zadrozny B. Efficient Variography with Partition Variograms. *Comput Geosciences* (2019) 131:52–9. doi:10.1016/j.cageo.2019.06.013
39. Alabert F. The Practice of Fast Conditional Simulations through the LU Decomposition of the Covariance Matrix. *Math Geol* (1987) 19:369–86. doi:10.1007/BF00897191
40. Hoffimann J, Scheidt C, Barfod A, and Caers J. Stochastic Simulation by Image Quilting of Process-Based Geological Models. *Comput Geosciences* (2017) 106: 18–32. doi:10.1016/j.cageo.2017.05.012
41. Mariethoz G, Renard P, and Straubhaar J. The Direct Sampling Method to Perform Multiple-point Geostatistical Simulations. *Water Resour Res* (2010) 46. doi:10.1029/2008WR007621
42. K Mogensen P, and N Riseth A. Optim: A Mathematical Optimization Package for Julia. *Joss* (2018) 3:615. doi:10.21105/joss.00615
43. Dunning I, Huchette J, and Lubin M.JuMP: A Modeling Language for Mathematical Optimization. *SIAM Rev* (2017) 59:295–320. arXiv:1508. doi:10.1137/15M1020575
44. Gutjahr A, Bullard B, and Hatch S. General Joint Conditional Simulations Using a Fast Fourier Transform Method. *Math Geol* (1997) 29:361–89. doi:10.1007/BF02769641
45. Hoffimann J. GeoStats.jl - High-Performance Geostatistics in Julia. *Joss* (2018) 3:692. doi:10.21105/joss.0069210.21105/joss.00692

18

Generation of a Dynamical Logic Gate from Unstable Dissipative Systems of Type 1

Roberto R. Rivera-Durón [1*]***, Ricardo Sevilla-Escoboza*** [2] ***and Qui-Ling Wang*** [3]

[1] *Unmanned System Research Institute, Northwestern Polytechnical University, Xi'an, China,* [2] *Centro Universitario de los Lagos, Universidad de Guadalajara, Lagos de Moreno, Mexico,* [3] *College of Transportation, Chang'an University, Xi'an, China*

Correspondence:
Roberto R. Rivera-Durón
rrrdrn@gmail.com

The obtainment of a dynamical logic gate (DLG), which is a device capable of implementing several logic functions using the same model, has been one of the goals of the scientific community. Dynamical systems, specifically those that display chaotic behavior, have been widely used to emulate different logic gates which are the basis of general-purpose computing. In this study, we present a methodology based on unstable dissipative systems of type 1 (UDS-1), a kind of dynamical system capable of generating multi-scrolls and multi-stability. Using these two features, we codify inputs, subsequently, we get the adequate output, developing in this way a dynamical (reconfigurable) logic gate that performs any of the sixteen possible logic functions of two inputs. A highlight of the proposed methodology is that the selection of the desired logic gate is realized just by varying a couple of parameters.

Keywords: unstable dissipative systems, multi-stability, reconfigurable computing, reconfigurable logic gate, dynamical logic gate

1. INTRODUCTION

In the last decades, a vast quantity of budget and effort has been invested to design and construct a unique device capable of implementing several logic gates into the same structure. In 1998, Sinha and Ditto showed the capacity of lattices of coupled logistic maps to emulate NOR gates, resulting in chaos computing [1]. From this pioneering study, several schemes have been exploited, these include chaotic continuous and discrete dynamical systems [2–7]; piece-wise linear (PWL) systems [8, 9]; resonators controlled by noise intensity [10]; cellular neuronal networks [11]; memristive devices [12, 13], and doping-free bipolar junction transistors controlled by polarity [14]. In chaos computing, chaotic elements are exploited to act as different logic functions by changing parameters so that these devices are more flexible than the silicon-based architectures.

On the other hand, multi-stability is intrinsically present in physics, chemistry, biology, among other fields [15]. It is defined as the coexistence of multiple possible final stable states. The final state to which the system will converge depends on the initial conditions [16]. In the dynamical systems field, multi-stability is an important feature related to dissipative systems. Unstable dissipative systems are those that have a focus-saddle equilibrium point responsible for stable and unstable manifolds, but also, the sum of their eigenvalues is negative [17].

In most of the previously presented approaches related to chaos computing, the sensitivity to the initial conditions of chaotic elements is exploited to obtain logic gates; however, it could be a disadvantage when an experimental implementation is realized due to small variations in the voltages or a little differences in the tolerance of components. In this study, we present a methodology based on the capacity of displaying multi-stability of unstable dissipative systems of type 1 (UDS-1) to implement a dynamical logic gate (DLG), also known as a reconfigurable logic gate. In our proposed method, logic zeros and logic ones are codified through one of the clearly distinguishable possible final states of the multi-stable USD-1. Although we obtain logic gates using multi-stability, which is closely related to the initial conditions, we take advantage of the concept of the basin of attraction, so that a vast set of initial conditions will produce the same response in our system. This gives the advantage to our model of being easily reliable and repeatable. An important aspect of our methodology is that by just varying two parameters, we can get the complete spectrum of two-input logic functions (16 logic gates), which represents an advantage in terms of time and resources for the future electronic implementation of the model.

The remaining of this article is structured as follows: In Section 2, we provide the fundamental theory of UDS-1. Our proposed methodology is explained in Section 3. Results for the DLG are shown and discussed in Section 4. Finally, in Section 5, some conclusions about this article are given.

2. UNSTABLE DISSIPATIVE SYSTEMS FUNDAMENTALS

In the same spirit of Campos-Cantón et al. [18], let us consider the following dynamical system:

$$\dot{x} = Ax, \tag{1}$$

where $x = [x_1, x_2, x_3]^T \in \mathbb{R}^3$ is the state vector, $A = [a_{ij}] \in \mathbb{R}^{3\times 3}$ denotes a linear operator and it is a non-singular matrix. Also, let $\Lambda = \{\lambda_1, \lambda_2, \lambda_3\}$ be the set of eigenvalues of matrix A.

The system given by Equation (1) is called an UDS-1, if the following two statements are satisfied:

1. Focus-saddle equilibrium condition. The matrix A must possess one negative pure real eigenvalue λ_1, whereas $\lambda_{2,3}$ are complex conjugate with a positive real part.
2. Dissipativity condition. The system is dissipative, if $\sum_{i=1}^{3} \mathrm{Re}(\lambda_i) < 0$.

Moreover, a UDS-1 is capable of displaying multi-scrolls if an adequate commutation control law is applied to it. A simple way to generate multi-scrolls consists of applying a PWL function to modify the system dynamics by changing the position of equilibrium points. Thus, if an additive term B is applied to Equation (1), it can be rewritten as:

$$\dot{x} = Ax + B, \tag{2}$$

where $B = [b_1, b_2, b_3]^T \in \mathbb{R}^3$ is a real vector, and it works as a discrete commutation function dependent on the state x. B changes depending on which domain $\mathcal{D}_i \subset \mathbb{R}^3$ the trajectory is located. The main idea is dividing the full phase space into domains, in other words, $\mathbb{R}^3 = \cup_{i=1}^{k} \mathcal{D}_i$. With this last regard, and supposing $B = [0, 0, b_3]^T$, then the switching function is given by:

$$b_3 = \begin{cases} \beta_1, & \text{if } x \in \mathcal{D}_1; \\ \beta_2, & \text{if } x \in \mathcal{D}_2; \\ \vdots & \vdots \\ \beta_k, & \text{if } x \in \mathcal{D}_k. \end{cases} \tag{3}$$

Because A is a non-singular matrix, the equilibrium point of the system given by Equation (2) is located at $x^* = -A^{-1}B$. Specifically, equilibrium points are $x_i^* = -A^{-1}\beta_i$ with $i = 1, 2, \ldots, k$. In this way, the system will have as equilibrium points as domains $\mathcal{D}_i$ are defined.

3. METHODS

In order to design a DLG, we start considering the following system in its canonical form:

$$\dot{x} = Ax = \begin{bmatrix} 0 & 1 & 0 \\ 0 & 0 & 1 \\ -0.5 & -0.7 & -0.5 \end{bmatrix} \begin{bmatrix} x_1 \\ x_2 \\ x_3 \end{bmatrix}, \tag{4}$$

whose eigenvalues are $\Lambda = \{\lambda_1 = -0.6358, \lambda_2 = 0.0679 + 0.8842i, \lambda_3 = 0.0679 + 0.8842i\}$; and $\sum_{i=1}^{3} \mathrm{Re}(\lambda_i) = -0.5 < 0$. Thus, system of Equation (4) satisfies the two conditions mentioned in Section 2 to classify it as a USD-1.

The next step in our methodology consists of forcing the system in Equation (4) to generate n scrolls. It is important to mention that the number of scrolls n to be generated can be arbitrarily chosen and increased, the only requirement is that $n \geq 2$. In our case, we decide to generate three scrolls. Therefore, we add the commutation vector B to Equation (4) and we can rewrite it as:

$$\dot{x} = Ax + B = \begin{bmatrix} 0 & 1 & 0 \\ 0 & 0 & 1 \\ -0.5 & -0.7 & -0.5 \end{bmatrix} \begin{bmatrix} x_1 \\ x_2 \\ x_3 \end{bmatrix} + \begin{bmatrix} 0 \\ 0 \\ b_3 \end{bmatrix}, \tag{5}$$

as we desire to generate three scrolls, then the same quantity of equilibrium points are necessary and we want them arbitrarily located at $x_{e1}^* = (-3, 0, 0)$, $x_{e2}^* = (0, 0, 0)$, $x_{e3}^* = (3, 0, 0)$, which are equispaced only along the x_1 plane. Also, let us remember that each equilibrium point is located in $x_i^* = -A^{-1}\beta_i$, this leads to $x_i^* = -2\beta_i$. Hence, we need to define the switching function b_3 and each β_i, which are described as:

$$b_3 = \begin{cases} 1.5 & \text{if } \quad x_1 < -1.5, \\ 0 & \text{if } -1.5 \leq x_1 \leq 1.5, \\ -1.5 & \text{if } \quad x_1 > 1.5 \end{cases} \tag{6}$$

where the commutation surfaces among domains are located at -1.5 and 1.5 to preserve the shape and symmetry of the scrolls.

Up to now, we have constructed a UDS-1 with the capability of generating three scrolls. In **Figure 1** are plotted projections of the states of the system given by Equations (5) and

(6) for several planes. The plot in **Figure 1A** corresponds to the x_1x_2 plane, where it is possible to distinguish the three scrolls clearly. **Figure 1B** is the projection onto the x_1x_3 plane; whereas **Figure 1C** shows the projection in the x_2x_3 plane.

The following step is controlling the system described by Equations (5) and (6) with the aim of transforming it into a multi-stable system. To achieve this goal, first, let us define the negative reciprocal of element a_{33} as $\mu = -1/a_{33}$, and second, let us multiply the last row of matrix A and the switching function b_3 times μ.

$$\dot{x} = Ax + B = \begin{bmatrix} 0 & 1 & 0 \\ 0 & 0 & 1 \\ -\frac{0.5}{0.5} & -\frac{0.7}{0.5} & -\frac{0.5}{0.5} \end{bmatrix} \begin{bmatrix} x_1 \\ x_2 \\ x_3 \end{bmatrix} + \begin{bmatrix} 0 \\ 0 \\ -\frac{1}{0.5} b_3 \end{bmatrix}, \quad (7)$$

The parameter μ induces multi-stability to the system but leaves the dissipativity unchanged. Now, each scroll we previously generated with the system described by Equations (5) and (6) have become a possible final stable state to which the system will converge depending on the initial conditions. In other words, the system will converge to one of these three scrolls depending on whether its initial condition belongs to the basin of attraction of the scroll. **Figure 2** shows these possible final states to which the system can converge. **Figure 2A** corresponds to the initial condition $x(0) = (-2.5, -1, 0)$; **Figure 2B** to $x(0) = (1.1, -1, 0)$; and finally, **Figure 2C** to $x(0) = (2.5, -1, 0)$.

To develop a system capable of emulating all possible two-input logic functions, we have built a simple three-node network in which each node is a multi-stable UDS-1 governed by Equation (7). The topology of this network is shown in **Figure 3**. Node 1 and node 2 act as inputs, whereas node 3 works as the output of the DLG. In the example, we are using to explain our methodology, the number of scrolls matches with the number of nodes in the network, but there is no relationship between these two quantities.

As it was previously explained, the multi-stable UDS-1 of Equation (7) can converge to three final attractors depending on the initial conditions, in such a way that we can choose two of these attractors to codify logical zeros and ones. Arbitrarily, we decide that if the multi-stable UDS-1 is converging to the left attractor (**Figure 2A**), then it will represent a logical zero. On the other hand, if the multi-stable UDS-1 is converging to the right attractor (**Figure 2C**), then this will be coded as a logical one. The initial condition $(x_0, y_0, z_0) = (-3, 0, 0)$ and $(x_0, y_0, z_0) = (3, 0, 0)$ belong to the basin of attraction of the left and right attractor, respectively. Thus, we can define the target point $(x, y, z) = (I_{1,2} \in \{-3, 3\}, 0.001, 0)$ to be reached through feedback control. Taking these assumptions into consideration, the dynamics of node 1 acting as the first input is described by:

$$\dot{x}_1 = y_1 - k(x_1 - I_1), \quad (8)$$
$$\dot{y}_1 = z_1 - k(y_1 - 0.001), \quad (9)$$
$$\dot{z}_1 = -0.5\mu x_1 - 0.7\mu y_1 - 0.5\mu z_1 + \mu b_{n1}, \quad (10)$$

where the switching function b_{n1} is:

$$b_{n1} = \begin{cases} 1.5 & \text{if } x_1 < -1.5, \\ 0 & \text{if } -1.5 \leq x_1 \leq 1.5, \\ -1.5 & \text{if } x_1 > 1.5 \end{cases} \quad (11)$$

The dynamics of node 2 which works as the second input is given by:

$$\dot{x}_2 = y_2 - k(x_2 - I_2), \quad (12)$$
$$\dot{y}_2 = z_2 - k(y_2 - 0.001), \quad (13)$$
$$\dot{z}_2 = -0.5\mu x_2 - 0.7\mu y_2 - 0.5\mu z_2 + \mu b_{n2}, \quad (14)$$

whose commutation function is:

$$b_{n2} = \begin{cases} 1.5 & \text{if } x_2 < -1.5, \\ 0 & \text{if } -1.5 \leq x_2 \leq 1.5, \\ -1.5 & \text{if } x_2 > 1.5 \end{cases} \quad (15)$$

How input node 1 and node 2 interconnect with the output node 3 considers the following linear affined system:

$$h(I_1, I_2) = \boldsymbol{\alpha} \cdot \mathbf{i} + \gamma, \quad (16)$$

where $\mathbf{i} = (I_1, I_2)^T$ is a column vector whose elements $I_{1,2} \in \{-3, 3\}$; $\boldsymbol{\alpha} = (\alpha_1, \alpha_2) \in \mathbb{R}^2$ and $\gamma \in \mathbb{R}$ are system parameters to be adjusted to obtain the desired logic gate. Function h results from summing the scalar product $\boldsymbol{\alpha} \cdot \mathbf{i}$ plus an offset given by γ. The output of the system is ruled by:

$$I_3(h) = \begin{cases} 3, & \text{if } |h| < \kappa; \\ -3, & \text{otherwise}. \end{cases} \quad (17)$$

where $\kappa \in \mathbb{R}$ is defined as a threshold.

Therefore, the dynamics of node 3 behaving as output is governed by:

$$\begin{aligned} \dot{x}_3 &= y_3 - k(x_3 - I_3), \\ \dot{y}_3 &= z_3 - k(y_3 - 0.001), \\ \dot{z}_3 &= -0.5\mu x_3 - 0.7\mu y_3 - 0.5\mu z_3 + \mu b_{n3}, \end{aligned} \quad (18)$$

with the function b_{n3} governed by:

$$b_{n3} = \begin{cases} 1.5 & \text{if } x_3 < -1.5, \\ 0 & \text{if } -1.5 \leq x_3 \leq 1.5, \\ -1.5 & \text{if } x_3 > 1.5 \end{cases} \quad (19)$$

The networked system described by Equations (10)–(19) emulates any of the possible sixteen two-input logic functions whose truth tables appear in **Table 1**. $\perp$ represents the contradiction or null; I_1I_2, AND; I_1I_2', inhibition of I_2; I_1, transfer of I_1; $I_1'I_2$, inhibition of I_1; I_2, transfer of I_2; $I_1 \oplus I_2$, XOR; $I_1 + I_2$, OR; $(I_1 + I_2)'$, NOR; $(I_1 \oplus I_2)'$, XNOR; I_2', complement

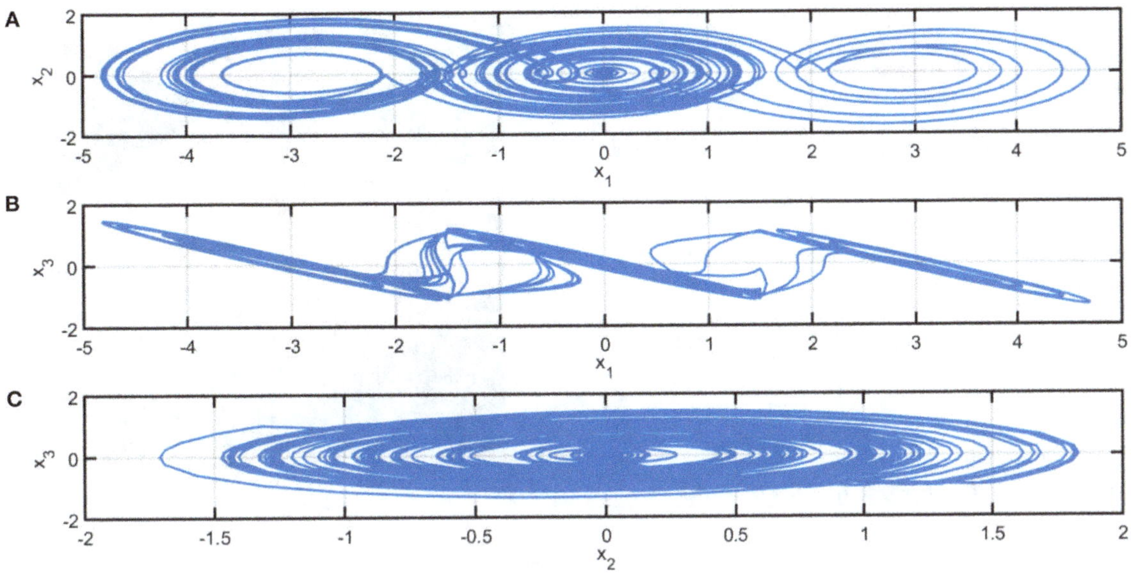

FIGURE 1 | Projections of the system given by equations. axplusbmethod and b3method onto the planes of R^3. The plot **(A)** corresponds to x_1x_2 plane; **(B)** is the projection onto x_1x_3 plane; and **(C)** is the projection in x_2x_3 plane.

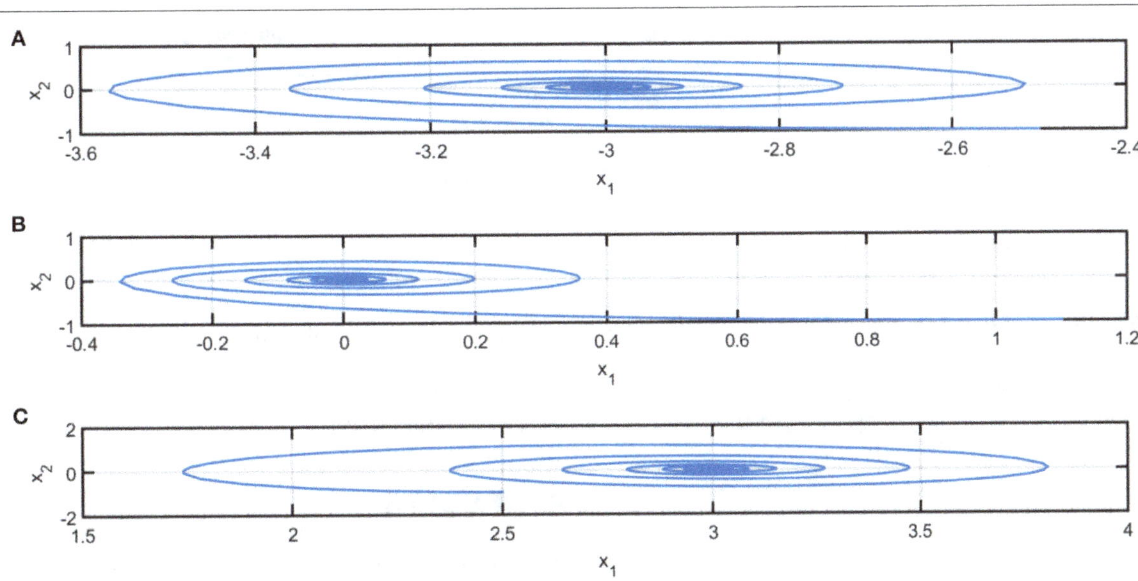

FIGURE 2 | Projections onto the x_1x_2 plane of the three final possible attractors of the multi-stable system of equation (7). axplusbmultimethod. Plot **(A)** corresponds to the initial condition $x(0) = (-2.5, -1, 0)$; Plot **(B)** to $x(0) = (1.1, -1, 0)$; and Plot **(C)** to $x(0) = (2.5, -1, 0)$.

of I_2; $I_1 + I_2'$, implication (I_2 implies I_1); I_1', complement of I_1; $I_1' + I_2$, implication (I_1 implies I_2); $(I_1 I_2)'$, NAND; $\top$, tautology or identity.

The selection of logic gate functionality is realized by adjusting system parameters α_1, α_2, γ, and κ so that Equations (16) and (17) are satisfied simultaneously. Depending on the values of I_1 and I_2, Equation (16) will have one of the results shown in the third column of **Table 2**. These results will fall or will not be inside the interval $(-\kappa, \kappa)$ according to the truth table of the desired logic function we desire to obtain. If h of Equation (16) falls in the open interval $(-\kappa, \kappa)$ defined in Equation (17), then $I_3 = 3$ which represents a logical one and the output node 3 will converge to the left attractor; otherwise, $I_3 = -3$ what is defined as a logical zero and the output node 3 will converge to the right attractor.

In the following lines, we briefly explain the selection of parameters for the case of the AND ($I_1 I_2$) gate, but an analog

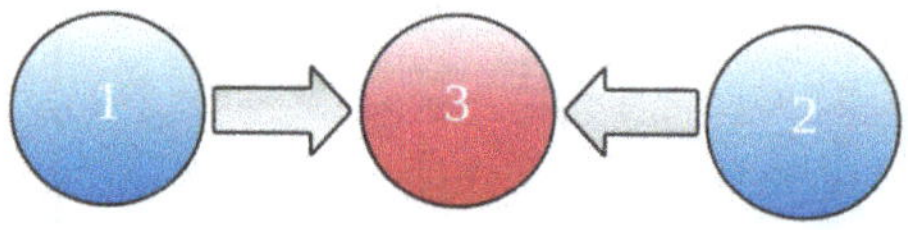

FIGURE 3 | Topology of the three-node network used to develop a dynamical logic gate (DLG).

TABLE 1 | Truth tables for all the sixteen possible two-input logic functions expressed in Boolean variables (0 and 1).

I_1	I_2	$\perp$	$I_1 I_2$	$I_1 I_2'$	I_1	$I_1' I_2$	I_2	$I_1 \oplus I_2$	$I_1 + I_2$
0	0	0	0	0	0	0	0	0	0
0	1	0	0	0	0	1	1	1	1
1	0	0	0	1	1	0	0	1	1
1	1	0	1	0	1	0	1	0	1
I_1	I_2	$(I_1 + I_2)'$	$(I_1 \oplus I_2)'$	I_2'	$I_1 + I_2'$	I_1'	$I_1' + I_2$	$(I_1 I_2)'$	$\top$
0	0	1	1	1	1	1	1	1	1
0	1	0	0	0	0	1	1	1	1
1	0	0	0	1	1	0	0	1	1
1	1	0	1	0	1	0	1	0	1

TABLE 2 | Values of $h(I_1, I_2)$ in Equation 16.

I_1	I_2	$h(I_1, I_2)$
-3	-3	$-3\alpha_1 - 3\alpha_2 + \gamma$
-3	3	$-3\alpha_1 + 3\alpha_2 + \gamma$
3	-3	$3\alpha_1 - 3\alpha_2 + \gamma$
3	3	$3\alpha_1 + 3\alpha_2 + \gamma$

TABLE 3 | List of the system parameters α_2 and γ used to emulate each logic function.

Logic function	$\perp$	$I_1 I_2$	$I_1 I_2'$	I_1	$I_1' I_2$	I_2	$I_1 \oplus I_2$	$I_1 + I_2$
α_2	0.0	0.5	-0.5	0.0	-0.5	0.5	0.8	0.5
γ	4.5	-4.0	-4.0	-3.0	4.0	-3.0	0.0	-2.0
Logic function	$(I_1 + I_2)'$	$(I_1 \oplus I_2)'$	I_2'	$I_1 + I_2'$	I_1'	$I_1' + I_2$	$(I_1 I_2)'$	$\top$
α_2	0.5	-0.8	0.5	-0.5	0.0	-0.5	0.5	0
γ	4.0	0.0	3.0	-2.0	3.0	1.0	2.0	0

procedure is necessary for the rest of the logic gates. First, we fix the threshold $\kappa = 3$. According to the truth table of AND gate shown in the fourth upper column of **Table 1**, only when $I_1 = 3$ and $I_2 = 3$, the sum h must fall inside the interval $(-3, 3)$; the remaining combinations of I_1 and I_2 will fall outside $(-3, 3)$. In such a way, the following inequalities must be accomplished simultaneously:

$$\begin{aligned} -3\alpha_1 - 3\alpha_2 + \gamma < -3 \quad &\vee \quad 3 < -3\alpha_1 - 3\alpha_2 + \gamma, \\ -3\alpha_1 + 3\alpha_2 + \gamma < -3 \quad &\vee \quad 3 < -3\alpha_1 + 3\alpha_2 + \gamma, \\ 3\alpha_1 - 3\alpha_2 + \gamma < -3 \quad &\vee \quad 3 < \ \ 3\alpha_1 - 3\alpha_2 + \gamma, \\ 3\alpha_1 + 3\alpha_2 + \gamma > -3 \quad &\wedge \quad 3 > \ \ 3\alpha_1 + 3\alpha_2 + \gamma. \end{aligned} \tag{20}$$

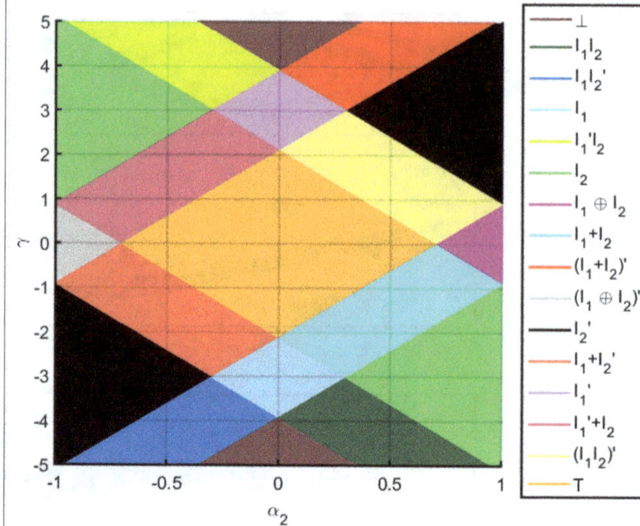

FIGURE 4 | Map of regions for parameters α_2 and γ of DLG.

TABLE 4 | Comparative frame among several approaches already presented and our proposal of DLG.

Reference	Number of obtained logic gates	Number of parameters to configure
Li et al. [7]	11	1
Peng et al. [8]	7	3
Peng et al. [9]	16	2
Guerra et al. [10]	2	1
Rivera-Durón *et al.*	16	2

After algebraic calculations, it is possible to determine that $\alpha_1 = 0.3$, $\alpha_2 = 0.5$, and $\gamma = -4.0$ are one of the several combinations that satisfy the corresponding inequalities in the system Equation (20).

4. RESULTS

The parameter selection is not unique, there are several combinations of $\alpha_1, \alpha_2, \gamma$, and κ that can satisfy Equations (16) and (17). For this reason, we utilized a computer code to determine the set of system parameters to obtain each of the sixteen two-input logic functions for a constant value of $\kappa = 3$. After inspecting the results, we noticed that the value of $\alpha_1 = 0.3$ appeared in all the cases, therefore, we can assume that α_1 is constant too. This last represents an advantage because the functionality of the DLG only depends on the values of α_2 and γ, which can result in the optimization of time and resources in the future experimental realization of the DLG. A map for parameters α_2 and γ to select functionality of DLG is shown in **Figure 4**. The complete list of the system parameters that we used to emulate each logic function is shown in **Table 3**, it is important to remark that in all the cases $\kappa = 3$ and $\alpha_1 = 0.3$.

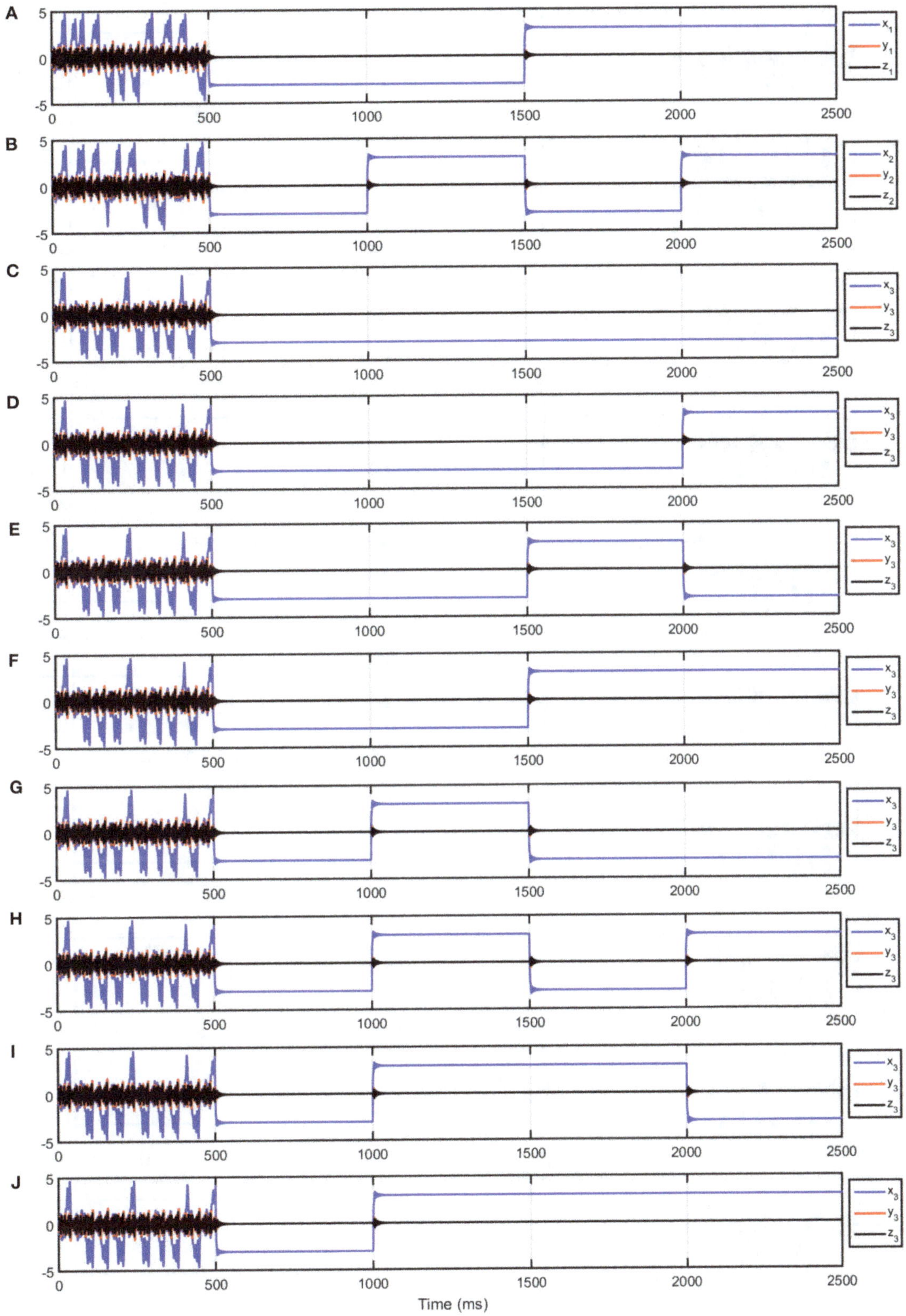

FIGURE 5 | Temporal evolution of the input nodes and for the output node emulating the logic functions shown in the upper part of Table Truth Tables. The plot in **(A)** corresponds to the input node 1. Plot **(B)** corresponds to the temporal evolution of input node 2. bot is represented in plot **(C)**; $I_1 I_2$ in plot **(D)**; $I_1 I_2'$ in plot **(E)**; I_1 in plot **(F)**; $I_1' I_2$ in plot **(G)**; I_2 in plot **(H)**; I_1 plus I_2 in plot **(I)**; I_1+I_2 in plot **(J)**.

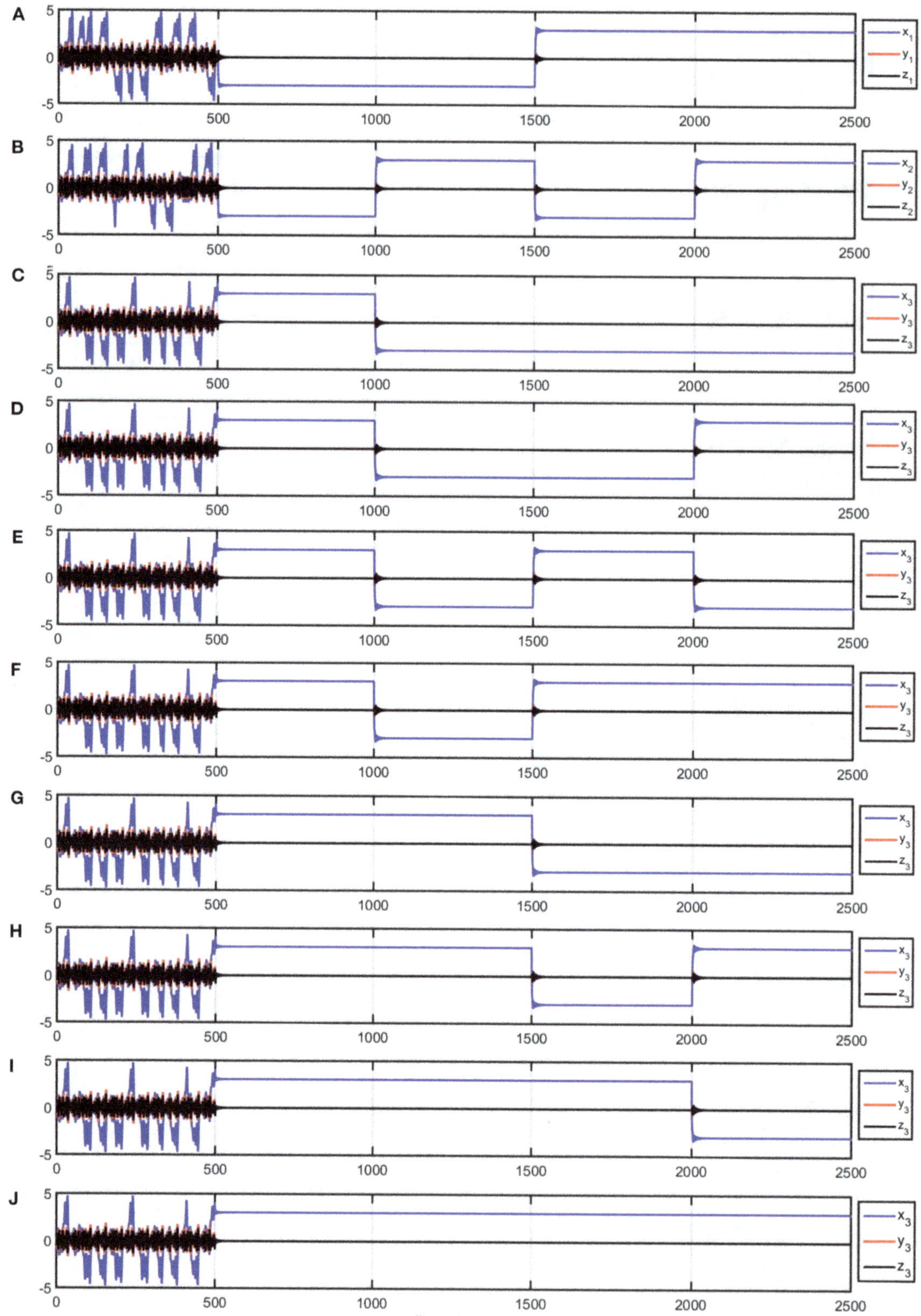

FIGURE 6 | Temporal evolution of the input nodes and for the output node emulating the logic functions shown in the lower part of Table TruthTables. The plot in **(A)** corresponds to the input node 1. Plot **(B)** corresponds to the temporal evolution of input node 2. $(I_1+I_2)'$ in plot **(C)**; $(I_1 \oplus I_2)'$ in plot **(D)**; I_2' in plot **(E)**; $I_1 + I_2'$ in plot **(F)**; I_1' in plot **(G)**; $I_1' + I_2$ in plot **(H)**; $(I_1/I_2)'$ in plot **(I)**; top in plot **(J)**.

TABLE 5 | The truth table for the full adder.

C_{in}	I_1	I_2	C_{out}	S
0	0	0	0	0
0	0	1	0	1
0	1	0	0	1
0	1	1	1	0
1	0	0	0	1
1	0	1	1	0
1	1	0	1	0
1	1	1	1	1

The three first columns correspond to the inputs, whereas the remaining ones, to the outputs.

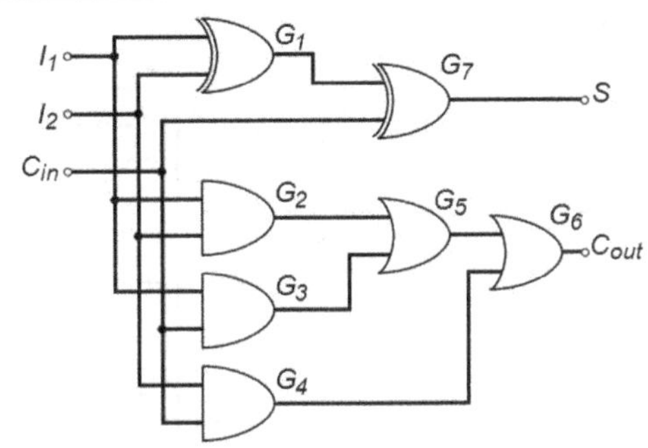

FIGURE 7 | Logic diagram of the full adder whose truth table is shown in **Table 5**.

We can compare our proposal with respect to other approaches to measure the benefits between the achieved logic gates and the number of parameters to be configured. In this sense, **Table 4** shows a comparative frame among previously presented studies and our proposal of DLG. From this table, it is possible to observe that [10] achieves only a pair of logic gates (AND, OR) by varying the resonator's operation parameters. The study in Peng et al. [8] obtains seven logic gates ($\perp$, AND, OR, NAND, NOR, XOR, $\top$) by tuning three parameters. Eleven logic gates ($\perp$, AND, OR, NAND, NOR, XNOR, $I_2, I_2', I_1'+I_2, I_1+I_2'$, $\top$) were achieved in Li et al. [7] by varying a single parameter. Special mention deserves the work done in Peng et al. [9], in which authors got all the possible two-input logic gates as well as we did in our approach through the tuning of a pair of parameters.

Numerical simulations were realized to prove the correct performance of the developed DLG. We started simulations letting the three nodes behave freely so that $\mu = 1$ and they are not in a multi-stable regime for time $0 \le t \le 500$. Then, we set $\mu = -1/0.5$ to get multi-stability in all the nodes, and we configured I_1 of node 1 to be $I_1 = -3$ for $500 < t \le 1500$ and $I_1 = 3$ for $1500 < t \le 2500$; in analogous way, I_2 of the node 2 was configured to be $I_2 = -3$ for $t \in \{(500,1000] \cup (1500,2000]\}$, and $I_2 = 3$ for $t \in \{(1000,1500] \cup (2000,2500]\}$. In this way, the four possible combinations of I_1 and I_2 shown in the first two columns of **Table 2** were accomplished. The plots of the temporal evolution of the input node 1 are shown in **Figures 5A, 6A**, whereas the plots in **Figures 5B, 6B** correspond to the temporal evolution of input node 2. The behavior of the output node 3 is plotted in **Figures 5, 6**. $\perp$ is represented in **Figure 5C**; I_1I_2 in **Figure 5D**; I_1I_2' in **Figure 5E**; I_1 in **Figure 5F**; $I_1'I_2$ in **Figure 5G**; I_2 in **Figure 5H**; $I_1 \oplus I_2$ in **Figure 5I**; I_1+I_2 in **Figure 5J**; $(I_1+I_2)'$ in **Figure 6C**; $(I_1 \oplus I_2)'$ in **Figure 6D**; I_2' in **Figure 6E**; $I_1 + I_2'$ in **Figure 6F**; I_1' in **Figure 6G**; $I_1' + I_2$ in **Figure 6H**; $(I_1I_2)'$ in **Figure 6I**; $\top$ in **Figure 6J**. From **Figures 5, 6** is possible to note the fast response time of the output node, this is because of the control law we added which yields the system to an initial condition into the basin of attraction of the desired attractor but also, this control law avoids falling in any of the equilibrium points of the system.

Now, let us implement a full adder to prove the performance of the DLG when it is executing compound functions. The full adder is a combinational circuit that realizes the arithmetical sum of three bits. I_1 and I_2 are the bits to be added, whereas C_{in} is the input carry bit coming from a previous sum. Due to the sum of three bits varies from 0 to 3, the circuit needs two bits to correctly represent the addition; these bits are S and C_{out}, which are the sum and the output carry, respectively. The truth table for the full adder is shown in **Table 5**. From the truth table is possible to determine the logic functions $S = C_{in} \oplus I_1 \oplus I_2$ and $C_{out} = C_{in}I_1 + C_{in}I_2 + I_1I_2$. The logic diagram to configure the full adder consists of seven logic gates (G_1 to G_7), and it is displayed in **Figure 7**. Again, we launched the numerical simulations with all the nodes behaving freely so that $\mu = 1$ and they are not in a multi-stable regime for time $0 \le t \le 250$, after this time, we set $\mu = -1/0.5$ to get multi-stability in all the nodes. To achieve the eight possible combinations in the inputs of the full adder, we proceeded as follows: first, we configured C_{in} at node 1 to be $C_{in} = -3$ for $250 < t \le 1250$ and $C_{in} = 3$ for $1250 < t \le 2250$; second, I_1 at the node 2 was configured to be $I_1 = -3$ for $t \in \{(250,750] \cup (1250,1750]\}$, and $I_1 = 3$ for $t \in \{(750,1250] \cup (1750,2250]\}$; finally, I_2 at node 3 was fixed to be $I_2 = -3$ for $t \in \{(250,500] \cup (750,1000] \cup (1250,1500] \cup (1750,2000]\}$, and $I_2 = 3$ for $t \in \{(500,750] \cup (1000,1250] \cup (1500,1750] \cup (2000,2250]\}$. The temporal evolution of the full adder is plotted in **Figure 8**. The plots in **Figures 8A–C** correspond to C_{in}, I_1, and I_2, respectively. **Figure 8D** displays the behavior of G_1 which is configured as XOR gate ($I_1 \oplus I_2$). **Figure 8E** shows the evolution of the AND gate G_2 (I_1I_2). In **Figure 8F** is plotted the AND gate G_3 (I_1C_{in}). **Figure 8G** is showing the evolution of the AND gate G_4 (I_2C_{in}). **Figure 8H** shows the temporal evolution of G_5 configured as OR gate and which receives in its inputs the signals coming from G_2 and G_3. The signal coming from G_4 and G_5 are received by G_6 and whose output corresponds to C_{out}, this is shown in **Figure 8I**. Finally, G_7 processes the signal coming from G_1 and C_{in} and its output results in the sum S, which is plotted in **Figure 8J**.

Our proposal of DLG can be taken into an electronic realization through currently available electronic components. Since the core of DLG is composed of three ordinary differential equations, they can be implemented using operational amplifiers (OP-AMP) in the basic configurations (integrator, inverted adder, and inverted), resistors, and capacitors. From an engineering viewpoint, the computational complexity and hardware cost to implement a DLG through a three-node network could be elevated compared with the current logic devices. For this reason, we have as future work to achieve the same results here reported

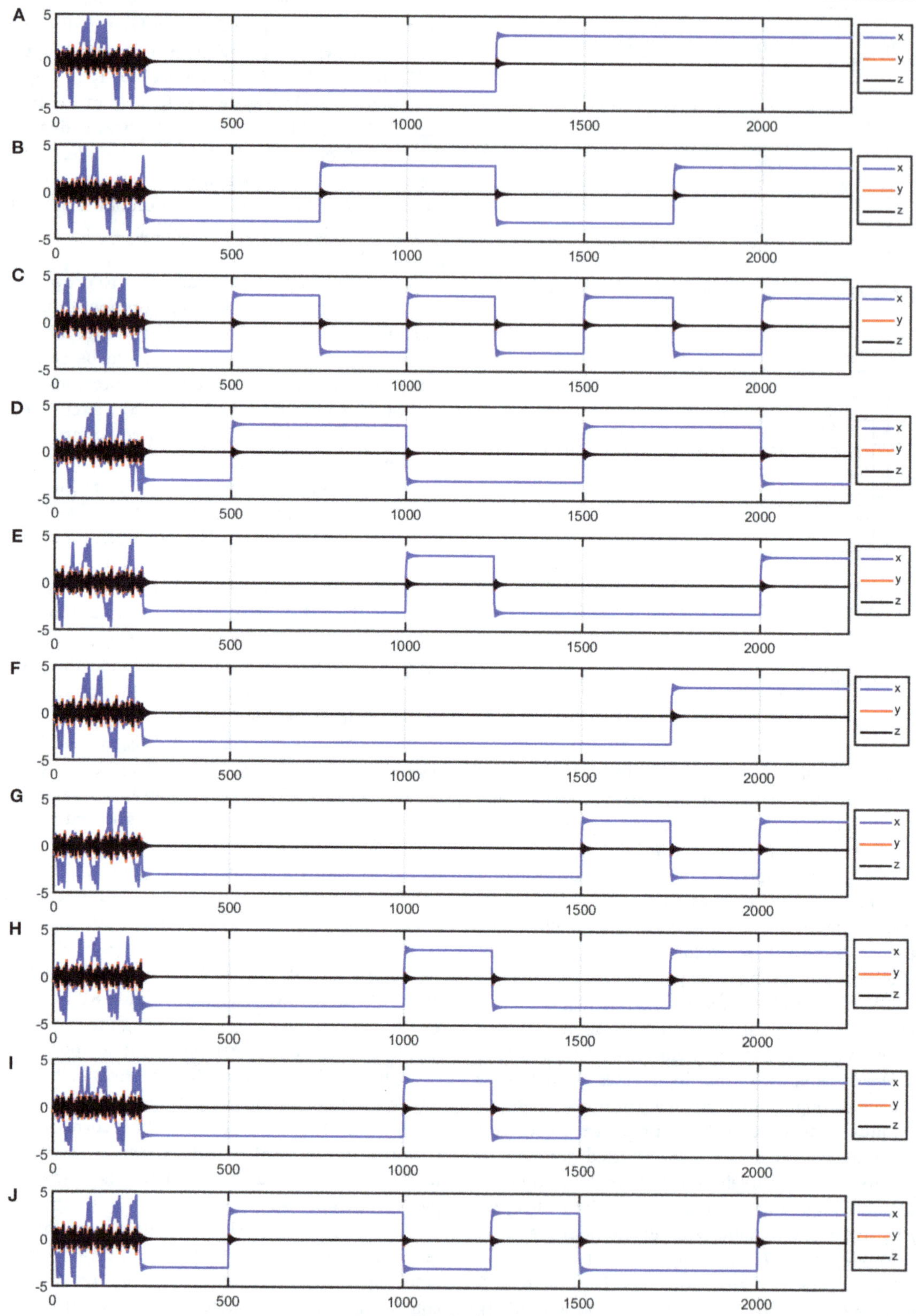

FIGURE 8 | Temporal evolution of the nodes in the full adder. The plot **(A)** corresponds to C_{in}; **(B)** to I_1; **(C)** to I_2; **(D)** to G_1; **(E)** to G_2; **(F)** to G_3; **(G)** to G_4; **(H)** to G_5; **(I)** to G_6 which is C_{out}; and **(J)** to G_7 which is the sum S.

but just using a single UDS-1 for each DLG, which will decrease the prices and the design tasks. However, our approach has the advantage that the programming to implement several functions can be quite quick and it can be done on the fly. Also, another advantage we can elucidate is related to a research issue; whereas in the traditional digital logic devices, as the FPGAs, the states updating is governed by a master clock so that the updating occurs in a synchronous way; in our device, we can add a delay element to investigate autonomous Boolean networks, a kind of dynamical system where the state updating happens when there exists a transition in any input. In this way, we can study the effect of delays in autonomous Boolean networks.

5. CONCLUSION

We presented a methodology to design a DLG using a multi-stable UDS-1. We constructed a three-node network of multi-stable UDS-1. In this topology, a couple of nodes act as inputs of the logic gate, whereas the remaining node is the output or response. Using a pair of the different final states (attractors) of the multi-stable system, we were able to codify Boolean ones and zeros, and subsequently, we obtain the adequate response to emulate all the possible two-input logic functions (16 logic gates). We are sure that our results are relevant because we just need to adjust a pair of parameters to select the functionality of the DLG, this could be important in terms of time and cost in the future experimental realization of this DLG.

AUTHOR CONTRIBUTIONS

RR-D was involved in methodology conceptualization, numerical simulations, writing the original draft, reviewing and editing the manuscript. RS-E is responsible for numerical simulations, validation, reviewing and editing the manuscript. Q-LW responsible for numerical simulations, validation, data visualization, and reviewing and editing the manuscript. All authors contributed to the article and approved the submitted version.

REFERENCES

1. Sinha S, Ditto W. Dynamics based computation. *Phys Rev Lett.* (1998) 81:2156–9.
2. Sinha S, Ditto W. Computing with distributed chaos. *Phys Rev E.* (1999) 60:363–77. doi: 10.1103/PhysRevE.60.363
3. Munakata T, Sinha S, Ditto W. Chaos computing: implementation of fundamental logical gates by chaotic elements. *IEEE Trans Circ Syst I Fundam Theory Appl.* (2002) 49:1629–33. doi: 10.1109/TCSI.2002.804551
4. Murali K, Sinha S, Ditto W. Implementation of NOR gate by a chaotic chua's circuit. *Int J Bifurcat Chaos.* (2003) 13:2669–72. doi: 10.1142/S0218127403008053
5. Murali K, Sinha S, Ditto W. Construction of a reconfigurable dynamic logic cell. *Pramana.* (2005) 64:433–41. doi: 10.1007/BF02704569
6. Murali K, Miliotis A, Ditto W, Sinha S. Logic from nonlinear dynamical evolution. *Phys Lett A.* (2009) 373:1346–51. doi: 10.1016/j.physleta.2009.02.026
7. Li L, Yang C, Hui S, Yu W, Kurths J, Peng H, et al. A reconfigurable logic cell based on a simple dynamical system. *Math Probl Eng.* (2013) 2013:735189. doi: 10.1155/2013/735189
8. Peng H, Yang Y, Li L, Luo H. Harnessing piecewise-linear systems to construct dynamic logic architecture. *Chaos Interdiscipl J Nonlin Sci.* (2008) 18:033101. doi: 10.1063/1.2953494
9. Peng H, Liu F, Li L, Yang Y, Wang X. Dynamic logic architecture based on piecewise-linear systems. *Phys Lett A.* (2010) 374:1450–6. doi: 10.1063/5.0046968
10. Guerra D, Bulsara A, Ditto W, Sinha S, Murali K, Mohanty P. A noise-assisted reprogrammable nanomechanical logic gate. *Nano Lett.* (2010) 10:1168–71. doi: 10.1021/nl9034175
11. Yuan X, Liu W. Implementation and improvement of dynamic logic gates based on cellular neural networks. In: *2012 Fifth International Workshop on Chaos-fractals Theories and Applications.* (Dalian) (2012). p. 99–103.
12. Rothenbuhler A, Tran T, Smith E, Saxena V, Campbell K. Reconfigurable threshold logic gates using memristive devices. In: *2012 IEEE Subthreshold Microelectronics Conference (SubVT).* Waltham, MA (2012). p. 1–3.
13. James A, Krestinskaya O, Maan A. Recursive threshold logica bioinspired reconfigurable dynamic logic system with crossbar arrays. *IEEE Trans Biomed Circ Syst.* (2020) 14: 1311–22. doi: 10.1109/TBCAS.2020.3027554
14. Sahu A, Kumar A, Tiwari S. Performance investigation of universal gates and ring oscillator using doping-free bipolar junction transistor. In: *2020 IEEE Silicon Nanoelectronics Workshop (SNW).* Honolulu, HI (2020). p. 125–6.
15. Hens C, Dana S, Feudel U. Extreme multistability: attractor manipulation and robustness. *Chaos Interdiscipl J Nonlin Sci.* (2015) 25:053112. doi: 10.1063/1.4921351
16. Pisarchik A, Feudel U. Control of multistability. *Phys Rep.* (2014) 540:167–218. doi: 10.1016/j.physrep.2014.02.007
17. Campos-Cantón E, Femat R, Chen G. Attractors generated from switching unstable dissipative systems. *Chaos: Interdis J Nonlin Sci.* (2012) 22:033121. doi: 10.1063/1.4742338
18. Campos-Cantón E, Barajas-Ramírez J, Solís-Perales G, Femat R. Multiscroll attractors by switching systems. *Chaos: Interdis J Nonlin Sci.* (2010) 20:013116. doi: 10.1063/1.3314278

Permissions

All chapters in this book were first published by Frontiers; hereby published with permission under the Creative Commons Attribution License or equivalent. Every chapter published in this book has been scrutinized by our experts. Their significance has been extensively debated. The topics covered herein carry significant findings which will fuel the growth of the discipline. They may even be implemented as practical applications or may be referred to as a beginning point for another development.

The contributors of this book come from diverse backgrounds, making this book a truly international effort. This book will bring forth new frontiers with its revolutionizing research information and detailed analysis of the nascent developments around the world.

We would like to thank all the contributing authors for lending their expertise to make the book truly unique. They have played a crucial role in the development of this book. Without their invaluable contributions this book wouldn't have been possible. They have made vital efforts to compile up to date information on the varied aspects of this subject to make this book a valuable addition to the collection of many professionals and students.

This book was conceptualized with the vision of imparting up-to-date information and advanced data in this field. To ensure the same, a matchless editorial board was set up. Every individual on the board went through rigorous rounds of assessment to prove their worth. After which they invested a large part of their time researching and compiling the most relevant data for our readers.

The editorial board has been involved in producing this book since its inception. They have spent rigorous hours researching and exploring the diverse topics which have resulted in the successful publishing of this book. They have passed on their knowledge of decades through this book. To expedite this challenging task, the publisher supported the team at every step. A small team of assistant editors was also appointed to further simplify the editing procedure and attain best results for the readers. Apart from the editorial board, the designing team has also invested a significant amount of their time in understanding the subject and creating the most relevant covers. They scrutinized every image to scout for the most suitable representation of the subject and create an appropriate cover for the book.

The publishing team has been an ardent support to the editorial, designing and production team. Their endless efforts to recruit the best for this project, has resulted in the accomplishment of this book. They are a veteran in the field of academics and their pool of knowledge is as vast as their experience in printing. Their expertise and guidance has proved useful at every step. Their uncompromising quality standards have made this book an exceptional effort. Their encouragement from time to time has been an inspiration for everyone.

The publisher and the editorial board hope that this book will prove to be a valuable piece of knowledge for researchers, students, practitioners and scholars across the globe.

List of Contributors

Nana Cabo Bizet and Octavio Obregón
Departamento de Física, División de Ciencias e Ingenierías, Universidad de Guanajuato, León, Mexico

César Damián
Departamento de Ingeniería Mecánica, Universidad de Guanajuato, Salamanca, Mexico

Roberto Santos-Silva
Departamento de Ciencias Naturales y Exactas, CUValles, Universidad de Guadalajara, Jalisco, Mexico

Farai Julius Mhlanga and Lazarus Rundora
Department of Mathematics and Applied Mathematics, University of Limpopo, Turfloop Campus, Sovenga, South Africa

Jiwei Zhang
Key Lab of Statistical Modeling and Data Analysis of Yunnan Province, School of Mathematics and Statistics, Yunnan University, Kunming, China

Jing Lu
Key Laboratory of Applied Statistics of MOE, School of Mathematics and Statistics, Northeast Normal University, Changchun, China

Hang Du
Key Laboratory of Applied Statistics of MOE, School of Mathematics and Statistics, Northeast Normal University, Changchun, China
School Affiliated to Longhua Institute of Educational Science, Shenzhen, China

Zhaoyuan Zhang
Key Laboratory of Applied Statistics of MOE, School of Mathematics and Statistics, Northeast Normal University, Changchun, China
School of Mathematics and Statistics, Yili Normal University, Yili, China

Teerawat Kamnardsiri
Department of Digital Game, College of Arts, Media and Technology, Chiang Mai University, Chiang Mai, Thailand

Pattaraporn Khuwuthyakorn and Worawit Janchai
College of Arts, Media and Technology, Chiang Mai University, Chiang Mai, Thailand

Sirinun Boripuntakul
Department of Physical Therapy, Faculty of Associated Medical Sciences, Chiang Mai University, Chiang Mai, Thailand

Yu Jia and Yu Ding
School of Civil Engineering, Central South University, Changsha, China

Xuan Wang, Jiasheng Zhang and Xiaobin Chen
School of Civil Engineering, Central South University, Changsha, China
National Engineering Research Center of High-Speed Railway Construction Technology, Central South University, Changsha, China

Muhammad Samraiz and Zahida Perveen
Department of Mathematics, University of Sargodha, Sargodha, Pakistan, Bhutto University, Upper Dir, Pakistan, Abdulaziz University, Wadi Aldawaser, Saudi Arabia

Gauhar Rahman
Department of Mathematics, Shaheed Benazir

Kottakkaran Sooppy Nisar
Department of Mathematics, College of Arts and Sciences, Prince Sattam Bin

Devendra Kumar
Department of Mathematics, University of Rajasthan, Jaipur, India

Nail Akhmediev
Department of Theoretical Physics, Research School of Physics, The Australian National University, Canberra, NSW, Australia

Zhanwei Liu, Luchao She and Guangming Fan
Fundamental Science on Nuclear and Simulation Technology Laboratory, Harbin Engineering University, Harbin, China

Yanwei Yue
Research and Development Center, China Nuclear Power Engineering Co, Ltd, Beijing, China

Tricia R. Prokop and Michael Wininger
Department of Rehabilitation Sciences, University of Hartford, West Hartford, CT, United States
Department of Biostatistics, Yale School of Public Health, New Haven, CT, United States
Cooperative Studies Program, Department of Veterans Affairs, West Haven, CT, United States

Zhen Wang, Lifeng Yang, Rui Gao, Zhe Liu, Shaoyuan Mo, Meng Fan and Xin Wang
CNPC Key Laboratory of Oil and Gas Reservoir Stimulation, Langfang, China
Research Institute of Petroleum Exploration and Development, Petro China, Beijing, China

Guanshui Xu
University of California, Riverside, Riverside, CA, United States

Joshua Wafula, Yusuf Karimjee, Yvonne Tamba, Geoffrey Malava, Caroline Muchiri, Grace Koech, Josephat Nyongesa and Keith Shepherd
World Agroforestry Centre (ICRAF), Nairobi, Kenya

Jan De Leeuw
World Agroforestry Centre (ICRAF), Nairobi, Kenya
ISRIC World Soil Information, Wageningen University, Wageningen, Netherlands

Eike Luedeling
World Agroforestry Centre (ICRAF), Nairobi, Kenya
Department of Horticultural Sciences, University of Bonn, Bonn, Germany

Koffi Messan Agbavon and Herve Michel Tenkam
Department of Mathematics and Applied Mathematics, North-West University Potchefstroom Campus, Potchefstroom, South Africa

Appanah Rao Appadu
Department of Mathematics and Applied Mathematics, Nelson Mandela University, Gqeberha, South Africa

Bilge Inan
Department of Basic Sciences of Engineering, Faculty of Engineering and Natural Sciences, ˙Iskenderun Technical University, Hatay, Turkey

Rachel Johnson
Department of Communication Disorders and Special Education, Old Dominion University, Norfolk, VA, United States

Jennifer Matthews
Department of Mathematics, U.S. Naval Academy, Annapolis, MD, United States

Norou Diawara
Department of Mathematics and Statistics, Old Dominion University, Norfolk, VA, United States

Rachel Carroll
Department of Mathematics and Statistics, University of North Carolina Wilmington, Wilmington, NC, United States

Zhi Wang and Jingchen Liu
Department of Statistics, Columbia University, New York, NY, United States

Xueying Tang
Department of Mathematics, University of Arizona, Tucson, AZ, United States

Salvador Cruz Rambaud and María José Muñoz Torrecillas
Departamento de Economía y Empresa, Universidad de Almería, Almería, Spain

Adriana Garcia
Faculty of Economics and Business, University of Groningen, Groningen, Netherlands

Muzaffer Ayvaz and Lieven De Lathauwer
Department of Electrical Engineering (ESAT), KU Leuven, Leuven, Belgium
Group Science, Engineering and Technology, KU Leuven Kulak, Kortrijk

Júlio Hoffimann, Maciel Zortea, Breno de Carvalho and Bianca Zadrozny
IBM Research, Rio de Janeiro, Brazil

Roberto R. Rivera-Durón
Unmanned System Research Institute, Northwestern Polytechnical University, Xi'an, China

Ricardo Sevilla-Escoboza
Centro Universitario de los Lagos, Universidad de Guadalajara, Lagos de Moreno, Mexico

Qui-Ling Wang
College of ransportation, Chang'an University, Xi'an, China

Index

A
Amplitude, 2, 12, 75-76, 78-81, 150

B
Bayesian Inference, 24, 28
Bayesian Model, 26, 35, 37-38, 40

C
Computer Vision, 42-44, 51, 203
Controlled Vertical Stress, 53, 64

D
Decision Outcomes, 120, 123-125
Demographics Profile, 47
Discount Rate, 173, 176-177
Dynamic Viscosity, 55

E
Education, 22, 43, 51, 93, 107-108, 122, 129, 148
Elastic Modulus, 55
Elementary Particles, 74, 79
Entropies, 1-2, 12-14
Equation, 1, 6, 12-16, 19, 22, 41, 53, 55, 57, 61, 64, 66, 73-76, 79-81, 84, 86, 92, 96, 105, 131-133, 136, 138, 144-147, 152, 161, 173-176, 184-185, 189-194, 220-222
Evolution Curves, 55, 59, 62

F
Fisher's Equation, 131-133, 146-147
Fractional Calculus, 66-67, 73
Fractional Derivative, 66-67, 73

G
Generalized Linear Model, 148-150, 152, 181
Geometric Brownian Motion, 15, 22-23

H
Harmonic Oscillator, 1-2, 6, 10, 12-13
High-value Variables, 123, 125-128
Hydraulic Conductivity, 53, 64
Hydraulic-fracturing, 109, 119
Hydrostatic Pressure, 59-60
Hypothesis Testing, 99, 102

I
Image Processing, 42-43, 50, 147, 201
Internal Erosion, 52-53, 63-65

L
Laser Equation, 66
Logarithm, 24

M
Metropolis-hastings Algorithm, 27, 40
Mittag-leffler Function, 67, 73
Mixed Effects, 106, 162
Monetary Income, 124-125, 127
Multivariate, 27, 103, 179-181, 183-185, 190, 192-193, 195, 199-201

N
Net Income, 121-122, 130
Non-informative Priors, 28, 35, 40
Nonextensive Entropies, 1-2
Nonlinear Diffusion, 2, 132, 147
Nonlinear Quantum Systems, 1-2
Nonlinearity, 2, 73-74
Numerical Simulation, 53, 82, 89, 91, 109, 118

O
Optical Fibers, 74-75, 80
Outcome Distribution, 125-128
Output Variables, 123-124

P
Parameter Recovery, 26, 32, 34-35
Path Integrals, 1, 14
Peregrine Wave, 74-76, 78, 81
Permeability, 52-53, 55, 58-59, 61-64
Porosity, 52-53, 55, 58-59, 61-63, 215
Probabilistic Projection, 120
Probability, 1-2, 4, 6, 13, 16, 19, 22-24, 26-27, 32-33, 37-40, 120, 123, 163-164, 166, 169, 204-205
Propagators, 1-2, 6, 12-13
Proppant Transport, 109, 118-119

Q
Quantropy, 1-2, 6, 10, 12-13
Quantum Mechanics, 1-2, 6, 12-14, 66

R
Rogue Waves, 74-76, 78-81

S
Scale Function, 15, 19, 22
Schrodinger Equation, 74
Seepage Field, 53, 59-60
Seepage Velocity, 52, 55-56, 60-61
Slice Sampling Algorithm, 24-28, 32, 34-35, 39-40
Statistical Analyses, 47

Statistical Mechanics, 1-2, 13-14

Statistical Physics, 2, 6

Statistics, 1-2, 4, 6, 10, 12-13, 23-24, 32, 37-38, 40-41, 47-48, 64, 93, 96-97, 102, 104, 148, 156, 163, 201, 205-206, 216

Stress Contrast, 109, 113, 115-118

Stress Interference, 109-110, 118

Superstatistics, 1-2, 13

T

Thermodynamic, 2, 53

Theta Functions, 74-75

Tsallis Entropy, 1-2

V

Value Chains, 120-121, 126, 129-130

Velocity, 42-43, 45-48, 52, 55-56, 60-61, 63, 75-76, 84-85, 87-90, 132

W

Wave Function, 1-2, 4, 6, 12

Z

Zipper Fracturing, 109-110, 112-115, 117-119

www.ingramcontent.com/pod-product-compliance
Lightning Source LLC
LaVergne TN
LVHW082203200926
848484LV00008B/2205